Diverse Quasiparticle Properties of Emerging Materials

Diverse Quasiparticle Properties of Emerging Materials: First-Principles Simulations thoroughly explores the rich and unique quasiparticle properties of emergent materials through a VASP-based theoretical framework. Evaluations and analyses are conducted on the crystal symmetries, electronic energy spectra/wave functions, spatial charge densities, van Hove singularities, magnetic moments, spin configurations, optical absorption structures with/without excitonic effects, quantum transports, and atomic coherent oscillations.

Key Features

- Illustrates various quasiparticle phenomena, mainly covering orbital hybridizations and spin-up/spin-down configurations
- Mainly focuses on electrons and holes, in which their methods and techniques could be generalized to other quasiparticles, such as phonons and photons
- Considers such emerging materials as zigzag nanotubes, nanoribbons, germanene, plumbene, bismuth chalcogenide insulators
- Includes a section on applications of these materials

This book is aimed at professionals and researchers in materials science, physics, and physical chemistry, as well as upper-level students in these fields.

Diverse Quasiparticle Properties of Emerging Materials

First-Principles Simulations

Edited by
Tran Thi Thu Hanh
Vo Khuong Dien
Ngoc Thanh Thuy Tran
Ching-Hong Ho
Thi Dieu Hien Nguyen
Ming-Fa Lin

CRC Press
Taylor & Francis Group
Boca Raton London

CRC Press is an imprint of the
Taylor & Francis Group, an **informa** business

First edition published 2023
by CRC Press
6000 Broken Sound Parkway NW, Suite 300, Boca Raton, FL 33487–2742

and by CRC Press
4 Park Square, Milton Park, Abingdon, Oxon, OX14 4RN

CRC Press is an imprint of Taylor & Francis Group, LLC

ISBN: 978-1-032-32305-3 (hbk)
ISBN: 978-1-032-34519-2 (pbk)
ISBN: 978-1-003-32257-3 (ebk)

DOI: 10.1201/9781003322573

Typeset in Times
by Apex CoVantage, LLC

Contents

**Chapter 5   Feature-Rich Quasiparticle Properties of Halogen-Adsorbed
                Silicene Nanoribbons 97**

*Duy Khanh Nguyen, Vo Duy Dat, Vo Van On, and
Ming-Fa Lin*

**Chapter 6   Essential Properties of Metals/Transition Metals-Adsorbed
                Graphene Nanoribbons..123**

Ngoc Thanh Thuy Tran, Shih-Yang Lin, and Ming Fa-Lin

Preface

This book *Diverse Quasiparticle Properties of Emerging Materials: First-Principles Simulations* is completed with intensive cooperation in scientific research between research groups from Taiwan and Vietnam. This book comprises 20 comprehensive chapters on the theoretical framework of quasiparticle properties. There are two approaches of quasiparticle viewpoints dominating the theoretical developments, namely, first-principles simulations and phenomenological models. This work is focused on the first-principles simulations.

The calculated results include the total ground state energies/the chemical modification energies, the optimal Moiré superlattices/normal unit cells, the atom- and spin-dominated band structures/wave functions, the spatial charge/spin density distributions, the atom- orbital- and spin-decomposed van Hove singularities, the net magnetic moments, the single-particle and many-body reflectance, absorption, transmission and energy loss spectra, the ballistic electrical conductivities, the Hall quantum ones, and the vibration phonons. They are sufficient to identify the various quasiparticle behaviors and to further generalize the previous theoretical framework.

Most of the research results in this book are carried out by research groups at prestigious universities such as National Cheng Kung University, Taiwan; Ho Chi Minh City University of Technology (HCMUT)—Vietnam National University Ho Chi Minh City, Vietnam; National Kaohsiung University of Science and Technology, Taiwan; Can Tho University, Vietnam; Can Tho University of Medicine and Pharmacy, Vietnam; Thu Dau Mot University, Vietnam. We are grateful to all the authors for their excellent contributions.

This book will hopefully be of great interest to the scientific community, and it will contribute to the development of research of emergent materials.

Acknowledgments

This work was financially supported by the Hierarchical Green-Energy Materials (Hi-GEM) Research Center, from The Featured Areas Research Center Program within the framework of the Higher Education Sprout Project by the Ministry of Education (MOE) and the Ministry of Science and Technology (Grant number MOST 108–2112-M-006–016-MY3) in Taiwan.

Editors

Tran Thi Thu Hanh is a lecturer at the Ho Chi Minh City University of Technology, Vietnam National University, Ho Chi Minh City, Vietnam. Dr. Hanh graduated with bachelor's and master's degrees from Voronezh State University, Russia, and a doctorate from Tokyo University, Japan. Her main area of research is computational physics using molecular dynamics and density functional theory methods. Dr. Hanh focuses on studying the interaction, the hydrogen storage of fuel cells, and the electronic properties, structural deformation of 2D, 3D materials, with many quality articles published in international journals. Find her on the web at https://fas.hcmut.edu.vn/personnel/thuhanhsp.

Vo Khuong Dien is a Ph.D. student at the Department of Physics, National Cheng Kung University. His research interests include the electronic and optical properties of low-dimensional group materials and first-principles calculations.

Ngoc Thanh Thuy Tran obtained her Ph.D. in physics in 2017 from the National Cheng Kung University (NCKU), Taiwan. Afterward, she worked as a postdoctoral researcher and then was promoted to assistant researcher at Hierarchical Green-Energy Materials (Hi-GEM) Research Center, NCKU. Her scientific interest is focused on the fundamental (electronic, magnetic, and thermodynamic) properties of 2D materials and rechargeable battery materials by means of the first-principles calculations.

Ching-Hong Ho is a postdoctoral researcher in the Department of Physics, National Cheng Kung University, Taiwan, where he received his Ph.D. in 2011. His research area of interest is theoretical condensed matter physics with a focus on topological aspects in the past years.

Thi Dieu Hien Nguyen achieved her Ph.D. in physics in July 2021 from National Cheng Kung University (NCKU), Taiwan Currently, she works as a postdoctoral researcher at NCKU. Her academic research focuses on essential properties of 1D, 2D materials and anode, cathode, electrolyte battery materials using first-principles calculations.

Ming-Fa Lin is a distinguished professor in the Department of Physics, National Cheng Kung University, Taiwan. He received his Ph.D. in physics in 1993 from the National Tsing-Hua University, Taiwan. His main scientific interests focus on essential properties of carbon related materials and low-dimensional systems.

Contributors

Vo Duy Dat
Institute of Applied
Technology
Thu Dau Mot University
Binh Duong, Vietnam

Vo Khuong Dien
Department of Physics
National Cheng Kung University
Tainan, Taiwan

Nguyen Thi Han
Department of Physics
National Cheng Kung University
Tainan, Taiwan

Tran Thi Thu Hanh
Laboratory of Computational Physics,
Faculty of Applied Science
Ho Chi Minh City University of
Technology (HCMUT)
Ho Chi Minh City, Vietnam
and
Vietnam National University Ho Chi
Minh City
Ho Chi Minh City, Vietnam

Ching-Hong Ho
Department of Physics
National Cheng Kung University
Tainan, Taiwan

Nguyen Van Hoa
Laboratory of Computational Physics,
Faculty of Applied Science
Ho Chi Minh City University of
Technology (HCMUT)
Ho Chi Minh City, Vietnam
and
Vietnam National University Ho Chi
Minh City
Ho Chi Minh City, Vietnam

Thi My Duyen Huynh
Department of Physics
National Cheng Kung University
Tainan, Taiwan

Phuoc Huu Le
Department of Physics and Biophysics,
Faculty of Basic Sciences
Can Tho University of Medicine and
Pharmacy
Can Tho, Vietnam

Ming-Fa Lin
Department of Physics
National Cheng Kung University
Tainan, Taiwan

Chiun-Yan Lin
Department of Physics
National Cheng Kung University
Tainan, Taiwan

Shih-Yang Lin
Department of Physics
National Cheng Kung University
Tainan, Taiwan

Wei-Bang Li
Core Facility Center
National Cheng Kung University
Tainan, Taiwan

Hsin-Yi Liu
Center for Micro/Nano Science and
Technology
National Cheng Kung University
Tainan, Taiwan

Thi Dieu Hien Nguyen
Department of Physics
National Cheng Kung University
Tainan, Taiwan

Duy Khanh Nguyen
Institute of Applied Technology
Thu Dau Mot University
Binh Duong, Vietnam

Vo Van On
Institute of Applied Technology
Thu Dau Mot University
Binh Duong, Vietnam

Jheng-Hong Shih
Department of Physics
National Cheng Kung
    University
Tainan, Taiwan

Pham Thi Bich Thao
Department of Physics
College of Natural Sciences
Can Tho University
Can Tho, Vietnam

Nguyen Thanh Tien
Department of Physics
College of Natural Sciences
Can Tho University
Can Tho, Vietnam

Ngoc Thanh Thuy Tran
Hierarchical Green-Energy Materials
    Research Center
National Cheng Kung University
Tainan, Taiwan

Ming-Hsiu Tsai
Department of Physics
National Cheng Kung University
Tainan, Taiwan

Le Thi Cam Tuyen
Department of Chemical Engineering,
    College of Engineering Technology
Can Tho University
Can Tho City, Vietnam

Yu-Ming Wang
Department of Physics
National Cheng Kung University
Tainan, Taiwan

Jhao-Ying Wu
Center of General Studies
National Kaohsiung University of
    Science and Technology
Kaohsiung, Taiwan

1 Introduction

*Tran Thi Thu Hanh, Jhao-Ying Wu,
Vo Khuong Dien, Thi Dieu Hien Nguyen,
Thi My Duyen Huynh, and Ming-Fa Lin*

A lot of emergent materials, which have been/will be successfully generated by various physical and chemical methods, are outstanding candidates in exploring diverse phenomena of quasiparticle properties both theoretically and experimentally. Such unusual materials cover graphene-related systems (diamond, bulk graphites, layered graphene, carbon nanotubes, graphene nanoribbons, fullerenes, onions, and chains) [1–8], few-layer group-IV and group-V ones (silicene/germanene/tinene/plumbene/ phosphorene/bismuthene/antimonene [9–15]), core anodes, electrolytes, and cathodes of lithium-ion-based batteries [e.g., the ternary lithium titanium/silicon/ iron compounds [16–19]], perovskite solar cells [20], transition-/rare-earth-metal disulfide-related compounds [21, 22], and quantum topological insulators [23]. Furthermore, they are easily modulated by chemical adsorptions/substitutions [24, 25], temperatures [26], mechanical strains [27], gate voltages [electric fields] [28], uniform/non-uniform magnetic fields [29, 30], time-dependent/static Coulomb fields [31, 32], and electromagnetic waves [33]. The intrinsic and extrinsic mechanism are very sufficient for creating diversified quasiparticle behaviors. This is clearly illustrated in crystal symmetries (Moiré superlattices or not [34, 35]), electronic energy spectra and wave functions, spatial charge densities [36], van Hove singularities [37], net magnetic moments [38], and atom- and orbital-induced spin configurations [39]. Both theoretical calculations and experimental measurements are developed to examine and verify various essential properties, being attributed to the dynamic and/or static responses of the same/composite quasiparticles (e.g., electrons/polarons [40]). This book is focused on the former, which is based on first-principles simulations [41]. The critical mechanisms in determining rich and unique phenomena are thoroughly explored from consistent physical quantities, especially for the orbital hybridizations of chemical bonds and the atom- and orbital-induced magnetic configurations [42]. Most importantly, the framework of quasiparticle viewpoints could be achieved through the systematic investigations. The predicted results are compared with the measured ones in detail [43].

In general, there are two approaches of quasiparticle viewpoints in dominating the theoretical developments (all the details in Chapter 2), namely, first-principles simulations (e.g., proposed orbital hybridizations and spin configurations in Ref [43]) and the phenomenological models (e.g., the generalized tight-binding model, the modified random-phase approximation and self-energy method; [44–46]). Each strategy must have plenty of merits and drawbacks. For example, the former/the latter can/ cannot successfully deal with optimal geometries, complicated band structures of

DOI: 10.1201/9781003322573-1

Moiré superlattices [34], spatial multi-orbital hybridizations [47], spin density distributions [48], the net magnetic and electric moments [49], complex excitonic effects [50], the low-dimensional quantum transports [51], and the collective atom vibrations (phonon spectra and polarization displacements [52]), while they are unable/able to explore the various magnetic quantization phenomena (e.g., the featured Landau levels/magneto-optical selection rules/magneto-electronic Coulomb excitations/quantum Hall effects [53–56]), the static/dynamic charge screening abilities (such as, the unusual Friedel oscillations due to charged impurities, plasma waves arising from the external ion beam, and the quasiparticle lifetimes of few-layer graphene/coaxial carbon nanotubes [57–59]). These have been clearly identified in systematic studies, as done for graphene-related emergent materials [60]. Apparently, how to promote their direct combinations would be very useful in understanding the same/composite quasiparticles and thus would achieve much progress of basic sciences. In this work, the VASP calculations are chosen for a theoretical development, being partially supported by the model discussions. The concise motivations of each book chapter are stated in the following paragraphs.

The various high-precision experiments are developed to detect the unusual physical/chemical/materials properties in emergent materials [the details are in Chapter 3], covering the delicate examinations and analyses about the geometric, electronic, magnetic, optical, Coulomb-excitation, and transport properties. The X-ray diffraction is frequently utilized to measure the crystal structures of bulk materials since the first theory and experiment by Bragg et al., such as the different Moiré super lattices in ternary lithium-titanium [61], lithium-silicon [62] and lithium-iron [63] oxide compounds (anode/electrolyte/cathode materials of lithium-ion-based batteries [64–66]), and stage-n graphite intercalation compounds (n corresponding to the number of graphitic layers between two intercalant ones;[67]). How to evaluate the reliable charge densities from the measured patterns are very interesting challenges. Specifically, the elastic scattering of the incident electron beams is available in observing the three-zero dimensional condensed-matter systems, in which both reflection low-energy electron diffraction (RLEED [68]) and tunneling electron microscopy (TEM [69]) are capable of providing the top and side views of surface-related structures, respectively. These two methods have clearly verified the low-dimensional crystal symmetries within the coaxial, few-layer, and deformed composite structures, such as single-/multi-walled carbon nanotubes [70], layered graphene systems with the different layers [71], normal stacking configurations [72] and twisted angles [73], planar/folded/curved/scrolled graphene nanoribbons [74–77], and buckled monolayer/bilayer silicene/germanene/tinene/plumbene [the significant coupling effects of stacking and buckling [78–81]]. As to the nano-scaled crystal structures, scanning tunneling microscopy (STM [82]) can reveal the periodical atom arrangements [83] and the local defects (vacancy, adatom intercalation and guest-atom substitution [84–86]). It is well known that STM is very successful in identifying the chirality and radius of a single-walled carbon nanotube [87], the edge boundary of an achiral/chiral graphene nanoribbon [88], the honeycomb lattices of monolayer group-IV systems [89], and the non-hexagonal phosphorene [90]. In addition, the spin-polarized STM is enhanced for its spatial resolution and thus is very useful in identifying the prominent ferromagnetism related to the atomic spin configurations [48].

The theoretical predictions on electronic energy spectra and wave functions are directly verified from scanning tunneling spectroscopy (STS [91]) and angle resolved photoemission spectroscopy (ARPES [92]). STS measurements can fully examine the dimension-dependent van Hove singularities due to the band-edge states. The very successful cases cover the geometry-determined symmetric peaks and the metallic or semiconducting behaviors in single-walled carbon nanotubes, the chirality- and width-dependent energy gaps of 1D graphene nanoribbons, the layer-number-, stacking-, twist-angle-, and doping-enriched band overlaps, band gaps, energy dispersion relations in few-layer graphene systems, and the greatly modified band properties across the Fermi level from 2D group-IV and group-V systems on distinct substrates [93]. Specifically, the spin-polarized STS measurements can distinguish the spin-split density of states [94]. On the ARPES side, their measurements can clearly reveal the quasiparticle energy spectra and lifetimes of occupied electronic states. The up-to-now works show that they have shown the diverse band dispersion for 2D materials, e.g., the linear, parabolic, partially flat, and oscillatory energy bands in AB- and ABC-stacked graphene systems [95]. The 1D graphene nanoribbons are observed to exhibit the parabolic bands in the presence of semiconducting behaviors [96]. In addition, no published papers are found about the ARPES spectra of 1D carbon nanotubes and 3D lithium titanium/silicon/iron oxides. The difficulty in defining vectors/transferred momenta and too many valence subbands should be the critical factors. Specifically, the wave-vector-dependent distribution width of ARPES spectra are available in determining the quasiparticle lifetimes, e.g., the enhanced Coulomb decay rates in monolayer alkali-doped graphene [97].

Four kinds of optical spectroscopies are able to measure the frequency-dependent reflectance [98], absorption [99], transmission [100], and photoluminescence spectra [101]. Specifically, the last ones are designed for the clear identifications of many-body effects (the greatly reduced threshold frequency, the extra-prominent absorption peaks of excitonic bound states, and the strongly modified features of single-particle vertical transitions [102]). Which kinds of measurements are suitable strongly depends on the sample thickness. For example, reflectance, transmission, and absorption spectra have been successfully measured for AB-stacked graphite [103], few-layered graphene systems with the different stacking configurations and carbon nanotubes, respectively. Furthermore, these examinations are able to clarify the low-energy π-electronic excitations and middle-energy σ-electronic ones [104], the layer-number- and stacking-enriched absorption structures [105], and the excitonic/Aharnov-Bohm effects [106]. It should be noted that the optical reflectance and photoluminescence spectra are detected for the multi-component lithium oxides [107], in which the measured results are too rough to achieve important conclusions. This is attributed to Moiré superlattices in creating a lot of valence and conduction subbands. However, the VASP simulations in this book will clearly specify the close relations between the active orbital hybridizations and the prominent absorption structures [108]. While the dynamic cases are recovered to the static ones [109] (the long wavelength limit is extended to any moment transfers [110]), the measured transport properties can clarify the semiconducting or metallic behaviors (the electron energy loss spectra are able to comprehend the single-particle and collective charge excitations [111]), especially for the quantum Hall conductivities of layered material [56] (the unusual plasmon modes in low-dimensional systems [112]). Very interestingly, experimental measurements are frequently utilized to fully explore the screening abilities of quasiparticle charges [113].

Numerous 2D materials have been synthesized and predicted since the discovery of graphene [114]. As a result, many studies for the structures and properties of 2D materials are available [115–117]. Hexagons are basic building blocks of the crystal structures for most 2D materials. Following the structure–property relationships that have been commonly explored to discover new materials [118], we expect that the properties of 2D materials can be modified if the building blocks for these nanomaterials are changed from hexagons to pentagons. Very recently, significant efforts have focused on stabilization of the pentagonal structure based on carbons, that is monolayer penta-graphene [119]. Penta-graphene (PG) is extracted from bulk T12-carbon phase. This phase is obtained by heating an interlocking-hexagon-based metastable carbon phase at high temperature [120]. It is found that the monolayer PG is an indirect band-gap semiconductor with a band gap of ~3.25 eV [121] which is smaller than SiC [122], BN [123], and BeO [124] nanostructures. Those studies showed that this structure has obtained dynamical, thermal, and mechanical stability. In a similar way to graphene, the PG sheets can be cut along typical crystallographic orientations in order to construct various penta-graphene nanoribbons (PGNRs) to obtain quasi-one-dimensional materials. Their electronic properties were systemically investigated [125] including confinement effects and quasiparticle phenomena. The resulting four typical nanoribbons, with different edge configurations, are denoted as zigzag-zigzag penta-graphene nanoribbon (ZZ-PGNR), zigzag-armchair penta-graphene nanoribbon (ZA-PGNR), zigzag-armchair penta-graphene nanoribbon (AA-PGNR), and sawtooth-sawtooth penta-graphene nanoribbon (SS-PGNR). This study confirmed that SS-PGNR is the most stable structure when compared with the other three types of PGNRs with similar width. SS-PGNR possesses semiconductor properties. Electronic and transport properties of the sawtooth-sawtooth penta-graphene nanoribbons were systematically investigated by using the density-functional theory (DFT) in combination with the non-equilibrium Green's function (NEGF) formalism in this chapter. Quasiparticle related electronic diversity of many SS-PGNR structures is investigated in Chapter 4. This is a very important basis to find the way to realize electronic devices based on this emergent material.

A new era of low-dimensional materials has indeed opened since a two-dimensional (2D) monolayer of layered graphite was successfully isolated by Geim and Novoselov through the mechanical exfoliation method in 2004 [114, 126, 127]. This first 2D monolayer graphitic system is widely known as graphene. Graphene is made of sp^2 hybridized carbon atoms packed in a highly symmetric hexagonal lattice [128]. The honeycomb network of graphene can be extended to create the basic building block of other carbon allotropes, in which it can be stacked to form 3D graphite [129], rolled to form (1D) nanotubes [130], cut to form 1D nanoribbons [131], and wrapped to form 0D fullerenes [132]. The orbital hybridization mechanism in graphene is that the C-(2s, $2p_x$, and $2p_y$) orbitals are hybridized to create strong σ bonds to hold the planar 2D sheet, while C-$2p_z$ orbitals remain freestanding to form weak π bonding along the z-direction. This evidences that σ and π bonds in graphene are separated, in which the π orbitals mainly contribute to a Dirac cone structure at low-lying energy

[133]. Specifically, the long-range π conjugation in graphene leads to many novel quasiparticle properties that have been interested in many recent studies [134]. To date, graphene has been utilized in various applications such as flexible devices [135], transparent conductors [136], high-speed devices, and batteries [137]. Unfortunately, graphene displays many disadvantages for nanoelectronic applications due to its zero-gap feature [138]. To overcome the critical drawbacks of graphene, various approaches have been used to open a band gap in graphene, including chemical functionalizations [139], atom dopings [140], mechanical strains [141], bilayer structures [142], finite-size confinements [143], inducing defects [144], and applying external fields [145]. Beyond graphene, many efforts have been strongly focused on graphene-like 2D materials and other emergent 2D systems, including silicene [146], germanene [147], stanine [148], phosphorene [149], antimonene [150], bismuthine [151], transition metal dichalcogenides (TMDs) [152], topological insulators (TIs) [153], metal-organic frameworks (MOFs) [154], and Mxenes [155], in which silicene, a 2D analog of graphene, is made of silicon atoms packed in a low buckled honeycomb lattice. Silicene possesses many graphene-like quasiparticle features [156]; however, silicene presents better compatibility in silicon-based electronic devices than graphene so that silicene has stirred studies to extend its potential for practical applications [157]. Unlike graphene, silicene can only be synthesized through bottom-up methods due to a lack of graphite-like layered silicon structure. The most common method to synthesize the monolayer silicene is to deposit silicon atoms on the metallic substrates [158–160] that provide the experimental evidence for the presence of the 2D silicon sheet, which was theoretically predicted in 1994 [161]. Up to now, silicene has been extended in many applications, including room-temperature field-effect transistors (FETs) [162], gas sensors [163], and batteries [164]. Nevertheless, the critical disadvantage of silicene for electronic devices is its small gap feature [165]. Thus, a lot of studies have been conducted in opening band gap for silicene, including chemical modifications [166], quantum confinements [167], stacking configurations [43], mechanical strains [168], and applying external fields [169]. Among these methods, creating the finite size-quantum confinements is the most powerful way to create a band gap that can remain the low-buckled honeycomb lattice of 2D host silicene without any modification in the chemical hybridization mechanism in the honeycomb networks. The finite size confinements of 2D silicene result in 1D silicene nanoribbons (SiNRs) with armchair (ASiNR) and zigzag (ZSiNR) edges [170]. SiNRs show the middle-gap quasiparticle properties that can fully overcome the main obstacle of 2D host silicene for electronic devices [171]. On the experimental side, SiNRs have been successfully synthesized from both top-down and bottom-up methods. The top-down method is to cut the 2D host silicene to create 1D SiNRs [172], while the bottom-up approach is to grow 1D SiNRs on metallic substrates or an insulating thin film [173, 174]. SiNRs with their dominant quasiparticle features and their compatibility in silicon-based electronic devices have attracted much attention from the scientific community recently [175]. On the other hand, a wide range of applications requires materials having greater diverse quasiparticle properties such that enriching the essential quasiparticle properties of SiNRs is an interesting issue for many studies. To diversify the essential quasiparticle properties of SiNRs, various methods have been applied, including chemical dopings [176], edge passivations

[177], stacking configurations [178], generating lattice defects [179], applying external fields [180], and forming heterostructures [181], in which atom doping is the most effective way to dramatically diversify the essential quasiparticle properties. Up to now, many kinds of atoms have been successfully doped in SiNRs to result in their diversified quasiparticle properties [182–186]. However, halogen adsorptions on SiNRs have not yet been revealed in detail, while halogen adatoms with very strong electron affinity can create a strong bonding with silicon atoms to greatly complicate in chemical hybridization mechanism that can result in significant diversified quasiparticle properties. Therefore, the diverse quasiparticle properties of halogen-adsorbed SiNRs are worthy of further investigation in Chapter 5. Furthermore, the developed first-principles theoretical framework in this chapter can be fully generalized to many other emergent layered materials.

Graphene nanoribbon (GNR), a one-dimensional (1D) narrow strip of graphene [187–189], has recently attracted much attention due to its remarkable properties. 1D quantum confinement effects of a GNR can greatly diversify the essential properties, which can overcome the limitation of application in 2D graphene with its zero-gap electronic structure. Nanoribbon width and edge structure play critical roles in the essential properties of GNRs. According to the edge structure, there are two typical GNRs, armchair and zigzag ones (AGNRs and ZGNRs) [190, 191]. The former belongs to non-magnetic semiconductors, while the latter are anti-ferromagnetic middle-gap semiconductors. Up to now, GNRs have been successfully synthesized by various experimental methods such as lithographic [192, 193], sonochemical breaking [194], oxidization reaction [195], chemical vapor deposition [196], unzipping CNTs using plasma etching [197], and so on. Recently, GNRs are promising in the fields of energy storages, e.g., field-effect transistors [198], lithium-ion batteries [199, 200], and fuel cells [201]. To further expand the range of application, GNRs' properties can be modulated by changing the geometric structures [202, 203], doping [204, 205], and applying external electric/magnetic fields [206, 207]. Chapter 6 aims to provide a systematic study on the fundamental properties of the metal/ transition metal adatom-adsorbed GNRs. The various Al-/Fe-/Co-/Ni-adsorption structures, critical multi-orbital hybridizations, significant non-magnetism (NM)/ anti-ferromagnetism (AFM)/ferromagnetism (FM), and metallic/semiconducting behaviors will be clearly illustrated.

A single-wall carbon nanotube as well as silicon nanotube can be regarded as a rolled honeycomb lattice of graphene and silicene, respectively. The successful systematic studies of carbon nanotubes (CNTs) have been synthesized by means of arc-discharge evaporation in 1991 [208]. Later, other developments [209–213] such as characterization [214–217], property [218–220], and applications [221–223] sprang up like mushrooms. Similarly, silicon nanotubes (SiNTs) were successful initially synthesized in 2000 via ozone to remove the tubular meso- and nanoporous silicate templates [223]. Soon after that, a lot of investigations have been reported such as different growth process [224–226], features [226, 227], and applications [228–230]. The electronic properties of the planar graphene nanoribbons exhibit semiconducting behavior. On the other hand, the cylindrical carbon nanotubes are metals or direct-gap semiconductors sensitive to the chirality and radius. Metallic nanotubes are exclusively comprised of either armchair nanotubes or very small zigzag nanotubes with radii $< \sqrt{3}b$ (b is the C-C bond length) [231]. The cylindrical silicon nanotubes exhibit the same behaviors as carbon nanotubes. They can be either metallic or semi-conducting depending on the radii and chiral vectors. The Metallic silicon nanotubes are only comprised of zigzag types, especially for the small size of tubes ((m,0), m $\leqq 9$), and others are semiconducting such as armchair silicon nanotubes and large size of zigzag ones. Large curvature effect enhances the σ and π mixing in the smaller tubes, leading to the metallic property. The theoretical calculations [232, 233] and experimental measurement have confirmed the curvature effects, the misorientations of $2p_z$ orbitals and hybridizations of carbon (2s, $2p_x$, $2p_y$, $2p_z$) and silicon (3s, $3p_x$, $3p_y$, $3p_z$) four orbitals, on a cylindrical surface, leading to the geometry dependent energy gaps. Chapter 7 introduces single-wall carbon and silicon nanotubes with different diameters and chiralities. The geometric structure, energy bands, spatial charge distributions, and orbital-projected density of states are discussed in detail. Silicon nanotube is characterized by sp^3 hybridization and the gear-like structure. The ground state energy E_0 obviously decreases with the increasing diameter, owing to the reduction of bond length, buckling distance (for SiNTs), and curvature effect. The variation of the band structure and PDOSs with the curvature is investigated thoroughly. The calculated results clearly indicate the unusual features of the energy band, such as energy gap, energy dispersions, band-edge states, mixing bands, band overlap, and state degeneracy. The total and local DOSs exhibit a plenty of prominent asymmetric peaks in the inverse of the square-root form. The zigzag carbon and silicon nanotubes are quite different from each other, mainly owing to the curvature effect, unsymmetrical structure, and the open/periodical boundary condition. These could be directly verified by the STS measurements.

Chapter 8 offers an analysis of electronic, optical properties of pristine silicene and substituted-silicene by B, C, and N atoms using density functional theory. Such guest-atoms possess three, four, and five electrons in the outermost cell, being suitable for a deep understanding of the quasiparticle properties of the substitutional silicene systems. The optical coefficients such as the real and imaginary dielectric function, dielectric function, electron loss function, absorption coefficient, refractivity, and reflectivity are calculated for both in-plane light polarization (perpendicular) and out of plane (parallel) polarization. The electronic and optical properties of the guest-substituted silicene systems become so different compare with the pristine ones. Our computational results present the p-type doping metallic behavior in boron-substituted silicene while in the carbon- and nitrogen-substituted cases, the systems become semiconductoring phenomena. The absorption intensity in the case of carbon-substituted silicene is highest in both polarization directions, but in the case of boron and nitrogen, it is almost unchanged compared to pristine. The comparison between the guest-atoms substitution and the pristine systems will be discussed in detail in all the properties.

Binary compounds, fully B-/C-/N-substituted germanenes, exhibit the diversified phenomena through the different chemical bondings presented in Chapter 9. The delicate first-principles calculations can present the buckling/planar honeycomb lattices, the atom-dominated band structures, the spatial charge densities, the spin density distributions, and the atom-, orbital-, and spin-decomposed density of states, being very useful in determining the critical orbital hybridizations and magnetic configurations. The concise pictures, the strong competition between sp^2 and sp^3 bondings, and the guest-dependent spin states, are responsible for the geometric symmetries, the metallic/wide-/narrow-gap behaviors, the modification/destruction of Dirac-cone structures, the nonmagnetic or ferromagnetic properties, the crossings/anti-crossings of π and σ bands, or the pure sp^3 energy bands.

Plumbene, the latest cousin of graphene, has been mentioned as a candidate material for topological insulator (TI) and room-temperature operations [12, 234] due to its rich and unique geometric and electronic properties. In 2019, Yuhara and his coworkers reported the successful fabrication of the single layer of lead atoms by molecular beam epitaxy (MBE) [235]. This work has prompted the development of related research, e.g., chemical decoration and/or hydrogenate of monolayer Plumbene. The chemical modifications, as revealed in the experimental and theoretical investigations [236, 237], are one of the most efficient approaches in dramatically changing the geometric, electronic, and optical properties through orbital hybridization modification. Very interestingly, the Hydrogen atom with 1s orbital in the electronic configuration exhibits the extremely strong chemical bonding with the Pb atom in Plumbene. The critical quasiparticle features include the significant orbital hybridizations in Pb-H chemical bonds, the significant change of the electronic properties in double and single side adsorption, the modify of optical spectrum in case of with/without excitonic effect, and very importantly, the effect of spin-orbital couplings on the electronic and optical properties of the hydrogenated systems are thoroughly examined from the first-principles simulations in Chapter 10. The current study is very useful in comprehending the crucial properties of 2D materials with chemical functionality.

Graphite is one of the mainstream materials in basic science research and potential applications [238]. Apparently, this system stirred plenty of theoretical and experimental [239, 240] studies more than one hundred years ago. Its layered structure, which consists of carbon mb lattices, exhibits the unusual crystal symmetries [241] and thus the unusual phenomena, such as, the AA [242], AB [242], ABC [243], and turbostratic stackings [244]. The graphitic spacing, being determined by the Van der Waals interactions [245], provides a very active environment in creating the chemical intercalations or de-intercalations for the various atoms/molecules/ions [246], especially for the charging and discharging processes in ion-based batteries [247]. The chemical modifications are capable of generating the n-, p-type dopings [248] or even the zero-gap semiconducting behaviors [249], the drastic changes of band structures and van Hove singularities through the zone-folding effects and significant intercalant-related interactions [13], the featured optical reflectance and absorption spectra in the presence/absence of quasi-stable excitons [250], the diverse (momentum, frequency)-dependent Coulomb excitations under distinct free carriers (the rich electron-hole and collective excitations), the very high electrical conductivities comparable to those metals, and a great enhancement of the superconducting transition temperature [250]. Very interestingly, this rather stable system is frequently utilized as the anode/cathode materials of lithium/aluminum-based batteries. The rich essential properties have been studied for the Li- and Li^+-related graphite intercalation compounds [249]. The critical quasiparticle properties, the significant orbital hybridizations in various intralayer and interlayer chemical bonds, are thoroughly examined from the first-principles simulations in Chapter 11. The intercalations and de-intercalations of large molecules are expected to become more complicated, mainly owing to the enlarged Moiré superlattices [251]. This study is very useful in comprehending aluminum-based batteries [252], certain important differences among the different graphite intercalation compounds, and close relations between the numerical methods and the phenomenological models [253].

Batteries [254], which store and release energy in terms of chemical energy, have become one of the mainstream items in research recently. Compared with other energy store systems, lithium-ion-based batteries (LIBs) have received a great deal of attention since they process desirable features, such as light weight, long life cycle, fast charging time, and ability to provide a sizable electronic current for electronic devices [255, 256]. Generally speaking, the commercial LIB is a complex combination of the electrolyte with the negative (cathode) and the positive (anode) electrodes [255, 256]. Furthermore, the physical/chemical pictures in each component are rather complicated and directly related to the performance of storage systems. The previous few theoretical studies are conducted on the geometric and electronic properties of LIBs' components through the first-principles calculations. However, the delicate results and analyses have been thoroughly absent up to now. That is to say, the calculated results are insufficient, and there are no critical mechanisms (concise physical pictures) in comprehending the diversified phenomena. The theoretical framework is based on the numerical calculations and delicate analyses were developed and applied for the layered $LiFeO_2$-a candidate for cathode compound in Chapter 12. The fundamental features, the critical quasiparticle properties, and the significant orbital hybridizations in various chemical bonds are thoroughly examined from the first-principles simulations. The charging and discharging of LIBs are expected to be complicated owing to the variation of chemical bonds and thus, orbital hybridizations. Our predictions provide certain meaningful information about the critical physical/chemical pictures in LIBs. Such state-of-the-art analysis is very useful for fully comprehending the diversified properties in anode/cathode/electrolyte and other emerging materials.

Beyond graphene, atomically thin TMDs have become a new flatform owing to their rich and unique properties [257–260]. Especially, the change in properties from monolayer to bilayer is more significant than that resulting from multilayers [261–264]. Bilayer TMDs reveal the interesting and unique properties compared to their monolayers such as higher density of states and carrier mobility [265–268]. This phenomenon anticipates superior performance in thin-film transistors and sensors. In addition, stacking orders in bilayer exhibit an alternative method in investigating their effects. Varying stacking modes in structural engineering can manipulate the electronic properties of bilayer TMDs as reported in MoS_2 [269–271] and WS_2 [272]. Among TMDs, HfX_2 (X = S, Se, or Te) is a group candidate that promises opportunities for investigation and applications based on their emergent and satisfactory findings. In the attempt to vary layered materials for reduction in the size of devices, layered structures such as monolayer or bilayer have been concerned. Besides, bulk and monolayer of HfX_2 had been explored and used in some electronic devices [273–276]. Bilayer HfX_2 should be therefore analyzed in order to enhance this group. Although bilayer $HfSe_2$ [277] has been studied and found promising for thermal conductivity, the perspective of all these materials in bilayer is still limited. Using VASP calculations, the quasiparticle problems related to electrons as formerly mentioned are resolved, indicating the close relations between theoretical framework and quasiparticles. In Chapter 13, we focus on the electronic properties of these materials constructed in bilayer to provide further information about their features.

Lithium-ion batteries have become popular and dominate in commercial purposes. They possess many high-performance characteristics such as high power, energy density, long life cycle, and friendly to the environment, as well as affordable prices. Many candidates are investigated further as potential cathode, anode, and electrolyte components. Our study in this chapter focuses on an excellent anode material with the zero-strain property of the volume during the lithium intercalation or deintercalation process. The ternary compound possesses a lot of advantages, e.g., the safety and long cycling life for lithium-ion batteries. Currently, $Li_4Ti_5O_{12}$ enters into the commercial anode product for Li^+-ion based batteries [278, 279]. Lithium titanate material presents rich and unique geometric, electronic properties under the quasiparticle framework [39, 278, 279]. The primitive cell contains a huge number of atoms, which is called a Moiré superlattice [39]. The geometric structure performs a non-uniform environment, which fundamentally comes from the Li-O and Ti-O bonds. Many significant electronic quasiparticle properties are presented such as band structures, atom-dominated energy spectrum, spatial charge density distributions, and the atom- or the orbital-decomposed density of states [279]. The theoretical quasiparticle properties could be tested under the high-resolution experimental measurements. Many experimental examinations can be used for investigating the whole structures, e.g., X-ray diffractions for measuring the lattice parameters, transmission electron microscopy (TEM) for morphology [280], angle-resolved photoemission spectroscopy (ARPES [281–283]), and scanning tunneling spectroscopy (STS [283–285]) for band structures examination along with van Hove singularities. Also, the theoretical development of a quasiparticle framework in geometry and the electronic in terms of multi-hybridizations is worthy to thoroughly investigate in Chapter 14.

In Chapter 15, the theoretical calculations for the low-lying vibrational H atoms adsorbed on the Pt(110) surface are presented. We use the H/Pt(110) model with the conventional ultrahigh vacuum (UHV) and the density functional theory (DFT) to study the phonon frequency (the quasiparticle frequency). The nature of hydrogen atoms, which were adsorbed on the four different sites of the Pt(110) surface, is shown. The most stable site of the short bridge is in agreement with previous studies. The highest stretching frequency of 2200 cm^{-1} and the zero-point energy (ZPE) of the H atom on the top site ~140 meV are calculated. Our results convincingly demonstrate the need to study the local oscillation to understand the dynamics of this system.

Bismuth chalcogenides are of great interest because of the exciting properties of topological insulators (TIs) and their potential applications in low power dissipation electronic devices, spintronics, and quantum computing. TIs are exotic materials with an insulating bulk and topologically protected surface states (TSSs) that exhibits Dirac linear energy dispersion inside the bulk gap, spin-polarization by spin-momentum locking nature. In bismuth chalcogenide TIs (e.g., Bi_2Te_3, Bi_2Se_3,

Bi_2Se_2Te, etc.), the dominant bulk conduction arising from naturally occurring crystal imperfections and residual carrier doping has greatly hindered the detection of Dirac fermions by means of weak anti-localization effect (WAL) and quantum oscillations at low temperatures. Regardless of such challenges, the transport method has been great success in probing the TSSs and studying its properties. The WAL effect agrees well with the Hikami-Larkin-Nagaoka model that allows to obtain the number of conduction channel and phase coherent length. However, in TIs, since the WAL reflects both the 3D contribution of spin–orbit coupling in bulk and the Dirac nature of the 2D TSSs, a detailed study of magnetoconductance (ΔG (θ, B)) in tilted magnetic fields ($\theta = 0$–$90°$) is essential to get insight into the origin of the observed WAL. If all the ΔG curves coincide with each other in the plot of ΔG (θ, B) versus the perpendicular component of the magnetic field, then the observed WAL effect is 2D in nature. In addition, TIs with sufficiently high surface electron mobility can present pronounced Shubnikov–de Haas (SdH) oscillations. The analysis of SdH oscillations leads to elucidating the Dirac nature of TSS with finite Berry phase and 2D Fermi surface; it also enables us to extract the carrier concentration, effective mass, Dingle temperature, and the Berry phase of TIs. Chapter 16 presents the recent advances in magnetotransport method to study on bismuth chalcogenide TIs and their most fascinating results.

Current emerging materials provide us with various significant applications in industry, particularly in energy storage and electronic equipment. In battery applications, $LiFeO_2$ can be served as a cathode material, while silicon-carbon nanotubes and $Li_4Ti_5O_{12}$ are popular for the anode side. Aluminum-chloride-graphene intercalated compounds are known as an abundant and friendly environment, which can contribute to the development of ion-based batteries. Other materials have significant applications in electronic and photoelectronic devices such as transition metal dichalcogenides (TMDs) material group HfX_2 (X = S, Se, or Te), Bismuth chalcogenide topological insulators (BiCh-TIs). In addition, penta-graphene nanoribbons metals/transition metals and halogen-adsorbed silicene nanoribbons can be developed for certain heterojunction devices and spintronics, respectively. Remarkably, spintronics is used to monitor the spin properties, which are based on the natural characteristics of electrons. Furthermore, hydrogenated absorption systems, for example, the adsorbed hydrogen on the Pt(110) surface, plumbene adsorption hydrogen, and hydrogen adsorption on two-dimensional germanene are developed to enhance hydrogen technology and battery applications. substituted silicene systems germanene and silence systems can be useful for light and lasers due to their wide band gap. Chapter 17 will present diverse practical contributions of these materials in industrial and daily applications.

In summary, Chapter 2 covers the theoretical frameworks of quasiparticle particles from both viewpoints of first-principles simulations and phenomenological models [286], as generalized from the precious developments [287]. The high-resolution experimental measurements are thoroughly characterized in Chapter 3. By delicate VASP calculations and analyses, the diverse quasiparticle phenomena clearly reveal in penta-graphene nanographene nanoribbons [Chapter 4], halogenated silicene nanoribbons [Chapter 5], metals/transition metals-adsorbed graphene nanoribbons [Chapter 6], zigzag silicon nanotubes [Chapter 7], boron-/carbon-/

nitrogen-substituted silicene [Chapter 8], adatom-substituted on germanene systems [Chapter 9], hydrogen-chemisorption plumbenes [Chapter 10], stage-1/stage-2/stage-3/stage-4 $AlCl_4$ graphite intercalation compounds [Chapter 11], ternary lithium iron oxides [Chapter 12], different stacking in bilayer HfX_2 (X=S, Se, or Te) [Chapter 13], lithium titanium oxides [Chapter 14], H-adsorbed Pt(110) surfaces [Chapter 15], and bismuth Chalcogenide topological Insulators [Chapter 16]. In addition to the theoretical analysis, the diverse related practical applications of these emerging materials will be covered in this book [Chapter 17].

In concluding remarks, the calculated results include the total ground state energies/the chemical modification energies [288], the optimal Moiré superlattices/normal unit cells [289], the atom- and spin-dominated band structures/wave functions [25], the spatial charge/spin density distributions [36], the atom- orbital- and spin-decomposed van Hove singularities [37], the net magnetic moments [205], the single-particle and many-body reflectance [290], absorption [291], transmission [292] and energy loss spectra [293], the ballistic electrical conductivities [294], the Hall quantum ones [295], and the vibration phonons [296]. They are very sufficient in identifying the various quasiparticle behaviors and further generalize the previous theoretical framework [297]. Concluding remarks, open issues, and obvious problems are, respectively, illustrated in Chapter 18–20.

REFERENCES

[1] Erohin S V, Ruan Q Y, Sorokin P B and Yakobson B I 2020 Nano-thermodynamics of chemically induced graphene-diamond transformation *Small* **16**

[2] Lee S M, Kang D S and Roh J S 2015 Bulk graphite: materials and manufacturing process *Carbon letters* **16** 135–46

[3] Weibel A, Flaureau A, Pham A, Chevallier G, Esvan J, Estournes C and Laurent C 2020 One-step synthesis of few-layered-graphene/alumina powders for strong and tough composites with high electrical conductivity *J Eur Ceram Soc* **40** 5779–89

[4] Stefan-van Staden R I and Comnea-Stancu I R 2021 Chiral single-walled carbon nanotubes as chiral selectors in multimode enantioselective sensors *Chirality* **33** 51–8

[5] Pawlak R, Liu X S, Ninova S, D'Astolfo P, Drechsel C, Sangtarash S, Haner R, Decurtins S, Sadeghi H, Lambert C J, Aschauer U, Liu S X and Meyer E 2020 Bottom-up synthesis of Nitrogen-doped porous graphene nanoribbons *J Am Chem Soc* **142** 12568–73

[6] Artyukh A A and Chernozatonskii L A 2020 Simulation of the formation and mechanical properties of layered structures with polymerized fullerene-graphene components *Jetp Lett* **111** 109–15

[7] Liu Y Y and Kim D Y 2015 Ultraviolet and blue emitting graphene quantum dots synthesized from carbon nano-onions and their comparison for metal ion sensing *Chem Commun* **51** 4176–9

[8] Gao E L, Li R S and Baughman R H 2020 Predicted confinement-enhanced stability and extraordinary mechanical properties for carbon nanotube wrapped Chains of linear carbon *Acs Nano* **14** 17071–9

[9] An X T, Zhang Y Y, Liu J J and Li S S 2013 Quantum spin hall effect induced by electric field in silicene *Appl Phys Lett* **102**

[10] Zhao F L, Wang Y, Zhang X, Liang X J, Zhang F, Wang L, Li Y, Feng Y Y and Feng W 2020 Few-layer methyl-terminated germanene-graphene nanocomposite with high capacity for stable lithium storage *Carbon* **161** 287–98

[11] Chen R B, Chen S C, Chiu C W and Lin M F 2017 Optical properties of monolayer tinene in electric fields *Sci Rep-Uk* **7**

[12] Mahmud S and Alam M K 2019 Large band gap quantum spin Hall insulator in methyl decorated plumbene monolayer: A first-principles study *Rsc Adv* **9** 42194–203

[13] Suragtkhuu S, Bat-Erdene M, Bati A S R, Shapter J G, Davaasambuu S and Batmunkh M 2020 Few-layer black phosphorus and boron-doped graphene based heteroelectrocatalyst for enhanced hydrogen evolution *J Mater Chem A* **8** 20446–52

[14] Zhong W, Zhao Y, Zhu B B, Sha J J, Walker E S, Bank S, Chen Y F, Akinwande D and Tao L 2020 Anisotropic thermoelectric effect and field-effect devices in epitaxial bismuthene on Si (111) *Nanotechnology* **31**

[15] Zou H, Zhang H, Yang Z X and Zhang Z H 2021 Magneto-electronic and spin transport properties of transition metal doped antimonene nanoribbons *Physica E* **126**

[16] Srout M, El Kazzi M, Ben Youcef H, Fromm K M and Saadoune I 2020 Improvement of the electrochemical performance by partial chemical substitution into the lithium site of titanium phosphate-based electrode materials for lithium-ion batteries: LiNi0.25Ti1.5 Fe-0.5(PO4)(3) *J Power Sources* **461**

[17] Tang F Q, Jiang T T, Tan Y, Xu X Y and Zhou Y K 2021 Preparation and electrochemical performance of silicon@graphene aerogel composites for lithium-ion batteries *J Alloy Compd* **854**

[18] Zhu Y, He X and Mo Y 2017 Strategies based on nitride materials chemistry to stabilize Li metal anode *Adv Sci* **4**(8) 1600517

[19] Rowsell J L C, Pralong V and Nazar L F 2001 Layered lithium iron nitride: A promising anode material for li-ion batteries *J Am Chem Soc* **123** 8598–9

[20] Haque F, Yi H M, Lim J, Duan L P, Pham H D, Sonar P and Uddin A 2020 Small molecular material as an interfacial layer in hybrid inverted structure perovskite solar cells *Mat Sci Semicon Proc* **108**

[21] Dong S J, Du J T, Lu Y L, Li J S, Wang L and Zhao H 2021 Ab initio identification of two-dimensional square-octagonal bismuthene doped with 3d transition metals as potential spin gapless semiconductor, bipolar magnetic semiconductor, and quantum anomalous Hall insulator *Physica E* **126**

[22] Zhang Y, Li K D and Liao J 2020 Facile synthesis of reduced-graphene-oxide/rare-earth-metal-oxide aerogels as a highly efficient adsorbent for rhodamine-B *Appl Surf Sci* **504**

[23] Tahir M, Manchon A, Sabeeh K and Schwingenschlogl U 2013 Quantum spin/valley hall effect and topological insulator phase transitions in silicene *Appl Phys Lett* **102**

[24] Luo Y, Ito Y, Zhong H F, Endou A, Kubo M, Manogaran S, Imamura A and Miyamoto A 2004 Density functional theory and tight-binding quantum chemical molecular dynamics calculations on Ce1-xCuxO2-delta catalyst and the adsorptions of CH3OH and CH3O on Ce1-xCuxO2-delta *Chem Phys Lett* **384** 30–4

[25] Pham H D, Su W P, Nguyen T D H, Tran N T T and Lin M F 2020 Rich p-type-doping phenomena in boron-substituted silicene systems *Roy Soc Open Sci* **7**

[26] Wu J Y, Chen S C and Lin M F 2014 Temperature-dependent Coulomb excitations in silicene *New J Phys* **16**

[27] Sun J, Yuan K P, Zhou W Y, Zhang X L, Onoe J, Kawazoe Y and Wang Q 2020 Low thermal conductivity of peanut-shaped carbon nanotube and its insensitive response to uniaxial strain *Nanotechnology* **31**

[28] Dutta R, Paitya N and Subash T D 2020 Electric field and surface potential analytical modeling of novel Double Gate Triple Material PiN Tunneling Graphene Nano Ribbon FET (DG-TM-PiN-TGNFET) *Silicon-Neth*

[29] Ebrahimr F, Nouraeil M, Dabbagh A and Civalek O 2019 Buckling analysis of graphene oxide powder-reinforced nanocomposite beams subjected to non-uniform magnetic field *Struct Eng Mech* **71** 351–61

[30] de Souza J F O, Ribeiro C A D and Furtado C 2014 Bound states in disclinated graphene with Coulomb impurities in the presence of a uniform magnetic field *Phys Lett A* **378** 2317–24

[31] Sindona A, Pisarra M, Gomez C V, Riccardi P, Falcone G and Bellucci S 2017 Calibration of the fine-structure constant of graphene by time-dependent density-functional theory *Phys Rev B* **96**

[32] Despoja V, Mowbray D J, Vlahovic D and Marusic L 2012 TDDFT study of time-dependent and static screening in graphene *Phys Rev B* **86**

[33] Ma Y, Lv C, Tong Z, Zhao C F, Li Y S, Hu Y Y, Yin Y H, Liu X B and Wu Z P 2020 Single-layer copper particles integrated with a carbon nanotube film for flexible electromagnetic interference shielding *J Mater Chem C* **8** 9945–53

[34] Gao Y, Lin X Q, Smart T, Ci P H, Watanabe K, Taniguchi T, Jeanloz R, Ni J and Wu J Q 2020 Band engineering of large-twist-angle graphene/h-BN Moire superlattices with pressure *Phys Rev Lett* **125**

[35] Dean C R, Wang L, Maher P, Forsythe C, Ghahari F, Gao Y, Katoch J, Ishigami M, Moon P, Koshino M, Taniguchi T, Watanabe K, Shepard K L, Hone J and Kim P 2013 Hofstadter's butterfly and the fractal quantum Hall effect in moire superlattices *Nature* **497** 598–602

[36] Pham H D, Gumbs G, Su W P, Tran N T T and Lin M F 2020 Unusual features of nitrogen substitutions in silicene *Rsc Adv* **10** 32193–201

[37] Pham H D, Lin S Y, Gumbs G, Khanh N D and Lin M F 2020 Diverse properties of carbon-substituted silicenes *Front Phys-Lausanne* **8**

[38] Muhammad I K, Swera K and Abdul M 2021 Computational study of 4d transition metals doped bismuthene for spintronics *Physica E* **126**

[39] Nguyen T D H, Pham H D, Lin S Y and Lin M F 2020 Featured properties of Li+-based battery anode: Li4Ti5O12 *Rsc Advances* **10** 14071–9

[40] Peithmann K, Korneev N, Flaspohler M, Buse K and Kratzig E 2000 Investigation of small polarons in reduced iron-doped lithium-niobate crystals by non-steady-state photocurrent techniques *Physica Status Solidi a-Applied Research* **178** R1–R3

[41] Chepkasov I V, Ghorbani-Asl M, Popov Z I, Smet J H and Krasheninnikov A V 2020 Alkali metals inside bi-layer graphene and MoS2: Insights from first-principles calculations *Nano Energy* **75**

[42] Lin S Y, Chang S L, Tran N T T, Yang P H and Lin M F 2015 H-Si bonding-induced unusual electronic properties of silicene: A method to identify hydrogen concentration *Phys Chem Phys* **17** 26443–50

[43] Lin S Y L H Y, Nguyen D K, Tran N T T, Pham H D, Chang S L, Lin C Y and Lin M F 2020 *Silicene-Based Layered Materials* (IOP Publishing)

[44] Lee S M and Niehaus T A 2020 Simulation of structural evolution using time-dependent density-functional based tight-binding method *J Nanosci Nanotechno* **20** 7206–9

[45] Paier J, Ren X, Rinke P, Scuseria G E, Gruneis A, Kresse G and Scheffler M 2012 Assessment of correlation energies based on the random-phase approximation *New J Phys* **14**

[46] Shishkin M and Kresse G 2007 Self-consistent GW calculations for semiconductors and insulators *Phys Rev B* **75**

[47] Dien V K, Han N T, Nguyen T D H, Huynh T M D, Pham H D and Lin M F 2020 Geometric and electronic properties of Li2GeO3 *Front Mater* **7**

[48] Zhao P and Chen G 2020 Spin-polarized transport properties and spin molecular Boolean logic gates in planar four-coordinate Fe complex-based molecular devices with carbon nanotube bridges and electrodes *J Magn Magn Mater* **493**

[49] Liu P T, Franchini C, Marsman M and Kresse G 2020 Assessing model-dielectric-dependent hybrid functionals on the antiferromagnetic transition-metal monoxides MnO, FeO, CoO, and NiO *J Phys-Condens Mat* **32**

[50] Liu P T, Kim B, Chen X Q, Sarma D D, Kresse G and Franchini C 2018 Relativistic GW plus BSE study of the optical properties of Ruddlesden-Popper iridates *Phys Rev Mater* **2**

[51] Yin Y H, Shao C, Zhang C, Zhang Z F, Zhang X W, Robertson J and Guo Y Z 2020 Anisotropic transport property of antimonene MOSFETs *Acs Appl Mater Inter* **12** 22378–86

[52] Bridges A, Yacoot A, Kissinger T and Tatam R P 2020 Polarization-sensitive transfer matrix modeling for displacement measuring interferometry *Appl Optics* **59** 7694–704

[53] Garcia-Flores A F, Terashita H, Granado E and Kopelevich Y 2010 Landau levels in bulk graphite by Raman spectroscopy (vol 79, 113105, 2009) *Phys Rev B* **81**

[54] Lin C Y, Wu J Y, Chang C P and Lin M F 2014 Magneto-optical selection rules of curved graphene nanoribbons and carbon nanotubes *Carbon* **69** 151–61

[55] Fandan R, Pedros J, Guinea F, Bosca A and Calle F 2019 Effect of quasiparticle excitations and exchange-correlation in Coulomb drag in graphene *Commun Phys-Uk* **2**

[56] Bao H R, Liao W H, Zhang X C, Yang H, Yang X X and Zhao H P 2017 Photoinduced quantum spin/valley hall effect and its electrical manipulation in silicene *J Appl Phys* **121**

[57] Affleck I, Borda L and Saleur H 2008 Friedel oscillations and the Kondo screening cloud (vol 77, art no 180404, 2008) *Phys Rev B* **78**

[58] Li H Y, Utama M I B, Wang S, Zhao W Y, Zhao S H, Xiao X, Jiang Y, Jiang L L, Taniguchi T, Watanabe K, Weber-Bargioni A, Zettl A and Wang F 2020 Global control of stacking-order phase transition by doping and electric field in few-layer graphene *Nano Lett* **20** 3106–12

[59] Chan Y 2020 A continuum study of ionic layer analysis for single species ion transport in coaxial carbon nanotubes *Epl-Europhys Lett* **131**

[60] Lin C Y C R B, Ho Y H and Lin F L 2020 *Electronic and Optical Properties of Graphite Related Systems* ([S.l.]: CRC Press)

[61] Abraham K M, Holleck G L, Nguyen T, Pasquariello D M and Schwartz D A 1989 The lithium titanium disulfide secondary battery *J Power Sources* **26** 313–17

[62] Uxa D, Huger E, Dorrer L and Schmidt H 2020 Lithium-silicon compounds as electrode material for Lithium-Ion batteries *J Electrochem Soc* **167**

[63] Wang S H, Yang Y and Guo K H 2020 An improved recursive total least squares estimation of capacity for electric vehicle lithium-iron phosphate batteries *Math Probl Eng* **2020**

[64] Li M, Liu T C, Bi X X, Chen Z W, Amine K, Zhong C and Lu J 2020 Cationic and anionic redox in lithium-ion based batteries *Chem Soc Rev* **49** 1688–705

[65] Chen J M, Xiong J W, Ji S M, Huo Y P, Zhao J W and Liang L 2020 All solid polymer electrolytes for lithium batteries *Prog Chem* **32** 481–96

[66] Ramzan M, Lebegue S and Ahuja R 2009 Ab initio study of lithium and sodium iron fluorophosphate cathodes for rechargeable batteries *Appl Phys Lett* **94**

[67] Li W B, Lin S Y, Tran N T T, Lin M F and Lin K I 2020 Essential geometric and electronic properties in stage-ngraphite alkali-metal-intercalation compounds *Rsc Adv* **10** 23573–81

[68] Ahmed R, Nakagawa T and Mizuno S 2020 Structure determination of ultra-flat stanene on Cu(111) using low energy electron diffraction *Surf Sci* **691**

[69] Khramov E V, Privezentsev V V, Palagushkin A N, Shcherbachev K D and Tabachkova N Y 2020 XAFS and TEM investigation of nanocluster formation in (64)Zn(+) ion-implanted and thermo-oxidized SiO(2) film *J Electron Mater* **49** 7343–8

[70] Kamedulski P, Zielinski W, Nowak P, Lukaszewicz J P and Ilnicka A 2020 3D hierarchical porous hybrid nanostructure of carbon nanotubes and N-doped activated carbon *Sci Rep-Uk* **10**

[71] Chittari B L, Chen G R, Zhang Y B, Wang F and Jung J 2019 Gate-tunable topological flat bands in trilayer graphene boron-nitride moire superlattices *Phys Rev Lett* **122**

[72] Ho Y H, Chiu Y H, Lin D H, Chang C P and Lin M F 2010 Magneto-optical selection rules in Bilayer Bernal Graphene *Acs Nano* **4** 1465–72

[73] Wang J, Bo W, Ding Y, Wang X and Mu X 2020 Optical, optoelectronic, and photoelectric properties in moire superlattices of twist bilayer graphene *Mater Today Phys* **14**

[74] Talantsev E F, Mataira R C and Crump W P 2020 Classifying superconductivity in moire graphene superlattices *Sci Rep-Uk* **10**

[75] He Y, Xiang K X, Zhou W, Zhu Y R, Chen X H and Chen H 2018 Folded-hand silicon/carbon three-dimensional networks as a binder-free advanced anode for high-performance lithium-ion batteries *Chem Eng J* **353** 666–78

[76] Slepchenkov M M and Glukhova O E 2020 Improving the sensory properties of layered phospholipid-graphene films due to the curvature of graphene layers *Polymers-Basel* **12**

[77] Sidorov A, Mudd D, Sumanasekera G, Ouseph P J, Jayanthi C S and Wu S Y 2009 Electrostatic deposition of graphene in a gaseous environment: A deterministic route for synthesizing rolled graphenes? *Nanotechnology* **20**

[78] Shih P H, Do T N, Gumbs G, Pham H D and Lin M F 2019 Electric-field-diversified optical properties of bilayer silicene *Opt Lett* **44** 4721–4

[79] Pham H D, Nguyen T D H, Vo K D, Huynh T M D and Lin M F 2020 Rich essential properties of boron, carbon, and nitrogen substituted germanenes *Appl Phys Express* **13**

[80] Matthes L, Pulci O and Bechstedt F 2014 Optical properties of two-dimensional honeycomb crystals graphene, silicene, germanene, and tinene from first principles *New J Phys* **16**

[81] Yuhara J and Le Lay G 2020 Beyond silicene: Synthesis of germanene, stanene and plumbene *Jpn J Appl Phys* **59**

[82] Bezerra L G P, Souza A L P, Lago A E A, Campos L B, Nunes T L, Paula V V, Oliveira M F and Silva A R 2019 Addition of equex STM to extender improves post-thawing longevity of collared peccaries' sperm *Biopreserv Biobank* **17** 143–7

[83] Song S K, Kim T H and Yeom H W 2019 Atomic structures of self-assembled epitaxially grown GdSi2 nanowires on Si(001) by STM *Sci Rep-Uk* **9**

[84] Hu Y H, Shu T, Mao C X, Xue L, Yan Z and Wu Y Y 2019 Arsenene and antimonene doped by group-VA atoms: First-principles studies of the geometric structures, electronic properties and STM images *Physica B* **553** 195–201

[85] Wang H X, Wu C H, Eren B, Hao Y B, Feng B M, Fang H T and Salmeron M 2019 Operando STM study of the interaction of imidazolium-based ionic liquid with graphite *Energy Storage Mater* **20** 139–45

[86] Haga T, Fujimoto Y and Saito S 2019 STM visualization of carbon impurities in sandwich structures consisting of hexagonal boron nitride and graphene *Jpn J Appl Phys* **58**

[87] Hong L, Gao D M, Wang J R and Zheng D 2020 Adsorption simulation of open-ended single-walled carbon nanotubes for various gases *Aip Adv* **10**

[88] Gao J, Uribe-Romo F J, Saathoff J D, Arslan H, Crick C R, Hein S J, Itin B, Clancy P, Dichtel W R and Loo Y L 2016 Ambipolar transport in solution-synthesized graphene nanoribbons *Acs Nano* **10** 4847–56

[89] Grzybowski G, Ware M E, Kiefer A and Claflin B 2020 Design of a remote plasma-enhanced chemical vapor deposition system for growth of tin containing group-IV alloys *J Vac Sci Technol B* **38**

[90] Arcudia J, Kempt R, Cifuentes-Quintal M E, Heine T and Merino G 2020 Blue phosphorene bilayer is a two-dimensional metal and an unambiguous classification scheme for buckled hexagonal bilayers *Phys Rev Lett* **125**

[91] Liu R Z, Ye Z Y, Wang J H and Liu L M 2019 Lifthitz transition and shadow gap in Li(Fe1-xCox) as investigated by STM/STS *J Supercond Nov Magn* **32** 3789–95

[92] Iwasawa H, Dudin P, Inui K, Masui T, Kim T K, Cacho C and Hoesch M 2019 Buried double CuO chains in YBa2Cu4O8 uncovered by nano-ARPES *Phys Rev B* **99**

[93] Zhao Y B, Li Y, Chen Y, Chen Y H, Zhou D Q and Zhao Z Q 2020 Ion implantation assisted synthesis of graphene on various dielectric substrates *Nano Res*

[94] Sundarayya Y, Mishra A K, Brand R A, Hahn H, Bansal C and Sunandana C S 2013 Magnetic phase transition and relaxation effects in lithium iron phosphate *Phys Status Solidi B* **250** 1599–605

[95] Yelgel C 2016 Electronic structure of ABC-stacked multilayer graphene and trigonal warping: A first principles calculation *Journal of Physics: Conference Series* 012022

[96] Meng S J, Shi H Y, Sun X D and Gao B 2020 Tuning ultrafast charge carrier dynamics of monolayer graphene using substrates *J Phys Chem C* **124** 21147–54

[97] Khademi A, Sajadi E, Dosanjh P, Bonn D A, Folk J A, Stohr A, Starke U and Forti S 2016 Alkali doping of graphene: The crucial role of high-temperature annealing *Phys Rev B* **94**

[98] Wang Y N, Zhang L, Yang W, Lv S S, Su C H, Xiao H, Zhang F Y, Sui Q M, Jia L and Jiang M S 2020 An in situ reflectance spectroscopic investigation to monitor two-dimensional MoS2 flakes on a sapphire substrate *Materials* **13**

[99] Olimov K, Falk M, Buse K, Woike T, Hormes J and Modrow H 2006 X-ray absorption near edge spectroscopy investigations of valency and lattice occupation site of Fe in highly iron-doped lithium niobate crystals *J Phys-Condens Mat* **18** 5135–46

[100] Qi Y P, Wang Y, Zhang X W, Liu C Q, Hu B B, Bai Y L and Wang X X 2019 A theoretical study of optically enhanced transmission characteristics of subwavelength metal Y-shaped arrays and its application on refractive index sensor *Results Phys* **15**

[101] Liu J Q, Li X L, Han Y D, Wu J B, Zhang X, Wang Z P and Xu Y 2020 Synergetic effect of tetraethylammonium bromide addition on the morphology evolution and enhanced photoluminescence of rare-earth metal-organic frameworks *Inorg Chem* **59** 14318–25

[102] Wu J Y, Chen S C, Do T N, Su W P, Gumbs G and Lin M F 2018 The diverse magneto-optical selection rules in bilayer black phosphorus *Sci Rep-Uk* **8**

[103] Chiu C W, Shyu F L, Chang C P, Chen R B and Lin M F 2003 Optical spectra of AB- and AA-stacked nanographite ribbons *J Phys Soc Jpn* **72** 170–7

[104] Rotenberg E and Bostwick A 2015 Super lattice effects in graphene on SiC(0001) and Ir(111) probed by ARPES *Synthetic Met* **210** 85–94

[105] Trainer D J, Putilov A V, Wang B K, Lane C, Saari T, Chang T R, Jeng H T, Lin H, Xi X X, Nieminen J, Bansil A and Iavarone M 2019 Moire superlattices and 2D electronic properties of graphite/MoS2 heterostructures *J Phys Chem Solids* **128** 325–30

[106] Smirnov D, Schmidt H and Haug R J 2012 Aharonov-Bohm effect in an electron-hole graphene ring system *Appl Phys Lett* **100**

[107] Andrianov A V, Zakhar'in A O, Zhukavin R K, Shastin V N, Abrosimov N V and Bobylev A V 2015 Terahertz intracenter photoluminescence of silicon with lithium at interband excitation *Jetp Lett+* **100** 771–5

[108] Houshmand F, Friedman R, Jalili S and Schofield J 2020 Exciton effect in new generation of carbon nanotubes: Graphdiyne nanotubes *J Mol Model* **26**

[109] Kresse G, Marsman M, Hintzsche L E and Flage-Larsen E 2012 Optical and electronic properties of Si3N4 and alpha-SiO2 *Phys Rev B* **85**

[110] Prakash S, Sharma G and Singh V 2019 Ultra-fast tuning of refractive index in lithium niobate slab by GHz acoustic wave *Optik* **178** 256–62

[111] Torbatian Z, Alidoosti M, Novko D and Asgari R 2020 Low-loss two-dimensional plasmon modes in antimonene *Phys Rev B* **101**

[112] Men N V and Phuong D T K 2020 Plasmon modes in double-layer gapped graphene at zero temperature *Phys Lett A* **384**

[113] Van Men N 2020 Coulomb bare interactions in inhomogeneous 4-layer graphene structures *Phys Lett A* **384**

[114] Novoselov K S, Geim A K, Morozov S V, Jiang D, Katsnelson M I, Grigorieva I V, Dubonos S V and Firsov A A 2005 Two-dimensional gas of massless Dirac fermions in graphene *Nature* **438** 197–200

[115] Kahlert J U, Rawal A, Hook J M, Rendina L M and Choucair M 2014 Carborane functionalization of the aromatic network in chemically-synthesized graphene *Chem Commun* **50** 11332–4

[116] Lee K W and Lee C E 2020 Strain and doping effects on the antiferromagnetism of AB-stacked bilayer silicene *Physica B* **577**

[117] Chen X P, Yang Q, Meng R S, Jiang J K, Liang Q H, Tan C J and Sun X 2016 The electronic and optical properties of novel germanene and antimonene heterostructures *J Mater Chem C* **4** 5434–41

[118] Novoa-De Leon I C, Johny J, Vazquez-Rodriguez S, Garcia-Gomez N, Carranza-Bernal S, Mendivil I, Shaji S and Sepulveda-Guzman S 2019 Tuning the luminescence of nitrogen-doped graphene quantum dots synthesized by pulsed laser ablation in liquid and their use as a selective photoluminescence on-off-on probe for ascorbic acid detection *Carbon* **150** 455–64

[119] Parvaiz M S, Shah K A, Dar G N, Chowdhury S, Farinre O and Misra P 2020 Electronic transport in penta-graphene nanoribbon devices using carbon nanotube electrodes: A computational study *Nanosyst-Phys Chem M* **11** 176–82

[120] Sathishkumar N, Wu S Y and Chen H T 2020 Charge-modulated/electric-field controlled reversible CO2/H-2 capture and storage on metal-free N-doped penta-graphene *Chem Eng J* **391**

[121] Rajbanshi B, Sarkar S, Mandal B and Sarkar P 2016 Energetic and electronic structure of penta-graphene nanoribbons *Carbon* **100** 118–25

[122] Matos M 2002 Electronic structure of several polytypes of SiC: A study of band dispersion from a semi-empirical approach *Physica B* **324** 15–33

[123] Wickramaratne D, Weston L and Van de Walle C G 2018 Monolayer to bulk properties of hexagonal boron nitride *J Phys Chem C* **122** 25524–9

[124] Wu W Z, Lu P, Zhang Z H and Guo W L 2011 Electronic and magnetic properties and structural stability of BeO sheet and nanoribbons *Acs Appl Mater Inter* **3** 4787–95

[125] Yuan P F, Zhang Z H, Fan Z Q and Qiu M 2017 Electronic structure and magnetic properties of penta-graphene nanoribbons *Phys Chem Chem Phys* **19** 9528–36

[126] Novoselov K S, Geim A K, Morozov S V, Jiang D, Zhang Y, Dubonos S V, Grigorieva I V and Firsov A A 2004 Electric field effect in atomically thin carbon films *Science* **306** 666–9

[127] Geim A K 2012 Graphene prehistory *Phys Scripta* **T146**

[128] Zhang C and Liu T X 2012 A review on hybridization modification of graphene and its polymer nanocomposites *Chinese Sci Bull* **57** 3010–21

[129] Ahuja R, Auluck S, Trygg J, Wills J M, Eriksson O and Johansson B 1995 Electronic-structure of graphite—effect of hydrostatic-pressure *Phys Rev B* **51** 4813–19

[130] Li K Y, Eres G, Howe J, Chuang Y J, Li X F, Gu Z J, Zhang L T, Xie S S and Pan Z W 2013 Self-assembly of graphene on carbon nanotube surfaces *Sci Rep-Uk* **3**

[131] Wu Z S, Ren W C, Gao L B, Liu B L, Zhao J P and Cheng H M 2010 Efficient synthesis of graphene nanoribbons sonochemically cut from graphene sheets *Nano Res* **3** 16–22

[132] Tang C, Oppenheim T, Tung V C and Martini A 2013 Structure-stability relationships for graphene-wrapped fullerene-coated carbon nanotubes *Carbon* **61** 458–66

[133] Kariyado T and Hatsugai Y 2015 Manipulation of Dirac cones in mechanical graphene *Sci Rep-Uk* **5**

[134] Phiri J, Johansson L S, Gane P and Maloney T 2018 A comparative study of mechanical, thermal and electrical properties of graphene-, graphene oxide- and reduced graphene oxide-doped microfibrillated cellulose nanocomposites *Compos Part B-Eng* **147** 104–13

[135] Han T H, Kim H, Kwon S J and Lee T W 2017 Graphene-based flexible electronic devices *Mat Sci Eng R* **118** 1–43

[136] Wassei J K and Kaner R B 2010 Graphene, a promising transparent conductor *Mater Today* **13** 52–9

[137] Zhang F, Zhang T F, Yang X, Zhang L, Leng K, Huang Y and Chen Y S 2013 A high-performance supercapacitor-battery hybrid energy storage device based on graphene-enhanced electrode materials with ultrahigh energy density *Energ Environ Sci* **6** 1623–32

[138] Sahu S and Rout G C 2017 Band gap opening in graphene: A short theoretical study *Int Nano Lett* **7** 81–9

[139] Xu X Z, Liu C, Sun Z H, Cao T, Zhang Z H, Wang E G, Liu Z F and Liu K H 2018 Interfacial engineering in graphene band gap *Chem Soc Rev* **47** 3059–99

[140] Fan X F, Shen Z X, Liu A Q and Kuo J L 2012 Band gap opening of graphene by doping small boron nitride domains *Nanoscale* **4** 2157–65

[141] Gui G, Li J and Zhong J X 2008 Band structure engineering of graphene by strain: First-principles calculations *Phys Rev B* **78**

[142] Tang S B, Wu W H, Xie X J, Li X K and Gu J J 2017 Band gap opening of bilayer graphene by graphene oxide support doping *Rsc Adv* **7** 9862–71

[143] Villamagua L, Carini M, Stashans A and Gomez C V 2016 Band gap engineering of graphene through quantum confinement and edge distortions *Ric Mat* **65** 579–84

[144] Iyakutti K, Kumar E M, Thapa R, Rajeswarapalanichamy R, Surya V J and Kawazoe Y 2016 Effect of multiple defects and substituted impurities on the band structure of graphene: A DFT study *J Mater Sci-Mater El* **27** 12669–79

[145] Mak K F, Lui C H, Shan J and Heinz T F 2009 Observation of an electric-field-induced band gap in bilayer graphene by infrared spectroscopy *Phys Rev Lett* **102**

[146] Oughaddou H, Enriquez H, Tchalala M R, Yildirim H, Mayne A J, Bendounan A, Dujardin G, Ali M A and Kara A 2015 Silicene, a promising new 2D material *Prog Surf Sci* **90** 46–83

[147] Acun A, Zhang L, Bampoulis P, Farmanbar M, van Houselt A, Rudenko A N, Lingenfelder M, Brocks G, Poelsema B, Katsnelson M I and Zandvliet H J W 2015 Germanene: The germanium analogue of graphene *J Phys-Condens Mat* **27**

[148] Zhu F F, Chen W J, Xu Y, Gao C L, Guan D D, Liu C H, Qian D, Zhang S C and Jia J F 2015 Epitaxial growth of two-dimensional stanene *Nat Mater* **14** 1020–5

[149] Khandelwal A, Mani K, Karigerasi M H and Lahiri I 2017 Phosphorene—the two-dimensional black phosphorous: Properties, synthesis and applications *Mater Sci Eng B-Adv* **221** 17–34

[150] Singh D, Gupta S K, Sonvane Y and Lukacevic I 2016 Antimonene: A monolayer material for ultraviolet optical nanodevices *J Mater Chem C* **4** 6386–90

[151] Akturk E, Akturk O U and Ciraci S 2016 Single and bilayer bismuthene: Stability at high temperature and mechanical and electronic properties *Phys Rev B* **94**

[152] Manzeli S, Ovchinnikov D, Pasquier D, Yazyev O V and Kis A 2017 2D transition metal dichalcogenides *Nat Rev Mater* **2**

[153] Moore J E 2010 The birth of topological insulators *Nature* **464** 194–8

[154] Saraci F, Quezada-Novoa V, Donnarumma P R and Howarth A J 2020 Rare-earth metal-organic frameworks: From structure to applications *Chem Soc Rev* **49** 7949–77

[155] Gogotsi Y and Anasori B 2019 The rise of MXenes *Acs Nano* **13** 8491–4

[156] Akbari E, Buntat Z, Afroozeh A, Pourmand S E, Farhang Y and Sanati P 2016 Silicene and graphene nano materials in gas sensing mechanism *Rsc Adv* **6** 81647–53

[157] Kharadi M A, Malik G F A, Khanday F A, Shah K R A, Mittal S and Kaushik B K 2020 Review-silicene: From material to device applications *Ecs J Solid State Sc* **9**

[158] Meng L, Wang Y L, Zhang L Z, Du S X, Wu R T, Li L F, Zhang Y, Li G, Zhou H T, Hofer W A and Gao H J 2013 Buckled silicene formation on Ir(111) *Nano Lett* **13** 685–90

[159] Fleurence A, Friedlein R, Ozaki T, Kawai H, Wang Y and Yamada-Takamura Y 2012 Experimental evidence for epitaxial silicene on diboride thin films *Phys Rev Lett* **108**

[160] Vogt P, De Padova P, Quaresima C, Avila J, Frantzeskakis E, Asensio M C, Resta A, Ealet B and Le Lay G 2012 Silicene: Compelling experimental evidence for graphene like two-dimensional silicon *Phys Rev Lett* **108**

[161] Takeda K and Shiraishi K 1994 Theoretical possibility of stage corrugation in Si and Ge analogs of graphite *Phys Rev B* **50** 14916–22

[162] Lima M P, Fazzio A and da Silva A J R 2018 Silicene-based FET for logical technology *Ieee Electr Device L* **39** 1258–61

[163] Aghaei S M, Monshi M M and Calizo I 2016 A theoretical study of gas adsorption on silicene nanoribbons and its application in a highly sensitive molecule sensor *Rsc Adv* **6** 94417–28

[164] Galashev A Y and Ivanichkina K A 2020 Silicene anodes for Lithium-Ion batteries on metal substrates *J Electrochem Soc* **167**

[165] Ye M, Quhe R, Zheng J X, Ni Z Y, Wang Y Y, Yuan Y K, Tse G, Shi J J, Gao Z X and Lu J 2014 Tunable band gap in germanene by surface adsorption *Physica E* **59** 60–5

[166] Nakano H, Tetsuka H, Spencer M J S and Morishita T 2018 Chemical modification of group IV graphene analogs *Sci Technol Adv Mat* **19** 76–100

[167] Aghaei S M and Calizo I 2015 Band gap tuning of armchair silicene nanoribbons using periodic hexagonal holes *J Appl Phys* **118**

[168] Jia T T, Zheng M M, Fan X Y, Su Y, Li S J, Liu H Y, Chen G and Kawazoe Y 2015 Band gap on/off switching of silicene superlattice *J Phys Chem C* **119** 20747–54

[169] Drummond N D, Zolyomi V and Fal'ko V I 2012 Electrically tunable band gap in silicene *Phys Rev B* **85**

[170] Aghaei S M, Torres I and Calizo I 2016 Structural stability of functionalized silicene nanoribbons with normal, reconstructed, and hybrid edges *J Nanomater* **2016**

[171] Chegel R and Hasani M 2020 Electronic and thermal properties of silicene nanoribbons: Third nearest neighbor tight binding approximation *Chem Phys Lett* **761**

[172] De Padova P, Kubo O, Oivieri B, Quaresima C, Nakayama T, Aono M and Le Lay G 2012 Multilayer silicene nanoribbons *Nano Lett* **12** 5500–3

[173] van den Broek B, Houssa M, Lu A, Pourtois G, Afanas'ev V and Stesmans A 2016 Silicene nanoribbons on transition metal dichalcogenide substrates: Effects on electronic structure and ballistic transport *Nano Res* **9** 3394–406

[174] Quertite K, Enriquez H, Trcera N, Tong Y F, Bendounan A, Mayne A J, Dujardin G, Lagarde P, El Kenz A, Benyoussef A, Dappe Y J, Kara A and Oughaddou H 2021 Silicene nanoribbons on an insulating thin film *Adv Funct Mater* **31**

[175] Li H, Wang L, Liu Q H, Zheng J X, Mei W N, Gao Z X, Shi J J and Lu J 2012 High performance silicene nanoribbon field effect transistors with current saturation *Eur Phys J B* **85**

[176] Chen A B, Wang X F, Vasilopoulos P, Zhai M X and Liu Y S 2014 Spin-dependent ballistic transport properties and electronic structures of pristine and edge-doped zigzag silicene nanoribbons: Large magnetoresistance *Phys Chem Chem Phys* **16** 5113–18

[177] Yao Y, Liu A P, Bai J H, Zhang X M and Wang R 2016 Electronic structures of silicene nanoribbons: Two-edge-chemistry modification and first-principles study *Nanoscale Res Lett* **11**

[178] Deng X Q and Sheng R Q 2019 Spin transport investigation of two type silicene nanoribbons heterostructure *Phys Lett A* **383** 47–53

[179] An R L, Wang X F, Vasilopoulos P, Liu Y S, Chen A B, Dong Y J and Zhai M X 2014 Vacancy effects on electric and thermoelectric properties of zigzag silicene nanoribbons *J Phys Chem C* **118** 21339–46

[180] Fang D Q, Zhang Y and Zhang S L 2014 Silicane nanoribbons: Electronic structure and electric field modulation *New J Phys* **16**

[181] Shen M, Zhang Y Y, An X T, Liu J J and Li S S 2014 Valley polarization in graphene-silicene-graphene heterojunction in zigzag nanoribbon *J Appl Phys* **115**

[182] Lu D B, Song Y L, Huang X Y and Wang C 2018 Optical properties of a single carbon chain-doped silicene nanoribbon *J Electron Mater* **47** 4585–93

[183] Zheng F B, Zhang C W, Yan S S and Li F 2013 Novel electronic and magnetic properties in N or B doped silicene nanoribbons *J Mater Chem C* **1** 2735–43

[184] Xu R F, Han K and Li H P 2018 Effect of isotope doping on phonon thermal conductivity of silicene nanoribbons: A molecular dynamics study *Chinese Phys B* **27**

[185] Ma L, Zhang J M, Xu K W and Ji V 2014 Nitrogen and Boron substitutional doped zigzag silicene nanoribbons: Ab initio investigation *Physica E* **60** 112–17

[186] Zhang J M, Song W T, Xu K W and Ji V 2014 The study of the P doped silicene nanoribbons with first-principles *Comp Mater Sci* **95** 429–34

[187] Austin J, Way E M, Jacobberger R M, Goeltl F, Saraswat V, Mavrikakis M and Arnold M S 2019 Tightly Pitched Sub-10 Nm Nanoribbons Grown Via Seeded Anisotropic Synthesis on Ge(001) *ECS Meeting Abstracts*

[188] Hell M G, Senkovskiy B V, Fedorov A V, Nefedov A, Woll C and Gruneis A 2016 Facile preparation of Au(111)/mica substrates for high-quality graphene nanoribbon synthesis *Phys Status Solidi B* **253** 2362–5

[189] Jacobse P H, Kimouche A, Gebraad T, Ervasti M M, Thijssen M, Liljeroth P and Swart I 2017 Electronic components embedded in a single graphene nanoribbon *Nat Commun* **8**

[190] Owens F J 2008 Electronic and magnetic properties of armchair and zigzag graphene nanoribbons *J Chem Phys* **128**

[191] Singh S and Kaur I 2020 Band gap engineering in armchair graphene nanoribbon of zigzag-armchair-zigzag based nano-FET: A DFT investigation *Physica E* **118**

[192] Han M Y, Ozyilmaz B, Zhang Y B and Kim P 2007 Energy band-gap engineering of graphene nanoribbons *Phys Rev Lett* **98**

[193] Bai J W, Duan X F and Huang Y 2009 Rational fabrication of graphene nanoribbons using a nanowire etch mask *Nano Lett* **9** 2083–7

[194] Wang X R, Ouyang Y J, Li X L, Wang H L, Guo J and Dai H J 2008 Room-temperature all-semiconducting sub-10-nm graphene nanoribbon field-effect transistors *Phys Rev Lett* **100**

[195] Fujii S and Enoki T 2010 Cutting of oxidized graphene into nanosized pieces *J Am Chem Soc* **132** 10034–41

[196] Sprinkle M, Ruan M, Hu Y, Hankinson J, Rubio-Roy M, Zhang B, Wu X, Berger C and de Heer W A 2010 Scalable templated growth of graphene nanoribbons on SiC *Nat Nanotechnol* **5** 727–31

[197] Jiao L Y, Xie L M and Dai H J 2012 Densely aligned graphene nanoribbons at similar to 35 nm pitch *Nano Res* **5** 292–6

[198] Li X L, Wang X R, Zhang L, Lee S W and Dai H J 2008 Chemically derived, ultra-smooth graphene nanoribbon semiconductors *Science* **319** 1229–32

[199] Lin J, Raji A R O, Nan K W, Peng Z W, Yan Z, Samuel E L G, Natelson D and Tour J M 2014 Iron oxide nanoparticle and graphene nanoribbon composite as an anode material for high-performance li-ion batteries *Adv Funct Mater* **24** 2044–8

[200] Li Y S, Ao X, Liao J L, Jiang J J, Wang C D and Chiang W H 2017 Sub-10-nm graphene nanoribbons with tunable surface functionalities for lithium-ion batteries *Electrochim Acta* **249** 404–12

[201] Zou X L, Wang L Q and Yakobson B I 2018 Mechanisms of the oxygen reduction reaction on B- and/or N-doped carbon nanomaterials with curvature and edge effects *Nanoscale* **10** 1129–34

[202] Chang S L, Wu B R, Yang P H and Lin M F 2012 Curvature effects on electronic properties of armchair graphene nanoribbons without passivation *Phys Chem Chem Phys* **14** 16409–14

[203] Wang Y, Zhan H F, Yang C, Xiang Y and Zhang Y Y 2015 Formation of carbon nanoscrolls from graphene nanoribbons: A molecular dynamics study *Comp Mater Sci* **96** 300–5

[204] Kawai S, Saito S, Osumi S, Yamaguchi S, Foster A S, Spijker P and Meyer E 2015 Atomically controlled substitutional boron-doping of graphene nanoribbons *Nat Commun* **6**

[205] Nguyen D K, Tran N T T, Nguyen T T and Lin M F 2018 Diverse electronic and magnetic properties of chlorination-related graphene nanoribbons *Sci Rep-Uk* **8**

[206] Huang Y C, Chang C P and Lin M F 2007 Magnetic and quantum confinement effects on electronic and optical properties of graphene ribbons *Nanotechnology* **18**

[207] Son Y W, Cohen M L and Louie S G 2006 Half-metallic graphene nanoribbons *Nature* **444** 347–9

[208] Iijima S 1991 Helical microtubules of graphitic carbon *Nature* **354** 56–8

[209] Deheer W A, Bacsa W S, Chatelain A, Gerfin T, Humphreybaker R, Forro L and Ugarte D 1995 Aligned carbon nanotube films—production and optical and electronic-properties *Science* **268** 845–7

[210] Liu J, Dai H J, Hafner J H, Colbert D T, Smalley R E, Tans S J and Dekker C 1997 Fullerene "crop circles" *Nature* **385** 780–1

[211] Thess A, Lee R, Nikolaev P, Dai H J, Petit P, Robert J, Xu C H, Lee Y H, Kim S G, Rinzler A G, Colbert D T, Scuseria G E, Tomanek D, Fischer J E and Smalley R E 1996 Crystalline ropes of metallic carbon nanotubes *Science* **273** 483–7

[212] Ren Z F, Huang Z P, Xu J W, Wang J H, Bush P, Siegal M P and Provencio P N 1998 Synthesis of large arrays of well-aligned carbon nanotubes on glass *Science* **282** 1105–7

[213] Fan S S, Chapline M G, Franklin N R, Tombler T W, Cassell A M and Dai H J 1999 Self-oriented regular arrays of carbon nanotubes and their field emission properties *Science* **283** 512–14

[214] Rao A M, Richter E, Bandow S, Chase B, Eklund P C, Williams K A, Fang S, Subbaswamy K R, Menon M, Thess A, Smalley R E, Dresselhaus G and Dresselhaus M S 1997 Diameter-selective Raman scattering from vibrational modes in carbon nanotubes *Science* **275** 187–91

[215] Cowley J M, Nikolaev P, Thess A and Smalley R E 1997 Electron nano-diffraction study of carbon single-walled nanotube ropes *Chem Phys Lett* **265** 379–84

[216] Wildoer J W G, Venema L C, Rinzler A G, Smalley R E and Dekker C 1998 Electronic structure of atomically resolved carbon nanotubes *Nature* **391** 59–62

[217] Odom T W, Huang J L, Kim P and Lieber C M 1998 Atomic structure and electronic properties of single-walled carbon nanotubes *Abstr Pap Am Chem S* **216** U77–8

[218] Kataura H, Kumazawa Y, Maniwa Y, Umezu I, Suzuki S, Ohtsuka Y and Achiba Y 1999 Optical properties of single-wall carbon nanotubes *Synthetic Met* **103** 2555–8

[219] Jost O, Gorbunov A A, Pompe W, Pichler T, Friedlein R, Knupfer M, Reibold M, Bauer H D, Dunsch L, Golden M S and Fink J 1999 Diameter grouping in bulk samples of single-walled carbon nanotubes from optical absorption spectroscopy *Appl Phys Lett* **75** 2217–19

[220] Bursill L A, Stadelmann P A, Peng J L and Prawer S 1994 Surface-plasmon observed for carbon nanotubes *Phys Rev B* **49** 2882–7

[221] Burghard M, Klauk H and Kern K 2009 Carbon-based field-effect transistors for nano-electronics *Adv Mater* **21** 2586–600

[222] Tans S J, Verschueren A R M and Dekker C 1998 Room-temperature transistor based on a single carbon nanotube *Nature* **393** 49–52

[223] Martel R, Schmidt T, Shea H R, Hertel T and Avouris P 1998 Single- and multi-wall carbon nanotube field-effect transistors *Appl Phys Lett* **73** 2447–9

[224] Schmidt O G and Eberl K 2001 Nanotechnology—Thin solid films roll up into nano-tubes *Nature* **410** 168

[225] Jeong S Y, Kim J Y, Yang H D, Yoon B N, Choi S H, Kang H K, Yang C W and Lee Y H 2003 Synthesis of silicon nanotubes on porous alumina using molecular beam epitaxy *Adv Mater* **15** 1172–6

[226] De Crescenzi M, Castrucci P, Scarselli M, Diociaiuti M, Chaudhari P S, Balasubramanian C, Bhave T M and Bhoraskar S V 2005 Experimental imaging of silicon nanotubes *Appl Phys Lett* **86**

[227] Yari H, Pakizeh M and Namvar-Mahboub M 2019 Effect of silica nanotubes on characteristic and performance of PVDF nanocomposite membrane for nitrate removal application *J Nanopart Res* **21**

[228] Kasavajjula U, Wang C S and Appleby A J 2007 Nano- and bulk-silicon-based insertion anodes for lithium-ion secondary cells *J Power Sources* **163** 1003–39

[229] Zhu J, Yu Z F, Burkhard G F, Hsu C M, Connor S T, Xu Y Q, Wang Q, McGehee M, Fan S H and Cui Y 2009 Optical absorption enhancement in amorphous silicon nanowire and nanocone arrays *Nano Lett* **9** 279–82

[230] Feng K, Li M, Liu W W, Kashkooli A G, Xiao X C, Cai M and Chen Z W 2018 Silicon-based anodes for lithium-ion batteries: From fundamentals to practical applications *Small* **14**

[231] Miyake T and Saito S 2003 Quasiparticle band structure of carbon nanotubes *Phys Rev B* **68**

[232] Kane C L and Mele E J 1997 Size, shape, and low energy electronic structure of carbon nanotubes *Phys Rev Lett* **78** 1932–5

[233] Shyu F L and Lin M F 2002 Electronic and optical properties of narrow-gap carbon nanotubes *J Phys Soc Jpn* **71** 1820–3

[234] Zhao H, Zhang C W, Ji W X, Zhang R W, Li S S, Yan S S, Zhang B M, Li P and Wang P J 2016 Unexpected giant-gap quantum spin hall insulator in chemically decorated plumbene monolayer *Sci Rep-Uk* **6**

[235] Yuhara J, He B J, Matsunami N, Nakatake M and Le Lay G 2019 Graphene's latest cousin: Plumbene epitaxial growth on a "Nano WaterCube" *Adv Mater* **31**

[236] Elias D C, Nair R R, Mohiuddin T M G, Morozov S V, Blake P, Halsall M P, Ferrari A C, Boukhvalov D W, Katsnelson M I, Geim A K and Novoselov K S 2009 Control of graphene's properties by reversible hydrogenation: Evidence for graphane *Science* **323** 610–13

[237] Pumera M and Wong C H A 2013 Graphane and hydrogenated graphene *Chem Soc Rev* **42** 5987–95

[238] Brownson D A C and Banks C E 2010 Graphene electrochemistry: An overview of potential applications *Analyst* **135** 2768–78

[239] Abergel D S L, Apalkov V, Berashevich J, Ziegler K and Chakraborty T 2010 Properties of graphene: A theoretical perspective *Adv Phys* **59** 261–482

[240] Geim A K and MacDonald A H 2007 Graphene: Exploring carbon flatland *Phys Today* **60** 35–41

[241] Nakamura D, Suzumura A and Shigetoh K 2015 Sintered tantalum carbide coatings on graphite substrates: Highly reliable protective coatings for bulk and epitaxial growth *Appl Phys Lett* **106**

[242] Ho J H, Lu C L, Hwang C C, Chang C P and Lin M F 2006 Coulomb excitations in AA- and AB-stacked bilayer graphites *Phys Rev B* **74**

[243] Ling C Y, Lee M H and Lin M F 2018 Coulomb excitations in ABC-stacked trilayer graphene *Phys Rev B* **98**

[244] Shallcross S, Sharma S, Kandelaki E and Pankratov O A 2010 Electronic structure of turbostratic graphene *Phys Rev B* **81**

[245] Zhang F, Jia L Q, Sun X T, Dai X Q, Huang Q X and Li W 2020 Tuning Schottky barrier in graphene/InSe van der Waals heterostructures by electric field *Acta Phys Sin-Ch Ed* **69**

[246] Eklund P C 1981 Optical studies of the electronic and lattice dynamical properties of graphite-intercalation compounds *B Am Phys Soc* **26** 265

[247] Xu J T, Dou Y H, Wei Z X, Ma J M, Deng Y H, Li Y T, Liu H K and Dou S X 2017 Recent progress in graphite intercalation compounds for rechargeable metal (Li, Na, K, Al)-Ion batteries *Adv Sci* **4**

[248] Meng X Q, Tongay S, Kang J, Chen Z H, Wu F M, Li S S, Xia J B, Li J B and Wu J Q 2013 Stable p- and n-type doping of few-layer graphene/graphite *Carbon* **57** 507–14

[249] Toyoura K, Koyama Y, Kuwabara A, Oba F and Tanaka I 2008 First-principles approach to chemical diffusion of lithium atoms in a graphite intercalation compound *Phys Rev B* **78**

[250] Marinopoulos A G, Reining L, Rubio A and Olevano V 2004 Ab initio study of the optical absorption and wave-vector-dependent dielectric response of graphite *Phys Rev B* **69**

[251] Ryu Y K, Frisenda R and Castellanos-Gomez A 2019 Superlattices based on van der Waals 2D materials *Chem Commun* **55** 11498–510

[252] Gao Y R, Zhu C Q, Chen Z Z and Lu G 2017 Understanding ultrafast rechargeable aluminum-ion battery from first-principles *J Phys Chem C* **121** 7131–8

[253] Dresselhaus G and Leung S Y 1981 Phenomenological model for the electronic-structure of graphite-intercalation compounds *Physica B & C* **105** 495–500

[254] Christensen J, Albertus P, Sanchez-Carrera R S, Lohmann T, Kozinsky B, Liedtke R, Ahmed J and Kojic A 2011 A critical review of li/air batteries *Journal of the Electrochemical Society* **159** R1

[255] Pistoia G 2013 *Lithium-Ion Batteries* (Amsterdam: Elsevier)

[256] Yoshio M, Brodd R J and Kozawa A 2009 *Lithium-Ion Batteries* vol 1 (Berlin: Springer)

[257] Xu M S, Liang T, Shi M M and Chen H Z 2013 Graphene-like two-dimensional materials *Chemical Reviews* **113** 3766–98

[258] Kim S Y, Kwak J, Ciobanu C V and Kwon S Y 2019 Recent developments in controlled vapor-phase growth of 2D group 6 transition metal dichalcogenides *Adv Mater* **31**

[259] Liu G B, Xiao D, Yao Y G, Xu X D and Yao W 2015 Electronic structures and theoretical modelling of two-dimensional group-VIB transition metal dichalcogenides *Chemical Society Reviews* **44** 2643–63

[260] Xia J, Yan J X and Shen Z X 2017 Transition metal dichalcogenides: Structural, optical and electronic property tuning via thickness and stacking *Flatchem* **4** 1–19

[261] Puretzky A A, Liang L B, Li X F, Xiao K, Sumpter B G, Meunier V and Geohegan D B 2016 Twisted MoSe2 bilayers with variable local stacking and interlayer coupling revealed by low-frequency Raman spectroscopy *Acs Nano* **10** 2736–44

[262] Samad L, Bladow S M, Ding Q, Zhuo J Q, Jacobberger R M, Arnold M S and Jin S 2016 Layer-controlled chemical vapor deposition growth of MoS2 vertical heterostructures via van der Waals epitaxy *Acs Nano* **10** 7039–46

[263] Ye H, Zhou J D, Er D Q, Price C C, Yu Z Y, Liu Y M, Lowengrub J, Lou J, Liu Z and Shenoy V B 2017 Toward a mechanistic understanding of vertical growth of van der Waals stacked 2D materials: A multiscale model and experiments *Acs Nano* **11** 12780–8

[264] He Y M, Sobhani A, Lei S D, Zhang Z H, Gong Y J, Jin Z H, Zhou W, Yang Y C, Zhang Y, Wang X F, Yakobson B, Vajtai R, Halas N J, Li B, Xie E Q and Ajayan P 2016 Layer engineering of 2D semiconductor junctions *Advanced Materials* **28** 5126–32

[265] Kim S, Konar A, Hwang W S, Lee J H, Lee J, Yang J, Jung C, Kim H, Yoo J B, Choi J Y, Jin Y W, Lee S Y, Jena D, Choi W and Kim K 2012 High-mobility and low-power thin-film transistors based on multilayer MoS2 crystals *Nature Communications* **3**

[266] Liu X C, Qu D S, Ryu J J, Ahmed F, Yang Z, Lee D Y and Yoo W J 2016 P-type polar transition of chemically doped multilayer MoS2 transistor *Advanced Materials* **28** 2345–51

[267] Li H, Yin Z Y, He Q Y, Li H, Huang X, Lu G, Fam D W H, Tok A I Y, Zhang Q and Zhang H 2012 Fabrication of single- and multilayer MoS2 film-based field-effect transistors for sensing NO at room temperature *Small* **8** 63–7

[268] Das S, Chen H Y, Penumatcha A V and Appenzeller J 2013 High performance multilayer MoS2 transistors with scandium contacts *Nano Letters* **13** 100–5

[269] Yan A, Ong C S, Qiu D Y, Ophus C, Ciston J, Merino C, Louie S G and Zettl A 2017 Dynamics of symmetry-breaking stacking boundaries in bilayer MoS2 *The Journal of Physical Chemistry C* **121** 22559–66

[270] Nalin Mehta A, Gauquelin N, Nord M, Orekhov A, Bender H, Cerbu D, Verbeeck J and Vandervorst W 2020 Unravelling stacking order in epitaxial bilayer MX2 using 4D-STEM with unsupervised learning *Nanotechnology* **31** 445702

[271] Zeng F, Zhang W-B and Tang B-Y 2015 Electronic structures and elastic properties of monolayer and bilayer transition metal dichalcogenides MX₂ (M = Mo, W; X = O, S, Se, Te): A comparative first-principles study *Chinese Physics B* **24**

[272] Wang Y, Cong C, Shang J, Eginligil M, Jin Y, Li G, Chen Y, Peimyoo N and Yu T 2019 Unveiling exceptionally robust valley contrast in AA- and AB-stacked bilayer WS2 *Nanoscale Horiz* **4** 396–403

[273] Kanazawa T, Amemiya T, Ishikawa A, Upadhyaya V, Tsuruta K, Tanaka T and Miyamoto Y 2016 Few-layer HfS2 transistors *Scientific Reports* **6**

[274] Ding G Q, Gao G Y, Huang Z S, Zhang W X and Yao K L 2016 Thermoelectric properties of monolayer MSe2 (M = Zr, Hf): Low lattice thermal conductivity and a promising figure of merit *Nanotechnology* **27**

[275] Yin L, Xu K, Wen Y, Wang Z X, Huang Y, Wang F, Shifa T A, Cheng R, Ma H and He J 2016 Ultrafast and ultrasensitive phototransistors based on few-layered HfSe2 *Applied Physics Letters* **109**

[276] Xu K, Wang Z X, Wang F, Huang Y, Wang F M, Yin L, Jiang C and He J 2015 Ultrasensitive phototransistors based on few-layered HfS2 *Advanced Materials* **27** 7881–7

[277] Yan P, Gao G Y, Ding G Q and Qin D 2019 Bilayer MSe2 (M= Zr, Hf) as promising two-dimensional thermoelectric materials: A first-principles study *Rsc Advances* **9** 12394–403

[278] Liu Y, Lian J, Sun Z, Zhao M, Shi Y and Song H 2017 The first-principles study for the novel optical properties of LiTi2O4, Li4Ti5O12, Li2Ti2O4 and Li7Ti5O12 *Chem Phys Lett* **677** 114–19

[279] Özen S, Şenay V, Pat S and Korkmaz Ş 2016 Optical, morphological properties and surface energy of the transparent Li4Ti5O12 (LTO) thin film as anode material for secondary type batteries *Journal of Physics D: Applied Physics* **49** 105303

[280] Egerton R F 2008 Electron energy-loss spectroscopy in the TEM *Rep Prog Phys* **72** 016502

[281] Kordyuk A 2014 ARPES experiment in fermiology of quasi-2D metals *Low Temperature Physics* **40** 286–96

[282] Damascelli A 2004 Probing the electronic structure of complex systems by ARPES *Physica Scripta* **2004** 61

[283] Bussolotti F, Chi D, Goh K J, Huang Y L and Wee A T 2020 *2D Semiconductor Materials and Devices* (Amsterdam: Elsevier) pp 199–220

[284] Hipps K 2006 *Handbook of Applied Solid State Spectroscopy* (Berlin: Springer) pp 305–50

[285] Kano S, Tada T and Majima Y 2015 Nanoparticle characterization based on STM and STS *Chemical Society Reviews* **44** 970–87

[286] Nomoto T, Koretsune T and Arita R 2020 Local force method for the ab initio tight-binding model: Effect of spin-dependent hopping on exchange interactions *Phys Rev B* **102**

[287] Bae C S, Freeman D L, Doll J D, Kresse G and Hafner J 2000 Energetics of hydrogen chemisorbed on Cu(110): A first principles calculations study *J Chem Phys* **113** 6926–32

[288] Zhao S N and Larsson K 2016 First principle study of the attachment of graphene onto non-doped and doped diamond (111) *Diam Relat Mater* **66** 52–60

[289] Chen G R, Sui M Q, Wang D M, Wane S P, Jung J, Moon P Y, Adam S, Watanabe K, Taniguchi T, Zhou S Y, Koshino M, Zhang G Y and Zhane Y B 2017 Emergence of tertiary Dirac points in graphene moire superlattices *Nano Lett* **17** 3576–81

[290] Kamnev A A, Tugarova A V, Shchelochkov A G, Kovacs K and Kuzmann E 2020 Diffuse Reflectance Infrared Fourier Transform (DRIFT) and Mossbauer spectroscopic study of Azospirillum brasilense Sp7: Evidence for intracellular iron(II) oxidation in bacterial biomass upon lyophilisation *Spectrochim Acta A* **229**

[291] Liu L J, He P, Xia Y J, Song H, Chang L Y, Chen J L and Pao C W 2020 X-ray absorption fine structure measurements on Ru-Zn/ZSM-5 during heterogeneous catalysis using an in situ spectroscopic cell *Electron Struct* **2**

[292] Shi Z Q, Zhang J Y, Zhao Q L, Guo B and Wang H 2020 Transmission Electron Microscopy (TEM) study of anisotropic surface damages in micro-cutting Polycrystalline Aluminate Magnesium Spinel (PAMS) crystals *Ceram Int* **46** 20570–5

[293] Xu P, Zheng D G, Zhu C H, Zhang M, Tian H F, Yang H X and Li J Q 2019 Energy loss spectrum and surface modes of two-dimensional black phosphorus *J Phys-Mater* **2**

[294] Iihara Y, Kawai T and Nonoguchi Y 2020 Ionic Dopant-encapsulating single-walled carbon nanotube films with metal-like electrical conductivity *Chem-Asian J* **15** 590–3

[295] Geissler F, Budich J C and Trauzettel B 2013 Group theoretical and topological analysis of the quantum spin hall effect in silicene *New J Phys* **15**

[296] Zhu T S and Ertekin E 2016 Phonons, localization, and thermal conductivity of diamond nanothreads and amorphous graphene *Nano Lett* **16** 4763–72

[297] Ebrahimi M, Horri A, Sanaeepur M and Tavakoli M B 2020 Tight-binding description of graphene-BCN-graphene layered semiconductors *J Comput Electron* **19** 62–9

2 The Theoretical Frameworks

*Chiun-Yan Lin, Ching-Hong Ho, Jhao-Ying Wu,
Vo Khuong Dien, Wei-Bang Li, and Ming-Fa Lin*

CONTENTS

2.1 VASP-BASED THEORETICAL FRAMEWORK FOR QUASIPARTICLES

This book is devoted to investigating the diverse physical/chemical/material properties of emergent materials [1–4] that provide outstanding phenomena in developing the unified quasiparticle framework using first-principles calculations. Specific
quasiparticle characteristics such as charge distributions and spin configurations
are proposed to account for the various interacting quasiparticle behaviors [5, 6].
Particularly, the significant orbital hybridizations and atom- and orbital-created spin
distributions can fully comprehend the unusual lattice symmetries [5], electronic
structures [7], spatial charge/spin densities [8], van Hove singularities [9], and optical
absorption peaks [10]. Such development is thoroughly examined from the specific
chemical environments, e.g., the stacking configurations and layer numbers of few-
layer graphenes/silicenes [11, 12], the chemical modifications on them [13, 14], the
novel heterojunctions between monolayer/bilayer silicene on Ag (111) [15], and the
electrolyte materials of LiGeO in Li^+-related batteries [16].

Besides theoretical and experimental studies, computer modeling and simulation play indispensable roles in fundamental scientific research. The goal of numerical computation is to find the energy spectra and wave function of quasiparticles in
condensed-matter systems. However, getting an exact solution of the Schrödinger
equation with single- and many-particle interactions is rather challenging, especially
when considering complex systems with a non-uniform environment geometric
structure. Some approximated theories have been made to accomplish reliable solutions [17–19]. Currently, the first-principles simulation is an efficient method to deal
with the complicated electron-electron interactions by using a series of approximations and simplifications to get the eigenvalues and eigenfunctions [20, 21].

In this book, the essential properties of emerging materials are investigated by
first-principles calculations under the Vienna Ab initio Simulation Package (VASP)
codes [22, 23]. The basic methodology of these codes is built not only by the density

DOI: 10.1201/9781003322573-2

functional theory (DFT) but also by other ante/post-DFT corrections such as the molecular dynamics (MD) [24], the random phase approximation (RPA) [25], hybrid functionals (HSE) [26], and so on. In detail, VASP finds the approximate solution within the DFT by solving the Kohn-Sham equations [27] within the Hartree–Fock approximation [28] and evaluating the Roothaan equations [29]. Obviously, compared to the phenomenological methods (e.g., the tight-binding model), such numerical calculation is very effective in finding the optimal structure as well as the complicated orbital hybridizations in various chemical bonds in many body-particles systems. However, when the tight-binding model successfully simulates the first-principle calculation energy spectra, the various essential properties could be fully explored in future studies such as the magnetic quantization phenomena [30], the quantum spin Hall effect [31], Coulomb excitation [32], and magneto-optical properties [33].

According to the reliable VASP calculations, the phenomenological models could be further developed from the main features of essential physical properties. Their close relations are able to greatly promote theoretical progress. The optimal geometric symmetries and low-lying valence and conduction energy subbands of the former are available in establishing the significant Hamiltonian matrices. Their characteristics are mainly determined by the quasiparticle charges/orbitals and spin configurations, as clearly indicated from the thorough studies of the current book. All the intrinsic interactions cover the multi-/single-orbital hybridizations (the hopping integrals of neighboring orbitals [34]), the on-site Coulomb potentials (the different ionization energies [12]], the spin-orbital couplings [35]), and the same-site Coulomb interactions of spin-up and spin-down states (the Hubbard-like ones [36]). For example, the sp^3-bonding silicene/germanene/tinene/plumbene systems [37–40] possess the 8×8 Hermitian matrices. The strengths of different orbital mixings and the magnitudes of spin-orbital interactions will be examined from the well-fitting of the low-energy band structure ($E^{c,v} < 3$ eV). In general, the suitable parameters, which are required in the tight-binding models, could be obtained under the concise bonding cases, such as the intralayer and interlayer atomic interactions of the carbon-$2p_z$ for bulk graphites [41], layered graphenes [42], carbon nanotubes [43], and graphene nanoribbon [44].

Furthermore, the generalized tight-binding model is thoroughly explored for the various quantization phenomena due to a perpendicular uniform magnetic field (the systematic investigations in [45]). On the other hand, cathode, electrolyte, and anode materials of ion-based batteries might have many atoms and active orbitals in a primitive unit cell so that they present the complicated energy dispersions near the Fermi level/band gap. These cases might be very difficult to obtain reliable tight-binding models, mainly owing to the highly non-uniform chemical environments. Whether the phenomenological models are practicable in fully understanding the diversified quasiparticle phenomena needs to be evaluated by the chemical pictures of orbital hybridizations [46].

When the theoretical models and/or simulations on electronic properties are well consistent with each other, they can directly combine with the other modified single- and many-particle theories for fully understanding the diversified properties [30] and the potential applications [47]. For example, the generalized tight-binding model is able to present the featured Landau levels (LLs) [45], e.g., the well-defined, perturbed and undefined LL modes (three kinds of LLs [48]), and their sensitive dependences on the magnetic-field strength through the non-crossing/crossing/anti-crossing behaviors. Its combinations with the modified random-phase approximation are very useful in thoroughly exploring the magneto-electronic Coulomb excitations, such as the inter-LL electron-hole pairs and magneto-plasmon modes [49]. This theory has covered the intralayer and interlayer orbital hybridizations and Coulomb interactions simultaneously. And then the inelastic quasiparticle Coulomb decays can be investigated by the modified self-energy method (develops in [50]). The direct linking of the static dynamic Kubo formula is to exhibit the unusual Hall effects through the various quantum conductivities [51]. Scattering events between the initial and final LLs can be completely investigated by using their main features in the real space [52]. Moreover, the further associations of the dynamic Kubo formula lead to rich and unique absorption spectra, e.g., the normal, extra, and absence of magneto-optical selection rules [33]. How to develop the excitonic theories in the well-known [53], emergent [54], and green energy materials [55] become very interesting issues. Under the current investigations, the pristine and hydrogenated group-IV systems are very outstanding candidates in promoting the quasiparticle framework, since they possess the π and σ, sp^3, and s-sp^3 bondings, respectively for graphene, silicene/germanene/tinene/plumbene, and the H-chemisorption systems (the single- and double-side absorption cases [56]). The systematic studies, which will be done on the geometric symmetries, band structures/wave functions, and van Hove singularities, as well as the single-particle and many-particle optical scattering theories, need to be constructed from the unified phenomenological models. Very interestingly, the 2D excitonic theories, could be developed to further comprehend the Coulomb-field-related optical scatterings and their charge screenings in various layered materials.

A fitting question that might be asked is which kinds of electronic properties (finite-gap/zero-gap semiconductors, semimetals, or metals) and magnetic configurations (ferromagnetism, anti-ferromagnetism, or non-magnetism) the material belongs to. The answer lies in the way we deal with the delicate orbital hybridizations and spin configuration analysis. It is noteworthy that the first-principle method can be revealed from the optimal lattice symmetries [57], the atom-dominated energy bands [58], the atom- and orbital-projected density of states (DOS) [59], and the spatial charge distributions [60]. They could be used to identify the complex orbital hybridizations among different atom types of the studied material. Such chemical bonds play a critical role in the fundamental properties, accounting for rich and unique geometric structures and electronic properties. Apart from the orbital hybridizations, the spin configuration is important for creating diverse magnetic configurations [61]. By utilizing the polarized calculations, the magnetic configurations can be examined by using spin-split band structures, the net magnetic moments, spin-density distribution, and the spin-decomposed DOS [5].

Few- and multi-layer graphene systems present the unusual geometric and electronic properties through the accurate first-principles calculations. For example, carbon honeycomb lattices keep planar after the interlayer couplings, clearly indicating the weak but significant van der Waals interactions and the perpendicular π and σ bondings [sp^2] [62]. The former and the latter are, respectively, reflected in the geometry-symmetry-diversified low-energy phenomena and the well characterizations of the

π and σ sigma-electronic states. Moreover, the electronic energy spectra can exhibit the linear, parabolic, sombrero-shaped, oscillatory partially flat energy dispersions with/without the non-crossing, crossing, and anti-crossing behaviors, as well as the split state degeneracy [63].

These featured results arise from the various stacking symmetries, e.g., AAA, ABA, ABC, and AAB ones [64, 65]. That is, the interlayer $2p_z$-orbital hybridizations are responsible for the low-energy essential properties, which are supported by the spatial charge distribution and orbital-projected van Hove singularities. Furthermore, density of states can determine the specific energy ranges of the corresponding orbital hybridizations. There also exist the curvature-induced sp^3 bondings in curved graphene-related systems, such as the strong effects of semiconductor-metal transitions on carbon nanotubes [66] and folded graphene nanoribbons [67].

In the case of graphite or few-layer graphene, π-orbitals formed by $2p_z$ orbitals give rise to the weak Van der Waals interaction [68]. By adjusting the layer number and stacking configuration of graphene, this interlayer interaction will also be altered as a result of the distorted π-bond [69]. If a graphene sheet is rolled up into a structure like a carbon nanotube or fullerene, the orbitals of carbon are rehybridized owing to the changes in bond lengths and bond angles. The curvature effects will lead to the drastic changes in carbon sp^2 bonds, in which the disorientation of $2p_z$ orbitals and the hybridizations of $(2s, 2p_x, 2p_y, 2p_z)$ orbitals are induced and resulting in a mixed sp^2-sp^3 hybridization [70]. Similarly, the buckled effect of other group-IV materials, e.g. silicene, also leads to the mixture of sp^2 and sp^3 bonds [71]. Apart from this, chemical functionalization is emerging as an effective way to engineer new orbital hybridizations. It is well known that H atoms with one 1s orbital have rather strong chemical interactions with other atoms, which could hybridize with $(2s, 2p_x, 2p_y, 2p_z)$ orbitals of carbon and form the sp^3s bond in graphane [72]. In most cases, the σ-bond is very strong and hard to be modified while π-bond is weak and easy to be destroyed or distorted, leading to the disappearance or distortion of the Dirac-cone band structure together with the blue/red shift of Fermi level [58, 73].

Both absorption and substitution have greatly diversified quasiparticle behaviors in graphene-related systems. For example, alkali, hydrogen, and oxygen adatoms are, respectively, situated on the hollow, top, and bridge sites above a graphene surface [1, 74]. By the thorough analyses on the VASP results, the critical chemical absorptions are created by the ns-$2p_z$, sp^3-sp^3, and $1s$-sp^3 orbital hybridizations. The first kind leads to the almost unchanged honeycomb lattice and σ bondings but the rigid blueshift of the Fermi level [creation of high-density conduction electrons]. However, the other two kinds generate buckling crystals, breaking/serious distortion of Dirac-cone bands, finite gaps, more energy dispersions, and many van Hove singularities with the distinct forms, especially for the high-concentration absorption cases [75].

Nearly all 2D materials possess very active surface or edge structures, so chemical modifications, including the adatom chemisorptions and guest-atom substitutions, could be utilized to create the dramatic transitions of the fundamental properties. Recent theoretical and experimental studies on surface reconstructions have demonstrated diversified physical and chemical phenomena [76], such as the tunable band gaps [31], the transformation of semiconductor-metal behavior [77], the destruction and recuperation of the Dirac-cone bands [78], the spin-split energy bands under the ferromagnetic configurations of the specific adatoms [39], and the diverse van Hove singularities in density of state [57].

The position-dependent chemical bonds, specifically the fluctuations of bond lengths, which generate the highly non-uniform environment, are responsible for the critical orbital hybridizations and thus determine the various fundamental properties. Obviously, the active surface environment of materials is easily obtainable in graphene, silicene, germanene, and tinene systems due to possessing rich dangling bonds [79, 80]. Apparently, such 2D material properties are mainly determined by the outermost p_z orbitals. The chemical modification could be slightly modified or even totally destroyed pi-bondings, and consequently, the creation of the single-/multi-orbital hybridizations often appears in specific chemical bonds. We have had many systematic studies about chemical modifications in both graphene and silicene systems [81–83]. Plenty of positions in the absorption of alkali-adatoms on graphene such as bridge, hollow, and top sites were investigated. The hollow-site positions exhibit the weak but significant chemical bondings that lead the roughly rigid Dirac-cone structure with an observable blue-shift Fermi level. The well-behaved of π bonding on the planar geometric structure shows the almost full charge transfer in honeycomb lattices. Meanwhile, the bridge- and top-sites can thoroughly change the orbital hybridizations, e.g., the multi-orbital hybridizations in C-O bonds of graphene oxides and the sp^3-bonding hydrogenated graphene systems, respectively. Interestingly, using boron, carbon, and nitrogen as guest atoms in chemical substitutions that are studied by VASP calculation could provide many unusual properties. P-type doping can be generated in boron-substituted silicene that creates a huge amount of free valence holes whereas the N–Si systems present complicated four-orbital hybridization with semiconducting ferromagnetic configurations. Obviously, chemical modification is an effective method to create diverse properties or enhance the potential application.

$Li_8Ge_4O_{12}$, a well-behaved electrolyte of a Li^+-based battery, exhibits the unusual phenomena. Its Moiré superlattice, with many atoms in a unit cell, has created a highly non-uniform chemical environment and thus determines the rich features of electronic and optical properties [62]. The active orbital hybridizations in Li-O and Ge-O bonds are examined through the complicated crystal structure, atom-dominated electronic energy spectrum, spatial charge densities, atom- and orbital-project van Hove singularities, and prominent optical responses. The unusual optical transitions include the red-shift optical gap, 14 frequency-dependent absorption structures, and the most prominent plasmon mode in terms of the dielectric functions, energy loss functions, reflectance spectra, and absorption coefficients. Optical transitions, relying on the directions of electric polarization, are strongly affected by the excitonic effects. The close combinations of electronic and optical properties can identify a dominating orbital hybridization for each available excitation channel. The developed theoretical framework can account for the diverse quasiparticle behaviors in cathode/electrolyte/anode materials of ion-based batteries.

When an external electromagnetic field incident to the material surface, it can create significant coupling between charges and propagating waves. Electrons are vertically excited from the occupied states to the unoccupied ones with the same wave vector, since photons can only provide sufficient energies under the normal cases. The valence electrons due to specific orbitals will effectively screen this perturbation, create the induced current density, and thus build the electric polarization. The transverse dielectric functions, which are determined by the specific ratio of the electric field before and after screening under the momentum and frequency space, have efficiently described the vertical excitations pictures [12, 84]. As a result, all the optical physical quantities through the long-wavelength limit can be identified. Therefore, our focus will concentrate on how to calculate and get the suitable and reliable dielectric function in the absence/presence of many-body effects.

Under the single-particle picture, the intensity and the number of available excitations channels of the excitations (imaginary part of dielectric function) are mainly determined by the square of the electric dipole moment and the joint density of states associated with the valence and conduction subbands as mentioned in the previous reports. During this process, the valence holes and conduction electrons will simultaneously exist. Furthermore, the combination of them through the Coulomb potential under suitable conditions will largely reduce the onset of the excitation frequencies, modifying the scattering intensity of the excitation channels. Such physical properties are closely related to the critical orbital hybridizations and, therefore, worthy to get a deeper understanding of the multi-orbital hybridization in the novel materials.

For more than three decades, ab initio techniques have been successfully used to investigate many properties of materials [85, 86]. DFT has proven to be a very powerful tool for electronic ground-state properties [87, 88]. For the optical excitation process, one-body Green's function approaches based on the GW approximation (correction of Kohn-Sham wave functions) for the electron self-energy has turned out to be highly successful [89, 90]. Furthermore, the interaction between the valence holes and conduction electrons, the excitonic effect, can be achieved by solving the standard Bethe-Salpeter equation. For example, the 3D solid-state electrolyte compound $Li_8Ge_4O_{12}$ is especially interesting due to the rich and unique geometric and electronic properties [91]. In addition, it showed various frequency-dependent absorption peaks that related to the complex orbital hybridization, the prominent plasmon peaks, the unusual transmission spectra, the absorption coefficient, the highly anisotropic optical transitions, and the extraordinary redshift of the optical gap due to the extremely strong Coulomb interactions. These calculations could combine with the phenomenological model and the experimental measurement to efficiently explore the diversified excitation phenomena of emergent materials. However, in high-resolution optical spectroscopies, wide frequency ranges are required for estimating the frequency-dependent dielectric function, which can create full assistance of the empirical models.

Very interestingly, monolayer and AB-bt bilayer silicene on Ag (111) can display charge- and spin-diversified essential properties. That is, the rich orbital hybridizations and spin-up and spin-down distributions account for the unusual phenomena at heterojunction through the significant silicene-substrate chemical bondings [15, 92, 93]. Enough substrate layers in the VASP calculations are used to simulate the buckled crystal structures. Apparently, the highly non-uniform Si-Si, Si-Ag, and Ag-Ag bonds/the various interlayer distances come into existence under a specific Moiré superlattice, clearly illustrating the very complicated intralayer/interlayer hopping integrals of the neighboring atoms [94]. The multi-orbital hybridizations, being due to Si-[3s, $3p_x$, $3p_y$, $3p_z$] and $Ag - \left[3d_{x^2-y^2}, 3d_{z^2}, 3d_{xy}, 3d_{yz}, 3d_{zx} \right]$, are thoroughly checked to be the critical mechanism for the featured electronic properties. Their strong cooperation/competition with Ag-orbital-dominated spin configurations even creates prominent ferromagnetism under the bilayer case.

The critical roles of charges and spins, which are fully covered in the theoretical framework, can illustrate the various physical/chemical/materials phenomena. Similar studies are required for the other quasi particles, such as the coupling/separation among electrons, phonons, and photons.

2.2   FRAMEWORK OF PHENOMENOLOGICAL MODELS: QUASIPARTICLE PROPERTIES

The theoretical framework of phenomenological models, which is focused on the diverse quasiparticle properties (both charge and spin behaviors [45, 83, 95–98]), has been clearly illustrated by the systematic investigations on the layered emergent materials. For example, electronic, magnetic, optical, Coulomb-excitation, and transport properties are thoroughly studied for 3D graphites [83, 95], 2D few-layer graphenes [83, 95, 96], 1D carbon nanotubes [99], graphene nanoribbons [97], and group-IV and group-V monolayer/bilayer systems [35]. Most importantly, the generalized tight-binding model [35, 83, 96, 97], being under the strong modifications of the pristine one, can efficiently resolve magneto-electronic states even in the presence of various intrinsic interactions. Rich and unique magnetic quantization phenomena are presented in the recent predictions. Its direct linking with the frequency-dependent/static Kubo formula is available in thoroughly studying the unusual magneto-optical absorption spectra/quantum Hall effects [45]. Other modified models cover the random-phase approximation (RPA) and the screened exchange self-energy, in which the interlayer orbital hybridizations, interlayer Coulomb interactions, and external fields could be taken into account simultaneously [95, 98]. These two developed methods can provide the full information about the many-body (momentum, frequency)-excitation phase diagrams and Coulomb decay rates. When they are further combined with the generalized tight-binding model, the diversified magneto-Coulomb excitations will be fully explored for any layered systems and even for bulk materials. How to unify the other models for more understanding of quasiparticle behaviors is under current investigation, e.g., the coupling together of the electron-electron (e-e) and electron-phonon interactions.

The generalized tight-binding model, which is built from the layer-dependent sublattices [33, 52, 96], can cover all the significant interactions due to planar/buckled lattice symmetry [6, 12, 13], layer number, stacking configuration [33, 100], single-/multi-orbital hybridizations [101], defects [102], spin-orbital coupling [9, 101], gate voltage [103], uniform magnetic field [9, 104, 105], and spatially modulated/composite fields [106]. Very interestingly, the vector-potential-created Peierls phase can create a Moiré superlattice [3, 6], where the giant Hamiltonian matrix, with the complex components, is very efficiently solved under the exact diagonalization method. The effective momentum in the presence of **B** is changed into **P-**e**A/**c, where $\mathbf{A}$ is the vector potential and $\mathbf{P}$ the canonical momentum. A periodic Peierls phase $G_{R,R'} = \dfrac{2\pi}{\phi_0}\displaystyle\int_{R'}^{R}\mathbf{A}(r)\cdot d\mathbf{r}$ is introduced in the tight-binding functions, where $\phi_0 = 2\pi\hbar c/e$ (4.1356×10^{-15} [$T\cdot m^2$]) is the flux quantum and R and R' represent the positions of atoms. The Hamiltonian matrix element coupled with the Peierls phase factor is given by

$$H_{i,j}^{B} = H_{i,j}e^{iG_{ij}} = H_{i,j}e^{i\frac{2\pi}{\phi_0}\int_{R'}^{R}A(\mathbf{r})\cdot d\mathbf{r}}, \qquad (2.1)$$

where H_{ij} indicates the zero-field case. The enlarged primitive unit cell is determined by the commensurate period of the lattice structure and the Peierls phase. As shown in Figure 2.1, the periodic length in a uniform magnetic field $\mathbf{B} = B_0\hat{x}$ is expressed as $l = 3R_B b\hat{x}$, where the Landau gauge $\mathbf{A} = (0,B_0 x,0)$ is used and the dimensionless quantity R_B. is given by $\dfrac{\phi_0/\left(3\sqrt{3}b^2/2\right)}{B_0} \approx \dfrac{79000}{B_0}$. The atoms in the unit cell are calculated as 4RB, half of which belong to A atoms and the other half are B atoms.

In the generalized tight-binding model, the wave function is a linear combination of the subenvelope functions spanned over all bases in the magnetically enlarged unit cell. The wave function is decomposed into two components on account of A and B atoms, as shown here:

$$|\Psi_\mathbf{k}> = \sum_{m=1}^{2R_B-1}\left(A_o|A_{mk}> +B_o|B_{mk}>\right)+\sum_{m=1}^{2R_B}\left(A_e|A_{mk}> +B_e|B_{mk}>\right), \qquad (2.2)$$

where o and e represent the odd and even, respectively. The subenvelope functions $A_{o,e}$ and $B_{o,e}$ represent the probability amplitude of wave function contributed by each carbon atom. They are deduced to be an nth-order Hermite polynomial multiplied with a Gaussian function. In other words, quantum number n is determined

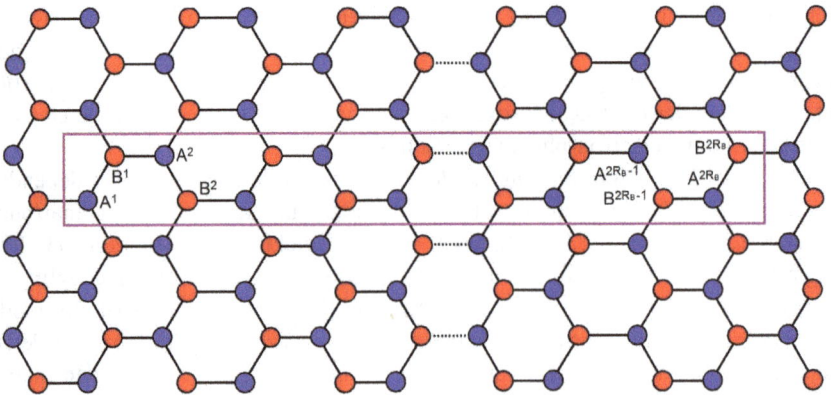

FIGURE 2.1 The enlarged primitive unit cell of graphene.

by the numbers of the zero points of the subenvelope function. This model can be applied to other condensed-matter materials. Systematic research has been completed for 3D graphites [1], 2D few-layer graphenes [83, 96], 1D carbon nanotubes [107–109], graphene nanoribbons [70, 82, 110], and group-IV and group-V monolayer/bilayer systems [30, 101, 111]. In addition to graphene, the currently known layered group-IV materials are composed of buckled honeycomb lattices, including few-layer silicene, germanene, tinene, and Pb adlayer, in which multi-orbital hybridization is further taken into account. The hopping integrals are modified due to the multi-orbital hybridization and interlayer and intralayer atomic interactions. Among them, the mis-orientation of orbitals also has a significant impact in non-planar systems. Regarding the magnetic field effect, we can obtain the explicit form of the Hamiltonian by the use of Peierls substitution. However, it is also necessary to consider boundary conditions in different dimensions, such as the periodicity along the stacking direction in graphites, the closed boundary in nanotubes, and the open boundary in ribbons [97]. A procedure for the band-like Hamiltonian matrix is introduced to efficiently solve the eigenvectors and eigenvalues by choosing an appropriate sequence for the bases.

(C, Si, Ge, Sn, Pb) atoms display four half-occupied orbitals; the electronic configurations are (2s, 2px, 2py, 2pz), (3s, 3px, 3py, 3pz), (4s, 4px, 4py, 4pz), (5s, 5px, 5py, 5pz), and (6s, 6px, 6py, 6pz), respectively. In general, these five 2D systems could be classified into three categories, C, Si/Ge, and Sn/Pb, according to the intensity of orbital hybridization; they, respectively, belong to the well separated π and σ bondings, the dominating sp^2 and sp^3 weak but significant sp^3 bondings, and the non-negligible sp^3 bondings [6]. The sharp contrast of atomic configuration between the condensed system and isolated status clearly manifests that the material surface can provide very active chemical environments. The buckled structures and the spin orbit coupling (SOC) are greatly enhanced with the increment of the atomic number. The significant orbital hybridization is predicted to cause the SOC strength to be a few orders of magnitude greater than that of planar graphene. Previously published papers have demonstrated that few-layer silicene and germanene present very obvious bottom-top and top-top stacking configurations, rather strong sp^3 bondings, and non-negligible spin-orbital interactions. Sn- and Pb-monolayer are predicted to have the important coupling effects between the sp^3 bondings and SOC, e.g., the σ-electronic states near the Fermi level.

A remarkable progress of this model has involved the aforementioned effects in the layered graphene/group-IV/group-V systems and revealed a variety of diverse magnetic quantization phenomena, such as the well-defined/perturbed/undefined Landau levels (LLs) [94, 96, 97], the normal or irregular field-strength dependences [3, 17, 18], the non-crossing/crossing/anti-crossing behaviors in the magneto-electronic energy spectra [100], and the valley-/spin-split behaviors [9]. These unusual quantization phenomena directly reflect the various band structures, with the linear, parabolic, oscillatory, partially at, sombrero-shaped, and mixing energy dispersions. The choice of r-space basis has advantages compared with k-space in calculating fundamental physical properties under complicated interactions. The calculation is difficultly achieved from the effective-mass approximation since the important hopping integrals and the magnetic-field effects are separated under the perturbation approximations.

The rich magneto-electronic properties are able to greatly diversify optical and transport properties. After the incidence of an electromagnetic wave, the vertical optical excitations are generated through the perturbation of an electric dipole moment. The coupling effects of dynamic charge screenings and magnetic quantization are fully explored through the direct combination about the generalized tight-binding model and the linear Kubo formula, mainly owing to their same bases [1, 8]. The spectral functions evaluated from the vertical transitions are expressed as

$$A(\omega) \propto \sum_{h,h'=c,v} \sum_{n,n'} \int_{1st\,BZ} \frac{dk}{(2\pi)^2} \left| \left\langle \Psi^{h'}(\mathbf{k},n') \left| \frac{\mathbf{E}\cdot\mathbf{P}}{m_e} \right| \Psi^h(\mathbf{k},n) \right\rangle \right|^2$$
$$\times Im\left[\frac{f\left(E^{h'}(\mathbf{k},n')\right) - f\left(E^h(\mathbf{k},n)\right)}{E^{h'}(\mathbf{k},n') - E^h(\mathbf{k},n) - \omega - i\Gamma} \right]$$

(2.3)

h represents the valence or conduction band, P is the momentum operator, $f\left(E^{h'}(\mathbf{k},n')\right)$ and $f\left(E^h(\mathbf{k},n)\right)$ the Fermi-Dirac distribution function, and Γ the broadening parameter due to the various deexcitation mechanisms. The two items in the integral represent velocity matrix elements due to dipole perturbation and the joint density of states. The former determines whether the transitions are available during the optical excitations.

The gradient approximation can be used to evaluate the velocity matrix elements for graphene-related systems

$$\left\langle \Psi^{h'}(\mathbf{k},n') \left| \frac{\hat{\mathbf{E}}\cdot\mathbf{P}}{m_e} \right| \Psi^h(\mathbf{k},n) \right\rangle = \frac{\partial}{\partial k_y} \left\langle \Psi^{h'}(\mathbf{k},n') | H | \Psi^h(\mathbf{k},n) \right\rangle$$

(2.4)

where the polarization direction is set along $\hat{y}$. Under this approximation, the matrix is given by the derivative of the Hamiltonian without directly calculating the left matrix product. Furthermore, in a uniform magnetic field, it is simply evaluated according to the inner product of the different subenvelope functions of the initial and final states due to the slowly change Peierls phase in the enlarged unit cell. Since the magneto-electronic subenvelope functions might possess the spatial symmetric/anti-symmetric probability distributions, the difference of quantum number between the initial and final LLs exhibits a specific rule.

Magneto-transport properties can be thoroughly examined in the framework of the generalized tight-binding model. Within the linear response, the transverse Hall conductivity is calculated from the static Kubo formula [7]

$$\sigma_{xy} = \frac{ie^2\hbar}{S} \sum_{\alpha} \sum_{\alpha\neq\beta} \left(f_\alpha - f_\beta\right) \frac{\langle \alpha | \dot{u}_x | \beta \rangle \langle \beta | \dot{u}_y | \alpha \rangle}{\left(E_\alpha - E_\beta\right)^2}$$

(2.5)

$|\alpha\rangle / |\beta\rangle$ is the initial/final LL state with energy E_α/E_β, S the area of the field-enlarged unit cell, f_α/f_β the Fermi-Dirac distribution functions, and u_x/u_y the velocity operator. With the subenvelope functions in Formula 2.5, the two matrix elements related to the velocity operator can be solved under the gradient approximation. The applicability of this model to calculate physical properties in the presence of external magnetic fields is obvious.

The normal $\Delta n = \pm 1$, extra, and vanishing selection rules are predicted to be, respectively, revealed by the well-behaved, perturbed, and undefined LLs [98, 106]. The similar scattering events are responsible for the main features of quantum Hall effects in layered graphenes [112, 113]. Monolayer, bilayer sliding systems, and trilayer AAA, ABA, ABC, and AAB stackings have clearly shown the specific non-integer conductivities, the integer ones with the different step structures, the splitting-induced reduction and diversity in electrical conductivity, a zero or non-zero conductivity at the Dirac point, and the well-like, staircase, composite, and irregular plateaus under the magnetic-dependencies. Such stacking- and layer-number-dependent characteristics arise from the previously mentioned LLs. The other group-IV and group-V 2D materials are also predicted to exhibit the diverse quantization behaviors [114].

Electron-electron interactions play critical roles in the fundamental properties, in which the band-structure effects dominate their excitation/deexcitation phenomena [70, 115–117]. To cover all the significant mechanisms of Hamiltonian, the RPA/self-energy method needs to be strongly modified for the consistency with the layer- and sublattice-based tight-binding model. The most important viewpoint lies in the sublattice-indexed electric polarization function tensors, being never present in the previous studies [118, 119]. In the presence of an external Coulomb potential, e-e interactions reflect the dynamic/static carrier screening. The electrons of the system will redistribute and create an induced potential to screen this perturbation [120]. The bare and screened response functions are thoroughly clarified from the dynamically inelastic scattering of electrons. Within the linear response, the dimensionless dielectric function in (q, ω) is defined as the ratio of the bare potential and the effective potential

$$\epsilon\left(q,\omega\right) = \lim_{V^{ex} \to 0} \frac{V^{ex}\left(q,\omega\right)}{V^{eff}\left(q,\omega\right)} \tag{2.6}$$

which can also be presented by the charge density and the longitudinal electric field, namely, ρ^{ex}/ρ^{tot} and D^{ex}/E^{tot} [121]. The bare polarization function (P) is used to describe the induced charge density in linear response to the effective Coulomb potential. By the Poisson equations and the self-consistent-field approach, the induced Coulomb potential is calculated as the product of the induced charge density and the bare Coulomb potential. As a result, the dielectric function is expressed as

$$\epsilon\left(q,\omega\right) = \epsilon_0 - V_q P\left(q,\omega\right), \tag{2.7}$$

Where ϵ_0 (= 2.4) is the background dielectric constant due to σ-electron excitations, and $V_q = 2\pi e^2/q$ for the 2D bare Coulomb potential [122]. Taking into account the band-structure effects, the bare polarization function is given by

$$P\left(q,\omega\right) = \sum_{h,h'=c,v}\sum_{n,n'} \left\langle \Psi^h\left(\mathbf{k},n\right) | e^{-i\mathbf{q}\cdot\mathbf{r}} | \Psi^{h'}\left(\mathbf{k}+\mathbf{q},n'\right) \right\rangle$$
$$\times \frac{f\left(E^{h'}\left(\mathbf{k},n'\right)\right) - f\left(E^h\left(\mathbf{k},n\right)\right)}{E^{h'}\left(\mathbf{k}+\mathbf{q},n'\right) - E^h\left(\mathbf{k},n\right) - \omega - i\Gamma}, \tag{2.8}$$

where $h = c/v$ stands for conduction/valence state. The Fermi-Dirac distribution is adjusted according to intrinsic and extrinsic cases at a finite temperature. The transferred momentum and frequency are conserved during the e-e Coulomb interaction, which is a necessary condition in describing the electronic excitations. The expressed dielectric function is also suitable to describe monolayer silicene and germanene with buckled honeycomb lattices; however, the electronic states are modified by the significant spin-orbital interactions, resulting in remarkable differences in rich and diversified electronic excitations [123].

The dielectric function of an N-layer graphene becomes more complicated [26]. The perturbed Coulomb potentials and induced charges arise from all the layers; that is, the effective potential needs to simultaneously take into account the external and induced Coulomb potentials due to each layer in the modified RPA. The stacking configuration and the number of layers determine the full band structure that can provide rich and unique electronic excitations in $(\mathbf{q}, \omega)$ -phase diagrams. When an electron beam is incident on a graphene plane, the potential is assumed to be uniform on each layer so the π electrons on different layers experience the similar bare Coulomb potential. Specifically, the excited electron and hole in each excitation pair due to the Coulomb perturbation frequently occur on distinct layers. The Feynman diagram of the Coulomb excitations is depicted in Figure 2.2. Characterized by the Dyson equation, the effective Coulomb potential for the two electrons on l-th and l'-th layers is expressed as

$$\epsilon_0 V_{ll'}^{eff}(q,\omega) = V_{ll'}(q) + \sum_{mm'} V_{lm}(q) P_{mm'}^{(1)}(q,\omega) V_{m'l'}^{eff}(q,\omega), \qquad (2.9)$$

in which each term is labeled by layer indexes and clearly reveals the Coulomb potential, induced charge density, and response function associated with any two layers. The first term is also useful in understanding the Coulomb decay rates in layered systems. The bilayer-like bare response function for the RPA bubble diagram is given by

$$P_{mm'}^{(1)}(q,\omega) = \sum_k \sum_{h,h'=c,v} \sum_{n,n'} \left(\sum_i u_{nmi}^h(\mathbf{k}) u_{n'm'i}^{*h'}(\mathbf{k+q}) \right) \times \left(\sum_{i'} u_{nmi'}^{*h}(\mathbf{k}) u_{n'm'i'}^{*h'}(\mathbf{k+q}) \right)$$
$$\times \frac{f\left(E^{h'}(\mathbf{k},n')\right) - f\left(E^h(\mathbf{k},n)\right)}{E^{h'}(\mathbf{k+q},n') - E^h(\mathbf{k},n) - \omega - i\Gamma}. \qquad (2.10)$$

Any electronic excitation, which satisfies the Pauli exclusion principle and the conservation of transferred momentum and frequency, can be decomposed into a number of layer-dependent contributions by analyzing their wave functions. This concept analysis performed on the tight-binding functions critically matches with

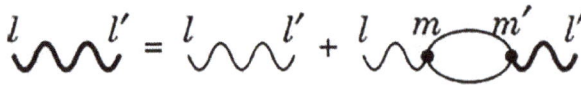

FIGURE 2.2   The Feynman diagram of the Coulomb excitations.

the layer-dependent Coulomb potential. This indicates that the modified RPA could be generalized in layered systems even in the presence of the external magnetic and electric fields. From the linear relationship between bare and screened potentials, we can write the layer-dependent dielectric function as

$$\epsilon_{ll'}\left(\mathbf{q},\omega\right)=\epsilon_0\delta_{ll'}\left(\mathbf{q}\right)-\sum_m V_{lm}\left(\mathbf{q}\right)P_{ml'}^{(1)}\left(\mathbf{q},\omega\right), \qquad (2.11)$$

showing that

$$V_{ll'}\left(\mathbf{q}\right)=\sum_{l''}\epsilon_{ll''}\left(\mathbf{q},\omega\right)V_{l''l'}^{eff}\left(\mathbf{q},\omega\right). \qquad (2.12)$$

The zero points of the dielectric tensor in ω space are available in understanding the energies of plasmon modes, while their intensities cannot be determined. In inelastic experimental measurements, the observed energy loss spectrum is associated with $\mathbf{q}$ and ω, which is necessary to be defined in the further formulas for describing the collective excitation of quasiparticles.

Very interestingly, the static charge screening phenomena are largely diversified by the atomic binding in condensed-matter systems. For the particular case of $\omega = 0$, the screening length is characterized by the static effective Coulomb potential between two charges, which is obtained by the Fourier transform from the momentum-space domain to the real-space domain [120]. The static screening effect is closely related to the Fermi momentum and density of free carriers in layered systems. Taking monolayer graphene as an example, the Fermi level is located at the Dirac point (the conduction/valence cone) in an intrinsic (extrinsic) case, so it belongs to a zero-gap semiconductor (a metal with the free electron/hole density roughly proportional to the square of the Fermi momentum). The static dielectric function used to describe the screening ability is zero/finite under the long wave-length limit $\mathbf{q}\to 0$ for the intrinsic/extrinsic graphene, which will affect the long-range coulomb behavior away from impurities [117, 124]. As a result, the effective Coulomb potential is predicted to remain the inverse distance decay function in the former case. However, for an extrinsic graphene, the e-e interaction close to the charged impurity will rapidly decrease and cause the well-known Friedel oscillations of the screening charge on the surface. The free carrier density and the Fermi surface or the energy gap are expected to play critical roles on the unusual screening behaviors. Whether the long-range effective Coulomb interactions decay quicker than the inverse of the characteristic momentum is the standard criterion for the charge screening ability. Further studies demonstrate the interesting phenomena in various systems including other important intrinsic factors, e.g., chirality, dimensionality, layer number, and stacking configuration, and extrinsic factors, e.g., temperature and external fields.

The effective energy loss function [the dimensionless screened response function] is suitably and reliably characterized from a dielectric function tensor [Dyson equation [83, 118]]. The inelastic scattering rate, corresponding to the probing electrons that transfer the specific momentum and frequency $(\mathbf{q}, \omega)$ to the system, can be delicately evaluated from the Born approximation [34]. Accounting for the experimental energy loss spectra, the dimensionless energy loss function is defined as follows according to biparticle scattering mechanism

$$\mathbf{Im}\left[-\frac{1}{\epsilon}\right] \equiv \frac{\sum_{l}\mathbf{Im}\left[-V_{ll}^{eff}\left(\mathbf{q},\omega\right)\right]}{\left[\sum_{lm}\frac{V_{lm}\left(\mathbf{q}\right)}{N}\right]} \tag{2.13}$$

which takes the average of all the external Coulomb potentials on the different layers. The dimensionless loss function is utilized to explore the various plasmon modes; all the methodology developed in the framework of modified RPA is suitable for a layered condensed-matter system, such as the layered graphene, silicene, germanene, tinene, phosphorene, antimonene, and bismuthene (the group-IV and group-V 2D materials). A unified theoretical framework, being conducted on the single- and many-particle properties, is very useful in thoroughly comprehending the diverse quasiparticle excitations and decays. For example, its successful applications on graphene-related systems clearly illustrate the lattice-symmetry-, layer-number-, stacking-configuration-, temperature-, magnetic-field-, and doping-enriched electron-hole excitation regions and plasmon modes [83]. In principle, the high-resolution angle-resolved photoemission spectroscopy can provide the full information of the quasiparticle. The e-e interactions, the orbital hybridizations (hopping integrals), and the external fields are unified together to calculate the quasiparticle self-energy, which can be used to characterize its particular status under the various scattering mechanisms of these effects. The well-defined real and imaginary parts of the self-energy can be used to define the energy shifts and widths of quasiparticles.

At low temperatures, the RPA self-energy, dominated by the e-e Coulomb interactions, is given by using the Matsubara Green's function [120]

$$\Sigma\left(\mathbf{k},h,ik_n\right) = -\frac{1}{\beta}\sum_{q,h',i\omega_m} V^{eff}\left(\mathbf{k},h,h',\mathbf{q},i\omega_m\right)G^{(0)}\left(\mathbf{k}+\mathbf{q},h',ik_n+i\omega_m\right), \tag{2.14}$$

where $G^{(0)}$ indicates the noninteracting Matsubara Green's function, $\beta = \left(k_B T\right)^{-1}$, complex fermion frequency $ik_n = \dfrac{i\left(2n+1\right)\pi}{\beta}$, and complex boson frequency $i\omega_m = \dfrac{i2m\pi}{\beta}$. The screened Coulomb potential is expressed as

$$V^{eff}\left(\mathbf{k},h,h',\mathbf{q},i\omega_m\right) = \frac{V\left(\mathbf{k},\mathbf{q},h,h'\right)}{\epsilon\left(\mathbf{q},i\omega_m\right)} = \frac{V_q\left|\left\langle\Psi^h\left(\mathbf{k},n\right)\left|e^{-i\mathbf{q}\cdot\mathbf{r}}\right|\Psi^{h'}\left(\mathbf{k}+\mathbf{q},n'\right)\right\rangle\right|^2}{\epsilon\left(\mathbf{q},i\omega_m\right)}, \tag{2.15}$$

where the band-structures effect needs to be taken into account as well as the intraband and the interband deexcitation channels.

Under the analytic continuation $ik_n \to E^h\left(k\right)$, the Coulomb decay rate of the $|\Psi^h\left(\mathbf{k},n\right)$ state is characterized as

$$\frac{-1}{2\tau(\mathbf{k},h)} = \frac{-1}{2\tau_e(\mathbf{k},h)} + \frac{-1}{2\tau_h(\mathbf{k},h)} = \sum_{q,h'}\mathbf{Im}\left[-V^{eff}(\mathbf{k},h,h',\mathbf{q},\omega_{de})\right]$$

$$\times\left\{n_B(-\omega_{de})\left[1-n_F\left(E^{h'}(\mathbf{k}+\mathbf{q})\right)\right]-n_F\left(E^{h'}(\mathbf{k}+\mathbf{q})\right)\right\}, \tag{2.16}$$

where n_B and n_F indicate the Bose-Einstein and Fermi-Dirac distribution functions, respectively, and the summation takes over all available deexcitions satisfying the Pauli exclusion principle and the conservations of energy and momentum (for detailed calculations, see [36]). In the Equation 2.16, the decay rate can be divided into electron decay and hole decay (indexed by subscript e and h) for the $|\Psi^h(\mathbf{k},n)$ state with deexcitation/decay energy $\omega_{de} = E^h(\mathbf{k}) - E^{h'}(\mathbf{k}+\mathbf{q})$ above or below the Fermi level. The diverse [momentum, frequency]-excitation diagrams become the available deexcitation channels of the quasiparticle Coulomb decay rates, leading to the very sensitive dependences on the Dirac points, the Fermi level, and the strong anisotropy.

By the detailed calculations, the Coulomb decay rates take the expression for the excited electrons and holes at zero temperature

$$\frac{1}{\tau_e(\mathbf{k},h)} + \frac{1}{\tau_h(\mathbf{k},h)} = -2\sum_{q,h'}\mathbf{Im}\left[-V^{eff}(\mathbf{k},h,h',\mathbf{q},\omega_{de})\right]$$

$$\times\left[-\Theta(\omega_{de})\Theta\left(E^{h'}(\mathbf{k}+\mathbf{q})-E_F\right)+\Theta(-\omega_{de})\Theta\left(E_F-E^{h'}(\mathbf{k}+\mathbf{q})\right)\right], \tag{2.17}$$

where E_F is the Fermi level and Θ the Heaviside step function that limits the available deexcitation channels. The derived Coulomb decay rate is just twice the energy width of the quasiparticle state. The theoretical framework of the modified self-energy equations could be applied with the layer-projection method for few-layer graphene-related systems and further generalized to cover spin-orbital effects in silicene and germanene [123, 125].

There also exist interesting charge screening behaviors, e.g., the beating Friedel oscillations in a doped monolayer graphene, the dipole-like/monopole induced charge distributions in a semiconducting/metallic carbon nanotubes, and the complicated wave-vector dependences of Coulomb scatterings in a double-walled armchair carbon nanotube [124]. In addition, the huge numerical calculations, being related to wave functions, have been overcome by a consistent framework. How to cover more quasiparticle phenomena is under current investigation, e.g., the dynamic charge density waves through strong fields, chemical modifications, spin configurations in biparticle interactions, and phonon-related inelastic scatterings. Whether the unification framework with the previously mentioned models could be achieved lies in lattice symmetries and scattering mechanisms.

REFERENCES

[1]   Lin Y T, Lin S Y, Chiu Y H and Lin M F 2017 Alkali-created rich properties in grapheme nanoribbons: Chemical bondings *Sci Rep-Uk* **7**

[2]   An X T, Zhang Y Y, Liu J J and Li S S 2013 Quantum spin hall effect induced by electric field in silicene *Appl Phys Lett* **102**

[3]   Acun A, Zhang L, Bampoulis P, Farmanbar M, van Houselt A, Rudenko A N, Lingenfelder M, Brocks G, Poelsema B, Katsnelson M I and Zandvliet H J W 2015 Germanene: The germanium analogue of graphene *J Phys-Condens Mat* **27**

[4]   Yu X L, Huang L and Wu J S 2017 From a normal insulator to a topological insulator in plumbene *Phys Rev B* **95**

[5]   Pham H D, Gumbs G, Su W P, Tran N T T and Lin M F 2020 Unusual features of nitrogen substitutions in silicene *Rsc Adv* **10** 32193–201

[6]   Pham H D, Lin S Y, Gumbs G, Khanh N D and Lin M F 2020 Diverse properties of carbon-substituted silicenes *Front Phys-Lausanne* **8**

[7]   Wu J Y, Chen S C, Do T N, Su W P, Gumbs G and Lin M F 2018 The diverse magneto-optical selection rules in bilayer black phosphorus *Sci Rep-Uk* **8**

[8]   Pham H D, Nguyen T D H, Vo K D, Huynh T M D and Lin M F 2020 Rich essential properties of boron, carbon, and nitrogen substituted germanenes *Appl Phys Express* **13**

[9]   Chen S C, Wu J Y and Lin M F 2018 Feature-rich magneto-electronic properties of bismuthene *New J Phys* **20**

[10] Chiu C W, Shyu F L, Chang C P, Chen R B and Lin M F 2003 Optical spectra of AB- and AA-stacked nanographite ribbons *J Phys Soc Jpn* **72** 170–7

[11] Ho J H, Lu C L, Hwang C C, Chang C P and Lin M F 2006 Coulomb excitations in AA- and AB-stacked bilayer graphites *Phys Rev B* **74**

[12] Shih P H, Do T N, Gumbs G, Pham H D and Lin M F 2019 Electric-field-diversified optical properties of bilayer silicene *Opt Lett* **44** 4721–4

[13] Li W Z, Liu M Y, Gong L, Chen Q Y, Cao C and He Y 2020 Emerging various electronic and magnetic properties of silicene by light rare-earth metal substituted doping *Superlattice Microst* **148**

[14] Nath P, Chowdhury S, Sanyal D and Jana D 2014 Ab-initio calculation of electronic and optical properties of nitrogen and boron doped graphene nanosheet *Carbon* **73** 275–82

[15] Arafune R, Lin C L, Kawahara K, Tsukahara N, Minamitani E, Kim Y, Takagi N and Kawai M 2013 Structural transition of silicene on Ag(111) *Surf Sci* **608** 297–300

[16] Lau J, DeBlock R H, Butts D M, Ashby D S, Choi C S and Dunn B S 2018 Sulfide solid electrolytes for lithium battery applications *Adv Energy Mater* **8**

[17] Blanc X, Cances E and Dupuy M S 2017 Variational projector augmented-wave method *Cr Math* **355** 665–70

[18] Forster A, Wagner C, Schuster J and Friedrich J 2017 Ab initio study of the trimethyl-aluminum atomic layer deposition process on carbon nanotubes: An alternative initial step *J Vac Sci Technol A* **35**

[19] Hurmach V V, Khrapatiy S V, Zavodovskyi D O, Prylutskyy Y I, Tauscher E and Ritter U 2020 Modeling of single-walled carbon nanotube binding to nitric oxide synthase and guanylate cyclase molecular structures *Neurophysiology* **52** 110–15

[20] Najafi F 2020 Thermodynamic studies of carbon nanotube interaction with Gemcitabine anticancer drug: DFT calculations *J Nanostructure Chem* **10** 227–42

[21] Kostelnik P, Seriani N, Kresse G, Mikkelsen A, Lundgren E, Blum V, Sikola T, Varga P and Schmid M 2007 The Pd (100)-(root 5 x root 5)R27 degrees-O surface oxide: A LEED, DFT and STM study *Surf Sci* **601** 1574–81

[22] Kresse G and Furthmuller J 1996 Efficient iterative schemes for ab initio total-energy calculations using a plane-wave basis set *Phys Rev B* **54** 11169–86

[23] Kresse G and Joubert D 1999 From ultrasoft pseudopotentials to the projector augmented-wave method *Phys Rev B* **59** 1758–75

[24] Cui S W, Wei J A, Li Q, Liu W W, Qian P and Wang X S 2021 Tolman length of simple droplet: Theoretical study and molecular dynamics simulation* *Chinese Phys B* **30**

[25] Jia F H, Kresse G, Franchini C, Liu P T, Wang J, Stroppa A and Ren W 2019 Cubic and tetragonal perovskites from the random phase approximation *Phys Rev Mater* **3**

[26] Flores E M, Moreira M L and Piotrowski M J 2020 Structural and electronic properties of bulk ZnX (X = O, S, Se, Te), ZnF2, and ZnO/ZnF2: A DFT investigation within PBE, PBE plus U, and Hybrid HSE functionals *J Phys Chem A* **124** 3778–85

[27] Kusakabe K 2001 A rigorous extension of the Kohn-Sham equation for strongly correlated electron systems *J Phys Soc Jpn* **70** 2038–48

[28] Plakhutin B N 2020 Brillouin's theorem in the Hartree-Fock method: Eliminating the limitation of the theorem for excitations in the open shell *J Chem Phys* **153**

[29] Rodriguez-Bautista M, Diaz-Garcia C, Navarrete-Lopez A M, Vargas R and Garza J 2015 Roothaan's approach to solve the Hartree-Fock equations for atoms confined by soft walls: Basis set with correct asymptotic behavior *J Chem Phys* **143**

[30] Shih P H, Do T N, Gumbs G, Huang D, Pham H D and Lin M F 2019 Rich magnetic quantization phenomena in AA bilayer silicene *Sci Rep-Uk* **9**

[31] Zhao H, Zhang C W, Ji W X, Zhang R W, Li S S, Yan S S, Zhang B M, Li P and Wang P J 2016 Unexpected giant-gap quantum spin hall insulator in chemically decorated plumbene monolayer *Sci Rep-Uk* **6**

[32] Ling C Y, Lee M H and Lin M F 2018 Coulomb excitations in ABC-stacked trilayer graphene *Phys Rev B* **98**

[33] Ho Y H, Chiu Y H, Lin D H, Chang C P and Lin M F 2010 Magneto-optical selection rules in bilayer bernal graphene *Acs Nano* **4** 1465–72

[34] Ho Y H, Chiu Y H, Su W P and Lin M F 2011 Magneto-absorption spectra of bernal graphite *Appl Phys Lett* **99**

[35] Lin M F C S C, Wu J Y and Lin C Y 2017 *Theory of Magnetoelectric Properties of 2D Systems* ([S.l.]: IOP Publishing Ltd.)

[36] Liu W V, Wilczek F and Zoller P 2004 Spin-dependent Hubbard model and a quantum phase transition in cold atoms *Phys Rev A* **70**

[37] Geissler F, Budich J C and Trauzettel B 2013 Group theoretical and topological analysis of the quantum spin hall effect in silicene *New J Phys* **15**

[38] Zhuang J C, Gao N, Li Z, Xu X, Wang J O, Zhao J J, Dou S X and Du Y 2017 Cooperative electron-phonon coupling and buckled structure in germanene on au(111) *Acs Nano* **11** 3553–9

[39] Chen R B, Chen S C, Chiu C W and Lin M F 2017 Optical properties of monolayer tinene in electric fields *Sci Rep-Uk* **7**

[40] Zhang L, Zhao H, Ji W X, Zhang C W, Li P and Wang P J 2018 Discovery of a new quantum spin hall phase in bilayer plumbene *Chem Phys Lett* **712** 78–82

[41] Ho Y H, Wang J, Chiu Y H, Lin M F and Su W P 2011 Characterization of Landau subbands in graphite: A tight-binding study *Phys Rev B* **83**

[42] Do T N, Shih P H, Chang C P, Lin C Y and Lin M F 2016 Rich magneto-absorption spectra of AAB-stacked trilayer graphene *Phys Chem Chem Phys* **18** 17597–605

[43] Marinopoulos A G, Reining L and Rubio A 2008 Ab initio study of the dielectric response of crystalline ropes of metallic single-walled carbon nanotubes: Tube-diameter and helicity effects *Phys Rev B* **78**

[44] Zschieschang U, Klauk H, Mueller I B, Strudwick A J, Hintermann T, Schwab M G, Narita A, Feng X L, Muellen K and Weitz R T 2015 Electrical characteristics of field-effect transistors based on chemically synthesized graphene nanoribbons *Adv Electron Mater* **1**

[45] Lin C-Y 2020 *Diverse Quantization Phenomena in Layered Materials* (CRC Press)

[46] Wang S Q 2011 A comparative first-principles study of orbital hybridization in two-dimensional C, Si, and Ge *Phys Chem Chem Phys* **13** 11929–38

[47] Molle A, Grazianetti C, Tao L, Taneja D, Alam M H and Akinwande D 2018 Silicene, silicene derivatives, and their device applications *Chem Soc Rev* **47** 6370–87

[48] Huang Y K, Chen S C, Ho Y H, Lin C Y and Lin M F 2014 Feature-rich magnetic quantization in sliding bilayer graphenes *Sci Rep-Uk* **4**

[49] Torbatian Z, Alidoosti M, Novko D and Asgari R 2020 Low-loss two-dimensional plasmon modes in antimonene *Phys Rev B* **101**

[50] Xu Z L, Ma M M and Liu P 2014 Self-energy-modified poisson-nernst-planck equations: WKB approximation and finite-difference approaches *Phys Rev E* **90**

[51] Ruzin I and Feng S C 1995 Universal relation between longitudinal and transverse conductivities in quantum hall-effect *Phys Rev Lett* **74** 154–7

[52] Ho Y H, Tsai S J, Lin M F and Su W P 2013 Unusual landau levels in biased bilayer bernal graphene *Phys Rev B* **87**

[53] Wang M and Li C M 2012 Excitonic properties of graphene-based materials *Nanoscale* **4** 1044–50

[54] Wei W and Jacob T 2013 Strong many-body effects in silicene-based structures *Phys Rev B* **88**

[55] Chang S H, Tseng P C, Chiang S E, Wu J R, Chen Y T, Chen C J, Yuan C T and Chen S H 2020 Structural, optical and excitonic properties of MA(x)Cs(1-x)Pb(IxBr1-x)(3) alloy thin films and their application in solar cells *Sol Energ Mat Sol C* **210**

[56] Yuan L H, Chen Y H, Kang L, Zhang C R, Wang D B, Wang C N, Zhang M L and Wu X J 2017 First-principles investigation of hydrogen storage capacity of Y-decorated porous graphene *Appl Surf Sci* **399** 463–8

[57] Tran N T T, Nguyen D K, Glukhova O E and Lin M F 2017 Coverage-dependent essential properties of halogenated graphene: A DFT study *Sci Rep-Uk* **7**

[58] Pham H D, Su W P, Nguyen T D H, Tran N T T and Lin M F 2020 Rich p-type-doping phenomena in boron-substituted silicene systems *Roy Soc Open Sci* **7**

[59] Nguyen T D H, Pham H D, Lin S Y and Lin M F 2020 Featured properties of li+-based battery anode: Li4Ti5O12 *Rsc Adv* **10** 14071–9

[60] Li W B, Lin S Y, Tran N T T, Lin M F and Lin K I 2020 Essential geometric and electronic properties in stage-ngraphite alkali-metal-intercalation compounds *Rsc Adv* **10** 23573–81

[61] Tran N T T, Dahal D, Gumbs G and Lin M F 2017 Adatom doping-enriched geometric and electronic properties of pristine graphene: A method to modify the band gap *Struct Chem* **28** 1311–18

[62] Lin S Y, Lin Y T, Tran N T T, Su W P and Lin M F 2017 Feature-rich electronic properties of aluminum-adsorbed graphenes *Carbon* **120** 209–18

[63] Tran N T T, Lin S Y, Glukhova O E and Lin M F 2016 Pi-bonding-dominated energy gaps in graphene oxide *Rsc Adv* **6** 24458–63

[64] Mohammadi Y, Moradian R and Tabar F S 2014 Effects of doping and bias voltage on the screening in AAA-stacked trilayer graphene *Solid State Commun* **193** 1–5

[65] Sugawara K, Yamamura N, Matsuda K, Norimatsu W, Kusunoki M, Sato T and Takahashi T 2018 Selective fabrication of free-standing ABA and ABC trilayer graphene with/without Dirac-cone energy bands *Npg Asia Mater* **10**

[66] Liu X, Pichler T, Knupfer M and Fink J 2003 Electronic and optical properties of alkali-metal-intercalated single-wall carbon nanotubes *Phys Rev B* **67**

[67] Gao Y, Liang X L, Han S P, Wu L, Zhang G F, Qin C B, Bao S X, Wang Q, Qi L L and Xiao L T 2020 High-efficiency adsorption for both cationic and anionic dyes using graphene nanoribbons formed by atomic-hydrogen induced single-walled carbon nanotube carpets *Carbon Lett* **30** 123–32

[68] Rui D R, Sun L Z, Kang N, Li J Y, Lin L, Peng H L, Liu Z F and Xu H Q 2020 Realization and transport investigation of a single layer-twisted bilayer graphene junction *Carbon* **163** 105–12

[69] de Vries F K, Zhu J H, Portoles E, Zheng G L, Masseroni M, Kurzmann A, Taniguchi T, Watanabe K, MacDonald A H, Ensslin K, Ihn T and Rickhaus P 2020 Combined minivalley and layer control in twisted double bilayer graphene *Phys Rev Lett* **125**

[70] Savin A V, Korznikova E A and Dmitriev S V 2020 Twistons in graphene nanoribbons on a substrate *Phys Rev B* **102**

[71] Oughaddou H, Enriquez H, Tchalala M R, Yildirim H, Mayne A J, Bendounan A, Dujardin G, Ali M A and Kara A 2015 Silicene, a promising new 2D material *Prog Surf Sci* **90** 46–83

[72] Lin S Y, Chang S L, Tran N T T, Yang P H and Lin M F 2015 H-Si bonding-induced unusual electronic properties of silicene: A method to identify hydrogen concentration *Phys Chem Chem Phys* **17** 26443–50

[73] Zheng X M, Chen W, Wang G, Yu Y Y, Qin S Q, Fang J Y, Wang F and Zhang X A 2015 The Raman redshift of graphene impacted by gold nanoparticles *Aip Adv* **5**

[74] Huang H C, Lin S Y, Wu C L and Lin M F 2016 Configuration- and concentration-dependent electronic properties of hydrogenated graphene *Carbon* **103** 84–93

[75] Nguyen D K, Tran N T T, Chiu Y H and Lin M F 2019 Concentration-diversified magnetic and electronic properties of halogen-adsorbed silicene *Sci Rep-Uk* **9**

[76] Shi P F, Wang C, Sun J Y, Lin P, Xu X T and Yang T 2020 Thermal conversion of polypyrrole nanotubes to nitrogen-doped carbon nanotubes for efficient water desalination using membrane capacitive deionization *Sep Purif Technol* **235**

[77] Guo Y, Pan F, Ye M, Sun X T, Wang Y Y, Li J Z, Zhang X Y, Zhang H, Pan Y Y, Song Z G, Yang J B and Lu J 2017 Monolayer bismuthene-metal contacts: A theoretical study *Acs Appl Mater Inter* **9** 23128–40

[78] Matthes L, Pulci O and Bechstedt F 2014 Optical properties of two-dimensional honeycomb crystals graphene, silicene, germanene, and tinene from first principles *New J Phys* **16**

[79] Lima M P, Fazzio A and da Silva A J R 2018 Silicene-based FET for logical technology *Ieee Electr Device L* **39** 1258–61

[80] Yuhara J and Le Lay G 2020 Beyond silicene: Synthesis of germanene, stanene and plumbene *Jpn J Appl Phys* **59**

[81] Lin S Y L H Y, Nguyen D K, Tran N T T, Pham H D, Chang S L, Lin C Y, Lin M F 2020 *Silicene-Based Layered Materials* (Bristol: IOP Publishing)

[82] Lin S Y T N T T, Chang S L, Su W P, Lin M F 2020 *Structure- and Adatom-Enriched Essential Properties of Graphene Nanoribbons* ([S.l.]: CRC Press)

[83] Lin C Y C R B, Ho Y H, Lin F L 2020 *Electronic and Optical Properties of Graphite Related Systems* ([S.l.]: CRC Press)

[84] Shih P H, Do T N, Gumbs G and Lin M F 2020 Electronic and optical properties of doped graphene *Physica E* **118**

[85] Wang S D, Wang W H and Zhao G J 2016 Thermal transport properties of antimonene: An ab initio study *Phys Chem Chem Phys* **18** 31217–22

[86] Marinopoulos A G, Reining L, Rubio A and Olevano V 2004 Ab initio study of the optical absorption and wave-vector-dependent dielectric response of graphite *Phys Rev B* **69**

[87] Bates S P, Kresse G and Gillan M J 1997 A systematic study of the surface energetics and structure of TiO2(110) by first-principles calculations *Surf Sci* **385** 386–94

[88] Kansara S, Shah J, Sonvane Y and Gupta S K 2019 Conjugation of biomolecules onto antimonene surface for biomedical prospects: A DFT study *Chem Phys Lett* **715** 115–22

[89] Liu P T, Kim B, Chen X Q, Sarma D D, Kresse G and Franchini C 2018 Relativistic GW plus BSE study of the optical properties of Ruddlesden-Popper iridates *Phys Rev Mater* **2**

[90] Klimes J, Kaltak M and Kresse G 2014 Predictive GW calculations using plane waves and pseudopotentials *Phys Rev B* **90**

[91] Dien V K, Han N T, Nguyen T D H, Huynh T M D, Pham H D and Lin M F 2020 Geometric and electronic properties of Li2GeO3 *Front Mater* **7**

[92] Pflugradt P, Matthes L and Bechstedt F 2014 Unexpected symmetry and AA stacking of bilayer silicene on Ag(111) *Phys Rev B* **89**

[93] Guo Z X and Oshiyama A 2014 Structural tristability and deep Dirac states in bilayer silicene on Ag(111) surfaces *Phys Rev B* **89**

[94] Do T N, Lin C Y, Lin Y P, Shih P H and Lin M F 2015 Configuration-enriched magneto-electronic spectra of AAB-stacked trilayer graphene *Carbon* **94** 619–32

[95] Lin C Y, Lin M F, Wu J Y, Chiu C W, Lin M F and Lin M F 2019 *Coulomb Excitations and Decays in Graphene-Related Systems* (Boca Raton: CRC Press)

[96] Lin C Y, Wu J Y, Ou Y J, Chiu Y H and Lin M F 2015 Magneto-electronic properties of multilayer graphenes *Phys Chem Chem Phys* **17** 26008–35

[97] Chung H C, Chang C P, Lin C Y and Lin M F 2016 Electronic and optical properties of graphene nanoribbons in external fields *Phys Chem Chem Phys* **18** 7573–616

[98] Lin C Y, Do T N, Huang Y K and Lin M F 2018 *Optical Properties of Graphene in Magnetic and Electric Fields* (Bristol: IOP publishing)

[99] Israr M, Raza F, Nazar N, Ahmad T, Khan M F, Park T J and Basit M A 2020 Rapid conjunction of 1D carbon nanotubes and 2D graphitic carbon nitride with ZnO for improved optoelectronic properties *Appl Nanosci* **10** 3805–17

[100] Lin C Y, Wu J Y, Chiu Y H, Chang C P and Lin M F 2014 Stacking-dependent magnetoelectronic properties in multilayer graphene *Phys Rev B* **90**

[101] Wu J Y, Chen S C, Gumbs G and Lin M F 2017 Field-induced diverse quantizations in monolayer and bilayer black phosphorus *Phys Rev B* **95**

[102] Pang H S, Wang H F, Li M L and Gao C H 2020 Atomic-scale friction on monovacancy-defective graphene and single-layer molybdenum-disulfide by numerical analysis *Nanomaterials-Basel* **10**

[103] Tsai S J, Chiu Y H, Ho Y H and Lin M F 2012 Gate-voltage-dependent Landau levels in AA-stacked bilayer graphene *Chem Phys Lett* **550** 104–10

[104] Shih P H, Do T N, Huang B L, Gumbs G, Huang D H and Lin M F 2019 Magnetoelectronic and optical properties of Si-doped graphene *Carbon* **144** 608–14

[105] Huang Y C, Chang C P and Lin M F 2007 Magnetic and quantum confinement effects on electronic and optical properties of graphene ribbons *Nanotechnology* **18**

[106] Ou Y C, Sheu J K, Chiu Y H, Chen R B and Lin M F 2011 Influence of modulated fields on the Landau level properties of graphene *Phys Rev B* **83**

[107] Anson-Casaos A, Ciria J C, Sanahuja-Parejo O, Victor-Roman S, Gonzalez-Dominguez J M, Garcia-Bordeje E, Benito A M and Maser W K 2020 The viscosity of dilute carbon nanotube (1D) and graphene oxide (2D) nanofluids *Phys Chem Chem Phys* **22** 11474–84

[108] Pan Z H, Yang J, Yang J, Zhang Q C, Zhang H, Li X, Kou Z K, Zhang Y F, Chen H, Yan C L and Wang J 2020 Stitching of Zn-3(OH)(2)V2O7 center dot 2H(2)O 2D nanosheets by 1D carbon nanotubes boosts ultrahigh rate for wearable quasi-solid-state zinc-ion batteries *Acs Nano* **14** 842–53

[109] Li N B 2020 Diffusion behaviors of energy and momentum for 1D single-walled carbon nanotubes *J Phys D Appl Phys* **53**

[110] Chen Y C, de Oteyza D G, Pedramrazi Z, Chen C, Fischer F R and Crommie M F 2013 tuning the band gap of graphene nanoribbons Synthesized from molecular precursors *Acs Nano* **7** 6123–8

[111] Zhu L J, Li L, Tao R, Fan X D, Wan X Y and Zeng C G 2020 Frictional drag effect between massless and massive fermions in single-layer/bilayer graphene heterostructures *Nano Lett* **20** 1396–402

[112] Xu Y Q, Chang J Q, Liang C, Sui X Y, Ma Y H, Song L T, Jiang W Y, Zhou J, Guo H B, Liu X F and Zhang Y 2020 Tailoring multi-walled carbon nanotubes into graphene quantum sheets *Acs Appl Mater Inter* **12** 47784–91

[113] Lee K W and Lee C E 2017 Half-metallic quantum valley Hall effect in biased zigzag-edge bilayer graphene nanoribbons *Phys Rev B* **95**

[114] Wu J Y, Su W P and Gumbs G 2020 Anomalous magneto-transport properties of bilayer phosphorene *Sci Rep-Uk* **10**

[115] Zhu J J, Badalyan S M and Peeters F M 2013 Plasmonic excitations in Coulomb-coupled N-layer graphene structures *Phys Rev B* **87**

[116] Hwang E H and Das Sarma S 2007 Dielectric function, screening, and plasmons in two-dimensional graphene *Phys Rev B* **75**

[117] Wunsch B, Stauber T, Sols F and Guinea F 2006 Dynamical polarization of graphene at finite doping *New J Phys* **8**

[118] Borghi G, Polini M, Asgari R and MacDonald A H 2009 Dynamical response functions and collective modes of bilayer graphene *Phys Rev B* **80**

[119] Sensarma R, Hwang E H and Das Sarma S 2010 Dynamic screening and low-energy collective modes in bilayer graphene *Phys Rev B* **82**

[120] Mahan G D 2000 *Many-Particle Physics* (New York: Kluwer Academic/Plenum Publisher)

[121] Kittel C 2005 *Introduction to Solid State Physics* (New York: Wiley)

[122] Shung K W K 1986 Dielectric function and plasmon structure of stage-1 intercalated graphite *Phys Rev B* **34** 979–93

[123] Shih P H, Chiu Y H, Wu J Y, Shyu F L and Lin M F 2017 Coulomb excitations of monolayer germanene *Sci Rep-Uk* **7**

[124] Chiu C W, Chung Y L, Yang C H, Liu C T and Lin C Y 2020 Coulomb decay rates in monolayer doped graphene *Rsc Adv* **10** 2337–46

[125] Wu J Y, Chen S C, Roslyak O, Gumbs G and Lin M F 2011 Plasma excitations in graphene: Their spectral intensity and temperature dependence in magnetic field *Acs Nano* **5** 1026–32

3 Experimental Measurements

*Jhao-Ying Wu, Nguyen Thanh Tien,
Phuoc Huu Le, Thi Dieu Hien Nguyen,
Vo Khuong Dien, and Ming-Fa Lin*

CONTENTS

3.1 INTRODUCTION

All the high-resolution experimental examinations can identify the quasiparticle properties, especially for the specific orbital hybridizations and spin configurations. In general, the electromagnetic waves [1], gate voltages [2], and electron beam Coulomb fields [3] are very popular in serving as the external perturbations, being able to clearly clarify the fundamental properties. The dynamic and static response abilities are responsible for their close relations [4]. This is the focus of the theoretical framework in Sections 2.1 and 2.2.

3.2 X-RAY SPECTROSCOPY, SCANNING TUNNELING MICROSCOPY, AND TUNNELING ELECTRON MICROSCOPY

X-ray diffraction, as shown in Figure 3.1 [5], is one of the very efficient methods in measuring the various crystal structures [6]. The most critical developments are initiated by Bragg [7] and Laue et al. [8] for their theoretical and experimental performances. The high-resolution measurements, which mainly arise from the elastic scatterings between charge densities and electromagnetic waves [9], are available for sufficiently large samples with observable responses. Without energy transfers, the prominent peak structures could be detected only under the Laue conditions, in which the charges of wave vectors are identical with the reciprocal lattice vectors [10]. The number, intensity, and form of them are very sensitive to the incident angles and wavelengths [11], since the important factors are mainly determined by

DOI: 10.1201/9781003322573-3

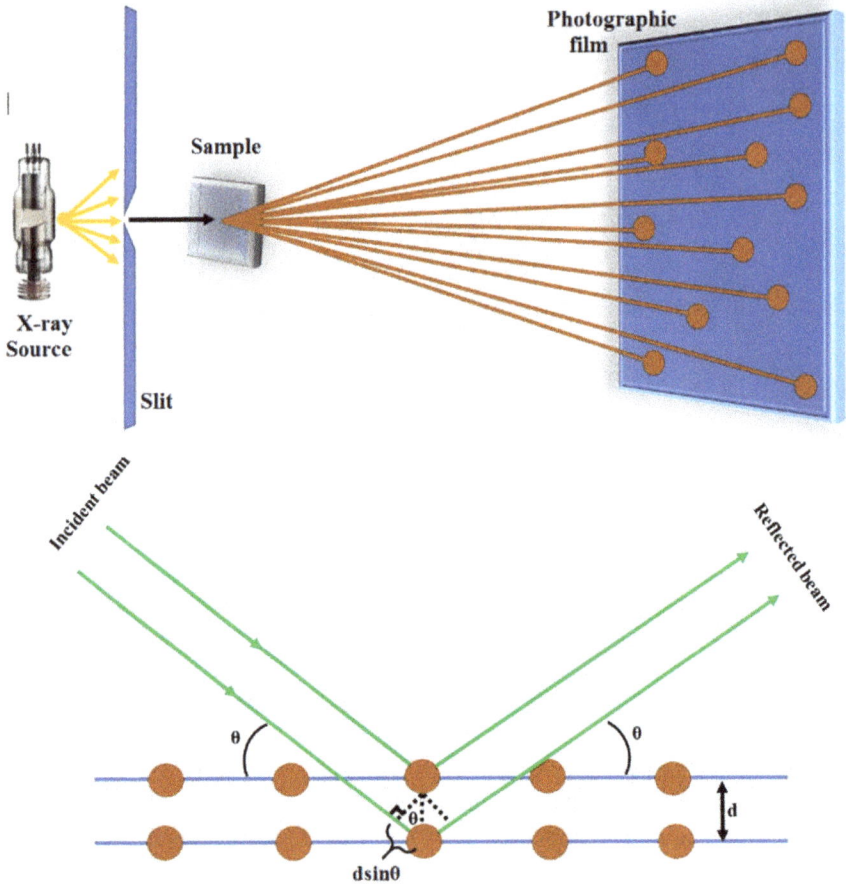

FIGURE 3.1   The elastic scatterings of charge density due to X-ray diffraction apparatus.

the intrinsic chemical environments. The featured diffraction patterns should be attributed to the chemical bonds/the orbital hybridizations in a primitive unit cell. By thorough analyses, the spatial charge density distributions could be estimated by the direct combinations of experimental measurements and theoretical calculations. Most previous studies do not provide such concise conclusions [12–14], leading to being almost mission impossible in examining the results of first-principles calculations [15–17]. In addition, the X-ray diffraction responses are too weak in nano-scaled materials to be accurately tested for the lattice symmetries [18–20].

The discrete X-ray diffraction peaks are capable of detecting the unique crystal symmetries of 3D Moiré superlattices [21], especially for the anode/electrolyte/cathode components of ion-based batteries [22] and the emergent bulk materials [23]. Very interestingly, during the charging or discharging processes for normal battery operations, the chemical reactions can drive the rapid variations of crystal structures.

These are clearly revealed as the intercalation or de-intercalation phenomena, e.g., the great reduction or increment of lithium-ion concentrations in multi-component lithium titanium/silicon/iron oxides of Li^+-based batteries [24–26]. A lot of quasi-stable or intermediate configurations come to exist as time gradually grows. In general, the initial and final lattice symmetries could be identified from the up-to-date X-ray measurements, such as the highly non-uniform environments due to various chemical bonds and observable bond-length fluctuations within a Moiré superlattice [27]. These two lattice symmetries are insufficient in determining the critical mechanisms of the ion transport properties. How to achieve more physical/chemical pictures of crystal structures must be the studying focuses in the near-future experimental observations, in which they are fully supported by the first-principles simulations [28].

Both scanning tunneling spectroscopy (STM [29]) and transmission electron microscopy (TEM [30]), as clearly illustrated in Figures 3.2 and 3.3, are very powerful in detecting the nano-scaled geometries with/without local defects. Their high-precision measurements directly provide an active environment of chemical bonds [31]. As for the former [STM], a gate voltage (V) is applied between the nanoprobe and material surface, so that a very weak quantum tunneling is greatly enhanced under the increment of V. The different probe heights, which correspond to the identical currents [32], are detected at the various positions. It is very famous that STM is able to identify both chiralities and radii of semiconducting or metallic single-walled carbon nanotubes [33], the chiral on non-chiral edge structures in graphene nanoribbon [34], the honeycomb structures of monolayer group-IV systems [35, 36], the various stackings in asymmetric bilayer graphene systems [37], and the non-hexagonal few-layer phosphorenes [38]. Specifically, the spin-polarized STM [39] could be largely modified for its spatial resolution and utilized to examine the ferromagnetic configurations associated with the atom and orbital-created spin distributions [40].

Tunneling electron microscopy, which is probed by an electron beam with suitable kinetic energy (Figure 3.3), is available in clarifying the side-view structures of the few-layer/very thin materials. Only the elastic scattering events, being distinguished from the inelastic Coulomb excitation spectra (electron energy loss spectra [41], belong to the diffraction patterns [42]). Their high-resolution examinations can detect the composite structures (the layered, coaxial, and closed composite structures), the normal and/or deformed crystals (the planar/folded/curved/scrolled lattices [43–46]), the volume spacing of two neighboring nanostructures [47], the number of layers/nanotubes [48, 49], the regular stacking configurations [50] and twisted cases [51], and the buckled lattices in the absence/presence of stacking couplings [52, 53]. These are fully supported by a lot of successful experimental syntheses [54, 55], thus leading to much progress of 2D emergent systems [56]. It should be noted that low-energy electron diffraction (LEED [57]) is another very powerful method in observing surface morphologies, such as monolayer/bilayer group-IV grown on various substrates [58, 59]. Obviously, STM, TEM, and LEED can present the diversified crystal symmetries of the low-dimensional materials due to the physical [60, 61] and chemical modifications [62–64]. Their combinations are able to present the full top and side views of various geometric symmetries [40, 65].

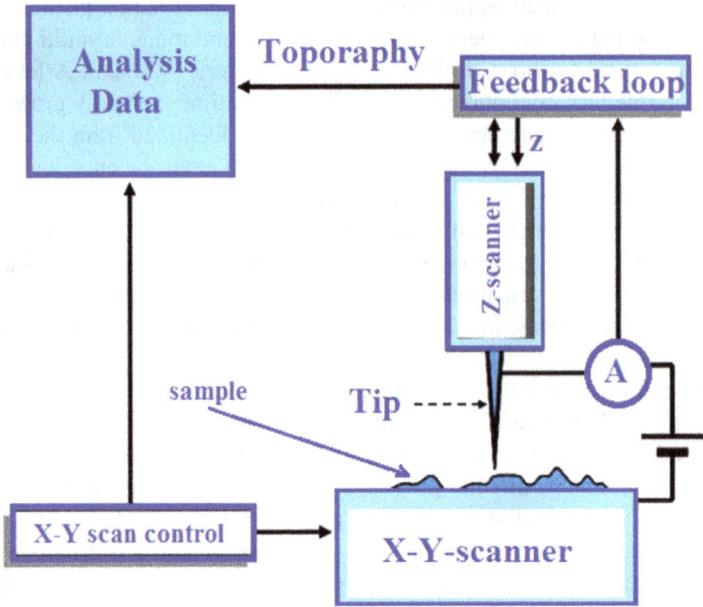

FIGURE 3.2   Setup of scanning tunneling microscopy (STM).

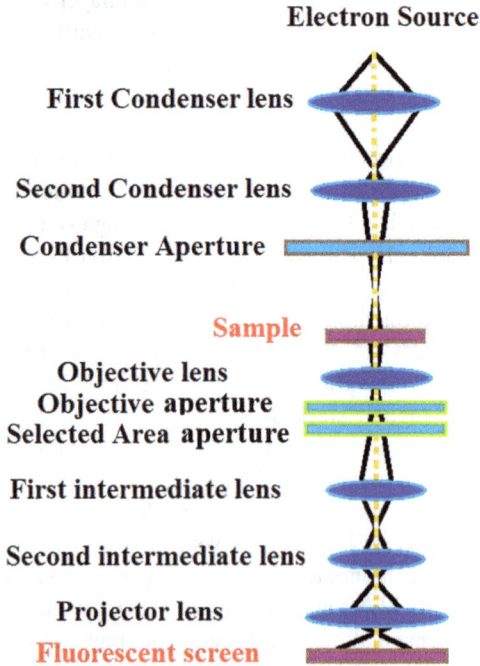

FIGURE 3.3   Transmission electron microscopy (TEM).

3.3 SCANNING TUNNELING SPECTROSCOPY AND ANGLE-RESOLVED PHOTOEMISSION SPECTROSCOPY

How to examine and verify the electronic properties is one of the mainstream research themes. Two kinds of experimental tools, scanning tunneling spectroscopy (STS in Figure 3.4 [66]) and angle-resolved photoemission (ARPES in Figure 3.5 [67]), are frequently used to detect the basic quasiparticle properties. Specifically, they have merits and drawbacks. STS is a very accurate extension of STM. Its high-precision measurements are done on the whole surface through the position-dependent tunneling quantum currents at the same probe height. By the delicate analyses on the differential conductance, the strong dependence with gate voltage is assumed to be roughly proportional to the density of states [68]. That is to say, the prominent structures in the measured results mainly arise from the van Hove singularities [69]. Only the band-edge state energies of valence and conduction energy spectra [70] could be tested in the STS experiments, while the information about the wave-vector dependences cannot be clarified through them.

Most importantly, the direct combinations of STS and STM [71] are very successful in precisely characterizing nanocrystal symmetries and identifying band properties simultaneously, especially for carbon-related honeycomb lattices (e.g., sp^2-bonding carbon nanotubes, graphene nanoribbons, and few-layer graphenes in Refs [72]). Their high-resolution examinations have clarified the geometry-diversified quasiparticle properties near the Fermi level. Very interestingly, both

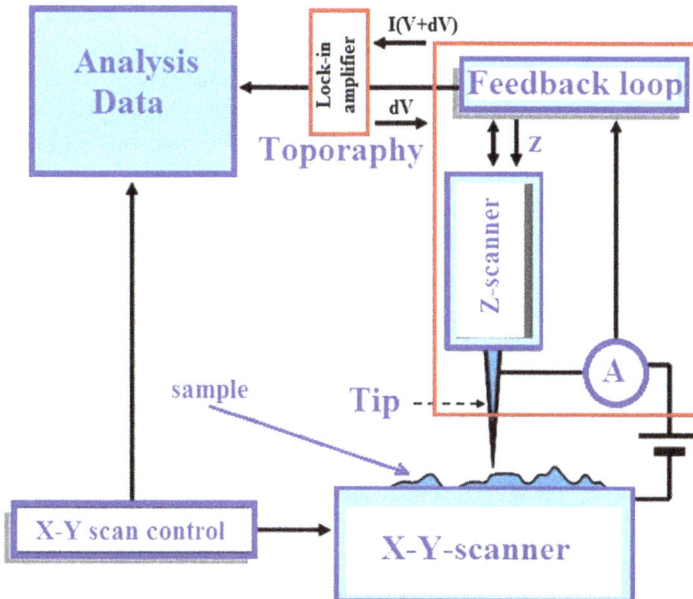

FIGURE 3.4   Scanning tunneling spectroscopy (STS).

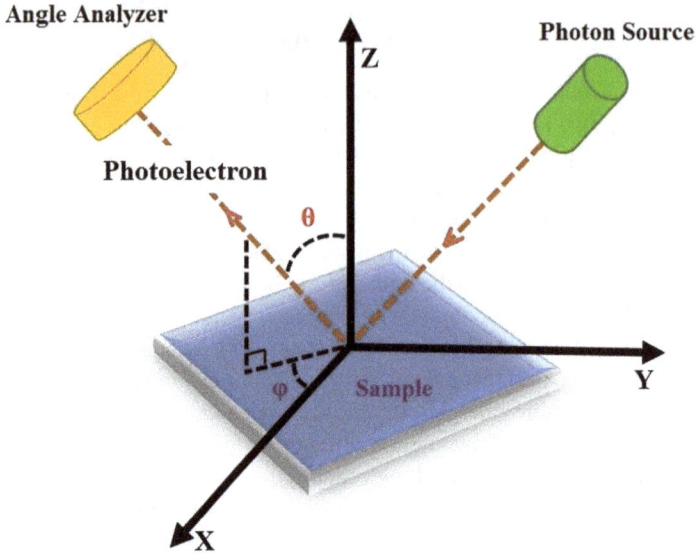

FIGURE 3.5   Angle-resolved photoemission spectroscopy.

experimental groups in 1998 clearly clarify the close relations between band properties (energy gaps and van Hove singularities [73]) and geometric symmetries (chiral
angles and radii [74]) for single-walled carbon nanotubes. The significant results
cover the 1D metallic/semiconducting behaviors in armchair systems/others [75, 76]
and the asymmetric peaks due to the 1D parabolic bands. Only the semiconducting
graphene nanoribbons are revealed in the simultaneous examinations of geometric
and electronic properties [77], mainly owing to the assistance of quantum confinement. It should be noticed that band properties become complex, e.g., the sensitive
dependences of energy gaps on edge structures [78], ribbon widths (the number of
armchair/zigzag lines in [79]), and spin configurations [80]. As to layered graphenes,
their semiconducting and semi-metallic behaviors are verified to strongly depend on
the substrate effects [81] and interlayer van der Waals interactions [82], respectively.
Furthermore, the low-energy quasiparticle properties are enriched by the number of
layers [83] and stacking configurations (AB/ABC/AAB ones [84–86]).

ARPES can directly examine the wave-vector-dependent energy spectra and lifetimes of the occupied quasiparticle states [87] but not for the unoccupied ones [88].
Part of the information about the latter could be supported by the STS observations
[66]. However, their measurements would become meaningless under the undefined
momentum transfers (the scattering cases of non-conserved momentum transfers,
e.g., uncertainties of momenta associated with surface boundaries [89]). When a suitable light with a well-defined wavelength [90] is illuminated on a specific sample,
the various elastic and inelastic scattering events come to exist within and outside
this system [91]. These complicated and unique phenomena are very sensitive to the
effective depth [92] and the intrinsic atomic interactions (the orbital- and spin-dependent
Hamiltonians [93]), since such critical mechanisms will determine the escape of

excited quasiparticles from the surface quantum confinement [94]. Generally speaking, quasiparticles, which are situated only below the fixed boundary about 10–20 Å, could make significant contributions to ARPES measurements. For example, the electron-electron Coulomb scatterings are very efficient [95]; therefore, the mean free paths are too short to break through the environmental bound [96]. As a result, the ARPES examinations are rather suitable and reliable in identifying electronic energy spectra and lifetimes of layered/2D emergent materials. However, their observations on 3D and 1D systems need to overcome the geometric issues, corresponding to the non-well-behaved momentum transfers during the scattering processes [50].

Very successfully, the high-precision ARPES measurements have verified the diverse quasiparticle energy spectra of few-layer group-IV materials [13, 97], especially for those of graphene-related systems [62, 98]. The clear identifications of diverse quasiparticle energy dispersions cover the 1D parabolic energy sub-bands about the high-symmetry point with width-declined band gaps for graphene nanoribbons [99], the linear valence Dirac-cone structure of monolayer graphene [100], two pairs of low-energy parabolic dispersions for bilayer AB stacking [101], the coexistence of linear and parabolic bands in twisted bilayer graphene [51], the linear and parabolic bands of tri-layer ABA stacking [85], the linear, partially flat, and Sombrero-shaped bands for tri-layer ABC stacking [85], the substrate-generated observable energy spacing between the separated valence and conduction bands in bilayer AB stacking, the substrate-created oscillatory bands for few-layer ABC stacking, the semimetal-semiconductor/semimetal-metal transitions and the tunable low-lying energy bands after the adatom/molecule chemisorption on graphene surfaces [102, 103], the 3D/semi-metallic electronic structure in Bernal graphite (the bilayer- and monolayer-like energy dispersions at $k_z = 0$ and 1, respectively; K and H corner points in the 3D first Brillouin zone of [70]), the separated valence and conduction Dirac points near the K/K' valleys for silicene sheets on silver [111]/[110] substrates $E_g \sim 0.6$ eV [104], the quasi-multi Dirac-cone band structures of few-layer germanenes on Au [111] [105], the hole-like metallic behavior near the Γ point in bilayer stanene on Bi_2Te_3 [111], and the property of metal for monolayer plumbene/Au [111] [106]. Specifically, the well-behaved 3D band structure of the AB-stacked graphite, which corresponds to the k_z-dependent one, is determined by using its features across the Fermi level. However, this characteristic is absent in a lot of 3F materials with Moiré superlattices, e.g., the complicated electronic energy spectra of anode/electrolyte/cathode components of lithium-ion-based batteries [107–109]. Such undefined momentum transfers are responsible for the absence of ARPES measurements in 3D battery materials [110].

3.4 REFLECTANCE/TRANSMISSION/ PHOTOLUMINESCENCE SPECTROSCOPIES

The different optical spectroscopies are available in examining the separate and/or composite quasiparticle properties, such as the collective excitations of electrons [111], phonons [112], photons [113], and their combined quasiparticles (plasmons/polarons/polaritons [114–116]). The external perturbations cover the electromagnetic waves, gate voltages, and electron beams. The experimental measurements can directly present the dynamic and/or static carrier responses [117], being closely related to the various elastic and inelastic scattering events. When they are combined with the Kramers-Kronig relations [118], the frequency-dependent transverse/longitudinal dielectric function could be estimated through certain approximations [119]. The prominent structures in Im[ε] and Re[ε] are mainly determined by the joint van Hove singularities [120], the matrix elements of optical/Coulomb [121, 122], and/or excitonic bound states [123]. The close relations between the strong responses and the active orbital hybridizations/the obvious spin configurations will be examined thoroughly in this work [124], i.e., this is included in the theoretical development of quasiparticle phenomena [125]. In general, a sufficiently strong light source, with frequency in the suitable range of $0\,eV \le \omega \le 40\,eV$, could be utilized to detect frequency-dependent reflectance spectra [126]. A PC-miniaturized spectrometer, which clearly corresponds to Figure 3.6(a), is available for the spectral range. An incident electromagnetic wave is initiated from a specific light source and then is guided to a reflection probe through the specific fibers of A1-A6 vertically transmitted into a condensed-matter sample [127]. There also exist silicon wafers and glass slides in serving as an appropriate reflective or transmitting substrate [128]. Through the simultaneous observations, a reflection probe detects all the reflected light beams by using fiber B in Figure 3.5(a). The continuous frequency-dependent reflectance spectrum is obtained from the high-resolution intensity ratio between the reflected and incident electric fields (details in Eq. (2.1.c4)). That is to say, a normally incident electromagnetic wave strongly interacts with valence and conduction carriers of an active material [129]. Furthermore, a reflectance signal is continuously observed by the optical spectrometer [130]. The previous measurements clearly show a lot of outstanding results in creating/verifying/clarifying the diverse optical properties, such as the diversified reflectance spectra in graphite [131], graphite intercalation compounds [132], layered graphene systems [133], carbon nanotube bundles [134], and carbon fullerene materials [135]. For example, the low-, middle-, and high-frequency vertical transitions, which, respectively, correspond to the ranges of $\omega < 1\,eV$, $\sim 5\text{–}7\,eV$ and $>10\,eV$, principally originate from the free electrons/holes (due to band overlaps or n-/p-type dopings [65, 136]), the π valence electrons [137], and $\pi+\sigma$ electrons [138]. The π and σ chemical bondings are well distinguished from each other in graphene-related materials, and their excitation behaviors are clearly diversified by the geometric symmetries [139], e.g., the main features of optical absorption structures [140]. Very interestingly, while the reflectance spectra are done for condensed-matter

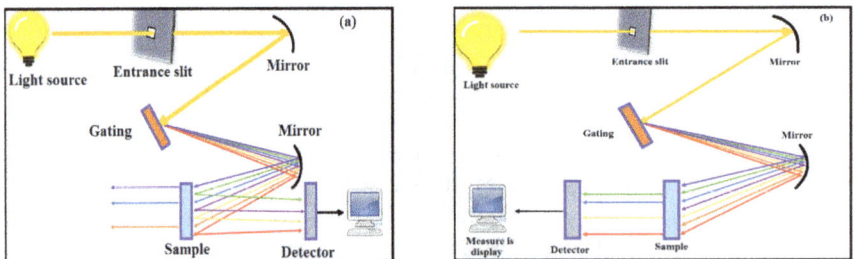

FIGURE 3.6 Optical measurements: (a) reflectance (b), and transmittance.

systems with Moiré superlattices, the delicate analyses on the complicated results almost become mission impossible. The anode/electrolyte/cathode materials of ion-based batteries provide outstanding information [22, 141–143]. Too many valence and conduction subbands, a lot of joint van Hove singularities, is the main reason. The direct combinations of the experimental observations and theoretical predictions might be capable of solving this open issue. The ω-dependent transmittance spectrum could be measured within a wide frequency range only for a thin enough sample. The transmission spectrometer is clearly illustrated in Figure 3.6(b), in which absorption spectrum will be achieved from the simultaneous observations of reflectance and transmittance. Most importantly, the transmission/absorption spectroscopy is rather suitable for thoroughly investigating the diverse optical behaviors of 2D/1D emergent materials. For example, their high-resolution measurements obviously display the diversified vertical transitions in group-IV [144]/group-V [145] 2D systems and 1D carbon nanotubes/graphene nanoribbons [146, 147]. The featured absorption structures, the frequency, number, form, and intensity of them, are verified to be very sensitive to dimensionalities [148], chiralities/boundary edges [149], layer numbers [150], stacking configurations [151], gate voltages [2], uniform/non-uniform magnetic fields [152, 153], and n-/p-type dopings [154]. Most experimental examinations are consistent with the theoretical predictions. However, such measurements would introduce extra challenges in terms of quantum-confinements effects (the dimension-crossover phenomena [155]). This is an outstanding near-future research theme for the layered emergent materials [156]. But for the pristine 3D systems, they are totally different from those of finite-width ones, since the essential quasi-particle properties are dramatically changed during the gradual variation of sample width. The measured absorption/transmittance spectra cannot be utilized to investigate the quasiparticle behaviors of pristine 3D materials, such as those of cathode/electrolyte/ anode materials in Li^+-related batteries [108, 157, 158]. Apparently, the ambiguous results [the random examinations] are unreliable and meaningless in understanding the critical mechanisms of optical excitations.

After the vertical transition, the excited electrons and holes can combine by the attractive Coulomb interactions at sufficiently low temperatures [159], decay during their propagations, and further create radiations in the narrow-frequency range [160]. The quasi-stable excitons belong to one of the mainstream many-particle interactions [161], in which they have strong effects on the featured optical properties, especially for the prominent absorption structures. Very interestingly, such effects play the critical mechanisms in optical devices [162]. The full quasiparticle dynamics could be thoroughly explored by the high-precision photoluminescence spectrometer under a specific frequency range near the optical threshold frequency (Figure 6.5 (c)). This is a contactless and non-destructive method, and it has successfully shown the observations of the main features of stable and metastable excitons [163]. The vertically optical transitions are initiated by an incident electromagnetic wave with a frequency above the threshold one. The weakly or strongly coupled excitons experience continuously inelastic scatterings, such as those due to electron-electron Coulomb (charge screenings in Refs [164]) and electron-phonon interactions (at enough high temperatures in Refs [165]). And then the npn-stable bound quasiparticles become the single-particle states by light emission or other non-radiative de-excitations [166].

The former could be collected by an optical detector. Both absorbance and photo-luminescence, which appear within a narrow-frequency range, are simultaneously examined for evaluating the bound-state energies, prominent absorption structures, and lifetimes of excitons. This is very reliable for the low-dimensional systems, e.g., the successful observations on the radius- and chirality-dependent excitonic effects in metallic and semiconducting carbon nanotubes [167]. Another direct combination of photoluminescence and reflectance spectra might be useful in distinguishing the single- and many-particle threshold absorption frequencies. Their high-resolution examinations are able to clarify the important differences among the quasi-state energies of excitons [168], the optical gap [169], and the band gap [170]. However, the previous experiments on the core materials of ion-based batteries cannot provide accurate results in doing the delicate analyses. These are urgently requested in the near-future research. When a time-resolved pump and probe, with the femtosecond resolutions [171], is applied to an emergent material [172], the specific optical transition channels could be excited and thus fully explored for their main features. This manner is useful in examining the theoretical viewpoints in Chapter 2, the distinct absorption peaks arising from the various multi-/single-orbital hybridization of chemical bonds [173].

REFERENCES

[1] Manzoor H U, Manzoor T and Hussain M 2020 Analysis of multiple surface electromagnetic waves on the planner interface of hyperbolic medium and rugate filter having sinusoidal refractive index profile *Plasmonics* **15** 287–91

[2] Tsai S J, Chiu Y H, Ho Y H and Lin M F 2012 Gate-voltage-dependent Landau levels in AA-stacked bilayer graphene *Chem Phys Lett* **550** 104–10

[3] de Souza J F O, Ribeiro C A D and Furtado C 2014 Bound states in disclinated graphene with Coulomb impurities in the presence of a uniform magnetic field *Phys Lett A* **378** 2317–24

[4] Batrudinov T M, Nekhoroshkova Y E, Paramonov E I, Zverev V S, Elfimova E A, Ivanov A O and Camp P J 2018 Dynamic magnetic response of a ferrofluid in a static uniform magnetic field *Phys Rev E* **98**

[5] Yashiro W, Shimizu K, Hirano K and Takahashi T 2002 Direct phase measurement of the X-ray specular reflection using modulation under the Bragg condition *Jpn J Appl Phys 2* **41** L592–4

[6] Cate J H, Yusupov M M, Yusupova G Z, Earnest T N and Noller H F 1999 X-ray crystal structures of 70S ribosome functional complexes *Science* **285** 2095–104

[7] Ushakov N M and Kolosov V V 2001 Bragg's reflection of the optical and acoustic waves from photoinduced periodic structures in iron-doped lithium niobate *Tech Phys Lett* **27** 1044–6

[8] Golovin A L and Imamov R M 1984 The investigation of X-Ray-scattering intensity for the case of Laue-diffraction under specular reflection conditions *Kristallografiya* **29** 410–12

[9] Simonsen I and Maradudin A A 1999 Numerical simulation of electromagnetic wave scattering from planar dielectric films deposited on rough perfectly conducting substrates *Opt Commun* **162** 99–111

[10] Zak J 1975 Lattice operators in crystals for bravais and reciprocal vectors *Phys Rev B* **12** 3023–6

[11] Wu X H, Zhang X L, Wang X L and Zeng Z 2016 Spin density waves predicted in zig-zag puckered phosphorene, arsenene and antimonene nanoribbons *Aip Adv* **6**

[12] Aghaei S M and Calizo I 2015 Band gap tuning of armchair silicene nanoribbons using periodic hexagonal holes *J Appl Phys* **118**

[13] Yuhara J and Le Lay G 2020 Beyond silicene: Synthesis of germanene, stanene and plumbene *Jpn J Appl Phys* **59**

[14] Meng L, Wang Y L, Zhang L Z, Du S X, Wu R T, Li L F, Zhang Y, Li G, Zhou H T, Hofer W A and Gao H J 2013 Buckled silicene formation on Ir(111) *Nano Lett* **13** 685–90

[15] Nagarajan V and Chandiramouli R 2018 First-principles investigation on structural and electronic properties of antimonene nanoribbons and nanotubes *Physica E* **97** 98–104

[16] Snehha P, Nagarajan V and Chandiramouli R 2018 Novel bismuthene nanotubes to detect NH3, NO2 and PH3 gas molecules: A first-principles insight *Chem Phys Lett* **712** 102–11

[17] Kremer L F and Baierle R J 2020 Graphene and silicene nanodomains in a ultra-thin SIC layer for water splitting and hydrogen storage: A first principle study *Int J Hydrogen Energ* **45** 5155–64

[18] Chen C N A, Zhang Q, Ma T and Fan W 2017 Synthesis and electrochemical properties of nitrogen-doped graphene/copper sulphide nanocomposite for supercapacitor *J Nanosci Nanotechno* **17** 2811–16

[19] Baldwin T O, Fraundorf P, Keefe G and Thomas J E 1975 X-Ray-diffraction and electron-microscopy studies of lithium precipitation in silicon and germanium *Acta Crystallographica Section A* **31** S266

[20] Chadwick B M, Jones D W, Wilde H J and Yerkess J 1985 X-Ray and Neutron-diffraction studies of the crystal-structures of the dicesium lithium hexacyanometallates of Iron(Iii) and Cobalt(Iii) *J Cryst Spectrosc* **15** 133–46

[21] Chen X F, Hu S and Luo Y 2000 Analysis of the chirped super Moire gratings based on acousto-optical superlattices *Int J Infrared Milli* **21** 939–44

[22] Cui J, Zheng H K and He K 2020 In Situ TEM study on conversion-type electrodes for rechargeable ion batteries *Adv Mater*

[23] Pardo E and Kapolka M 2017 3D computation of non-linear eddy currents: Variational method and superconducting cubic bulk *J Comput Phys* **344** 339–63

[24] Esaka T, Hayashi M, Sakaguchi H, Takai S and Matsubayashi M 2002 Analysis of lithium ion distribution in electrolyzed Li1.33Ti1.67O4 by neutron computed tomography *Solid State Ionics* **147** 107–14

[25] Kang D Y, Kim C, Gueon D, Park G, Kim J S, Lee J K and Moon J H 2015 3D woven-like carbon micropattern decorated with silicon nanoparticles for use in lithium-ion batteries *Chemsuschem* **8** 3414–18

[26] Yu M C, Bian X F, Liu S, Yuan C, Yang Y H, Ge X L, Guan R Z and Wang C 2019 3D hollow porous spherical architecture packed by iron-borate amorphous nanoparticles as high-performance anode for lithium-ion batteries *Acs Appl Mater Inter* **11** 25254–63

[27] Alexeev E M, Ruiz-Tijerina D A, Danovich M, Hamer M J, Terry D J, Nayak P K, Ahn S, Pak S, Lee J, Sohn J I, Molas M R, Koperski M, Watanabe K, Taniguchi T, Novoselov K S, Gorbachev R V, Shin H S, Fal'ko V I and Tartakovskii A I 2019 Resonantly hybridized excitons in Moiré superlattices in van der Waals heterostructures *Nature* **567** 81–6

[28] Kim H and Kim G 2020 Adsorption properties of dopamine derivatives using carbon nanotubes: A first-principles study *Appl Surf Sci* **501**

[29] Bezerra L G P, Souza A L P, Lago A E A, Campos L B, Nunes T L, Paula V V, Oliveira M F and Silva A R 2019 Addition of equex STM to extender improves post-thawing longevity of collared peccaries' sperm *Biopreserv Biobank* **17** 143–7

[30] Kaloshin V A and Le N T 2020 2D-periodic over-wave-range antenna array of TEM horns with a feeder *J Commun Technol El* **65** 1140–6

[31] Peng D Q, Chen Y X, Ma H L, Zhang L, Hu Y, Chen X N, Cui Y L, Shi Y L, Zhuang Q C and Ju Z C 2020 Enhancing the cycling stability by tuning the chemical bonding between phosphorus and carbon nanotubes for Potassium-Ion battery anodes *Acs Appl Mater Inter* **12** 37275–84

[32] Peres N M R, Tsai S W, Santos J E and Ribeiro R M 2009 Scanning tunneling microscopy currents on locally disordered graphene *Phys Rev B* **79**

[33] Marinopoulos A G, Reining L and Rubio A 2008 Ab initio study of the dielectric response of crystalline ropes of metallic single-walled carbon nanotubes: Tube-diameter and helicity effects *Phys Rev B* **78**

[34] Lin Y T, Lin S Y, Chiu Y H and Lin M F 2017 Alkali-created rich properties in grapheme nanoribbons: Chemical bondings *Sci Rep-Uk* **7**

[35] Shih P H, Chiu Y H, Wu J Y, Shyu F L and Lin M F 2017 Coulomb excitations of monolayer germanene *Sci Rep-Uk* **7**

[36] Gong X X, Ye Z L, Lu S, Liu K, Liu J Y and Liu Z L 2020 Tuning the structural and electronic properties of arsenene monolayers by germanene, silicene, and stanene domain doping *Physica E* **122**

[37] Shen B, Huang Z W, Ji Z, Lin Q, Chen S L, Cui D J and Zhang Z N 2019 Bilayer graphene film synthesized by hot filament chemical vapor deposition as a nanoscale solid lubricant *Surf Coat Tech* **380**

[38] Wu J Y, Su W P and Gumbs G 2020 Anomalous magneto-transport properties of bilayer phosphorene *Sci Rep-Uk* **10**

[39] Ara F, Oka H, Sainoo Y, Katoh K, Yamashita M and Komeda T 2019 Spin properties of single-molecule magnet of double-decker Tb(III)-phthalocyanine (TbPc2) on ferromagnetic Co film characterized by spin polarized STM (SP-STM) *J Appl Phys* **125**

[40] Pham H D, Gumbs G, Su W P, Tran N T T and Lin M F 2020 Unusual features of nitrogen substitutions in silicene *Rsc Adv* **10** 32193–201

[41] Chen C H and Silcox J 1974 Electron energy-losses in graphite—bulk plasmons, guided surface-wave, interband-transitions and Cherenkov radiation *B Am Phys Soc* **19** 462

[42] Cowley J M, Nikolaev P, Thess A and Smalley R E 1997 Electron nano-diffraction study of carbon single-walled nanotube ropes *Chem Phys Lett* **265** 379–84

[43] Al-Dirini F, Mohammed M A, Hossain F M, Nirmalathas T and Skafidas E 2016 All-graphene planar double-quantum-dot resonant tunneling diodes *Ieee J Electron Devi* **4** 30–9

[44] Kim K, Lee Z, Malone B D, Chan K T, Aleman B, Regan W, Gannett W, Crommie M F, Cohen M L and Zettl A 2011 Multiply folded graphene *Phys Rev B* **83**

[45] Xia S X, Zhai X, Wang L L, Sun B, Liu J Q and Wen S C 2016 Dynamically tunable plasmonically induced transparency in sinusoidally curved and planar graphene layers *Opt Express* **24** 17886–99

[46] Meng L Y, Xia Y X, Liu W G, Zhang L, Zou P and Zhang Y S 2015 Hydrogen micro-explosion synthesis of platinum nanoparticles/nitrogen doped graphene nanoscrolls as new amperometric glucose biosensor *Electrochim Acta* **152** 330–7

[47] Liu P and Xiang B 2017 2D hetero-structures based on transition metal dichalcogenides: Fabrication, properties and applications *Sci Bull* **62** 1148–61

[48] Kaskun S, Akinay Y and Kayfeci M 2020 Improved hydrogen adsorption of ZnO doped multi-walled carbon nanotubes *Int J Hydrogen Energ* **45** 34949–55

[49] Kumar A, Sachdeva G, Pandey R and Karna S P 2020 Optical absorbance in multilayer two-dimensional materials: Graphene and antimonene *Appl Phys Lett* **116**

[50] Do T N, Lin C Y, Lin Y P, Shih P H and Lin M F 2015 Configuration-enriched magneto-electronic spectra of AAB-stacked trilayer graphene *Carbon* **94** 619–32

[51] de Vries F K, Zhu J H, Portoles E, Zheng G L, Masseroni M, Kurzmann A, Taniguchi T, Watanabe K, MacDonald A H, Ensslin K, Ihn T and Rickhaus P 2020 Combined minivalley and layer control in twisted double bilayer graphene *Phys Rev Lett* **125**

[52] Ni Z Y, Liu Q H, Tang K C, Zheng J X, Zhou J, Qin R, Gao Z X, Yu D P and Lu J 2012 Tunable bandgap in silicene and germanene *Nano Lett* **12** 113–18

[53] Lee K W and Lee C E 2019 Quantum spin-valley hall effect in AB-stacked bilayer silicene *Sci Rep-Uk* **9**

[54] Kussmann D, Hoffmann R D and Pottgen R 1998 Syntheses and crystal structures of CaCuGe, CaAuIn, and CaAuSn—Three different superstructures of the KHg2 type *Z Anorg Allg Chem* **624** 1727–35

[55] Guan L, Jin H Z, Wang Y, Xiong X J, Chen Y X and Wang X 2020 Syntheses, structures and properties of two one-dimensional coordination polymers constructed by phenol-sulfonate ligands and rare earth metal ions *Main Group Chem* **19** 245–55

[56] Borisenko D P, Gusev A S, Kargin N I, Komissarov I V, Kovalchuk N G and Labunov V A 2019 Plasma assisted-MBE of GaN and AlN on graphene buffer layers *Jpn J Appl Phys* **58**

[57] McDougall D, Hattab H, Hershberger M T, Hupalo M, von Hoegen M H, Thiel P A and Tringides M C 2016 Dy uniform film morphologies on graphene studied with SPA-LEED and STM *Carbon* **108** 283–90

[58] Choi S H, Kim Y L and Byun K M 2011 Graphene-on-silver substrates for sensitive surface plasmon resonance imaging biosensors *Opt Express* **19** 458–66

[59] Suzuki S and Yoshimura M 2017 Chemical stability of graphene coated silver substrates for surface-enhanced raman scattering *Sci Rep-Uk* **7**

[60] Lee F, Tripathi M, Lynch P and Dalton A B 2020 Configurational effects on strain and doping at graphene-silver nanowire interfaces *Appl Sci-Basel* **10**

[61] Chen X B and Duan W H 2015 Quantum thermal transport and spin thermoelectrics in low-dimensional nano systems: Application of nonequilibrium Green's function method *Acta Phys Sin-Ch Ed* **64**

[62] Banerjee D, Sankaran K J, Deshmukh S, Ficek M, Bhattacharya G, Ryl J, Phase D M, Gupta M, Bogdanowicz R, Lin I N, Kanjilal A, Haenen K and Roy S S 2019 3D hierarchical boron-doped diamond-multilayered graphene nanowalls as an efficient supercapacitor electrode *J Phys Chem C* **123** 15458–66

[63] Vashchenko A V, Kuzmin A V and Shainyan B A 2020 Si-doped single-walled carbon nanotubes as potential catalysts for oxygen reduction reactions *Russ J Gen Chem+* **90** 454–9

[64] Iyakutti K, Kumar E M, Thapa R, Rajeswarapalanichamy R, Surya V J and Kawazoe Y 2016 Effect of multiple defects and substituted impurities on the band structure of graphene: A DFT study *J Mater Sci-Mater El* **27** 12669–79

[65] Pham H D, Su W P, Nguyen T D H, Tran N T T and Lin M F 2020 Rich p-type-doping phenomena in boron-substituted silicene systems *Roy Soc Open Sci* **7**

[66] Matsui T, Kambara H, Niimi Y, Tagami K, Tsukada M and Fukuyama H 2005 STS observations of landau levels at graphite surfaces *Phys Rev Lett* **94**

[67] Tan S N, Mou Y P, Liu Y Q and Feng S P 2020 ARPES autocorrelation in electron-doped cuprate superconductors *J Supercond Nov Magn* **33** 2305–11

[68] Isihara A and Smrcka L 1986 Density and magnetic-field dependences of the conductivity of two-dimensional electron-systems *J Phys C Solid State* **19** 6777–89

[69] Nguyen T D H, Pham H D, Lin S Y and Lin M F 2020 Featured properties of Li+-based battery anode: Li4Ti5O12 *Rsc Adv* **10** 14071–9

[70] Pham H D, Lin S Y, Gumbs G, Khanh N D and Lin M F 2020 Diverse properties of carbon-substituted silicenes *Front Phys-Lausanne* **8**

[71] Mechehoud F, Benaioun N E, Hakiki N E, Khelil A, Simon L and Bubendorff J L 2018 Thermally oxidized Inconel 600 and 690 nickel-based alloys characterizations by combination of global photoelectrochemistry and local near-field microscopy techniques (STM, STS, AFM, SKPFM) *Appl Surf Sci* **433** 66–75

[72] Lin S Y L H Y, Nguyen D K, Tran N T T, Pham H D, Chang S L, Lin C Y, Lin M F 2020 *Silicene-Based Layered Materials* (Bristol: IOP Publishing)

[73] Kim P, Odom T W, Huang J L and Lieber C M 1999 Electronic density of states of atomically resolved single-walled carbon nanotubes: Van Hove singularities and end states *Phys Rev Lett* **82** 1225–8

[74] Saito R, Takeya T, Kimura T, Dresselhaus G and Dresselhaus M S 1998 Raman intensity of single-wall carbon nanotubes *Phys Rev B* **57** 4145–53

[75] Kimouche A, Ervasti M M, Drost R, Halonen S, Harju A, Joensuu P M, Sainio J and Liljeroth P 2015 Ultra-narrow metallic armchair graphene nanoribbons *Nat Commun* **6**

[76] Inoue S and Akagi H 2007 A bidirectional isolated DC-DC converter as a core circuit of the next-generation medium-voltage power conversion system *Ieee T Power Electr* **22** 535–42

[77] Yazdi A Z, Roberts E and Sundararaj U 2014 Electrochemical synthesis of nitrogen doped graphene nanoribbons *Abstr Pap Am Chem S* **248**

[78] Han M Y, Ozyilmaz B, Zhang Y B and Kim P 2007 Energy band-gap engineering of graphene nanoribbons *Phys Rev Lett* **98**

[79] Austin J, Way E M, Jacobberger R M, Goeltl F, Saraswat V, Mavrikakis M and Arnold M S 2019 Tightly Pitched Sub-10 Nm Nanoribbons Grown Via Seeded Anisotropic Synthesis on Ge(001) *ECS Meeting Abstracts*

[80] Deng X Q and Sheng R Q 2019 Spin transport investigation of two type silicene nanoribbons heterostructure *Phys Lett A* **383** 47–53

[81] Wehling T O, Lichtenstein A I and Katsnelson M I 2008 First-principles studies of water adsorption on graphene: The role of the substrate *Appl Phys Lett* **93**

[82] Shirali K, Shelton W A and Vekhter I 2021 Importance of van der Waals interactions forab initiostudies of topological insulators *J Phys-Condens Mat* **33**

[83] Yang B, Zhang S H, Lv J, Li S, Shi Y Y, Hu D C and Ma W S 2021 Large-scale and green production of multi-layer graphene in deep eutectic solvents *J Mater Sci* **56** 4615–23

[84] Lu C L, Chang C P, Huang Y C, Chen R B and Lin M L 2006 Influence of an electric field on the optical properties of few-layer graphene with AB stacking *Phys Rev B* **73**

[85] Koshino M 2010 Interlayer screening effect in graphene multilayers with ABA and ABC stacking *Phys Rev B* **81**

[86] Lin C Y, Huang B L, Ho C H, Gumbs G and Lin M F 2018 Geometry-diversified Coulomb excitations in trilayer AAB stacking graphene *Phys Rev B* **98**

[87] Maschek M, Rosenkranz S, Heid R, Said A H, Giraldo-Gallo P, Fisher I R and Weber F 2015 Wave-vector-dependent electron-phonon coupling and the charge-density-wave transition in TbTe3 *Phys Rev B* **91**

[88] Miller T L, Arrala M, Smallwood C L, Zhang W, Hafiz H, Barbiellini B, Kurashima K, Adachi T, Koike Y, Eisaki H, Lindroos M, Bansil A, Lee D H and Lanzara A 2015 Resolving unoccupied electronic states with laser ARPES in bismuth-based cuprate superconductors *Phys Rev B* **91**

[89] Dammeier L, Schwonnek R and Werner R F 2015 Uncertainty relations for angular momentum *New J Phys* **17**

[90] Park G, Shchekin O B, Huffaker D L and Deppe D G 1998 Lasing from InGaAs/GaAs quantum dots with extended wavelength and well-defined harmonic-oscillator energy levels *Appl Phys Lett* **73** 3351–3

[91] Tougaard S and Sigmund P 1982 Influence of elastic and inelastic-scattering on energy-spectra of electrons emitted from solids *Phys Rev B* **25** 4452–66

[92] Tsubaki M and Mizoguchi T 2018 Fast and accurate molecular property prediction: Learning atomic interactions and potentials with neural networks *J Phys Chem Lett* **9** 5733–41

[93] Dyall K G 1994 An exact separation of the spin-free and spin-dependent terms of the Dirac-Coulomb-Breit Hamiltonian *J Chem Phys* **100** 2118–27

[94] Raty J Y, Galli G, Bostedt C, van Buuren T W and Terminello L J 2003 Quantum confinement and fullerenelike surface reconstructions in nanodiamonds *Phys Rev Lett* **90**

[95] Rana F 2007 Electron-hole generation and recombination rates for Coulomb scattering in graphene *Phys Rev B* **76**

[96] Chabot V, Higgins D, Yu A P, Xiao X C, Chen Z W and Zhang J J 2014 A review of graphene and graphene oxide sponge: Material synthesis and applications to energy and the environment *Energ Environ Sci* **7** 1564–96

[97] Mendoza-Sanchez B, Coelho J, Pokle A and Nicolosi V 2015 A 2D graphene-manganese oxide nanosheet hybrid synthesized by a single step liquid-phase co-exfoliation method for supercapacitor applications *Electrochim Acta* **174** 696–705

[98] Liu F, Zhang X L, Zhang X, Liu M M, Shao Q S, Dong L, Yan W and Zhang J J 2020 Novel Fe3C nanoparticles encapsulated in bamboo-like nitrogen-doped carbon nanotubes as high-performance electrocatalyst for Zinc-Air battery *J Electrochem Soc* **167**

[99] Gao J, Uribe-Romo F J, Saathoff J D, Arslan H, Crick C R, Hein S J, Itin B, Clancy P, Dichtel W R and Loo Y L 2016 Ambipolar transport in solution-synthesized graphene nanoribbons *Acs Nano* **10** 4847–56

[100] Badalyan S M and Jauho A P 2020 Coulomb drag between a carbon nanotube and monolayer graphene *Phys Rev Res* **2**

[101] Ho J H, Lu C L, Hwang C C, Chang C P and Lin M F 2006 Coulomb excitations in AA- and AB-stacked bilayer graphites *Phys Rev B* **74**

[102] Sluiter M H F and Kawazoe Y 2003 Cluster expansion method for adsorption: Application to hydrogen chemisorption on graphene *Phys Rev B* **68**

[103] Karlicky F, Lepetit B and Lemoine D 2014 Quantum modelling of hydrogen chemisorption on graphene and graphite *J Chem Phys* **140**

[104] Cahangirov S, Audiffred M, Tang P Z, Iacomino A, Duan W H, Merino G and Rubio A 2013 Electronic structure of silicene on Ag(111): Strong hybridization effects *Phys Rev B* **88**

[105] Zhuang J C, Gao N, Li Z, Xu X, Wang J O, Zhao J J, Dou S X and Du Y 2017 Cooperative electron-phonon coupling and buckled structure in Germanene on Au(111) *Acs Nano* **11** 3553–9

[106] Yuhara J, He B J, Matsunami N, Nakatake M and Le Lay G 2019 Graphene's latest cousin: Plumbene epitaxial growth on a "Nano WaterCube" *Adv Mater* **31**

[107] Jang J, Kang I, Yi K W and Cho Y W 2018 Highly conducting fibrous carbon-coated silicon alloy anode for lithium ion batteries *Appl Surf Sci* **454** 277–83

[108] Kizilaslan A, Cetinkaya T and Akbulut H 2020 2H-MoS(2) as an artificial solid electrolyte interface in all-solid-state Lithium-Sulfur batteries *Adv Mater Interfaces* **7**

[109] Pereira N, Badway F, Wartelsky M, Gunn S and Amatucci G G 2009 Iron oxyfluorides as high capacity cathode materials for Lithium batteries *J Electrochem Soc* **156** A407–16

[110] Li C, Chen K, Guan M X, Wang X W, Zhou X, Zhai F, Dai J Y, Li Z J, Sun Z P, Meng S, Liu K H and Dai Q 2019 Extreme nonlinear strong-field photoemission from carbon nanotubes *Nat Commun* **10**

[111] Chiu C W, Chung Y L, Yang C H, Liu C T and Lin C Y 2020 Coulomb decay rates in monolayer doped graphene *Rsc Adv* **10** 2337–46

[112] Zhu T S and Ertekin E 2016 Phonons, localization, and thermal conductivity of diamond nanothreads and amorphous graphene *Nano Lett* **16** 4763–72

[113] Kibis O V 2010 Metal-insulator transition in graphene induced by circularly polarized photons *Phys Rev B* **81**

[114] Sadhukhan K and Agarwal A 2017 Anisotropic plasmons, Friedel oscillations, and screening in 8-P mmn borophene *Phys Rev B* **96**

[115] Del Cima O M and Miranda E S 2018 Electron-polaron-electron-polaron bound states in mass-gap graphene-like planar quantum electrodynamics: S-wave bipolarons *Eur Phys J B* **91**

[116] Fali A, White S T, Folland T G, He M Z, Aghamiri N A, Liu S, Edgar J H, Caldwell J D, Haglund R F and Abate Y 2019 Refractive index-based control of hyperbolic phonon-polariton propagation *Nano Lett* **19** 7725–34

[117] Dumm M, Basov D N, Komiya S, Abe Y and Ando Y 2002 Electromagnetic response of static and fluctuating stripes in cuprate superconductors *Phys Rev Lett* **88**

[118] Dolganov P V, Ksyonz G S, Dmitrienko V E and Dolganov V K 2013 Description of optical properties of cholesteric photonic liquid crystals based on Maxwell equations and Kramers-Kronig relations *Phys Rev E* **87**

[119] Walter J P and Cohen M L 1970 Wave-vector-dependent dielectric function for Si, Ge, GaAs, and ZnSe *Phys Rev B* **2** 1821–6

[120] Jorio A, Souza A G, Dresselhaus G, Dresselhaus M S, Saito R, Hafner J H, Lieber C M, Matinaga F M, Dantas M S S and Pimenta M A 2001 Joint density of electronic states for one isolated single-wall carbon nanotube studied by resonant Raman scattering *Phys Rev B* **63**

[121] Read A J and Needs R J 1991 Calculation of optical matrix-elements with nonlocal pseudopotentials *Phys Rev B* **44** 13071–3

[122] Tsiper E V 2002 Analytic Coulomb matrix elements in the lowest landau level in disk geometry *J Math Phys* **43** 1664–7

[123] Pandey J and Soni A 2019 Unraveling biexciton and excitonic excited states from defect bound states in monolayer MoS2 *Appl Surf Sci* **463** 52–7

[124] Wang D C, Wei S R, Yuan A R, Tian F H, Cao K Y, Zhao Q Z, Zhang Y, Zhou C, Song X P, Xue D Z and Yang S 2020 Machine learning magnetic parameters from spin configurations *Adv Sci* **7**

[125] Fulde P 1997 Heavy-quasiparticle phenomena in metals and semimetals *Physica B* **230** 1–8

[126] Wu Y Q, Lin Y M, Bol A A, Jenkins K A, Xia F N, Farmer D B, Zhu Y and Avouris P 2011 High-frequency, scaled graphene transistors on diamond-like carbon *Nature* **472** 74–8

[127] Pujari S R, Kambale M D, Bhosale P N, Rao P M R and Patil S R 2002 Optical properties of pyrene doped polymer thin films *Mater Res Bull* **37** 1641–9

[128] Choumane H, Ha N, Nelep C, Chardon A, Reymond G O, Goutel C, Cerovic G, Vallet F, Weisbuch C and Benisty H 2005 Double interference fluorescence enhancement from reflective slides: Application to bicolor microarrays *Appl Phys Lett* **87**

[129] Zhou J, Chen Y J, Li H, Dugnani R, Du Q, UrRehman H, Kang H M and Liu H Z 2018 Facile synthesis of three-dimensional lightweight nitrogen-doped graphene aerogel with excellent electromagnetic wave absorption properties *J Mater Sci* **53** 4067–77

[130] Fantini S, Hueber D, Franceschini M A, Gratton E, Rosenfeld W, Stubblefield P G, Maulik D and Stankovic M R 1999 Non-invasive optical monitoring of the newborn piglet brain using continuous-wave and frequency-domain spectroscopy *Phys Med Biol* **44** 1543–63

[131] Taft E A and Philipp H R 1965 Optical properties of graphite *Phys Rev* **138** A197–A201

[132] Setton R, Beguin F and Piroelle S 1982 Graphite-intercalation compounds as reagents in organic-synthesis—an overview and some recent applications *Synthetic Met* **4** 299–318

[133] Xia S X, Zhai X, Wang L L and Wen S C 2018 Plasmonically induced transparency in double-layered graphene nanoribbons *Photonics Res* **6** 692–702

[134] Kis A, Csanyi G, Salvetat J P, Lee T N, Couteau E, Kulik A J, Benoit W, Brugger J and Forro L 2004 Reinforcement of single-walled carbon nanotube bundles by intertube bridging *Nat Mater* **3** 153–7

[135] Oku T, Kuno M, Kitahara H and Narita I 2001 Formation, atomic structures and properties of boron nitride and carbon nanocage fullerene materials *Int J Inorg Mater* **3** 597–612

[136] Matsoso B J, Ranganathan K, Mutuma B K, Lerotholi T, Jones G and Coville N J 2016 Time-dependent evolution of the nitrogen configurations in N-doped graphene films *Rsc Adv* **6** 106914–20

[137] Liu L and Shen Z X 2009 Bandgap engineering of graphene: A density functional theory study *Appl Phys Lett* **95**

[138] Huang H C, Lin S Y, Wu C L and Lin M F 2016 Configuration- and concentration-dependent electronic properties of hydrogenated graphene *Carbon* **103** 84–93

[139] Lin C Y C R B, Ho Y H, Lin F L 2020 *Electronic and Optical Properties of Graphite Related Systems* ([S.l.]: CRC PRESS)

[140] Lu D B, Song Y L, Huang X Y and Wang C 2018 Optical properties of a single carbon chain-doped silicene nanoribbon *J Electron Mater* **47** 4585–93

[141] Choi S and Manthiram A 2002 Synthesis and electrode properties of nanocrystalline lithium copper iron oxide cathodes *J Electrochem Soc* **149** A570–3

[142] Dubois V, Pecquenard B, Soule S, Martinez H and Le Cras F 2017 Dual cation- and anion-based redox process in lithium titanium oxysulfide thin film cathodes for all-solid-state lithium-ion batteries *Acs Appl Mater Inter* **9** 2275–84

[143] Zheng C, Wu J X, Li Y F, Liu X J, Zeng L X and Wei M D 2020 High-performance lithium-ion-based dual-ion batteries enabled by few-layer MoSe2/Nitrogen-doped carbon *Acs Sustain Chem Eng* **8** 5514–23

[144] Wang L S, Niu B, Lee Y T and Shirley D A 1990 Photoelectron-spectroscopy and electronic-structure of heavy group Iv-Vi diatomics *J Chem Phys* **92** 899–908

[145] Kovalevsky K A, Abrosimov N V, Zhukavin R K, Pavlov S G, Hubers H W, Tsyplenkov V V and Shastin V N 2015 Terahertz lasers based on intracentre transitions of group V donors in uniaxially deformed silicon *Quantum Electron* **45** 113–20

[146] Xu Q, Li W J, Ding L, Yang W J, Xiao H H and Ong W J 2019 Function-driven engineering of 1D carbon nanotubes and 0D carbon dots: Mechanism, properties and applications *Nanoscale* **11** 1475–504

[147] Li X L, Wang X R, Zhang L, Lee S W and Dai H J 2008 Chemically derived, ultra-smooth graphene nanoribbon semiconductors *Science* **319** 1229–32

[148] Fan Y S, Guo C C, Zhu Z H, Xu W, Wu F, Yuan X D and Qin S Q 2017 Monolayer-graphene-based perfect absorption structures in the near infrared *Opt Express* **25** 13079–86

[149] Joly Y 2001 X-ray absorption near-edge structure calculations beyond the muffin-tin approximation *Phys Rev B* **63**

[150] Koshino M and Ando T 2009 Electronic structures and optical absorption of multilayer graphenes *Solid State Commun* **149** 1123–7

[151] Lan Y Z 2018 First-principles studies of effects of layer stacking, opposite atoms, and stacking order on two-photon absorption of two-dimensional layered silicon carbide *Comp Mater Sci* **151** 231–9

[152] Dell'Anna L and De Martino A 2009 Multiple magnetic barriers in graphene *Phys Rev B* **79**

[153] Park S and Sim H S 2008 Magnetic edge states in graphene in nonuniform magnetic fields *Phys Rev B* **77**

[154] Fan S Q, Tang X D, Zhang D H, Hu X D, Liu J, Yang L J and Su J 2019 Ambipolar and n/p-type conduction enhancement of two-dimensional materials by surface charge transfer doping *Nanoscale* **11** 15359–66

[155] Wang Z Y and Zhuang P F 2016 Critical behavior and dimension crossover of pion superfluidity *Phys Rev D* **94**

[156] Sun Z P, Martinez A and Wang F 2016 Optical modulators with 2D layered materials *Nat Photonics* **10** 227–38

[157] Maximov M, Nazarov D, Rumyantsev A, Koshtyal Y, Ezhov I, Mitrofanov I, Kim A, Medvedev O and Popovich A 2020 Atomic layer deposition of lithium-nickel-silicon oxide cathode material for thin-film lithium-ion batteries *Energies* **13**

[158] Wilson A M, Zank G, Eguchi K, Xing W and Dahn J R 1997 Pyrolysed silicon-containing polymers as high capacity anodes for lithium-ion batteries *J Power Sources* **68** 195–200

[159] Mott N F 1975 Coulomb gap and low-temperature conductivity of disordered systems *J Phys C Solid State* **8** L239–40

[160] Luo J H, Wang J H, Fang Z J, Shao J W and Li J G 2018 Optimal design of a high efficiency LLC resonant converter with a narrow frequency range for voltage regulation *Energies* **11**

[161] Zvara M, Grill R, Hlidek P, Orlita M and Soubusta J 2002 Photoluminescence of biased GaAs/AlxGa1-xAs double quantum wells—many-body effects *Physica E* **12** 335–9

[162] Atwater H 2007 COLL 491-Plasmonics: A route to optical metamaterials and nanoscale optical devices *Abstr Pap Am Chem S* **234**

[163] Zoubi H and Ritsch H 2010 Metastability and directional emission characteristics of excitons in 1D optical lattices *Epl-Europhys Lett* **90**

[164] Terekhov I S, Milstein A I, Kotov V N and Sushkov O P 2008 Screening of Coulomb impurities in graphene *Phys Rev Lett* **100**

[165] Richardson C F and Ashcroft N W 1997 High temperature superconductivity in metallic hydrogen: Electron-electron enhancements *Phys Rev Lett* **78** 118–21

[166] Kiczek B and Ptok A 2017 Influence of the orbital effects on the Majorana quasi-particles in a nanowire *J Phys-Condens Mat* **29**

[167] Ma C, Clark S, Liu Z X, Liang L L, Firdaus Y L, Tao R, Han A, Liu X G, Li L J, Anthopoulos T D, Hersam M C and Wu T 2020 Solution-processed mixed-dimensional hybrid perovskite/carbon nanotube electronics *Acs Nano* **14** 3969–79

[168] Quintela M F C M and Peres N M R 2020 A colloquium on the variational method applied to excitons in 2D materials *Eur Phys J B* **93**

[169] Risko C, McGehee M D and Bredas J L 2011 A quantum-chemical perspective into low optical-gap polymers for highly-efficient organic solar cells *Chem Sci* **2** 1200–18

[170] Srikant V and Clarke D R 1998 On the optical band gap of zinc oxide *J Appl Phys* **83** 5447–51

[171] Zhang Y L, Guo L, Wei S, He Y Y, Xia H, Chen Q D, Sun H B and Xiao F S 2010 Direct imprinting of microcircuits on graphene oxides film by femtosecond laser reduction *Nano Today* **5** 15–20

[172] Das P, Ganguly S, Banerjee S and Das N C 2019 Graphene based emergent nanolights: A short review on the synthesis, properties and application *Res Chem Intermediat* **45** 3823–53

[173] Zhou W, Kapetanakis M D, Prange M P, Pantelides S T, Pennycook S J and Idrobo J C 2012 Direct determination of the chemical bonding of individual impurities in graphene *Phys Rev Lett* **109**

4 Electronic and Transport Properties of the Sawtooth-Sawtooth Penta-Graphene Nanoribbons

Nguyen Thanh Tien, Pham Thi Bich Thao, and Ming-Fa Lin

CONTENTS

4.1 INTRODUCTION

Low-dimensional materials are of great interest to both theorists and experimentalists, owing to their novel electronic properties, which arise mainly because of a host of quantum confinement effects and quasiparticle properties. Two-dimensional

DOI: 10.1201/9781003322573-4

nanomaterials have captivated enormous attention due to the surge in graphene research, which opens an avenue for the development of 2D semiconductors for future multifunctional electronic and optoelectronic applications [1–5]. Graphene presents a zero-energy gap and linear energy bands across the Fermi level so that the charge carriers resemble massless Dirac fermions [6, 7]. However, the application of graphene in semiconductor devices requires a band gap in order to switch the conductivity between on and off states. Therefore, physical and chemical approaches for creating an energy gap in graphene are a critical studying topic (under discussion). Many attempts have been made to improve their electronic properties by creating ribbon structures from the two-dimensional graphene [8–10]. The chemical functionalization of graphene is the most effective method to induce a band gap and dramatically change the fundamental physical properties [11–15]. Due to the powerful applications of carbon nanostructure, many studies have been devoted to exploring new carbon allotropes.

Very recently, significant efforts have focused on the stabilization of a novel carbon allotrope, named monolayer penta-graphene (MPG) [16–19]. Penta-graphene (PG) is extracted from the bulk T12-carbon phase. This phase is obtained by heating an interlocking-hexagon-based metastable carbon phase at high temperatures [20]. It is found that the MPG is an indirect band-gap semiconductor with a band gap of ~3.25 eV [16, 18, 21], which is smaller than SiC [22], BN [23, 24], and BeO [25] nanostructures. Those studies showed that this structure has obtained dynamical, thermal, and mechanical stability. Furthermore, Wu *et al.* [17] used first-principles lattice dynamics and iterative solution of the phonon Boltzmann transport equation (BTE) to investigate the thermal conductivity of MPG and its more stable derivative, hydrogenated monolayer penta-graphene (HMPG). They showed that in contrast to the hydrogenation of graphene, which leads to a dramatic decrease in thermal conductivity, HMPG shows a notable increase in thermal conductivity, which is much higher than that of MPG. Also, a theoretical study has reported the stability and electronic properties of one-dimensional penta-graphene nanoribbon (PGNR) [26]. Their calculations predict that PGNR is dynamically stable; it is mechanically flexible tolerating up to 11.5% of axial strain. They also explored the electronic properties of the PGNRs with different widths. It is found that the band gap of wider PGNRs changes only marginally with further increasing the ribbon width, which opens up the possibility of a stable nanoribbon with a large band gap [26]. Tien et. al. investigated the electronic structures and the I-V characteristics of the sawtooth-sawtooth PGNRs (SSPGNRs) under a sequence of uniaxial strains in a range from 10% compression to 10% stretch [27]. In this strained range, carbon atoms still keep a pentagon network, but with the changing bond lengths. The fundamental physical properties (band structure, I-V characteristic) of SSPGNRs seem to be more sensitive to compressive strain than the stretch strain. The current intensity of the compressive-SSPGNR is by 2 orders of magnitude compared to that of the tensile-SSPGNR at the same strain in a range from 6% to 10%.

It is clear that there have been many attempts to materialize the graphene and penta-graphene materials by sophisticated experimental procedures, and numerous theoretical explorations have shown a huge potential of these low-dimensional materials to revolutionize electronic and optic-electronic devices [28–31]. It is necessary

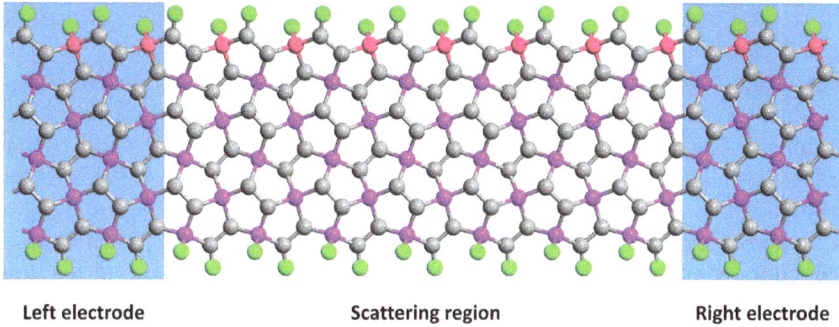

Left electrode   **Scattering region**   **Right electrode**

FIGURE 4.1   Schematic illustration of device model (top view). The transparent blue rectangles illustrate the left and right electrodes. C_1 (sp^3 hybridization) atoms and C_2 atoms (sp^2
hybridization) are purple and gray in turn. Green represents termination elements. Pink represents doping elements.

to study many problems related to the essential properties of low-dimensional systems based on graphene and penta-graphene materials for developing the next generation of electronic and optic-electronic devices [32]. Figure 4.1 shows a simple
device model consists of two electrodes and the active (scattering) region based on
the SSPGNR material.

This chapter will provide a reliable prediction about the novel and essential properties of the 1D PGNR nanostructured materials. Electronic and transport properties
of the SSPGNRs were systematically investigated by using the density-functional
theory in combination with the non-equilibrium Green's function formalism. We
pay attention to the diversity of the electronic properties and the electronic transport of SSPGNRs by doping and edge passivation of structures. We studied the
electronic and transport properties of an H-passivated SSPGNR, n-type (silicon-
Si, nitrogen-N, and phosphorus-P) substitutional doping SSPGNRs and p-type
(boron-B, aluminum-Al, and gallium-Ga) doping. We also studied the influence of
the edge configurations in terms of H-passivated edges (HH-SSPGNR) and edges
terminated by non-metallic atoms (H, P, Si) such as identical edge termination
(PP-SSPGNR and SiSi-SSPGNR) and alternate edge termination (PH-SSPGNR and
SiH-SSPGNR). All studied samples are optimized using DFT calculations within
the generalized gradient approximation (GGA) of Perdew-Burke-Ernzerhof (PBE).
The electronic and transport properties such as band structure (BS), the density of
states (DOS), the partial density of states (PDOS), transmission spectrum (T(E))
and Volt-Ampere characteristics (I-V curves) were computed. Our results show that
electronic structures and the currents of the studied samples depend on doped elements and doping type.

The doping dramatically affects the electronic properties and the I-V characteristic of studied samples. More specifically, the current intensity of N-SSPGNR and
P-SSPGNR increase by 8 orders of magnitude compared to that of SSPGNR, while
the one of Si-SSPGNR has negligible change. However, there are also considerable
differences in I-V curves of samples doping with N and P. The same effect has

occurred for p-type doping. It was found that SSPGNR band gap can be controlled through changing various passivation elements as well as termination forms, which leads to the transition from a semiconductor to a metal or a semimetal. For the influence on transport properties, P and Si atoms of alternate cases improve significantly the weak point (very poor current) of the bare and traditionally passivated models. The 9-order rise of current magnitude is observed in PH-SSPGNR and SiH-SSPGNR compared to HH-SSPGNR and PP-SSPGNR. Interestingly, the oscillation current-voltage characteristic appears when SSPGNR is functionalized identically by Si atoms. This study helps guide the selection of I-V rules in atom-doped SSPGNRs and the edges-terminated one. The electronic diversity of these structures shows the way to adjust the electronic properties of these promising materials in the different types of solid-state nanoelectronic devices. These calculations provide insight into the quasiparticle (added hole and electron) states of SSPGNRs.

4.2 COMPUTATIONAL METHODS

Electronic and transport properties of the SSPGNRs are explored by first-principles calculations based on DFT and the NEGF using the Atomistix ToolKit (ATK) software package [33, 34]. First-principle calculations are used to calculate the properties of SSPGNRs by means of the interactions between atoms, molecules, and groups of molecules, etc. The methods use the laws of physics and do not rely on approximations or fitting to experimental measurements.

The width of the studied structures is six and four sawtooth chains. First, the samples are optimized using DFT calculations within the generalized gradient approximation (GGA) of Perdew-Burke-Ernzerhof (PBE) with similar conditions: $1 \times 1 \times 21$ k-point sampling, a mesh cutoff energy of 680 eV, and electron temperature 300K. A periodic boundary condition along the z-direction was applied to the SSPGNRs, while vacuum spaces of 15 Å are imposed on the plane perpendicular to the z-axis of the SSPGNR layer to guarantee negligible interaction between periodic images.

To evaluate the stability of these structures and the energy value to form the various SSPGNRs, we calculate the binding energy (EB) and edge formation energy (EFE) by equations that are listed in detail in the following sections. To understand the electronic properties of SSPGNRs, we calculate the electronic BS, DOS, and corresponding PDOS. The investigation on device transport is carried out by NEGF formalism. The current through devices is calculated by Landauer Büttiker [35, 36] formula:

$$I(V_b) = \frac{2e}{h} \int_{-\infty}^{+\infty} T(E, V_b) \left[f(E - \mu_L) - f(E - \mu_R) \right] dE \qquad (4.1)$$

Here $T(E)$ is the transmission function at energy E under bias V_b can be obtained by the following formula:

$$T(E, V_b) = Tr[\Gamma_L(E, V_b)) G(E, V_b) \Gamma_R(E, V_b) G'(E, V_b) \qquad (4.2)$$

and $f(E - \mu_L)$, $f(E - \mu_R)$ are the Fermi distribution functions for left/right lead electrochemical potentials ($\mu_L = E_F + eV_b/2, \mu_R = E_F - eV_b/2$), respectively. $G^r(G^a)$ is the retarded (advanced) Green functions and $\Gamma_{L/R}$ is the coupling matrices.

4.3   RESULTS AND DISCUSSIONS

4.3.1   H-PASSIVATED SSPGNR

4.3.1.1   The Structural Stability and Electronic Properties

Figure 4.2 shows a unit cell of the H-passivated SSPGNR structure with SSPGNR before optimization (Figure 4.2(a)) and SSPGNR after optimization (Figure 4.2(b)).

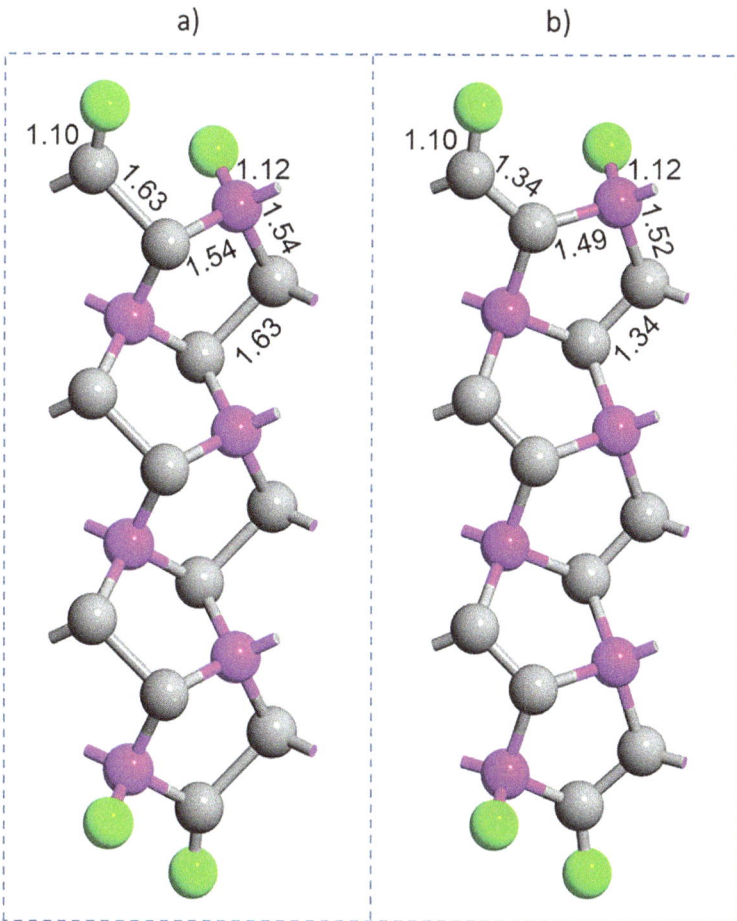

FIGURE 4.2   Schematic illustration (top view) of the H-passivated SSPGNR structures: (a) before optimization, (b) after optimization with purple balls (C_1: sp^3 hybridization), gray balls (C_2: sp^2 hybridization), and green balls (hydrogen atoms).

We first assessed the stability of the H-passivated SSPGNR structures by calculating the binding energy. The binding energy E_{BE} of the sample is calculated through the following formula:

$$E_{BE} = \frac{E_{total} - n_C E_C - n_H E_H}{n_C + n_H},$$
(4.3)

with E_{total} is the total energy for one supercell. E_C, E_H denote isolated energy for one C, H atom, and n_C, n_H, are the number of C, H atoms, respectively. Numerical results indicate that the binding energy E_{BE} of the sample is –7.4 eV. The binding energy of H-passivated SSPGNR is consistent with the research of Yuan et. al [37]. The binding energy of the H-passivated SSPGNR has a negative value shows that the one is thermodynamics stability.

In Figure 4.3, we present the calculated band structure of the H-passivated SSPGNR. It shows that the H-passivated SSPGNR possesses a semiconductor nature with a 2.6 eV band gap (E_g). It has a direct band gap with both the conduction band minimum and the valence band maximum at Γ point.

The total DOS and PDOS of the H-passivated SSPGNR were also drawn in Figure 4.4. It shows that C has a major contribution to most electronic states around the Fermi level.

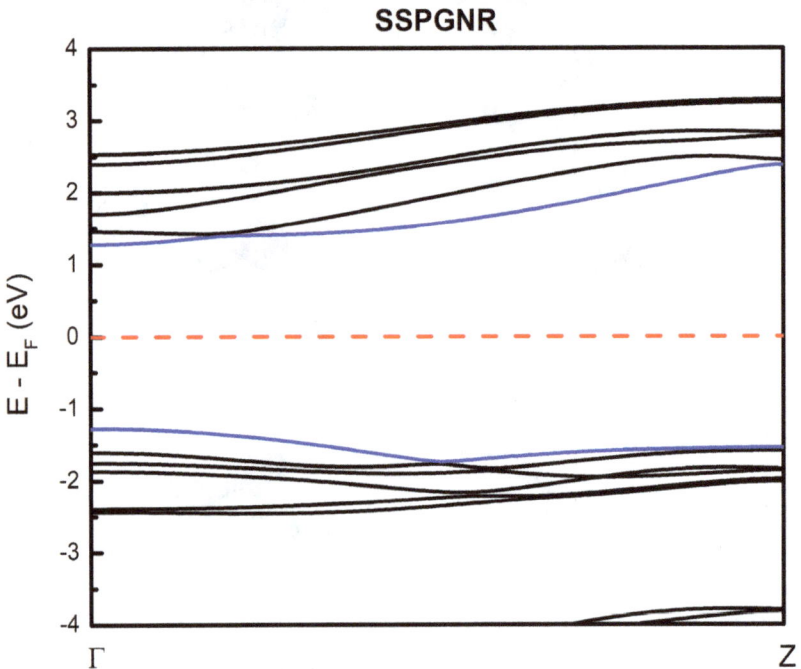

FIGURE 4.3   The band structure of the H-passivated SSPGN; the dashed line indicates the Fermi level.

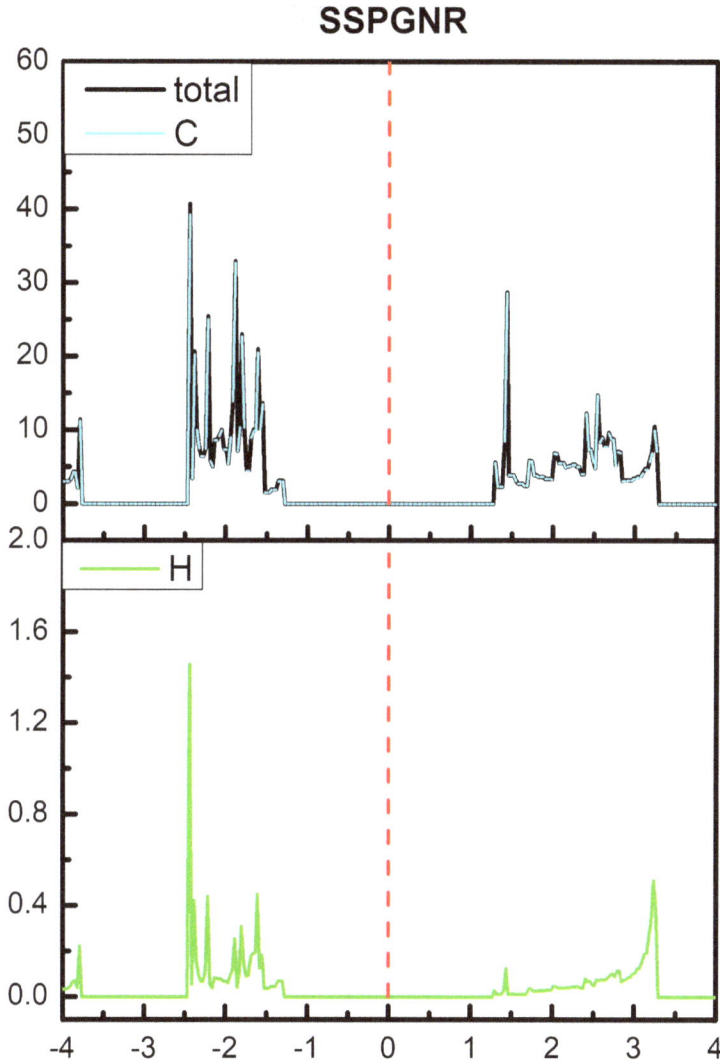

FIGURE 4.4   The density of state of the H-passivated SSPGNR. The above figure shows total DOS and PDOS of C, the below figure the PDOS of H.

4.3.1.2   Electronic Transport Properties

We show the correlation between current and bias voltage of the H-passivated SSPGNR in Figure 4.5. The current intensity is almost unchanged when the bias voltage increases from 0 V to 1.0 V. The current of the SSPGNR starts to increase at a bias voltage of 1.1 V, reaches its maximum at 1.5 V, and decreases to the minimum at 1.7 V. The second peak appears at 1.8 V and draws the reduction current between 1.8 V and 2 V. This shows the characteristics of a semiconductor, but when the high bias voltage, there are fluctuations in the current.

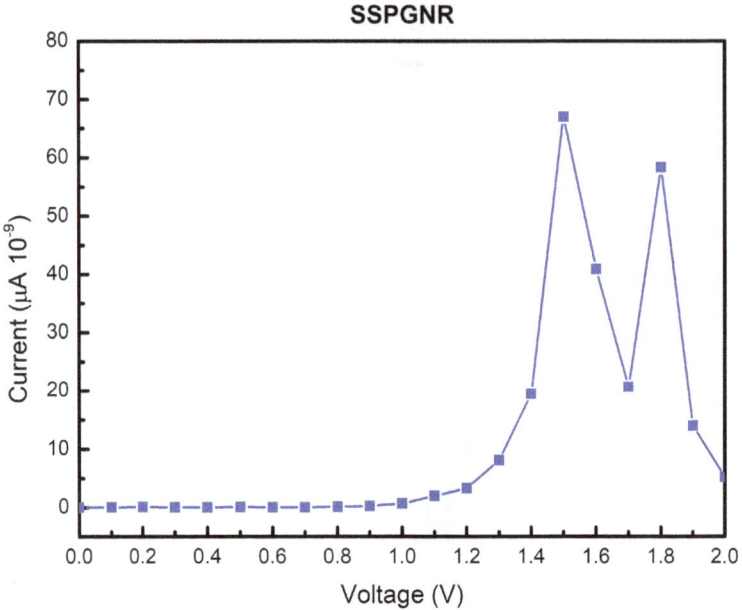

SSPGNR

FIGURE 4.5   The I-V curves of the H-passivated SSPGNR.

To further understand the I-V curves of the H-passivated SSPGNR, we examine the transmission spectra as a function of the applied bias and electron energy in Figure 4.6. T(E) almost remains equal to 0 with the increasing bias voltage to 1.0 V. Nevertheless, there exist higher coefficients at several bias windows from 1.0 V that leads to the current fluctuation.

4.3.2   N-TYPE DOPING SSPGNRS (SILICON-SI, NITROGEN-N, AND PHOSPHORUS-P)

To enhance the electrical conductivity, the doping method is often implemented [38]. We select the edge for dopant positions because the substitutional atom at the nanoribbon edges shows more special effects [39]. Figure 4.7 shows a unit cell of the H-passivated SSPGNR structures with doping one atom (Si, N, P) in a similar position.

4.3.2.1   The Structural Stability and Electronic Properties

We also consider the stability of the *n-type doping* H-passivated SSPGNR structures by calculating the binding energy. Binding energy E_{BE} of three doping samples is calculated the same as Eq. (4.3), through the following formula,

$$E_{BE} = \frac{(E_{total} - n_C E_C - n_H E_H - n_X E_x)}{n_C + n_H + n_X}, \tag{4.4}$$

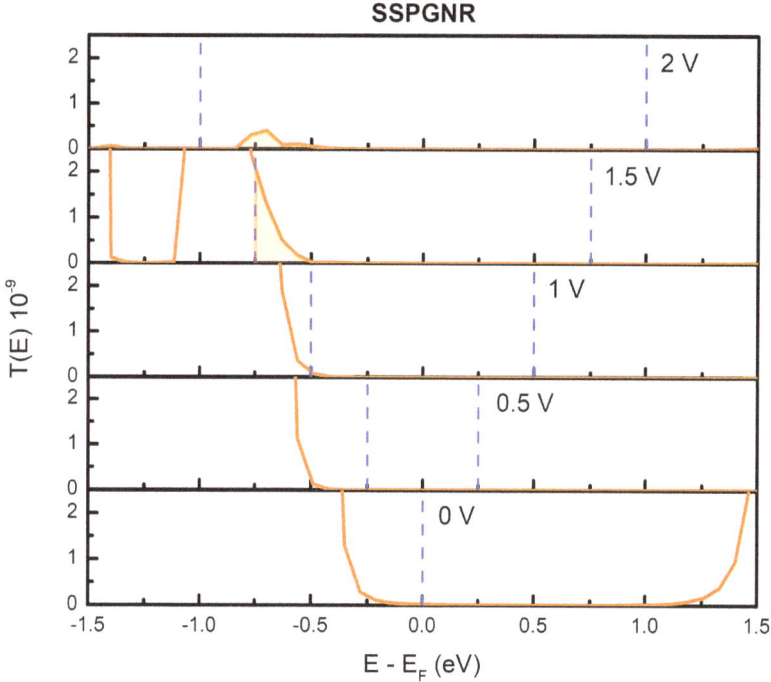

FIGURE 4.6 Transmission spectrum T(E) at different bias voltages for H-passivated SSPGNR. The vertical dashed lines show the bias windows.

with E_X denoting isolated energy for X doped-atom (X = Si, N, P) and n_X is the number of X atom ($n_X = 1$). Numerical results indicate that the binding energy of those samples has a negative value (between −6.9 and −7.2 eV). Those results are quite compatible with the results of Li et al [40]. In these doped structures, N-SSPGNR possesses the highest stability.

In Figure 4.8, we present the calculated band structures of three n-type doping SSPGNRs. It can be obviously seen that doping leads to significant variation in band structures. Replacing C atoms by Si atoms, the width of the band gap reduces about 15 percentages ($E_{g\_Si} = 2.2$ eV), changes into indirect form, and still keeps an intrinsic semiconductor. This is similar to the study of the band gap in the Si-doped 2D penta-graphene [41]. In contrast, two samples (N-SSPGNR, P-SSPGNR) exhibit an absolutely different feature. Specifically, both of them have a sub-band passing through the Fermi level, and the distribution of the conduction band and valence band states is nearly homogeneous. Moreover, the Fermi energy is shifted closer to the edge of the conduction band. Interestingly, replacing C by N and P varies the system from an intrinsic semiconductor to an n-type semiconductor. This can be explained by the fact that P and N are n-type dopants and belong to the VA group with the same electron number (5 electrons) in the outer layer. They may have an extra electron after binding to their neighbors (C atoms). This is different from substituting C by Si, the element has 4 electrons in its outer layer like C. This result illustrates that doping P

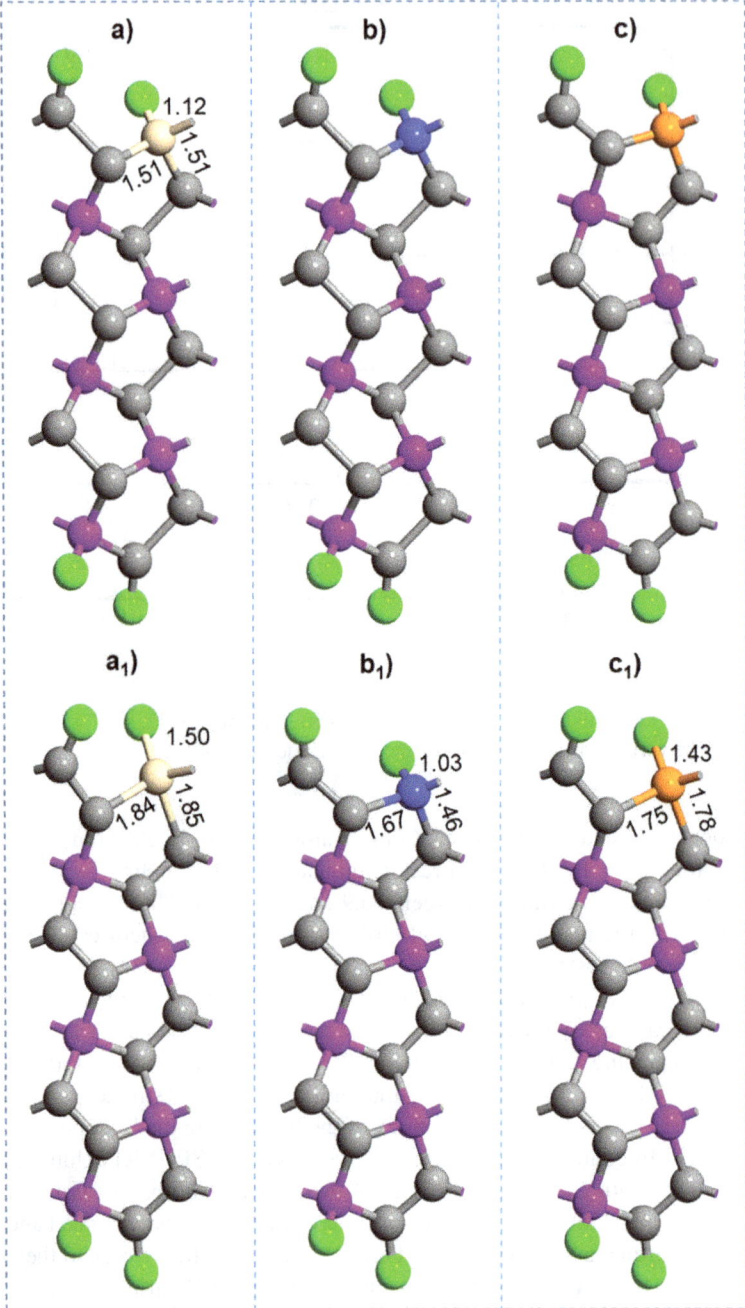

FIGURE 4.7 Schematic illustration (top view) of the n-type doping SSPGNR structures: (a) Si-doped, (b) N-doped, and (c) P-doped before optimization. (a1), (b1), and (c1) present structures after optimization. Si, N, P, and H are colored by cream, blue, orange, and green, respectively.

a) Si-SSPGNR

b) N-SSPGNR

c) P-SSPGNR

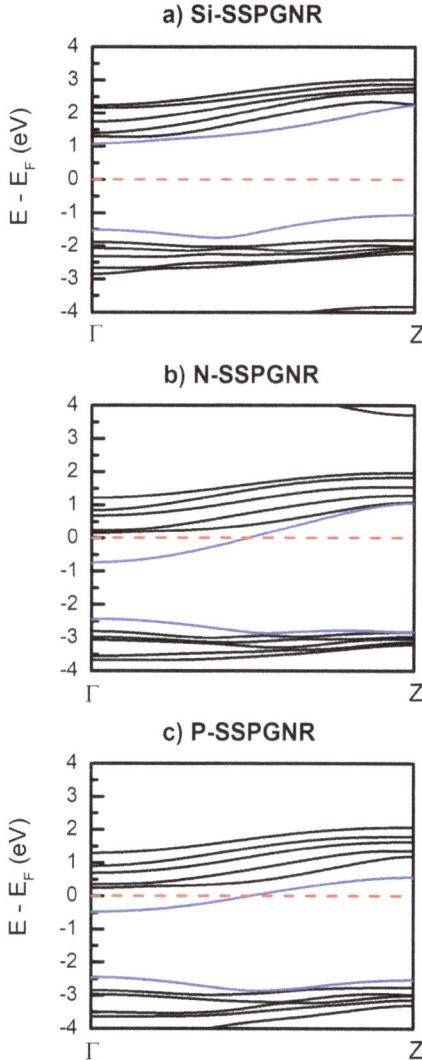

FIGURE 4.8 The band structures of the n-type doping SSPGNRs: (a) Si-doped, (b) N-doped, and (c) P-doped. The dashed line indicates the Fermi level.

and N in PGNRs changes the electronic transport properties compared to the pristine structure. Otherwise, Si-doped PGNR has nearly identical transport with the pristine case. This content will be detailed in the transport properties sub-section.

In order to clearly understand the effect of dopants on band structure, we consider the density of states (DOS) and the partial density of states (PDOS) of all examined systems in Figure 4.9. In the case of Si-SSPGNR, new electronic states of Si atoms appear in the energy zone corresponding to the band gap of the pristine SSPGNR (Figure 4.9(a)), leading to the reduction of the band gap. The presence of

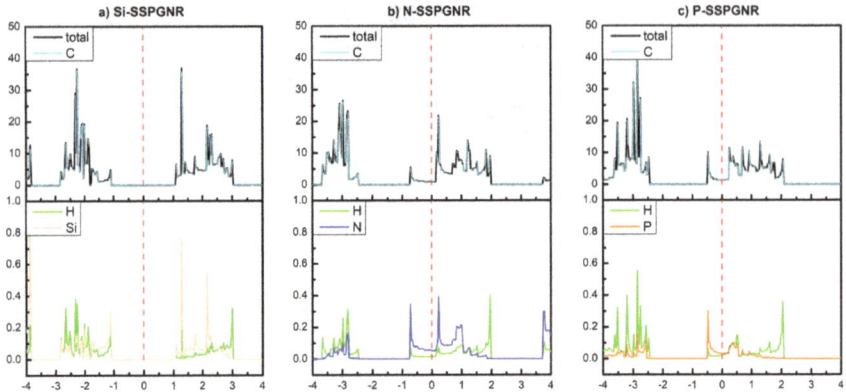

FIGURE 4.9   The density of state of the n-type doping SSPGNRs: (a) Si-doped, (b) N-doped, and (c) P-doped. The dashed line indicates the Fermi level.

electron states of N and P in the above energy area is also observed through PDOS in Figure 4.9(b) and Figure 4.9(c). This also helps to explain the reason for the change in band structure obtained earlier. From these analyses, it can be seen that the alteration in band structure depends mainly on the electronic distribution from the dopants [41]. We also note Si (the element of IV-A group as C) but N, P (the element of V-A group, different from C).

4.3.2.2   Electronic Transport Properties

Figure 4.10 shows the correlation between current and bias voltage of Si-SSPGNR, N-SSPGNR, and P-SSPGNR. For Si-SSPGNR, the current intensity is almost unchanged when the bias voltage increases from 0 V to 1.6 V. The current intensity increases from 1.6 V, reaches the maximum at 1.8 V, drops to a minimum at 1.9 V, and increases when the bias voltage is at 2 V. However, the current intensity is very small, and the peak intensity is about 6.0×10^{-8} μA. In contrast, the current intensity of N-SSPGNR and PSSPGNR rises linearly as bias voltage increases from 0 V to 0.5 V. While the N-SSPGNR current reaches its maximum at 0.8 V, then decreases as the bias voltage increases, the P-SSPGNR current reaches its maximum at 0.5 V then fluctuates considerably within the remaining bias voltage range.

To further understand the I-V curves of studied systems, we examine the transmission spectra as a function of the applied bias and electron energy in Figure 4.11. Numerical results in Figure 4.11(a) show T(E) for Si-SSPGNR. T(E) almost is equal to 0 with the increasing bias voltage to 1.0 V. Nevertheless, there exist higher coefficients at several bias windows from 1.0 V that leads to the current fluctuation. However, the transmission coefficients can be seen in the range of 10^{-9}, a very small value. Meanwhile, in N-SSPGNR and P-SSPGNR, we can observe the peaks of T(E) in all the displayed voltages is about 10^9 times higher than the one of SSPGNR and Si-SSPGNR. The outcomes show the strong influence of the transmitted spectrum on the I-V characteristics of the device. The filled zone bounded by T(E) curve and horizontal axis in the bias window can be used to explain changing trend of the I-V curves.

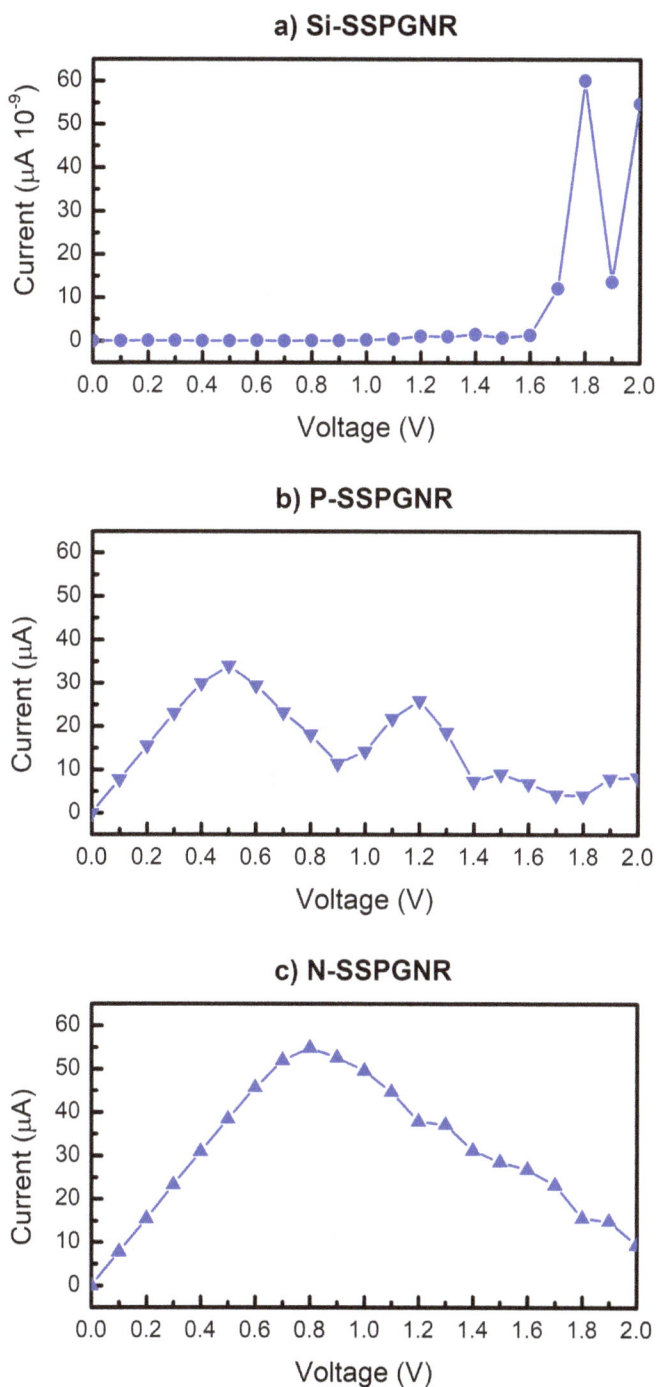

FIGURE 4.10 The I-V curves of the n-type doping SSPGNRs: (a) Si-doped, (b) N-doped, and (c) P-doped.

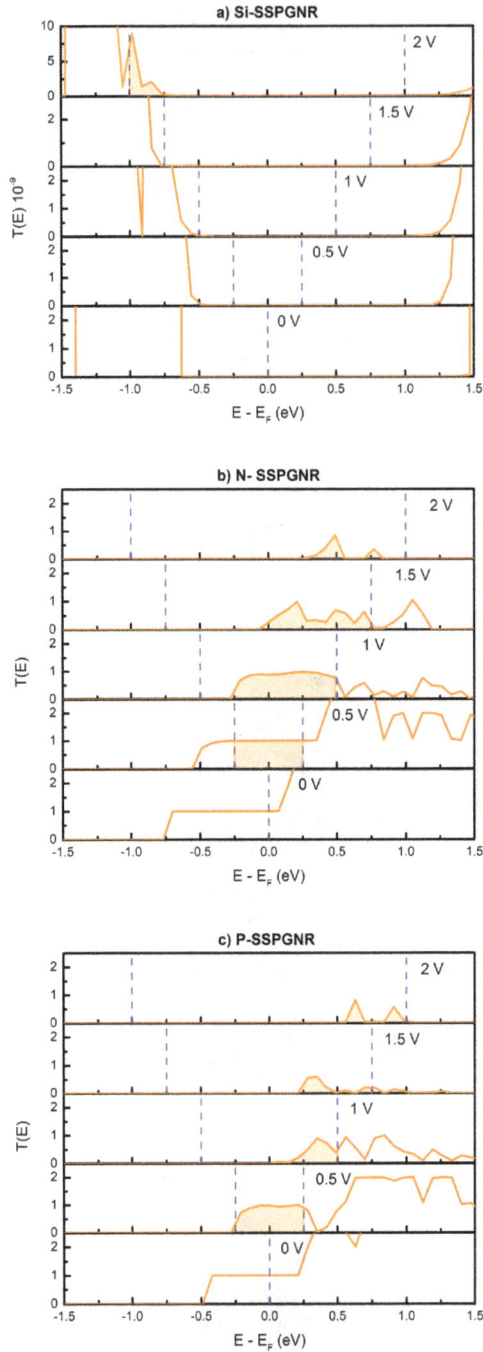

FIGURE 4.11 Transmission spectrum T(E) at different bias voltages for the n-type doping SSPGNRs: (a) Si-doped, (b) N-doped, and (c) P-doped. The vertical dashed lines show the bias windows.

4.3.3   P-Type Doping SSPGNRs (Boron-B, Aluminum-Al, and Gallium-Ga)

We also consider p-type doping. Figure 4.12 shows a unit cell of the H-passivated SSPGNR structures with doping one atom (Si, N, P) in a similar edge position.

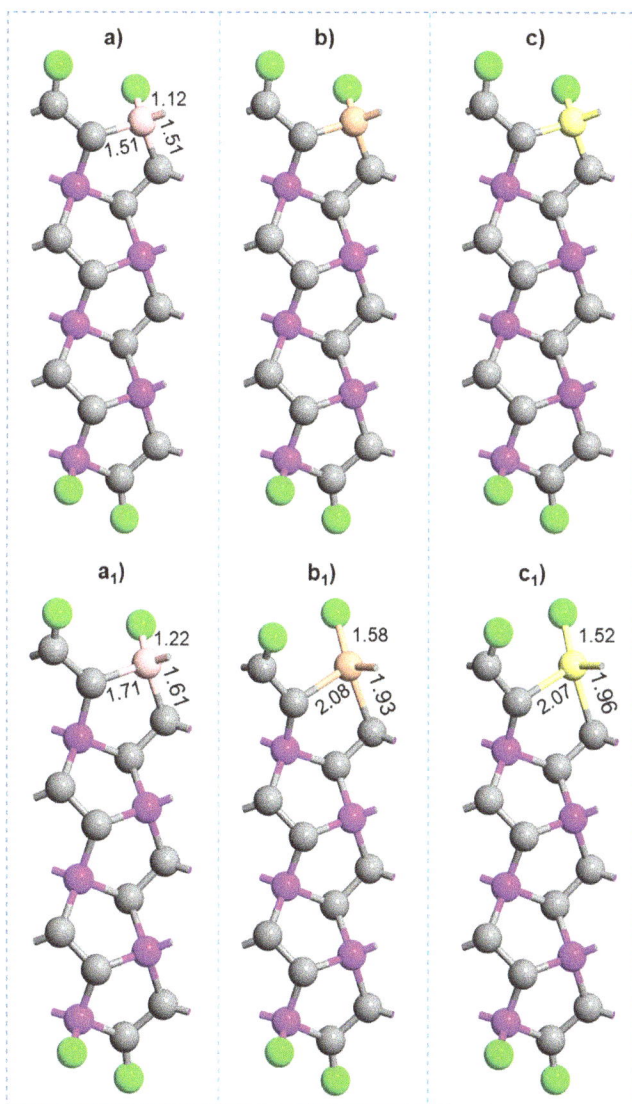

FIGURE 4.12 Schematic illustration (top view) of the p-type doping SSPGNR structures: (a) B-doped, (b) Al-doped, and (c) Ga-doped. The bottom view present structures (a1), (b1), (c1) before (after) optimization. B, Al, Ga, and H are colored by cream, orange, yellow, and green, respectively.

4.3.3.1   The Structural Stability and Electronic Properties

Binding energy E_{BE} of three p-type doping samples is calculated the same as Eq. (4.4) by replacing X = B, Al, Ga. Numerical results indicate that the binding energy of those samples has negative value (E_{BE_B} = −7.7 eV, E_{BE_Al} = −7.61 eV, E_{BE_Ga} = −3.64 eV). In these doped structures, B-SSPGNR possesses the highest stability.

In Figure 4.13, we present the calculated band structures of three p-type doping SSPGNRs. It can be obviously seen that doping leads to significant variation in band

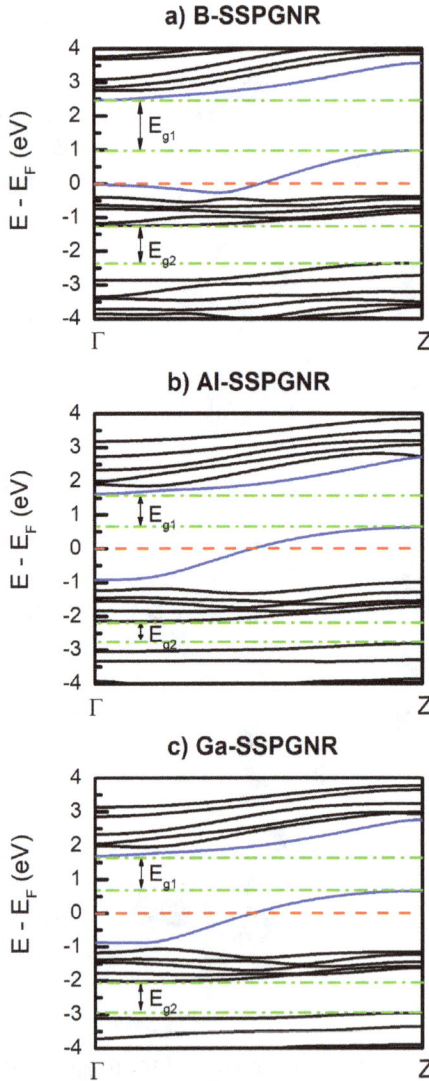

FIGURE 4.13   The band structures of the p-type doping SSPGNRs: (a) B-doped, (b) Al-doped, and (c) Ga-doped. The dashed line indicates the Fermi level.

structures. Replacing C atoms by B, Al, Ga atoms, the width of the band gap reduces significantly, e.g. E_{g_X} (X= B, Al, Ga) = 1.4, 0.9, 0.85 eV. The B, Al, Ga doping SSPGNRs change into indirect form and p-type semiconductor. This is also similar to the study of the band gap in the Ga-doped 2D penta-graphene [41]. All of them have a sub-band passing through the Fermi level, and their Fermi energy is shifted closer to the edge of the valence band. This can be explained by the fact that B, Al, and Ga are p-type dopants and belong to group-III with the same electron number (3 electrons) in the outer layer. The doping atoms lose an electron after binding to their neighboring C atoms. The shape of the valence band maximum of the B-doped SSPGNR is slightly different from the one of the Al, Ga-doped SSPGNRs.

In order to verify the effect of dopants on band structure, we also consider DOS and PDOS of all p-type doping systems examined in Figure 4.14. The presence of electron states of B, Al, and Ga in the above energy area is also observed through PDOS in Figure 4.14(a,b,c). This demonstrates the role of doping atoms that changes the band structure obtained above.

4.3.3.2   Electronic Transport Properties

Figure 4.15 presents the I-V characteristic of B-SSPGNR, Al-SSPGNR, and Ga-SSPGNR. Their progress is almost the same. The highest value of currents is about 42 µA. However, the shape of the I-V curve of B-SSPGNR is slightly different. The highest value of current of one almost keeps constant when the bias voltage increases from 0.65 V to 1.42 V.

Numerical results in Figure 4.16 show T(E) for the p-type doping SSPGNRs. The trend of change of T (E) in two cases (Al-doped SSPGNR and Ga-doped SSPGNR) is almost the same. Particularly in the case of B-doped SSPGNR, the changing trend of T(E) is different. At 1 V and 1.5 V bias potential, there are large conductance channels in the bias windows. This may clearly explain the changing trend of the I-V curves in the three research samples.

Graphene nanoribbons in the presence of impurities are a very good candidate for the next generation of nanoscale devices [42]. On the other hand, one can increase its applications in electronic components, including field effect transistors and supercapacitors, by controlling the quantum capacitance of them [43]. Research results on electronic properties and electronic transport of the doped PGNRs enrich the understanding of the physical properties of carbon-based electronic materials. This research result provides more opportunities to select new-generation electronic components.

4.3.4   EDGE-TERMINATED SSPGNRs

We consider four various terminated structures of SSPGNR with the width of four carbon chains [44]. The terminated structures are named in a general formula XY-SSPGNR (X/Y is H, P, and Si) as shown in Figure 4.17. They are classified into two forms: identical edge termination (where both X and Y are the same elements, namely HH-SSPGNR, PP-SSPGNR, and SiSi-SSPGNR), and alternate edge termination (X is either P or Si and Y remains H, specifically PH-SSPGNR and SiH-SSPGNR).

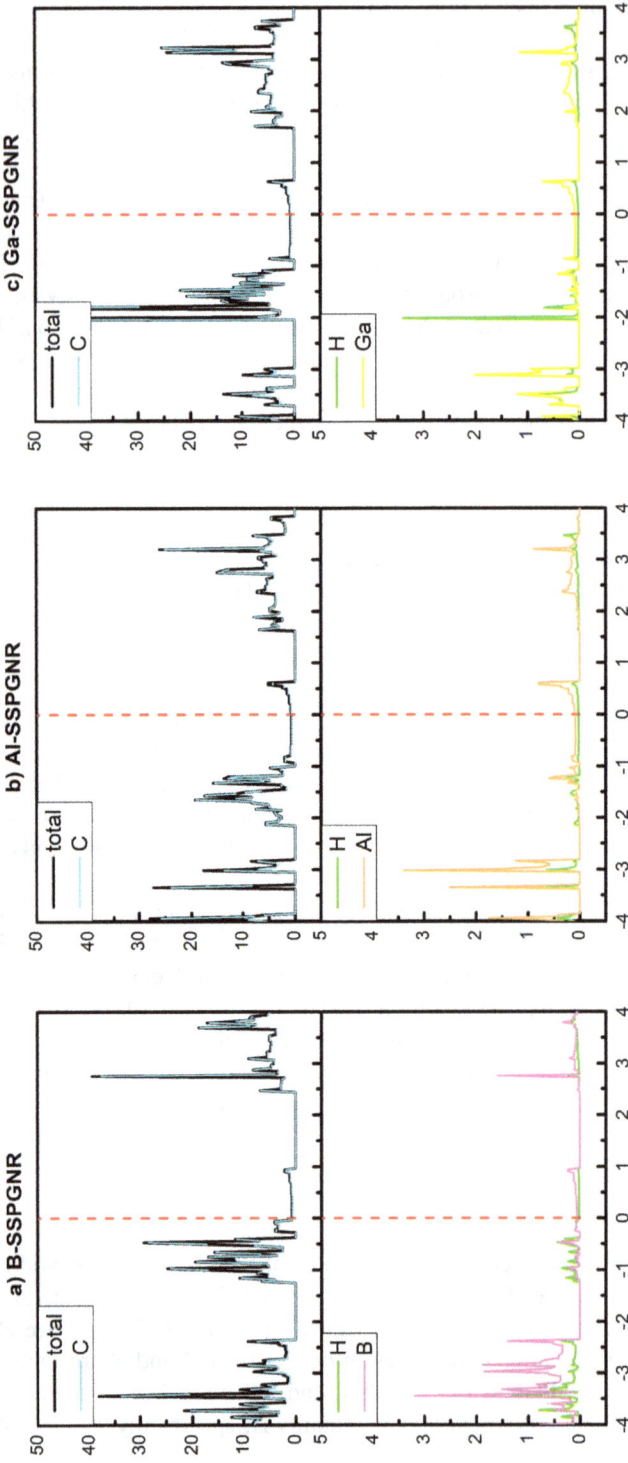

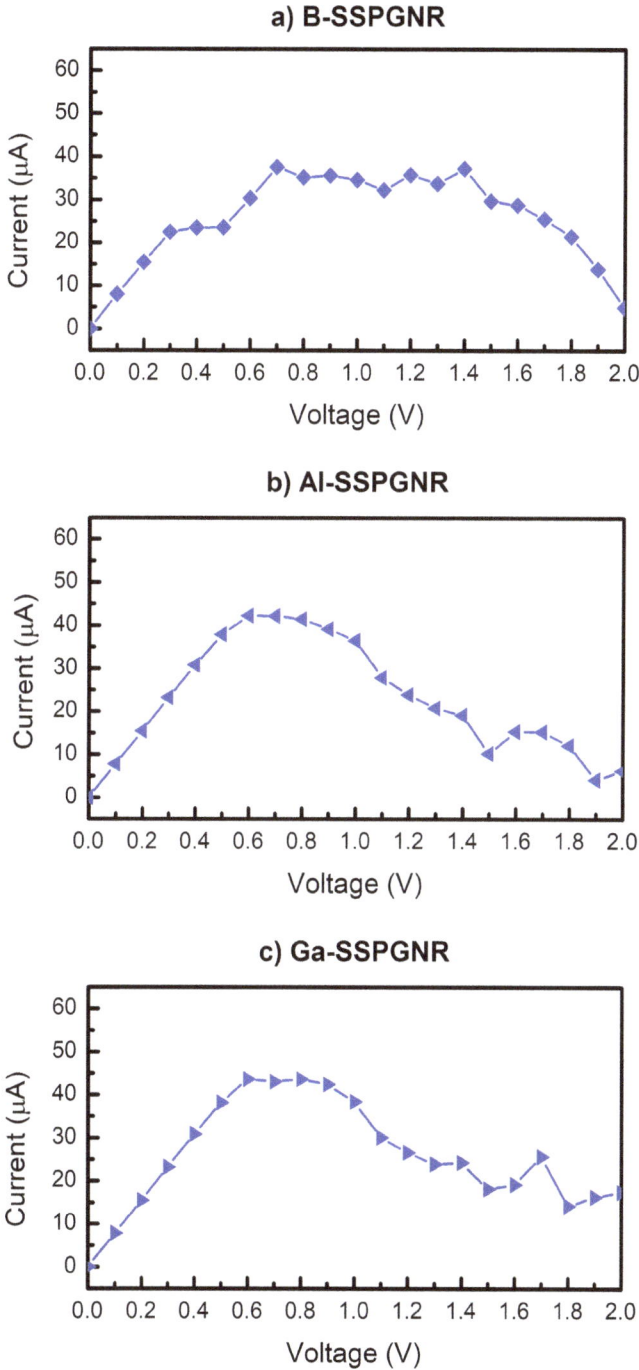

FIGURE 4.15   The I-V curves of the p-type doping SSPGNRs: (a) B-doped, (b) Al-doped, and (c) Ga-doped.

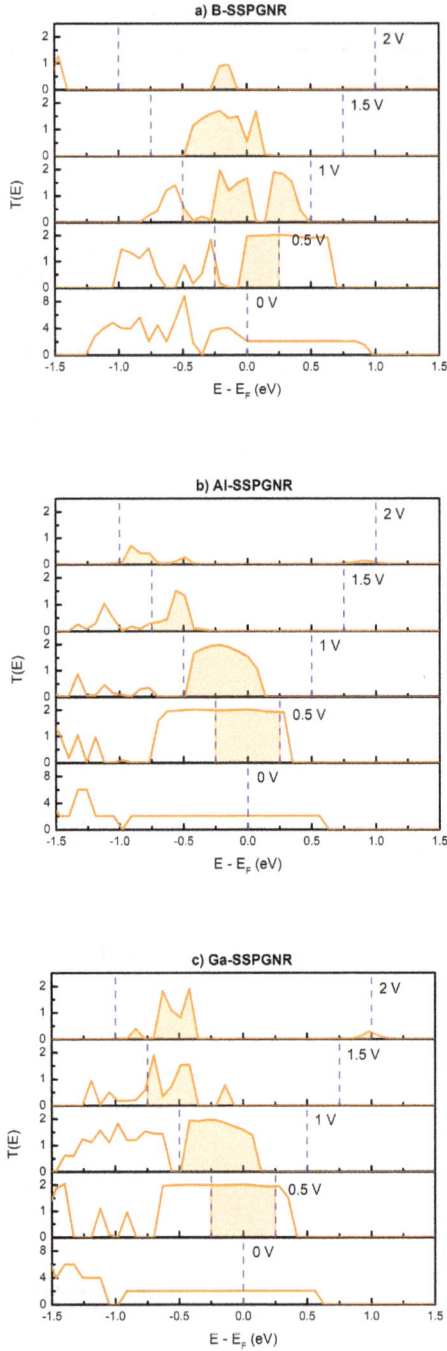

FIGURE 4.16   Transmission spectrum T(E) at different bias voltages for the p-type doping SSPGNRs: (a) B-doped, (b) Al-doped, and (c) Ga-doped. The vertical dashed lines show the bias windows.

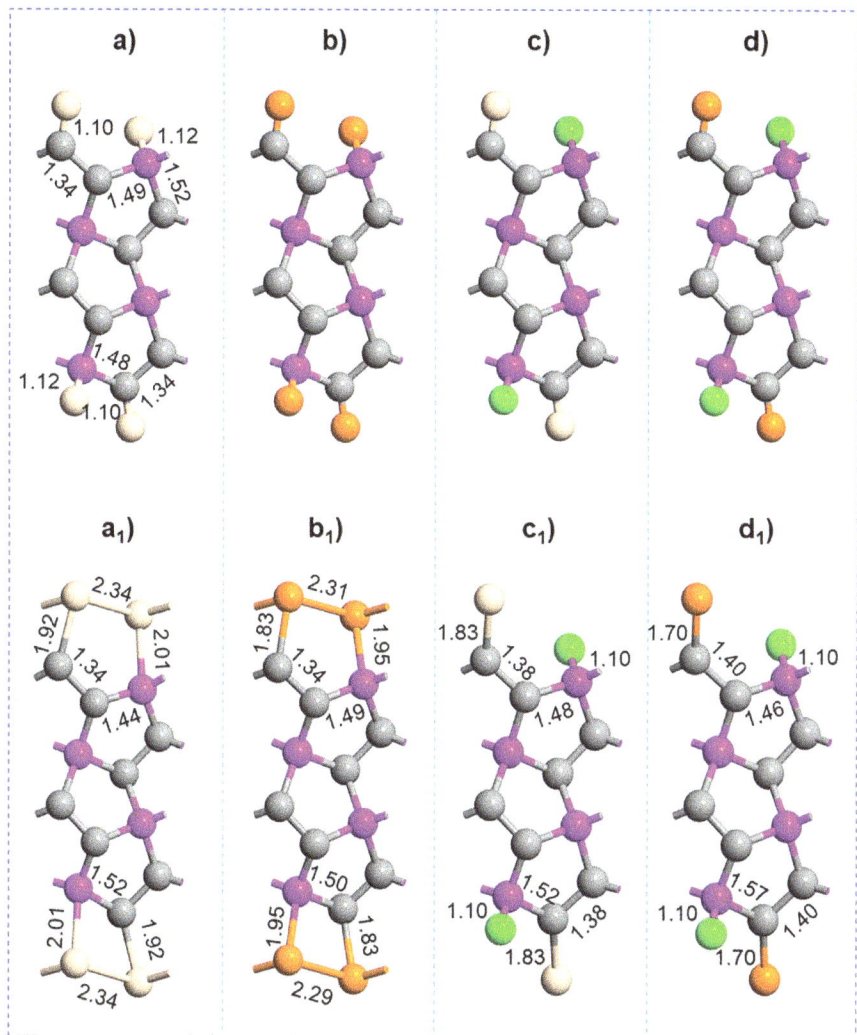

FIGURE 4.17 Schematic illustration (top view) of the edge terminated SSPGNRs: (a) SiSi-edge, (b) PP-edge, (c) SiH-edge, and (d) PH-edge before optimization. The bottom view (a1), (b1), (c1), and (d1) present structures after optimization. Si, P, and H are colored by cream, orange, and green, respectively.

4.3.4.1 The Structural Stability and Electronic Properties

To evaluate the stability of these structures and the energy value to form an edge from a bare SSPGNR, we calculate the binding energy (EB) by Eq. (4.4) and the edge formation energy (EFE) through the following equations.

$$E_{EFE} = \frac{(E_{total\_XY} - E_{total\_bare} - n_X E_x - n_y E_y)}{2L}, \tag{4.5}$$

Here, E_{total_XY}, E_{total_bare}, and $E_X(E_Y)$ denote the total electronic energy of a terminated-ribbon unit cell, a bare-ribbon unit cell, and one isolated edge-termination atom (X and Y can be H, P, or Si), respectively. $n_X(n_Y)$ is the total number of edge-termination atoms in the previously mentioned unit cell in turn. L is the lattice constant along the repeated nanoribbon axis.

Calculation results of E_{EB} and E_{EFE} are presented in Table 4.1 for comparison.

In E_{EB} data, the stability is ranked as follows: PP-SSPGNR > SiSi-SSPGNR > PH-SSPGNR > HH-SSPGNR > SiH-SSPGNR. This order shows that PP-SSPGNR and SiSi-SSPGNR are more dominant regarding energetic stability. Our assessments are compatible with Li et al.'s studies [40].

To understand the influence of edge effects on the electronic properties of SSPGNRs, we calculate the electronic band structure for each system in Figure 4.18. We turn our attention to the effects of other passivation atoms (P, Si) in the remaining four models: SiSi-SSPGNR (Figure 4.18a), PP-SSPGNR (Figure 4.18b), SiH-SSPGNR (Figure 4.18c), and PH-SSPGNR (Figure 4.18d) compared to the influence of H atoms in HH-SSPGNR (Figure 4.3). It is clear that there are significant differences in the electronic band structure of the structures' passivation by (P, Si) atoms. The change is very noticeable with respect to the valence band maximum (VBM) and the conduction band minimum (CBM).

In order to verify the effect of edges on band structure, we also consider DOS and PDOS of all studied systems examined in Figure 4.19. The DOS of HH-SSPGNR (Figure 4.4) displays the availability of electronic states beyond the energy range from 1.36 eV to -1.36 eV i.e., no state is found near the Fermi level. This confirms that HH-SSPGNR is a wide gap semiconductor. In combination of X = Si and Y = Si (Figure 4.19a), the VBM is localized at Γ-point, and the CBM lies at Z-point SiSi-SSPGNR, exhibiting semimetal behavior and coherent with the absence of a band gap in DOS. The DOS enhancement is also observed, and several peaks of DOS are contributed by Si. For X = P and Y = P (Figure 4.19b), a semiconducting behavior is observed with an energy gap of 2.388 eV and a significant increase of DOS, namely the highest peak at 38.266 eV in CB and several peaks in VB. The DOS change derives from the noticeable contribution of P atoms having more valance electrons than H.

Next, we discuss the situation of alternate edge termination. In consideration for X = P and Y = H (Figure 4.19c), we observe a very small energy gap of 0.192 eV

TABLE 4.1

Binding Energy and Edge Formation Energy of the Studied Structures

Studied Structures	E_{EB} (eV)	E_{EFE} (eV)
HH-SSPGNR	−7.56	−2.80
PP-SSPGNR	−7.98	−3.77
SiSi-SSPGNR	−7.75	−3.23
PH-SSPGNR	−7.61	−2.92
SiH-SSPGNR	−7.53	−2.73

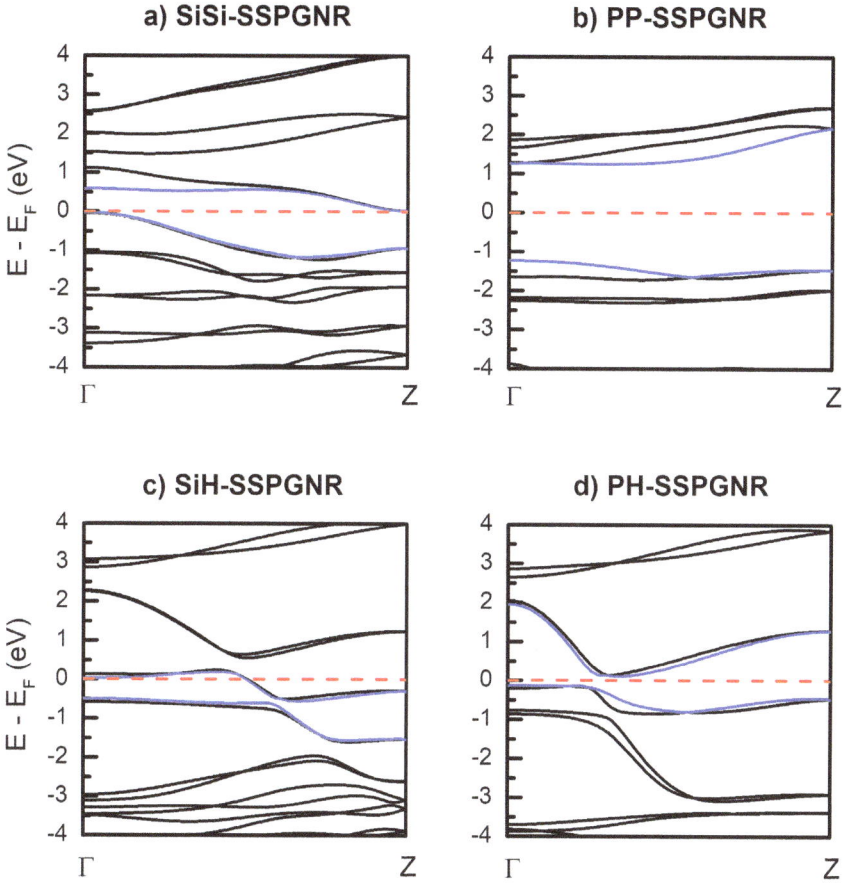

FIGURE 4.18 The band structures of the edge terminated SSPGNRs: (a) SiSi-edge, (b) PP-edge, (c) SiH-edge, and (d) PH-edge. The dashed line indicates the Fermi level.

corresponding to the absence of electronic states at the vicinity of the Fermi level. And the SiH-SSPGNR exhibits a metallic character with the existence of two bands crossing the Fermi level (Figure 4.19d).

4.3.4.2 Electronic Transport Properties

The calculated I-V characteristics for all four systems are shown in Figure 4.20. As can be seen, the magnitudes of the current under applied bias voltage follow the pattern PP-SSPGNR < HH-SSPGNR < SiSi-SSPGNR < PH-SSPGNR < SiH-SSPGNR. The extremely small currents belong to HH-SSPGNR (Figure 4.5) and PP-SSPGNR (Figure 4.20b), which is nearly 9 orders of magnitude smaller compared to SiH-SSPGNR (Figure 4.20c) and PH-SSPGNR (Figure 4.20d). There are three distinct types of current pictures in Figure 4.20 [41]. First, one can see that there is a clear tendency in Figure 4.20b and Figure 4.20d which presents the typical characteristic

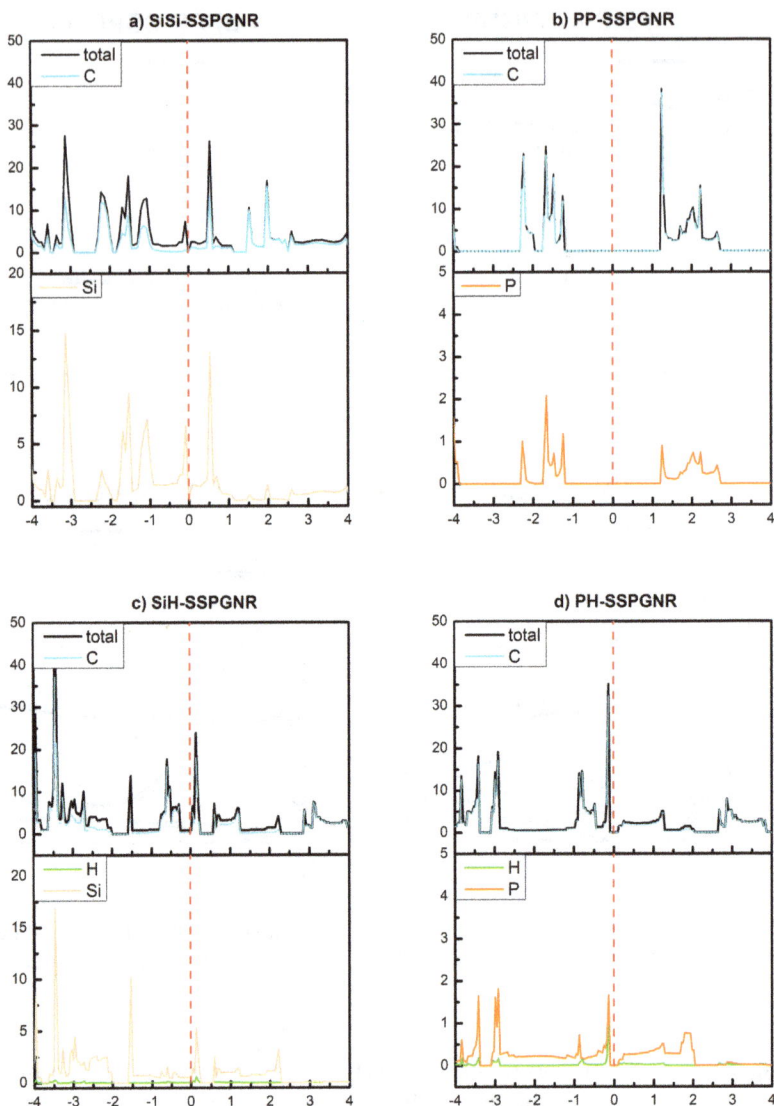

FIGURE 4.19 The density of state of the edge terminated SSPGNRs: (a) SiSi-edge, (b) PP-edge, (c) SiH-edge, and (d) PH-edge. The dashed line indicates the Fermi level.

of a semiconductor. Second, it is very special that the oscillation current appears in SiSi-SSPGNR (Figure 4.20a) with various peaks and valleys, revealing a negative differential resistance (NDR) effect. Finally, at bias voltage < 0.2 V the SiH-SSPGNR (Figure 4.20c) behaves ohmic with a linear I-V characteristic [45].

Numerical results in Figure 4.21 show T(E) for the terminated SSPGNRs. The changing tendency of T(E) accurately illustrates the changing tendency of I-V for all studied samples.

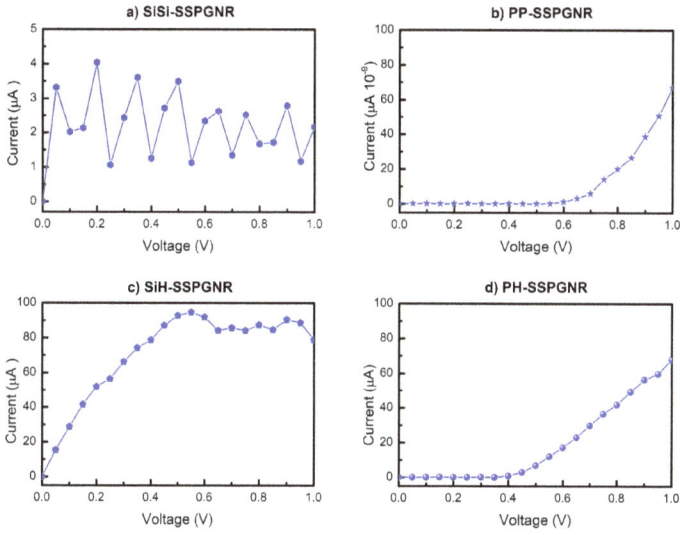

FIGURE 4.20　The I-V curves of the edge terminated SSPGNRs: (a) SiSi-edge, (b) PP-edge, (c) SiH-edge, and (d) PH-edge.

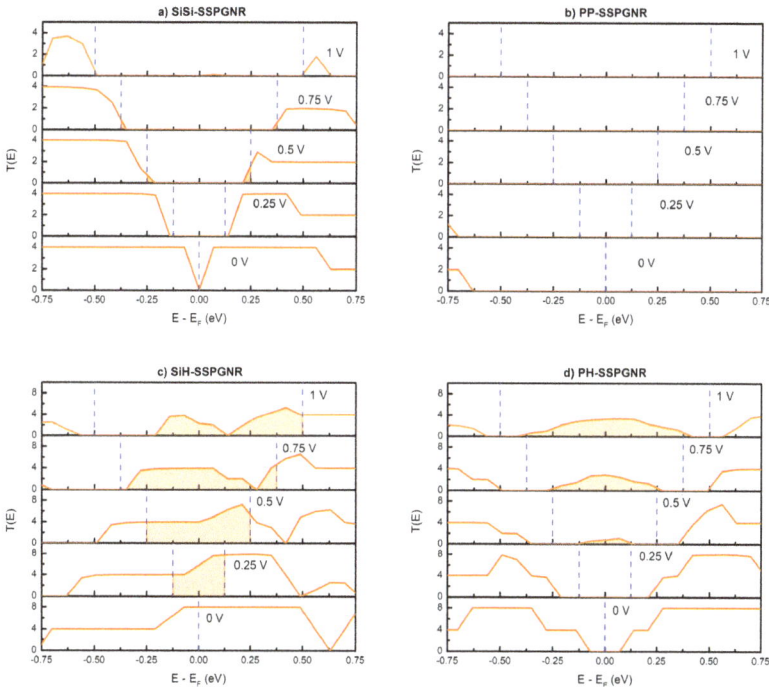

FIGURE 4.21　Transmission spectrum T(E) at different bias voltages for the edge terminated SSPGNRs: (a) SiSi-edge, (b) PP-edge, (c) SiH-edge, and (d) PH-edge. The vertical dashed lines show the bias windows.

4.4   CONCLUDING REMARKS

In this chapter, we have presented the study of penta-graphene nanoribbon, a pentagonal material, through the analysis of different physical phenomena by numerical simulation. This subgroup of the family of penta-materials presents remarkable properties under substitutional doping and edge termination. Firstly, the penta-graphene was found to be a wide band gap material with an optical response that may have applications in optoelectronic devices.

Also, the n-type doping elements (Si, N, and P) and p-type doping elements (B, Al, and Ga) significantly affect the band structure of the SSPGNR systems. Specifically, the band gap of the doped samples is reduced compared to the pristine sample. Current intensity of N-SSPGNR, P-SSPGNR, B-SSPGNR, Al-SSPGNR, and Ga-SSPGNR increases about 10^8 times due to the rising number of free electrons, number of free holes, and the number of transport channels at low temperature. Meanwhile, electronic and transport characteristics of SSPGNR doped Si are almost as similar as pristine SSPGNR by Si and C located in the same group of the periodic table. In addition, the behavior of the transmission completely matches the I-V characteristics. The NDR behavior is one intrinsic characteristic of the SSPGNR-based devices, especially with doped SSPGNR.

Additionally, the investigated ribbons are very sensitive to termination elements. Band structures can be semiconducting, metallic or semi-metallic, depending on appropriate termination. Alternate edge functionalization by P or Si together with H atoms not only enriches SSPGNR current by 10^9 times but also brings high feasibility to form functionalized ribbons, which provides good low-dimensional structures likely to exist in experiments. The enhancement of PH-SSPGNR and SiH-SSPGNR current is contributed by the number of transmission modes as well as charge carriers. The study found oscillation current appears in SiSi-SSPGNR. We suppose that this study provides a basic analysis to understand the impact of edge chemistry, resulting in guiding experimental efforts toward applications of PGNRs.

The results show that acceptors and donors regarding the quasiparticles, which typically act as strong potential wells for electrons and holes, form a previously unrecognized "impurity" band near VBM or CBM. In contrast, the edge termination groups are responsible for delocalized both lower (valence) and upper (conduction) conjugated bands. The change is very noticeable with respect to the VBM and the CBM.

ACKNOWLEDGMENTS

This work is funded by Vietnam National Foundation for Science and Technology Development (NAFOSTED) under grant number 103.01–2020.16.

REFERENCES

[1]   Wang Z Y, Dong F, Shen B, Zhang R J, Zheng Y X, Chen L Y, Wang S Y, Wang C Z, Ho K M, Fan Y J, Jin B Y and Su W S 2016 Electronic and optical properties of novel carbon allotropes *Carbon* **101** 77–85

[2] Coleman J N, Lotya M, O'Neill A, Bergin S D, King P J, Khan U, Young K, Gaucher A, De S, Smith R J, Shvets I V, Arora S K, Stanton G, Kim H Y, Lee K, Kim G T, Duesberg G S, Hallam T, Boland J J, Wang J J, Donegan J F, Grunlan J C, Moriarty G, Shmeliov A, Nicholls R J, Perkins J M, Grieveson E M, Theuwissen K, McComb D W, Nellist P D and Nicolosi V 2011 Two-dimensional nanosheets produced by liquid exfoliation of layered materials *Science* **331** 568–71

[3] Allen M J, Tung V C and Kaner R B 2010 Honeycomb carbon: A review of graphene *Chem Rev* **110** 132–45

[4] Tien N T, Thao D N, Thao P T B and Quang D N 2016 Key scattering mechanisms limiting the lateral transport in a modulation-doped polar heterojunction *J Appl Phys* **119**

[5] Tien N T, Hung N N T, Nguyen T T and Thao P T B 2017 Linear intersubband optical absorption in the semiparabolic quantum wells based on AlN/AlGaN/AlN under a uniform electric field *Physica B* **519** 63–8

[6] Novoselov K S, Geim A K, Morozov S V, Jiang D, Zhang Y, Dubonos S V, Grigorieva I V and Firsov A A 2004 Electric field effect in atomically thin carbon films *Science* **306** 666–9

[7] Haberer D, Vyalikh D V, Taioli S, Dora B, Farjam M, Fink J, Marchenko D, Pichler T, Ziegler K, Simonucci S, Dresselhaus M S, Knupfer M, Buchner B and Gruneis A 2010 Tunable band gap in hydrogenated quasi-free-standing graphene *Nano Lett* **10** 3360–6

[8] Chung H C, Chang C P, Lin C Y and Lin M F 2016 Electronic and optical properties of graphene nanoribbons in external fields *Phys Chem Chem Phys* **18** 7573–616

[9] Chung H C, Lin Y T, Lin S Y, Ho C H, Chang C P and Lin M F 2016 Magnetoelectronic and optical properties of nonuniform graphene nanoribbons *Carbon* **109** 883–95

[10] Chung H C, Lee M H, Chang C P and Lin M F 2011 Exploration of edge-dependent optical selection rules for graphene nanoribbons *Opt Express* **19** 23350–63

[11] Gierz I, Riedl C, Starke U, Ast C R and Kern K 2008 Atomic hole doping of graphene *Nano Lett* **8** 4603–7

[12] Ryu S, Han M Y, Maultzsch J, Heinz T F, Kim P, Steigerwald M L and Brus L E 2008 Reversible basal plane hydrogenation of graphene *Nano Lett* **8** 4597–602

[13] Kuila T, Bose S, Mishra A K, Khanra P, Kim N H and Lee J H 2012 Chemical functionalization of graphene and its applications *Prog Mater Sci* **57** 1061–105

[14] Nguyen D K, Tran N T T, Nguyen T T and Lin M F 2018 Diverse electronic and magnetic properties of chlorination-related graphene nanoribbons *Sci Rep-Uk* **8**

[15] Tien N T, Phuc V T and Ahuja R 2018 Tuning electronic transport properties of zigzag graphene nanoribbons with silicon doping and phosphorus passivation *AIP Adv* **8**

[16] Zhang S H, Zhou J, Wang Q, Chen X S, Kawazoe Y and Jena P 2015 Penta-graphene: A new carbon allotrope *P Natl Acad Sci USA* **112** 2372–7

[17] Wu X F, Varshney V, Lee J, Zhang T, Wohlwend J L, Roy A K and Luo T F 2016 Hydrogenation of penta-graphene leads to unexpected large improvement in thermal conductivity *Nano Lett* **16** 3925–35

[18] Stauber T, Beltran J I and Schliemann J 2016 Tight-binding approach to penta-graphene *Sci Rep-Uk* **6**

[19] Cranford S W 2016 When is 6 less than 5? Penta- to hexa-graphene transition *Carbon* **96** 421–8

[20] Einollahzadeh H, Dariani R S and Fazeli S M 2016 Computing the band structure and energy gap of penta-graphene by using DFT and G(o)W(o) approximations *Solid State Commun* **229** 1–4

[21] Avramov P, Demin V, Luo M, Choi C H, Sorokin P B, Yakobson B and Chernozatonskii L 2015 Translation symmetry breakdown in low-dimensional lattices of pentagonal rings *J Phys Chem Lett* **6** 4525–31

[22] Behzad S, Chegel R, Moradian R and Shahrokhi M 2014 Theoretical exploration of structural, electro-optical and magnetic properties of gallium-doped silicon carbide nanotubes *Superlattice Microst* **73** 185–92

[23] Moradian R, Shahrokhi M, Charganeh S S and Moradian S 2012 Structural, magnetic, electronic and optical properties of iron cluster (Fe-6) decorated boron nitride sheet *Physica E* **46** 182–8

[24] Naderi S, Shahrokhi M, Noruzi H R, Gurabi A and Moradian R 2013 Structural, electronic and magnetic properties of Fe and Co monatomic nanochains encapsulated in BN nanotube bundle *Eur Phys J-Appl Phys* **62**

[25] Shahrokhi M and Leonard C 2016 Quasi-particle energies and optical excitations of wurtzite BeO and its nanosheet *J Alloy Compd* **682** 254–62

[26] Rajbanshi B, Sarkar S, Mandal B and Sarkar P 2016 Energetic and electronic structure of penta-graphene nanoribbons *Carbon* **100** 118–25

[27] On V V, Thanh L N and Tien N T 2020 The electronic properties and electron transport of sawtooth penta-graphene nanoribbon under uniaxial strain: Ab-initio study *Philos Mag* **100** 1834–48

[28] Magda G Z, Jin X Z, Hagymasi I, Vancso P, Osvath Z, Nemes-Incze P, Hwang C Y, Biro L P and Tapaszto L 2014 Room-temperature magnetic order on zigzag edges of narrow graphene nanoribbons *Nature* **514** 608-+

[29] Han M Y, Ozyilmaz B, Zhang Y B and Kim P 2007 Energy band-gap engineering of graphene nanoribbons *Phys Rev Lett* **98**

[30] Li X L, Wang X R, Zhang L, Lee S W and Dai H J 2008 Chemically derived, ultra-smooth graphene nanoribbon semiconductors *Science* **319** 1229–32

[31] Tapaszto L, Dobrik G, Lambin P and Biro L P 2008 Tailoring the atomic structure of graphene nanoribbons by scanning tunnelling microscope lithography *Nat Nanotechnol* **3** 397–401

[32] Marmolejo-Tejada J M and Velasco-Medina J 2016 Review on graphene nanoribbon devices for logic applications *Microelectron J* **48** 18–38

[33] Taylor J, Guo H and Wang J 2001 Ab initio modeling of quantum transport properties of molecular electronic devices *Phys Rev B* **63**

[34] Brandbyge M, Mozos J L, Ordejon P, Taylor J and Stokbro K 2002 Density-functional method for nonequilibrium electron transport *Phys Rev B* **65**

[35] Landauer R 1957 Spatial variation of currents and fields due to localized scatterers in metallic conduction *IBM J Res Dev* **1** 223–31

[36] Destefani C F and Marques G E 2000 Electronic transport in quasi-1D mesoscopic systems: The correlated electron approach *Physica E* **7** 786–9

[37] Yuan P F, Zhang Z H, Fan Z Q and Qiu M 2017 Electronic structure and magnetic properties of penta-graphene nanoribbons *Phys Chem Chem Phys* **19** 9528–36

[38] Tien N T, Thao P T B, Phuc V T and Ahuja R 2019 Electronic and transport features of sawtooth penta-graphene nanoribbons via substitutional doping *Physica E* **114**

[39] Berdiyorov G R, Bahlouli H and Peeters F M 2016 Effect of substitutional impurities on the electronic transport properties of graphene *Physica E* **84** 22–6

[40] Li Y H, Yuan P F, Fan Z Q and Zhang Z H 2018 Electronic properties and carrier mobility for penta-graphene nanoribbons with nonmetallic-atom -terminations *Org Electron* **59** 306–13

[41] Berdiyorov G R, Dixit G and Madjet M E 2016 Band gap engineering in penta-graphene by substitutional doping: First-principles calculations *J Phys-Condens Mat* **28**

[42] Zhou Y H, Qiu N X, Li R W, Guo Z S, Zhang J, Fang J F, Huang A S, He J, Zha X H, Luo K, Yin J S, Li Q W, Bai X J, Huang Q and Du S Y 2016 Negative differential resistance and rectifying performance induced by doped graphene nanoribbons p-n device *Phys Lett A* **380** 1049–55

[43] Shylau A A, Klos J W and Zozoulenko I V 2009 Capacitance of graphene nanoribbons *Phys Rev B* **80**

[44] Tien N T, Thao P T B, Phuc V T and Ahuja R 2020 Influence of edge termination on the electronic and transport properties of sawtooth penta-graphene nanoribbons *J Phys Chem Solids* **146**

[45] Arjmand T, Tagani M B and Soleimani H R 2017 The effect of buckling on I-V characteristics of symmetric and asymmetric zigzag germanene nanoribbons: A first-principle calculation *J Phys D Appl Phys* **50**

5 Feature-Rich Quasiparticle Properties of Halogen-Adsorbed Silicene Nanoribbons

Duy Khanh Nguyen, Vo Duy Dat,
Vo Van On, and Ming-Fa Lin

CONTENTS

5.1   INTRODUCTION

Silicon, a rich element of group IV, possesses a unique electron configuration that can exist in various allotropes under various dimensions [1]. Similar to the development of carbon-based materials, the bulk [2], one-dimensional (1D) [3], and zero-dimensional (0D) allotropes [4] of silicon have been early identified, in which the traditional bulk structure of silicon was well known as the main components in the semiconductor industry. Under low-dimensional allotrope forms of silicon, 1D nanotubes and 0D fullerene have early come into light, while a basic 2D structure of silicon was only predicted in 1994 [5]. However, such a 2D silicon structure has just attracted much attention since the first successful synthesis of monolayer graphene through the top-down method of mechanical exfoliation from layered graphite in 2004 [6]. The 2D silicon nanosheet, termed silicene, is known as the closest 2D counterpart of graphene [7]. Silicene presents a low buckled one-atom-thickness honeycomb lattice that exhibits a mix of sp^2/sp^3 orbital hybridizations in Si-Si bonds [8]. This indicates that separation of σ and π bonds in silicene is not clear as sp^2 hybridized

DOI: 10.1201/9781003322573-5

graphene even though both silicene and graphene show a Dirac cone structure made of p_z orbitals in low-lying energy [9]. Apart from graphene, silicene can only be synthesized through the bottom-up method that has to be grown from silicon atoms on metallic substrates since it does not have a layered silicon structure in nature as graphite does [10, 11]. Silicene possesses many rich quasiparticle properties that are very potential for many practical applications, such as silicene-based field-effect transistors (FETs) operating at room temperature [12], 2D sensor [13], and energy storage devices [14]. Furthermore, many other unusual quasiparticle properties coming from the low-buckled one-atom thickness honeycomb structure of silicene have been discovered, including a large band gap induced by the spin-orbit coupling at the Dirac point [15], a quantum spin Hall effect [16], the transition from a topological insulating phase to a band insulator [17], giant magnetoresistance [18], presence of anomalous quantum Hall effects [19], or the emergence of a valley-polarized metal [20]. Such potential quasiparticle properties indicate that silicene can be a promising 2D candidate to replace graphene due not only to the graphene-like 2D features but also to its compatibility with current silicon-based electronic devices. Unfortunately, the mili-gap feature of silicene limits its potential in nanoelectronic applications [21]. Therefore, opening a band gap for silicene becomes a critical topic that has attracted much attention from the scientific community up to now. Various methods have been applied to create an energy gap in silicene, including the finite size termination of the 2D silicene nanosheet, atom dopings [22], chemical functionalizations [23], forming bilayer and multilayer structures [24], mechanical strains [25], and applying external fields [26], in which the chemical modification is one of the effective methods to enrich essential quasiparticle properties. Through recent first-principles predictions, top site adsorptions of hydrogen and halogen atoms on silicene could create large gap semiconductors under fully double-side adsorptions that is due to the full termination of π bonds of silicon atoms by adatoms, and these adatom-diversified systems are regarded as p-type doped materials since the strong electron affinity of the adatoms attracted electrons from the silicene to leave free holes [27, 28]. Also, the halogenated silicene systems have been verified by low-temperature scanning tunneling microscopy (STM) studies [29]. It should be mentioned that the low-buckled 2D silicene structures under the atom adsorptions become higher buckled 2D structures due to the serious breaking of π bonds and a significant weakening of σ bonds [30]. Contrary, substitutions of guest atoms in silicene could lead to feature-rich quasiparticle properties; however, it remains in lower buckled or planar 2D structures [31]. This indicates that the quasi π and σ bonds are formed in Si-adatom bonds. Next door to enriching host materials through chemical modifications, the development of novel materials with rich quasiparticle properties through reducing the dimension of host materials has also achieved great success with the emergence of many novel 1D materials that can effectively overcome the internal disadvantages of 2D host materials for many practical applications. Specifically, forming a 1D silicon structure from 2D host silicene is a good trend to induce band gap in silicene for applications in silicon-based electronic devices.

On the geometric aspect, the simplest approach to create silicon-based 1D materials from the 2D host silicene is to terminate the 2D silicene nanosheet at finite sizes that result in 1D silicene nanoribbons (SiNRs) [32]. Under the finite-size termination effects, two typical edges of SiNRs appear, namely armchair (ASiNR) and zigzag (ZSiNR) silicene nanoribbons. To reduce the dangling effects along the edges of SiNRs, various atoms or a group of atoms have been used to passivate the edges, including the passivation of hydrogen [33], halogen [34], oxygen [35], and OH [36]. As a result, SiNRs adopt the mix of sp^2/sp^3 orbital hybridizations in the low-buckled one-atom-thickness honeycomb lattice of 2D silicene. It should be noted that the representation of 2D silicene is to show a Dirac cone structure made of $3p_z$ orbitals of silicon atoms at low-lying energy. Within 1D SiNRs, this feature is to replace by the highest valence and lowest conduction energy bands symmetric via the Fermi level, and these low-lying energy bands mainly come from $3p_z$ orbitals of silicon atoms [37]. Both 1D ASiNR and ZSiNR belong to a direct middle-gap semiconductor, and the band gaps of SiNRs strongly depend on the ribbon widths [38]. On the other hand, ASiNR and ZSiNR own different magnetic configurations due to different localization of silicon atoms at edges, in which ASiNR is nonmagnetic, while ZSiNR displays the anti-ferromagnetic configuration with symmetry of spin-up and -down states across the ribbon [39]. The opened band gaps in 1D SiNRs are great advantages to solve the main obstacle in 2D silicene for applications in electronic devices. Besides, the unique magnetic configuration of ZSiNRs can be very potential 1D materials for spintronic devices. Parallel to the synthesis of 1D SiNRs from the top-down method of 2D host silicene, the bottom-up approach is also the possible method to synthesize 1D SiNRs with precision at an atomic level, in which the 1D SiNRs are grown on metallic substrates or an insulating thin film [40, 41]. The SiNRs, with their existing rich quasiparticle properties, have been widely utilized for many practical applications, including gas sensors [42], electronic [43], and spintronic [44] devices. However, to be suitable for a wide range of practical applications, the existing quasiparticle properties of SiNRs need to be further modified. Recently, how to diversify the essential quasiparticle properties of SiNRs has quickly become a hot spot for many studies. Various approaches have been used to enrich the essential quasiparticle properties of SiNRs, including edge functionalizations [45], atom substitutions or adsorptions [30], inducing defects [46], forming heterostructures [47], mechanical strains [48], and applying external fields [49].

From the chemical viewpoint, atom doping is the most effective method to dramatically alter the quasiparticle properties of SiNRs, in which diversifying the existing quasiparticle properties can be easily achieved through changes in adatom concentrations and distributions [50]. Through atom doping, it is very effective to accurately determine the semiconducting-metallic or nonmagnetic-magnetic-anti-ferromagnetic transitions at specific concentrations or distributions. Up to now, various kinds of atoms have been used to dope in SiNRs. Cu adsorptions on ZSiNRs have been studied in which the center of the nearest hexagon to the zigzag edge is the most favorable site for adsorbing Cu atoms. Charge transfer from Si atoms to Cu adatoms leads to the ferromagnetic state [51]. Also, adsorption effects of Ti atoms on ZSiNRs have been revealed, in which Ti atoms prefer being adsorbed at the hollow site inside the nanoribbon rather than on zigzag edges [52]. On the other hand, P-doped SiNRs have been investigated, where the P adatom preferentially substitutes at the edges of SiNRs [53]. For N- or B-doped SiNRs, substitutions of N or B for Si are preferentially at the edge sites [54]. However, the rich quasiparticle properties of SiNRs under halogen adsorptions remain unclear; therefore, halogen adsorption effects in SiNRs are worthy of further detailed studies.

In this chapter, the diverse structural, electronic, and magnetic properties of halogen-adsorbed ASiNR and ZSiNR have been explored in detail through the first-principles calculations. Detailed investigations in various adatom concentrations and distributions are considered, in which the semiconducting-metallic transitions in both ASiNR and ZSiNR are determined at specific halogen concentrations. On the magnetic aspect, halogen-adsorbed ASiNRs fully belong to nonmagnetic materials regardless of adatom concentrations and distributions, while diverse magnetic configurations in halogen-adsorbed ZSiNRs have mainly emerged at low adatom concentrations that are fully destroyed at specific high adatom concentrations. Due to the strong electron affinity, halogen adatoms attract electrons from SiNRs in which the free holes appear in the adsorbed systems. This indicates that halogen-adsorbed SiNRs can be regarded as p-type doping. The developed first-principles physical quantities, including the atom-dominated band structures, the orbital-projected density of states, magnetic moments, and spin density distributions, have been analyzed in a rigid logic to reveal the halogen-enriched quasiparticle properties. The theoretical framework developed under the first-principles calculations in this chapter can be fully generalized for other emergent systems.

5.2 COMPUTATIONAL DETAILS

In this chapter, diverse structural, electronic, and magnetic properties of halogen-adsorbed SiNRs are fully investigated using the spin-polarized density functional theory (DFT), implemented in the Vienna Ab Initio Simulation Package (VASP) [55]. VASP is accurate and reliable for most condensed-matter materials since the suitable crystal potentials, electron-electron Coulomb interactions, and spin configurations are covered in the evaluation processes. The many-body exchange and correlation energies, which come from the electron-electron Coulomb interactions, are calculated from the Perdew-Burke-Ernzerhof (PBE) functional under the generalized gradient approximation [56]. Furthermore, the projector-augmented wave (PAW) pseudopotentials can characterize the intrinsic electron-ion interactions. As to the complete set of plane waves, the kinetic energy cutoff is set to be 500 eV, being suitable for evaluating Bloch wave functions and electronic energy spectra. A vacuum space of 20 Å is inserted between periodic images to avoid any significant interaction. The first Brillouin zone of 1D SiNRs is sampled by $12 \times 1 \times 1$ and $100 \times 1 \times 1$ k-point meshes within the Monkhorst-Pack scheme for geometric optimizations and self-consistent calculations, respectively. The convergence for the ground state energy is equal to 10^{-5} eV between two consecutive steps, and the maximum Hellman-Feynman force acting on each atom is less than 0.01 eV/Å during the ionic relaxations. The first-principles physical quantities developed through the VASP calculations, including the binding energies, atom-dominated band structures, spin- and orbital-projected density of states, and spin density distributions that can provide the full information in comprehending the diverse quasiparticle properties of SiNRs under the atom doping effects.

5.3 RESULT AND DISCUSSION

5.3.1 STRUCTURAL PROPERTIES

Optimal geometric structures of halogen-adsorbed silicene nanoribbons (SiNRs) are obtained through the DFT calculations, in which they are displayed in Figures 5.1(a) and 5.1(b) for armchair and zigzag silicene nanoribbons (ASiNR and ZSiNR), respectively. The lattice constants of ASiNRs and ZSiNRs are evaluated as b = 3a

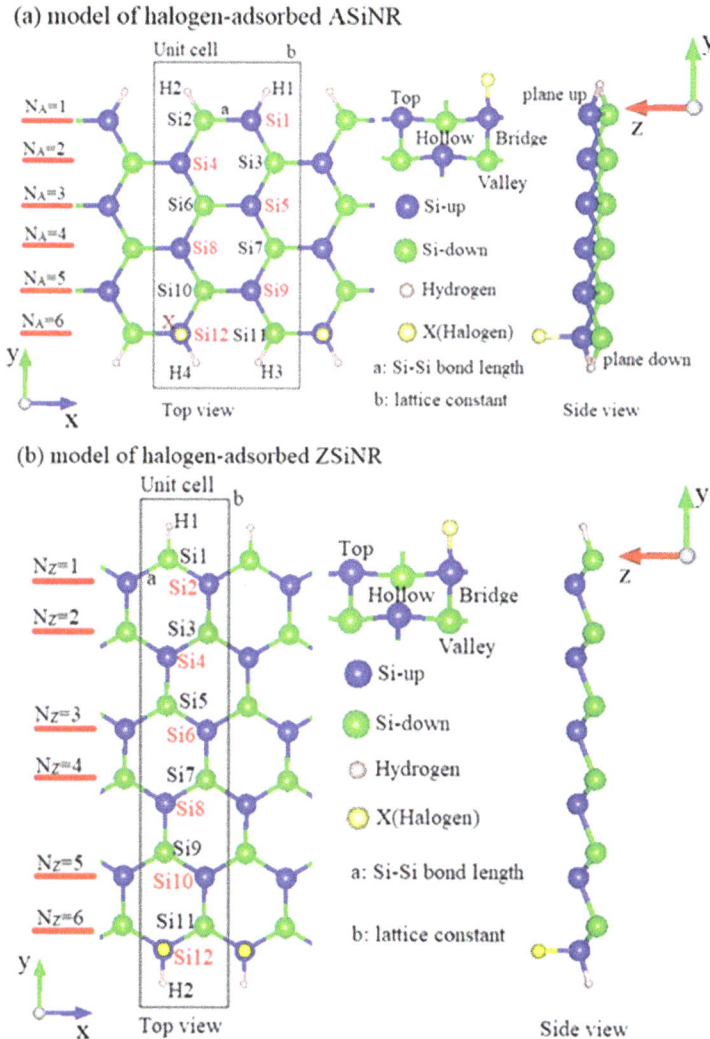

FIGURE 5.1 Model study of (a) halogen-adsorbed armchair silicene nanoribbon (ASiNR) and (b) halogen-adsorbed zigzag silicene nanoribbon (ZSiNR), in which the lattice constants of $b = 3a$ (Å) and $b = 2\sqrt{3}a$ (Å) and the widths of N_A and N_Z represent for ASiNR and ZSiNR systems, respectively.

and $b = 2\sqrt{3}\,a$, where a denotes for Si-Si bond lengths are shown in Table 5.1, respectively. The widths of ASiNRs and ZSiNRs can be described through the number of dimer lines illustrated by the red solid lines in Figure 5.1, in which the typical widths of 6- and 7-ASiNRs are included in the study, as shown in Figures 5.1(a) and 5.1(b), respectively. Due to the pseudo-Jahn-Teller effect, the symmetry of ASiNRs and ZSiNRs is broken to form the low-buckled honeycomb lattice, where it possesses a mix of sp^2/sp^3 hybridizations in Si-Si bonds. The buckling slightly depends on the width of SiNRs, which are 0.660 Å, and 0.669 Å for 7-ASiNRs, and 6-ASiNRs, as displayed in Table 5.1, respectively. However, the buckling of 6-ASiNRs is about 48% larger than the one of 6-ZSiNRs. This results from the strong coupling effect in ASiNRs that is due to a large number of uncoupled orbitals of Si atoms. Apart from the 2D silicene systems, the Si-Si bond lengths in SiNRs are ununiform across the nanoribbons due to effects of edges, where the difference in bond lengths of Si-edge and Si-non-edge is about 0.3%. Under various halogen adsorptions, the geometric structure is greatly diversified. The study model of halogen-adsorbed SiNRs is presented in Figure 5.1, where the halogen atoms are optimally adsorbed on the top site among the bridge, hollow, and valley sites regardless of any adatom concentrations and distributions. The structural stability can be evaluated via the binding energy of $E_b = E_T - E_P - nE_A/n$, where the E_T, E_P, and E_A and n account for the ground state energy of the total system, pristine SiNRs, isolated halogen adatoms, and the number of adatoms, respectively. As a result, the smaller binding energies result in greater geometric stability. The E_b gets smaller values under larger atomic numbers, as evidenced in Table 5.1, in which the E_b of F, Cl, Br, I, and At-adsorbed systems respectively is −5.30 eV, −3.31 eV, −2.68 eV, −2.08 eV, and −1.82 eV. This implies that the stronger chemical bonds should obtain greater structural stability. According to the calculated results presented in Table 5.1, the binding energies, Si-Si, and Si-adatom bond lengths are slightly affected by various adatom concentrations. Meanwhile, the buckling significantly increases as the halogen concentration increases. As for the F-adsorbed 6-ASiNRs, the buckling increases from 0.76 Å at the 8.33% adsorption to the critical value of 0.995 Å at 50% adsorption. The higher adatom concentrations induce the larger coupling effect of adatom-closed unpair Si-3p orbitals, leading to shortening of the adatom-Si-Si angle from 114.69° to 111.71°. Besides, further increasing of adatom concentration decreases the buckling, since the number of unpaired Si-3p orbitals is reduced. Especially, at the 100% adsorption systems, the effect of halogen adatoms on ZSiNRs and ASiNRs is similar, where the number of unpaired Si-$3p_z$ orbitals dramatically decreases due to full occupation of electrons from the adatoms that leads to the reduced buckling and adatom-Si-Si bond angle of 0.483 Å and 109.4°, respectively, as evidenced in Table 5.1. Furthermore, it is worthy to say that the magnitude of the E_b in the F, Cl, Br halogen-adsorbed SiNRs is always greater than 3 eV, indicating that these systems belong to the chemical adsorption, where its buckling and adatom-Si-Si bond angles are almost the same regardless of any adatom concentrations. Meanwhile, the magnitude of the E_b in the I and At halogen-adsorbed SiNRs is always less than 3 eV; therefore, these systems are regarded as the physical adsorption, where its charge transfer plays the main role to diversify the essential properties. Within these physical adsorption systems, the I and

TABLE 5.1

Optimized Structural Parameters of Pristine and Halogen-Adsorbed ASiNRs and ZSiNRs Under Various Adatom Concentrations and Distributions, Including Binding Energy (eV), First and Second Si-Si Bond Length (Å), Buckling (Å), Adatom-Si Bond Length (Å), and Adatom-Si-Si Bond Angle (°). The representation of (1F)$_{edge/non-edge}$ is shown for the number of adatoms and the distribution of adatoms, respectively

SiNRs	Adsorption configuration	Concentration (%)	Binding energy (eV)	1st Si-Si bond length (Å)	2nd Si-Si bond length (Å)	Buckling (Å)	Adatom-Si bond length (Å)	Adatom-Si-Si bond angle (°)
Pristine ASiNRs	7-ASiNR (Si:H = 14:4)	X	X	2.271	2.268	0.660	X	X
	6-ASiNR (Si:H = 12:4)	X	X	2.258	2.264	0.669	X	X
Halogen-adsorbed 6-ASiNRs	(1F)$_{edge}$	8.33%	−5.5278	2.331	2.266	0.76	1.634	114.69
	(1F)$_{non-edge}$	8.33%	−5.3006	2.321	2.274	0.73	1.636	116.11
	(2F)	16.66%	−5.3736	2.338	2.244	0.856	1.633	113.04
	(4F)	33.33%	−5.5108	2.326	2.267	0.837	1.640	112.33
	(6F)	50%	−5.4619	2.333	2.253	0.995	1.632	111.71
	(12F)	100%	−5.4992	2.368	X	0.583	1.623	114.47
	(1Cl)$_{non-edge}$	8.33%	−3.3175	2.325	2.275	0.688	2.096	116.82
	(12Cl)	100%	−3.4151	2.362	X	0.917	2.061	114.33
	(1Br)$_{non-edge}$	8.33%	−2.6842	2.324	2.274	0.734	2.274	117.02
	(12Br)	100%	−2.7420	2.366	X	0.921	2.239	117.37
	(1I)$_{non-edge}$	8.33%	−2.0843	2.324	2.275	0.827	2.509	116.82
	(12I)	100%	−1.9474	2.382	X	0.919	2.472	124.06
	(1At)$_{non-edge}$	8.33%	−1.8224	2.324	2.275	0.720	2.615	116.76
	(12At)	100%	−1.5562	2.390	X	0.922	2.586	126.19

(Continued)

TABLE 5.1
(Continued)

SiNRs	Adsorption configuration	Concentration (%)	Binding energy (eV)	1st Si-Si bond length (Å)	2nd Si-Si bond length (Å)	Buckling (Å)	Adatom-Si bond length (Å)	Adatom-Si-Si bond angle (°)
Pristine ZSiNR	6-ZSiNR (Si:H = 12:2)	X	X	2.221	2.292	0.452	X	X
Halogen-adsorbed 6-ZSiNRs	$(1F)_{edge}$	8.33%	−5.6557	2.328	2.258	0.832	1.631	115.64
	$(1F)_{non-edge}$	8.33%	−5.3270	2.366	2.227	1.083	1.637	106.98
	$(2F)_{edge-edge}$	16.66%	−5.5945	2.328	2.259	0.863	1.629	115.14
	$(2F)_{non-edge-non-edge}$	16.66%	−5.3976	2.331	2.278	1.007	1.633	118.94
	$(2F)_{edge-non-edge}$	16.66%	−5.4176	2.362	2.227	1.046	1.634	106.77
	$(3F)$	25%	−5.4643	2.356	2.229	1.018	1.632	107.90
	$(4F)$	33.33%	−5.3718	2.339	2.229	0.602	1.634	113.88
	$(6F)$	50%	−5.3956	2.340	2.230	0.607	1.633	113.77
	$(10F)$	83.33%	−5.3631	2.337	2.212	0.717	1.633	113.59
	$(12F)$	100%	−5.2454	2.312	X	0.483	1.626	109.41
	$(1Cl)_{edge}$	8.33%	−3.4629	2.331	2.262	0.858	2.082	116.24
	$(1Cl)_{non-edge}$	8.33%	−3.1551	2.361	2.227	1.083	2.098	105.74
	$(1Br)_{edge}$	8.33%	−2.6759	2.326	2.262	0.831	2.257	116.08
	$(1I)_{edge}$	8.33%	−1.7444	2.330	2.277	0.951	2.526	117.03
	$(1At)_{edge}$	8.33%	−1.3330	2.305	2.280	0.899	2.691	116.09

At adatoms attract electrons from surrounding Si atoms that create more unpaired Si-3p orbitals, resulting in higher buckling of 0.951 Å and 0.899 Å in the I-adsorbed SiNRs and At-adsorbed SiNRs, as demonstrated in Table 5.1, respectively. Generally, the geometric structures of ASiNRs, and ZSiNRs are affected by halogen adsorptions in different ways due to the different arrangements of unpaired Si-3p orbitals at the edge. Moreover, the large electronegativity of the F, Cl, and Br adatoms separate them from the I and At adatoms whose electronegativity is smaller. This results that the F, Cl, and Br-adsorbed SiNRs belong to the chemical adsorptions, while the I- and At-adsorbed SiNRs become the physical adsorptions.

5.3.2 ELECTRONIC PROPERTIES

5.3.2.1 Atom-Dominated Band Structure

The electronic band structure is known as a critical physical quantity in the condensed-matter physics that can provide full information on the electronic properties of materials. Energy gaps and energy dispersions are the main characteristics of electronic band structure to observe changes in electronic properties of various systems. The 1D band structure of pristine 7- and 6-ASiNRs are presented in Figures 5.2(a) and 5.2(b), respectively, in which the Fermi level (E_F) is set at zero energy to measure all electronic states. The 1D energy bands fully belong to anti-crossing weak dispersed bands, and the 1D valence and conduction bands are symmetric via Fermi level at low-lying energy that becomes asymmetric at deeper energies. This comes from the weak separation of π and σ bonds in low-buckled Si-Si bonds, in which π bands of Si-$3p_z$ orbitals mainly dominate at low-lying energy, while the σ bands of Si-($3p_x$, $3p_y$) orbitals fully contribute at deep energy. The pristine ASiNRs show a direct energy gap (E_g^d) that is determined by the highest occupied valence and lowest unoccupied conduction band at Γ point. The band gap strongly depends on the widths of ASiNRs that are governed by the rule of $E_g(2n+1) > E_g(2n) > E_g(2n + 2)$, where n is an integer. With n = 3, the widths of the studying systems are 7-ASiNR and 6-ASiNR, in which the band gap of the former is much larger than the latter as shown in Table 5.2. This rule remains unchanged for other widths. The rich 1D band structure of ASiNRs will be dramatically diversified under various halogen adsorptions.

Adjusting adatom concentrations and distributions is an effective way to greatly diversify the electronic properties of ASiNRs. The single halogen atom adsorption on the 6-ASiNR can be considered the possible lowest adsorption concentration of 8.3%. The atom-dominated electronic band structures of 8.3% systems are presented in Figures 5.2(c–h), in which Figures 5.2(c) and (d) show adsorption of a single F atom at the edge and non-edge positions, and Figures 5.2I, (f), (g), and (h) display for adsorption of Cl, Br, I, and At atoms at the non-edge position, respectively. The Fermi level in the 8.3% systems exhibits a redshift; this means the E_F moves downward on the valence band. The redshift of E_F results in generating free holes that come from intersecting of the weakly dispersed valence energy bands with E_F. This indicates that all 8.3% systems belong to p-type metals. Also, the p-type metallic behavior remains almost identical regardless of adatom distributions at edge or non-edge, as evidenced in Figures 5.2(c) and (d) for F adsorbed at the edge and non-edge positions, respectively. It should be mentioned that the atomic radius of halogen atoms

TABLE 5.2
Semiconducting/Metallic Behaviors [Band Gap $E_g^{d/i}$ (eV)/Metal], Magnetic Moment (μ_B), and Magnetism of Pristine and Halogen-Adsorbed ASiNRs and ZSiNRs under Various Adatom Concentrations and Distributions, in Which the $E_g^{d/i}$ (eV) Correspond to the Direct/Indirect Band Gap and the Abbreviation of NM, AFM, and FM Represent the Non-Magnetic, Anti-Ferromagnetic, and Ferromagnetic Features, Respectively

SiNRs	Adsorption configuration	Concentration (%)	Semiconductor $E_g^{d(i)}$ (eV)/metal	Magnetic moment (μ_B)	Magnetism (NM/FM/AFM)
Pristine ASiNRs	7-ASiNR (Si:H = 14:4)	X	$E_g^d = 0.61$	0	NM
	6-ASiNR (Si:H = 12:4)	X	$E_g^d = 0.27$	0	NM
Halogen-adsorbed 6-ASiNRs	$(1F)_{edge}$	8.33%	Metal	0	NM
	$(1F)_{non\text{-}edge}$	8.33%	Metal	0	NM
	(2F)	16.66%	Metal	0	NM
	(4F)	33.33%	Metal	0	NM
	(6F)	50%	$E_g^d = 0.43$	0	NM
	(12F)	100%	$E_g^d = 1.04$	0	NM
	$(1Cl)_{non\text{-}edge}$	8.33%	Metal	0	NM
	(12Cl)	100%	$E_g^d = 1.61$	0	NM
	$(1Br)_{non\text{-}edge}$	8.33%	Metal	0	NM
	(12Br)	100%	$E_g^d = 1.56$	0	NM
	$(1I)_{non\text{-}edge}$	8.33%	Metal	0	NM
	(12I)	100%	$E_g^d = 1.11$	0	NM
	$(1At)_{non\text{-}edge}$	8.33%	Metal	0	NM
	(12At)	100%	$E_g^d = 0.87$	0	NM
Pristine ZSiNR	6-ZSiNR (Si:H = 12:2)	X	$E_g^d = 0.29$	0	AFM
Halogen-adsorbed 6-ZSiNRs	$(1F)_{edge}$	8.33%	Metal	0.17	FM
	$(1F)_{non\text{-}edge}$	8.33%	Metal	0.23	FM
	$(2F)_{edge\text{-}edge}$	16.66%	Metal	0	NM
	$(2F)_{non\text{-}edge\text{-}non\text{-}edge}$	16.66%	Metal	0	NM
	$(2F)_{edge\text{-}non\text{-}edge}$	16.66%	Metal	0.19	FM
	(3F)	25%	Metal	0	NM
	(4F)	33.33%	$E_g^d = 0.42$	0	NM
	(6F)	50%	$E_g^i = 0.38$	0	NM
	(10F)	83.33%	$E_g^i = 0.21$	0	NM
	(12F)	100%	$E_g^d = 0.13$	0	NM
	$(1Cl)_{edge}$	8.33%	Metal	0.18	FM
	$(1Cl)_{non\text{-}edge}$	8.33%	Metal	0.24	FM
	$(1Br)_{edge}$	8.33%	Metal	0.19	FM
	$(1I)_{edge}$	8.33%	Metal	0.21	FM
	$(1At)_{edge}$	8.33%	Metal	0.22	FM

increases with its atomic number under the order of F < Cl < Br < I < At. This means that F has the strongest electron affinity among the others. As a result, F adatoms make the strongest bonding with Si atoms, and the adatom-Si bonds become weaker for another halogen. This implies that the F-Si bond length is shortest, and it becomes larger with an increase of atomic numbers, as indicated in Table 5.1. The feature is fully reflected in the atom-dominated band structures of halogen-adsorbed ASiNRs, in which the F adatoms mainly dominate in the deepest valence energy bands, as illustrated by red circles in Figures 5.2(c) and (d). Meanwhile, the dominance of other

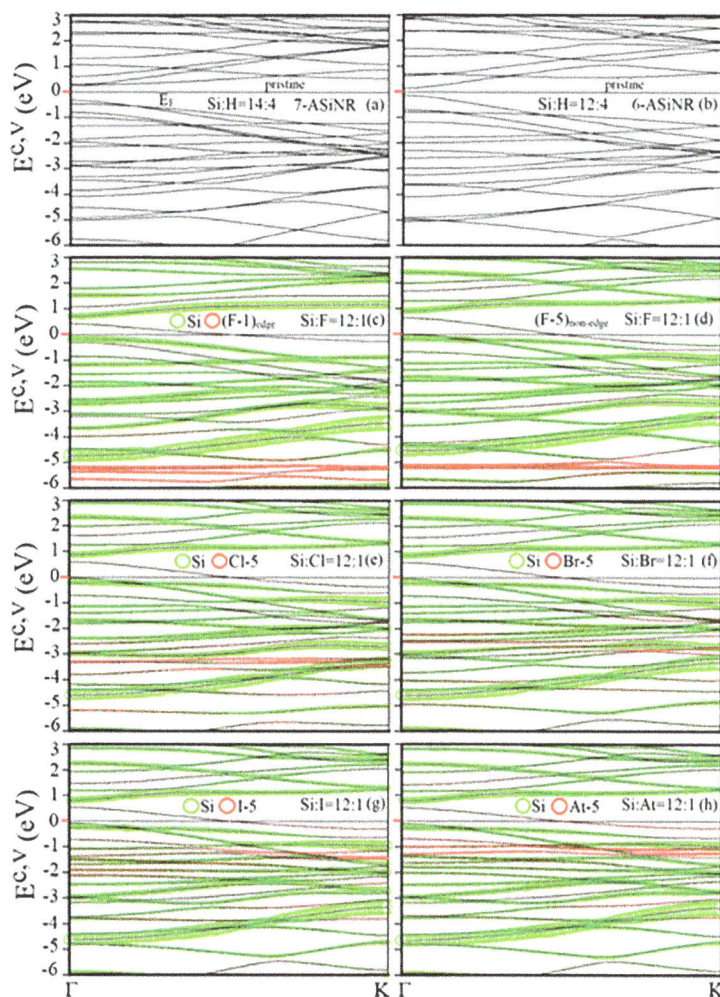

FIGURE 5.2 Atom-projected band structures of pristine ASiNRs under different widths for (a) 7-ASiNR, (b) 6-ASiNR and halogen-adsorbed ASiNRs under lowest-concentration of 8.3% for (c) $(1F)_{edge}$, (d) $(1F)_{non-edge}$, (e) $(1Cl)_{non-edge}$, (f) $(1Br)_{non-edge}$, (g) $(1I)_{non-edge}$, and $(1At)_{non-edge}$. Green and red circles display for dominance of silicon atoms and halogen adatoms, respectively.

halogen adatoms occurs at higher valence energy bands, as shown in Figures 5.2(e–h) for Cl, Br, I, and At adatoms, respectively. It can be concluded that the middle-gap feature of the pristine ASiNRs fully becomes the p-type metallic behavior under the single halogen adsorptions regardless of the adatom kinds and adatom distributions.

Under gradually increasing adatom concentrations from 8.3% to 100%, the electronic properties of ASiNRs are greatly diversified, as shown in Figure 5.3. At

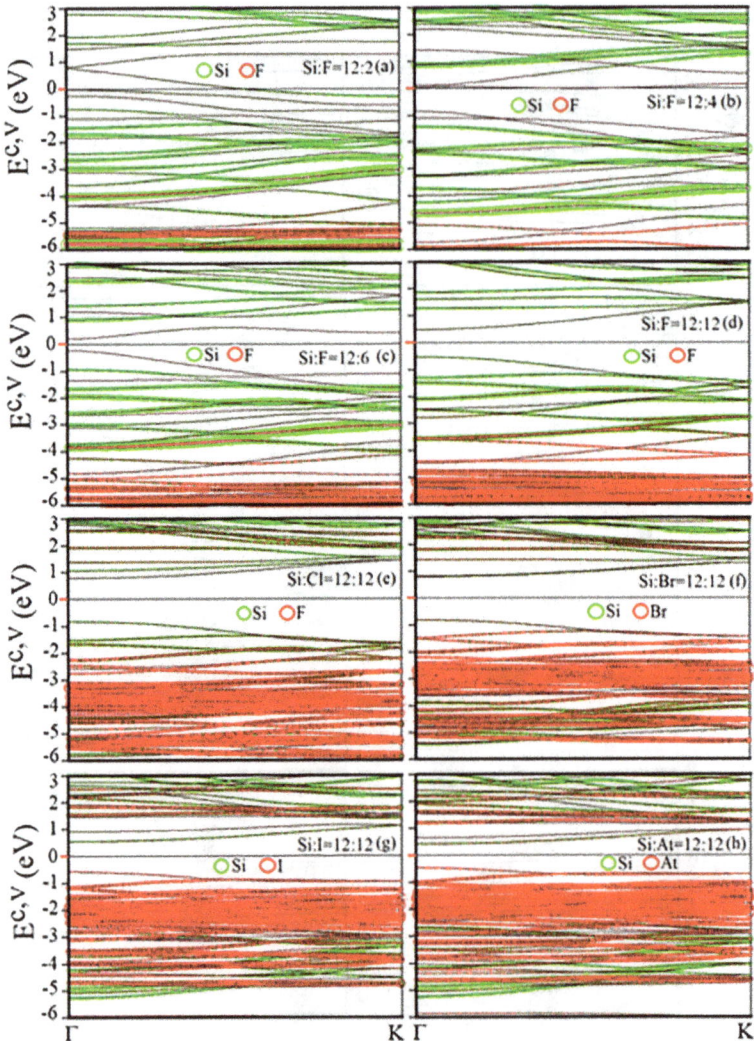

FIGURE 5.3 Atom-projected band structures of halogen-adsorbed ASiNRs under high concentrations (a) 16.6% of Si:F = 12:2, (b) 33.3% of Si:F = 12:4, (c) 50% of Si:F = 12:6, (d) 100% of Si:F = 12:12, (e) 100% of Si:Cl = 12:12, (f) 100% of Si:Br = 12:12, (g) 100% of Si:I = 12:12, and (h) 100% of Si:At = 12:12. Green and red circles display dominance of silicon atoms and halogen adatoms, respectively.

slightly increased concentrations of 16.6% and 33. 3%, the p-type metallic behavior remains; this is owing to the fact that π bonds in Si-Si significantly survive in the systems. However, the contribution of the free hole density is changed under these different low concentrations, as shown in Figures 5.3(a) and (b) for 16.6% and 33.3 % F adsorptions, respectively. As a result, the dominance of adatoms in energy bands becomes larger. Under the critical increase of adatom concentration at 50%, this system belongs to a p-type middle-gap semiconductor, in which a band gap of 0.16 eV is opened, as compared with the pristine system. The p-type metallic-semiconducting transition at 50% adsorption is because the π bonds are half terminated by the adatoms that are insufficient to dominate at low-lying energy. Beyond the critical concentration, the band gap is further opened with an increase of adatom concentrations, whereas the direct gap features of the band structures come to survive. The largest band gap of 1.04 eV is found at the highest adatom concentration of 100%, and the dominant intensity of adatoms is highest in the 100% system. Serious destroying or termination of π bonds is responsible for the increase of band gaps with the increase of adatom concentrations. The p-type metallic-semiconducting transitions under the sufficiently high adatom concentrations occur the same for other halogen adatoms. The induced band gaps in the other 100% halogen systems are 1.61 eV, 1.56 eV, 1.11 eV, and 0.87 eV for Cl, Br, I, and At adatoms, respectively, as shown in Figures 5.3(e), (f), (g), and (h). As a result, adatom dominant intensities in 100% systems of other halogen are denser and tend to dominate at higher energy bands.

The difference in edge terminations can lead to different diverse features. As a result, there remain certain differences in the electronic band structure of ASiNR and ZSiNR. The band structure of pristine 6-ZSiNR is shown in Figure 5.4(a), in which it presents a pair of partial flat valence and conduction bands nearest to E_F that corresponds to wave functions localized at the zigzag boundaries. These partial flat bands exhibit the spin degeneracy that will show the anti-ferromagnetic configuration across the 1D nanoribbon, as discussed in Figure 5.8(a). Within low-lying energies, it is mainly contributed by the partial flat bands that their energy dispersions become more obvious at deeper valence/higher conduction energies. Besides, ZSiNR displays a direct energy gap that is mainly determined by the highest occupied valence and lowest unoccupied conduction partial flat bands. This direct gap originates from the strong competition between quantum confinement and spin configuration. These partial flat bands at low-lying energies are made of weak π bonds so that they are very sensitive to halogen adsorptions.

With the single halogen adsorption at the concentration of 8.3% as shown in Figures 5.4(b–h), the redshift of E_F also comes to exist that leads to having one highest occupied partial flat band intersecting with the E_F, indicating that the 8.3% zigzag systems belong to the p-type metal. This p-type feature is due to the electron transfer from the Si atoms to halogen adatoms. Especially, there is a spin splitting in energy bands, in which it appears obvious at low-lying energy near to the K point, as illustrated by pink and black curves for spin-up and spin-down states in Figure 5.4(b–h), respectively. The spin-splitting bands become degeneracy at deeper valence/higher conduction energies. Specifically, the magnitude of the spin splitting strongly depends on the adatom distributions at the edge and non-edge positions, as shown in Figures 5.4(b) and 5.4(c) for F adatoms and Figures 5.4(d) and 5.4(e) for Cl

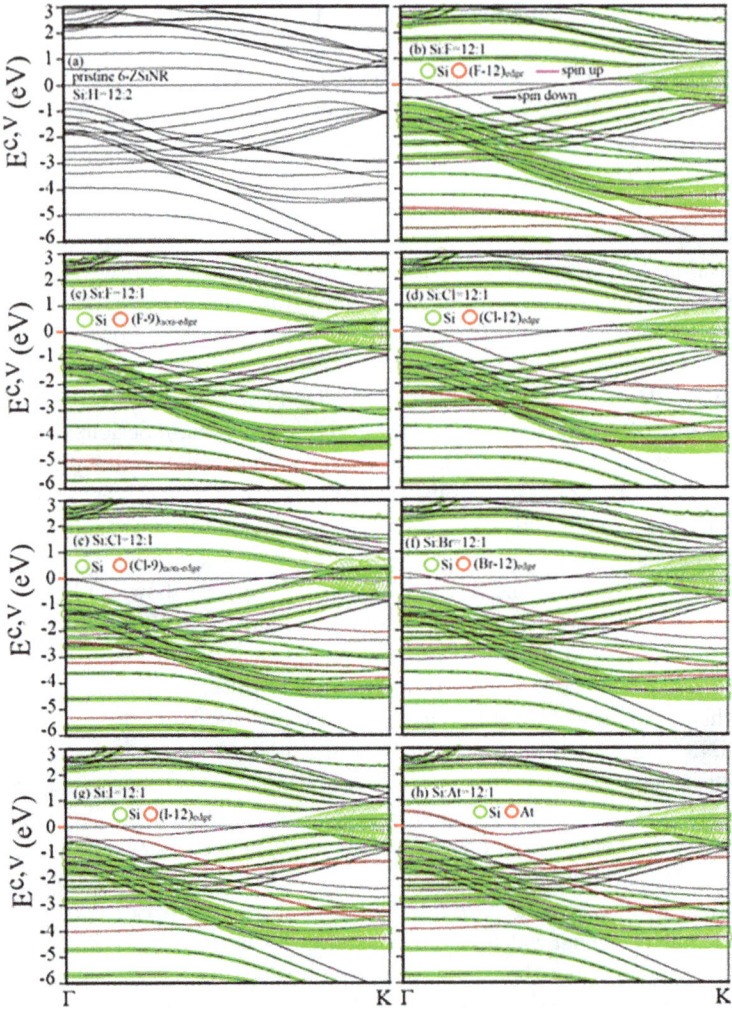

FIGURE 5.4 Atom-projected band structures of pristine ZSiNR for (a) 6-ZSiNR and halogen-adsorbed ZSiNRs under lowest concentration of 8.3% for (b) $(1F)_{edge}$, (c) $(1F)_{non-edge}$, (d) $(1Cl)_{edge}$, (e) $(1Cl)_{non-edge}$, (f) $(1Br)_{edge}$, (g) $(1I)_{edge}$, and (h) $(1At)_{edge}$. Green and red circles and pink and black curves display silicon atoms and halogen adatoms and spin-up and spin-down polarizations, respectively.

adatoms, respectively. This will result in different spin configurations at the edge and non-edge adsorptions (discussed in detail in Figure 5.8). This spin splitting feature is identical for Br, I, and At, as shown in Figures 5.4(f), (g), and (h), respectively. The p-type metallic and spin-splitting behaviors are dramatically changed under further increasing of adatom concentrations. At the double adatom adsorptions of 16.6%, these features are diversified under adatom distributions. The spin splitting is significantly diversified under the three typical adatom distributions of edge-edge, non-edge-non-edge, and edge-non-edge. The edge-edge and non-edge-non-edge distributions fully destroy the spin splitting that leads to non-spin-splitting bands, as evidenced in Figures 5.5(a) and (b), respectively. This is due to the two adatoms destroying the ferromagnetic spin orientations in each zigzag edge. Meanwhile, the edge-non-edge distribution results in ferromagnetic spin splitting bands, as displayed in Figure 5.5(c). Under the increased adatom concentration of 25% as presented in Figure 5.5(d), the spin splitting is fully destroyed regardless of adatom distributions so that this increased concentration can be regarded as the critical concentration in

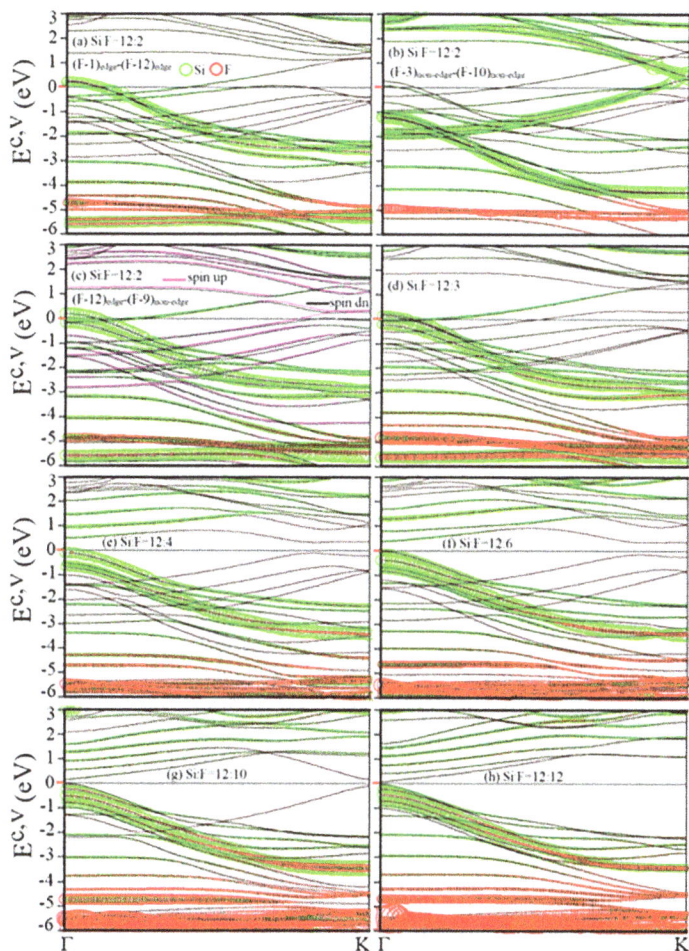

FIGURE 5.5 Atom-projected band structures of halogen-adsorbed ZSiNRs under different high-concentrations and distributions for 16.6% under distributions of (a) $(1F)_{edge}$-$(1F)_{edge}$, (b) $(1F)_{non-edge}$-$(1F)_{non-edge}$, (c) $(1F)_{edge}$-$(1F)_{non-edge}$, (d) 25% of Si:F = 12:3, (e) 33.3% of Si:F = 12:4, (f) 50% of Si:F = 12:6, (g) 83.3% of Si:F = 12:10, and (h) 100% of Si:F = 12:12. Green and red circles and pink and black curves display dominance of silicon atoms and halogen adatoms and spin-up and spin-down polarizations, respectively.

magnetic-non-magnetic transition. However, this system retains its p-type metallic behavior. As for critical high increased concentration at 33.3% shown in Figure 5.5(e), the p-type metallic-semiconducting transition occurs. Beyond the critical concentration, all the high-concentration systems exhibit the p-type semiconducting behavior, as illustrated in Figures 5.5(f), 5.5(g), and 5.5(h) for 50%, 83.3%, and 100% adsorptions, respectively.

5.3.2.2 Orbital-Projected Density of States

Atom- and orbital-projected density of states (DOSs) can verify all main features in the electronic band structures and further provide critical information on atomic orbitals-decomposed contributions in energy bands. The DOSs of 6-ASiNR without and with various halogen adsorptions is presented in Figure 5.6. For DOSs of pristine 6-ASiNR shown in Figure 5.6(a), it has a vacant region centered at the E_F that confirms the energy gap in its corresponding band structure. At low-lying energy, two prominent peaks symmetric via E_F appear that are mainly made of Si-$3p_z$ orbitals, as illustrated by solid red curves in Figure 5.6(a), resulting from weakly dispersed bands. It should be mentioned that the low-lying DOSs of this pristine system is mainly contributed by the Si-$3p_z$ orbital as a result of the dominance of π bands. At deeper energies, all the prominent peaks are mainly made of Si-$(3p_x, 3p_y)$ orbitals that verify for the contribution of weakly dispersed σ bands. It is further noted that there is a separation of π and σ orbitals at low-lying and deep energies, where the evidence for the separation of π and σ bands, coming from splitting of π and σ bonds in a low-buckled honeycomb lattice. within the middle-energy range of −1.2 eV to −3 eV, there exist hybridizations between the Si-$(3s, 3_x + 3p_y,$ and $3p_z)$ orbitals that result from weak sp³ hybridization mechanism in the low-buckled 1D honeycomb structure, as displayed by solid blue, green, and red curves in Figure 5.6(a), respectively. Generally speaking, the separation of π and σ orbitals at low-lying and deep energies and hybridizations in s, p_x, p_y, and p_z orbitals at middle energy is strong evidence for a mixed hybridization mechanism of sp²/sp³ in low-buckled ASiNRs. The DOSs of the pristine system is fully reshaped under various halogen adatom adsorptions. As for the single adatom adsorptions shown in Figures 5.6(b), (e), (f), (g), and (h) for F, Cl, Br, I, and At adatoms, DOSs comes to localize at the E_F that originates from the redshift of E_F to create free holes, indicating that the systems belong to p-type metals, respectively. Specifically, for the F case, F-$(2p_x$ and $2p_y)$ orbitals create a very strong single peak at the very deep energies, and there is a hybridization between the single peak with small peaks of Si-$(3s, 3p_x,$ and $3p_y)$ orbitals, in which F-$(2p_x$ and $2p_y)$ orbitals are shown by dashed pink curves in Figure 5.6(b). The strong single peak structure at the deep energy will decompose into several smaller sub-peaks, in which these small sub-peaks dominate at higher valence energies that rise from Cl, Br, I, and At, as shown in Figures 5.6(e), (f), (g), and (h) for Cl, Br, I, and At, respectively. The strong single peaks at deep energy and small sub-peaks at higher energies result from the strong dominances of halogen adatoms in deep and higher valence bands, and the hybridizations between orbitals of halogen adatoms and Si atoms can evidence for co-contributions of these atoms in energy bands. As for further increased concentration, higher DOSs locates at the E_F, and several prominent sub-peaks of F adatoms dominate at deep energies, as illustrated in Figure 5.6(d). Under the highest

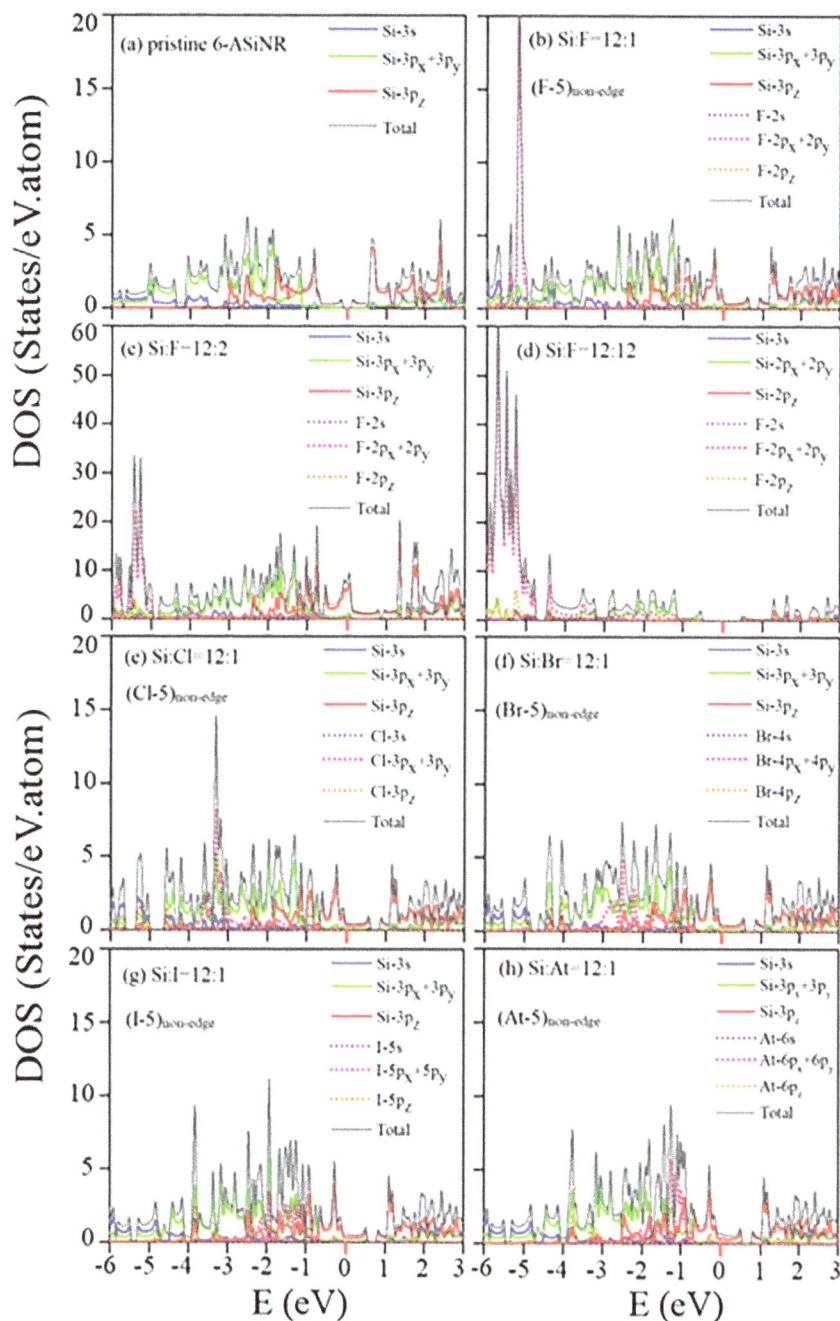

FIGURE 5.6 Atom- and orbital-projected density of states (DOSs) of pristine ASiNR and halogen-adsorbed ASiNRs under different concentrations for (a) pristine 6-ASiNR, (b) 8.3% of Si:F = 12:1, (c) 16.6% of Si:F = 12:2, (d) 100% of Si:F = 12:12, (e) 8.3% of Si:Cl = 12:1, (f) 8.3% of Si:Br = 12:1, (g) 8.3% of Si:I = 12:1, and (h) 8.3% of Si:At = 12:1.

concentration of 100%, the low-lying DOSs of π orbitals are seriously deformed, and many prominent peaks of F orbitals appear at deep valence energies that belong to sub-peak structure. It is noted that spin-splitting peaks are absent in DOSs of ASiNR systems since these systems belong to non-magnetic materials.

The fundamental difference between ASiNR and ZSiNR systems is due to unique ferromagnetic spin arrangements along each zigzag edge and anti-ferromagnetic ones across the ribbon center. This unique spin arrangement is greatly diversified under various adatom adsorptions that can be fully identified through symmetric and asymmetric prominent peaks in DOSs. The DOSs of pristine 6ZSiNR is shown in Figure 5.7(a), in which the spin-up and spin-down prominent peaks are identical and fully symmetric via the E_F in the whole energy range, resulting from spin degeneracy energy bands. Specifically, at low-lying energy, two spin-up prominent peaks of Si-3p$_z$ orbitals are thoroughly symmetric with the spin-down ones through the E_F that features from spin degeneracy partial flat bands, as illustrated by black triangles in Figure 5.7(a), where the gap states between these π prominent peaks are responsible for the energy gap. It should be mentioned that the pristine ZSiNR system adopts the mix of sp^2/sp^3 hybridizations so that it displays a weak separation of π and σ orbitals, in which π orbitals of Si-3p$_z$ mainly contribute at low-lying energy of −1 eV to 1 eV and σ ones of Si-(3s, 3p$_x$, and 3p$_y$) strongly dominate at deep energy of −6 eV to −3 eV, as illustrated by black, green, and red curves in Figure 5.7(a) for Si-(3s, 3p$_{x,y}$, and 3p$_z$) orbitals, respectively. Meanwhile, the weak sp^3 hybridization can be realized at middle energy of −3 eV to −1 eV such that it simultaneously occurs Si-(3s, 3p$_{x,y}$, and 3p$_z$) orbitals. The unique features of DOSs in the pristine ZSiNR are significantly reshaped under various halogen adsorptions. Under the single halogen adsorptions, the symmetric prominent peaks of Si-3p$_z$ orbitals at low-lying energy become asymmetric ones, and it appears DOSs at the E_F that comes from the spin-splitting bands intersecting with the E_F, as shown in Figures 5.7(b), 5.7(e), 5.7(f), 5.7(g), and 5.7(h) for F, Cl, Br, I, and At systems, respectively. Besides, the strong peaks made of F-2p$_{x,y}$ orbitals come to appear at deep energy and hybridize with Si-3p$_{x,y}$ orbitals that evidence for strong dominance of F adatoms at deep energy bands. The deep energy strong peaks of p$_{x,y}$ orbitals become higher ones for other halogen adatoms, as illustrated by dashed pink curves in Figures 5.7(e–h). As a further increase of concentrations of double adatoms, the spin-splitting configurations only exist at specific adatom distribution of edge-non-edge, as shown in Figure 5.7(c), in which more asymmetric sub-peaks of Si-3p$_z$ orbitals come to appear at low-lying energy. Especially, at the critical increased concentration of 25%, the spin-splitting feature of DOSs is fully vanished regardless of adatom distributions, as shown in Figure 5.7(d). However, the presence of DOSs at the E_F in the increased concentrations remains.

5.3.3 Magnetic Configurations

Magnetic properties of materials can be fully elucidated through their spatial spin density distribution. In this study, magnetic configurations only exist in ZSiNR systems. For pristine 6-ZSiNR, its spin density distribution is shown in Figure 5.8(a), in which spin-up and -down states are described by pink and yellow balls, respectively. Along each zigzag edge, it always presents a ferromagnetic spin arrangement, in

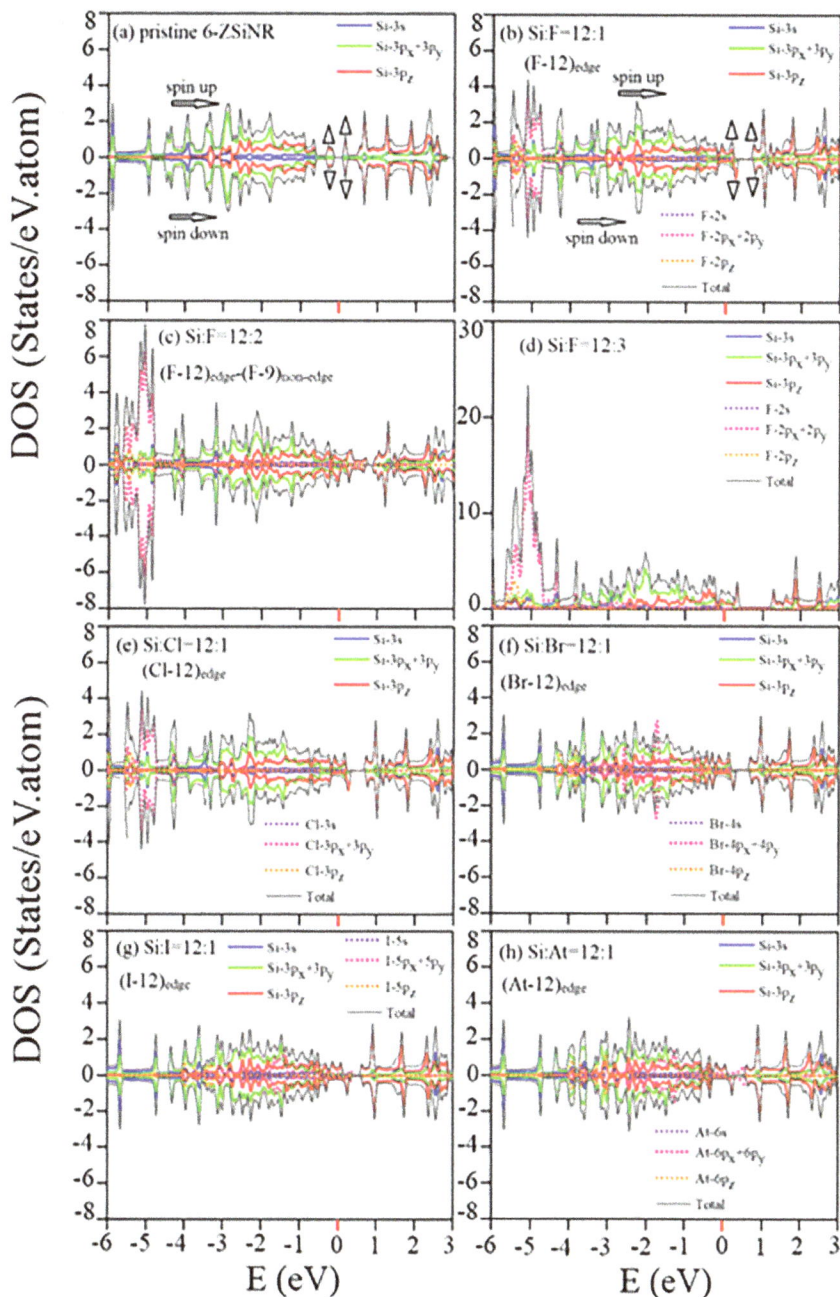

FIGURE 5.7 Atom- and orbital-projected density of states (DOSs) of pristine ZSiNR and halogen-adsorbed ASiNRs under different concentrations for (a) pristine 6-ZSiNR, (b) 8.3% of Si:F = 12:1, (c) 16.6% of Si:F = 12:2 under distribution of $(1F)_{edge}$-$(1F)_{non-edge}$, (d) 25% of Si:F = 12:3, I 8.3% of Si:Cl = 12:1, (f) 8.3% of Si:Br = 12:1, (g) 8.3% of Si:I = 12:1, and (h) 8.3% of Si:At = 12:1.

which spin-up and -down states tend to orientate at upper and lower Si atoms, respectively. This unique spin orientation is owing to its low-buckled honeycomb lattice. Across the ZSiNR center, spin up and down states are identically distributed and symmetric so that the net magnetic moment of this system has vanished, as shown in Table 5.1. As a result, the pristine ZSiNR can be regarded as an anti-ferromagnetic system, namely anti-ferromagnetic configuration. Under various halogen adsorptions, it can lead to diverse spin arrangements/configurations. As for the single halogen adsorptions at edge positions, the adatom will destroy spin orientations at the edge and the edge-close regions that the spin orientations only exists in the opposite

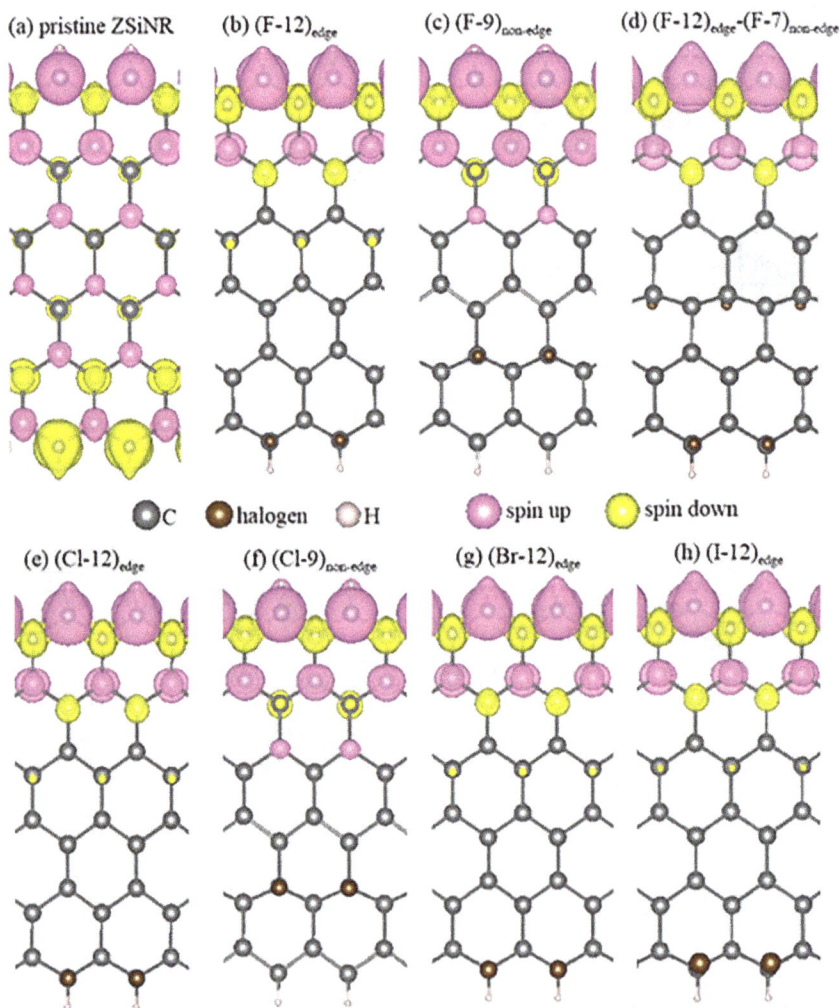

FIGURE 5.8 Spin density distribution of (a) pristine ZSiNR and halogen-adsorbed ZSiNRs of (b) (F-12)$_{edge}$, (c) (F-9)$_{non-edge}$, (d) (F-12)$_{edge}$-(F-7)$_{non-edge}$, (e) (Cl-12)$_{edge}$, (f) (Cl-9)$_{non-edge}$, (g) (Br-12)$_{edge}$, and (h) (I-12)$_{edge}$.

edge and the edge-closed regions, leading to the edge ferromagnetic configuration, as shown in Figures 5.8(b), (e), (g), and (h) for F, Cl, Br, and I adatoms, respectively. The adatom distributions at non-edge positions can also induce another spin configuration, as shown in Figures 5.8(c) and (f) for F and Cl adatoms, respectively. Under further increasing of concentration of double adatoms, the ferromagnetic spin configuration only remains at specific adatom distribution of edge-non-edge positions, as illustrated in Figure 5.8(d) for F adatoms, in which F adatoms will destroy spin orientations at the edge and edge-closed region such that this ferromagnetic configuration is mainly contributed by the opposite edge- and non-edge-atoms. This ferromagnetic spin arrangement at the specific adatom distribution also remains identical for other halogen adatoms of Cl, Br, I, and At.

5.4   CONCLUDING REMARKS

The DFT method implemented in the VASP package is a very useful tool to explore novel quasiparticle properties and has been extensively used for many studies, since the DFT predictions could lead to very accurate and reliable results. Furthermore, such accurate DFT predictions play a critical role to significantly reduce the random testing time for further experimental studies. Currently, the VASP calculations have been the main instrument in many theoretical research groups and laboratories because of their effective deployment in high-performance computing clusters (HPCC) to efficiently deal with large-scale study models. Specifically, many essential physical quantities can be developed through the VASP calculations, including the binding energies, phonon spectrum, optimal bond lengths, bond angles, atom-dominated band structures, orbital-projected density of states, charge density distributions, charge density differences, magnetic moments, and spin density distributions, dielectric functions, absorption spectrum, electron energy loss-function, in which the binding energies, optimal lattice parameters, electronic band structures, density of states, magnetic moments, and spin density distributions have been effectively utilized to determine many novel quasiparticle properties of many emergent layered materials, such as graphene [57], graphene nanoribbons [58], silicene [59], silicene nanoribbons [60], germanene [61], layered transition metal dichalcogenides [62], phosphorene [63], and emergent 2D heterostructures [64]. On the experimental side, the optimal geometric structures, electronic band structures, and density of states predicted through DFT calculations can be fully verified by the available experimental measurements of scanning tunneling microscopy (STM) [65], angle-resolved photoemission spectroscopy (ARPES) [66], and scanning tunneling spectroscopy (STS) [67], respectively.

   In this chapter, the diverse quasiparticle properties of halogen-adsorbed SiNRs are fully revealed in the spin-polarized DFT calculations. The essential physical quantities have been developed through the first-principles calculations, including the binding energies, optimal lattice parameters, atom-dominated band structures, atom- and orbital-projected DOS, and spin density distributions. These developed physical quantities are sufficient to clarify the rich quasiparticle phenomena induced by the halogenation effects. Specifically, the diversified geometric structures are evaluated through the binding energies and optimal lattice parameters. The feature-rich electronic properties are rigidly analyzed in atom-dominated band structures that can be thoroughly verified through atom- and orbital-projected DOSs. The diverse magnetic configurations are thoroughly comprehended in the spin density distributions. As for halogen-adsorbed ASiNR systems, under the single halogen adsorptions, the direct middle-gap semiconducting feature of the pristine system becomes the p-type metallic behavior for all the halogen adatoms. This is because the halogen adatoms attract electrons from ASiNRs to generate free holes in the systems. This is to say that creating a redshift of E_F leads to a p-type metallic behavior. When the adatom concentrations reach the critical value of 50% adsorption, the p-type metallic-semiconducting transition appears, and the band gaps are further opened under higher adatoms concentrations. The largest band gap is found at the highest concentration of 100% regardless of any halogen adatoms. On the other side, the unique anti-ferromagnetic configuration of the pristine ZSiNR system is greatly diversified under the various halogen concentrations and distributions. For the single halogen-adsorbed ZSiNR systems, it shows the Si-related ferromagnetic configurations for all halogen adatoms, in which the net magnetic moments are different in edge and one-edge adsorption positions. Under further increasing of concentrations of double adatoms, the Si-related ferromagnetic configuration only remains at specific adatoms distribution of edge and non-edge adsorption positions. It is worthy to notice that the diverse quasiparticle properties of halogen-adsorbed SiNRs can be suitable for a wide range of applications in electronic and spintronic devices. Furthermore, the developed first-principles theoretical framework in this chapter can be fully generalized to many other emergent layered materials.

ACKNOWLEDGMENTS

This research is funded by Thu Dau Mot University, Binh Duong Province, Vietnam. Also, this research used resources of the HPCC at Thu Dau Mot University, Vietnam, and National Cheng Kung University, Taiwan.

REFERENCES

[1] Kurakevych O O, Le Godec Y, Crichton W A and Strobel T A 2016 Silicon allotropy and chemistry at extreme conditions *Energy Procedia* **92** 839–44

[2] Yusufu A et al. 2014 Bottom-up nanostructured bulk silicon: A practical high-efficiency thermoelectric material *Nanoscale* **6** 13921

[3] Perim E, Paupitz R, Botari T and Galvao D S 2014 One-dimensional silicon and germanium nanostructures with no carbon analogs *Phys Chem Chem Phys* **16** 24570–4

[4] Tiwari J N, Tiwari R N and Kim K S 2012 Zero-dimensional, one-dimensional, two-dimensional, and three-dimensional nanostructured materials for advanced electrochemical energy devices *Prog Mater Sci* **57** 724–803

[5] Takeda K and Shiraishi K 1994 Theoretical possibility of stage corrugation in Si and Ge analogs of graphite *Phys Rev B* **50** 14916–22

[6] Novoselov K S et al. 2004 Electric field effect in atomically thin carbon films *Science* **306** 666–9

[7] Akbari E, Buntat Z, Afroozeh A, Pourmand S E, Farhang Y and Sanati P 2016 Silicene and graphene nanomaterials in gas sensing mechanism *RSC Adv* **6** 81647

[8]   Tang C, Oppenheim T, Tung V C and Martini A 2013 Structure—stability relationships for graphene-wrapped fullerene-coated carbon nanotubes *Carbon* **61** 458–66
[9]   Sadeddine S et al. 2017 Compelling experimental evidence of a Dirac cone in the electronic structure of a 2D Silicon layer *Sci Rep* **7** 44400
[10] Cherukara M J, Narayanan B, Chan H and Sankaranarayanan S K R S 2017 Silicene growth through island migration and coalescence *Nanoscale* **9** 10186–92
[11] Lin C-L et al. 2012 Structure of silicene grown on Ag(111) *Appl Phys Express* **5**, 45802
[12] Tao L et al. 2015 Silicene field-effect transistors operating at room temperature *Nat Nanotechnol* **10** 227–31
[13] Hu W, Xia N, Wu X, Li Z and Yang J 2014 Silicene as a highly sensitive molecule sensor for NH_3, NO, and NO_2 *Phys Chem Chem Phys* **16** 6957–62
[14] Galashev A Y and Ivanichkina K A 2020 Silicene anodes for lithium-ion batteries on metal substrates *J Electrochem Soc* **167** 50510
[15] Karpas M, Faria Junior P E, Gmitra M and Fabian J 2019 Spin-orbit coupling in elemental two-dimensional materials *Phys Rev B* **100** 125422
[16] Liu C-C, Feng W and Yao Y 2011 Quantum spin hall effect in silicene and two-dimensional germanium, *Phys Rev Lett* **107** 76802
[17] Ezawa M 2015 Monolayer topological insulators: Silicene, germanene, and stanene *J Phys Soc Japan* **84** 121003
[18] Xu C et al. 2012 Giant magnetoresistance in silicene nanoribbons *Nanoscale* **4** 3111–17
[19] Pan H, Li Z, Liu C-C, Zhu G, Qiao Z and Yao Y 2014 Valley-polarized quantum anomalous Hall effect in silicene *Phys Rev Lett* **112** 106802
[20] Ezawa M 2012 Valley-polarized metals and quantum anomalous Hall effect in silicene *Phys Rev Lett* **109** 55502
[21] Ni Z et al. 2014 Tunable band gap and doping type in silicene by surface adsorption: Towards tunneling transistors *Nanoscale* **6** 7609–18
[22] Hernández Cocoletzi H and Castellanos Águila J E 2018 DFT studies on the Al, B, and P doping of silicene, *Superlattices Microstruct* **114** 242–50
[23] Huang B, Xiang H J and Wei S-H 2013 Chemical functionalization of silicene: Spontaneous structural transition and exotic electronic properties *Phys Rev Lett* **111** 145502
[24] Padilha J E and Pontes R B 2015 Free-standing bilayer silicene: The effect of stacking order on the structural, electronic, and transport properties *J Phys Chem C* **119** 3818–25
[25] Qin R, Wang C-H, Zhu W and Zhang Y 2012 First-principles calculations of mechanical and electronic properties of silicene under strain *AIP Adv* **2** 22159
[26] Wu J-Y, Chen S-C, Gumbs G and Lin M-F 2016 Feature-rich electronic excitations of silicene in external fields *Phys Rev B* **94** 205427
[27] Johll H, Lee M D K, Ng S P N, Kang H C and Tok E S 2014 Influence of interconfigurational electronic states on Fe, Co, Ni-silicene materials selection for spintronics *Sci Rep* **4** 7594
[28] Lin X and Ni J 2012 Much stronger binding of metal adatoms to silicene than to graphene: A first-principles study *Phys Rev B* **86** 75440
[29] Resta A et al. 2013 Atomic structures of silicene layers grown on Ag(111): Scanning tunneling microscopy and noncontact atomic force microscopy observations *Sci Rep* **3** 2399
[30] Tran N T T, Gumbs G, Nguyen D K and Lin M-F 2020 Fundamental properties of metal-adsorbed silicene: A DFT study *ACS Omega* **5** 13760–9
[31] Lin S-Y, Liu H-Y, Nguyen D K, Tran N T T, Pham H D, Chang S-L, Lin C-Y and Lin M-F 2020 Carbon-, boron- and nitrogen-substituted silicene compounds *IOP Publishing* 9–43

[32] De Padova P, Perfetti P, Olivieri B, Quaresima C, Ottaviani C and Le Lay G 2012 1D graphene-like silicon systems: Silicene nano-ribbons *J Phys Condens Matter* **24** 223001

[33] Lu Y H et al. 2009 Effects of edge passivation by hydrogen on electronic structure of armchair graphene nanoribbon and band gap engineering *Appl Phys Lett* **94** 122111

[34] Cao L, Li X, Li Y and Zhou G 2020 Electrical properties and spintronic application of carbon phosphide nanoribbons with edge functionalization *J Mater Chem C* **8** 9313–21

[35] Sadeghi H, Sangtarash S and Lambert C J 2015 Enhanced thermoelectric efficiency of porous silicene nanoribbons *Sci Rep* **5** 9514

[36] Yao Y, Liu A, Bai J, Zhang X and Wang R 2016 Electronic structures of silicene nanoribbons: Two-edge-chemistry modification and first-principles study *Nanoscale Res Lett* **11** 371

[37] Nguyen D K, Tran N T T, Chiu Y-H and Lin M-F 2019 Concentration-diversified magnetic and electronic properties of halogen-adsorbed silicene *Sci Rep* **9** 13746

[38] Mehdi Aghaei S and Calizo I 2015 Band gap tuning of armchair silicene nanoribbons using periodic hexagonal holes *J Appl Phys* **118** 104304

[39] Dong H, Fang D, Gong B, Zhang Y, Zhang E and Zhang S 2015 Electronic and magnetic properties of zigzag silicene nanoribbons with Stone—Wales defects *J Appl Phys* **117** 64307

[40] Quertite K et al. 2020 Silicene nanoribbons on an insulating thin film *Adv Funct Mater* 2007013

[41] van den Broek B, Houssa M, Lu A, Pourtois G, Afanas'ev V and Stesmans A 2016 Silicene nanoribbons on transition metal dichalcogenide substrates: Effects on electronic structure and ballistic transport *Nano Res* **9** 3394–406

[42] Aghaei S M, Monshi M M and Calizo I 2016 A theoretical study of gas adsorption on silicene nanoribbons and its application in a highly sensitive molecule sensor *RSC Adv* **6** 94417–28

[43] Kharadi M A, Malik G F A, Khanday F A, Shah K A, Mittal S and Kaushik B K 2020 Review-silicene: From material to device applications *ECS J Solid State Sci Technol* **9** 115031

[44] Zhou J, Bournel A, Wang Y, Lin X, Zhang Y and Zhao W 2017 Silicene spintronics: Fe(111)/silicene system for efficient spin injection *Appl Phys Lett* **111** 182408

[45] Mehdi Aghaei S, Torres I and Calizo I 2016 Structural stability of functionalized silicene nanoribbons with normal, reconstructed, and hybrid edges *J Nanomater* 5959162

[46] Walia G K and Randhawa D K K 2018 First-principles investigation on defect-induced silicene nanoribbons: A superior media for sensing NH_3, NO_2 and NO gas molecules *Surf Sci* **670** 33–43

[47] Rosales L, Orellana P, Barticevic Z and Pacheco M 2008 Transport properties of graphene nanoribbon heterostructures *Microelectronics J* **39** 537–40

[48] Mahmoudi M, Fathipour M and Ahangari Z 2016 Improved double-gate armchair silicene nanoribbon field-effect-transistor at large transport band gap *Chinese Phys B* **25** 8

[49] Lu W-T, Sun Q-F, Tian H-Y, Zhou B-H and Liu H-M 2020 Band bending and zero-conductance resonances controlled by edge electric fields in zigzag silicene nanoribbons *Phys Rev B* **102** 125426

[50] Núñez C, Orellana P A, Rosales L, Römer R A and Domínguez-Adame F 2017 Spin-polarized electric current in silicene nanoribbons induced by atomic adsorption *Phys Rev B* **96** 45403

[51] Ghasemi N, Ahmadkhan Kordbacheh A and Berahman M 2019 Electronic, magnetic and transport properties of zigzag silicene nanoribbon adsorbed with Cu atom: A first-principles calculation *J Magn Magn Mater* **473** 306–11

[52] Xu L, Wang X-F, Zhou L and Yang Z-Y 2015 Adsorption of Ti atoms on zigzag silicene nanoribbons: Influence on electric, magnetic, and thermoelectric properties *J Phys D Appl Phys* **48** 215306

[53] Zhang J-M, Song W-T, Xu K-W and Ji V 2014 The study of the P doped silicene nanoribbons with first-principles *Comput Mater Sci* **95** 429–34

[54] Ma L, Zhang J-M, Xu K-W and Ji V 2014 Nitrogen and boron substitutional doped zigzag silicene nanoribbons: Ab initio investigation *Phys E Low-Dimensional Syst Nanostructures* **60** 112–17

[55] Kresse G and Furthmuller J 1996 Efficient iterative schemes for ab initio total-energy calculations using a plane-wave basis set *Phys Rev B* **54** 11169

[56] Perdew J-P, Burke K and Ernzerhof M 1996 Generalized gradient approximation made simple *Phys Rev Lett* **77** 3865

[57] Tran N-T-T, Nguyen D-K, Glukhova O-E and Lin M-F 2017 Coverage-dependent essential properties of halogenated graphene: A DFT study *Sci Rep* **7** 17858

[58] Nguyen D-K, Lin Y-T, Lin S-Y, Chiu Y-H, Tran N-T-T and Lin M-F 2017 Fluorination-enriched electronic and magnetic properties in graphene nanoribbons *Phys Chem Chem Phys* **19** 20667–76

[59] Nguyen D-K, Tran N-T-T, Chiu Y-H and Lin M-F 2019 Concentration-diversified magnetic and electronic properties of halogen-adsorbed silicene *Sci Rep* **9** 1–15

[60] Zhang X, Zhang D, Xie F, Zheng X, Wang H and Long M 2017 First-principles study on the magnetic and electronic properties of Al or P doped armchair silicene nanoribbons *Phys Lett A* **381** 2097–102

[61] Pang Q, Li L, Zhang C-L, Wei X-M and Song Y-L 2015 Structural, electronic and magnetic properties of 3d transition metal atom adsorbed germanene: A first-principles study *Mater Chem Phys* **160** 96–104

[62] Chen Q, Ding Q, Wang Y, Xu Y and Wang J 2020 Electronic and magnetic properties of a two-dimensional transition metal phosphorous chalcogenide TMPS₄ *J Phys Chem C* **124** 12075–80

[63] Tang Y, Zhou W, Hu C, Pan J and Ouyang F 2019 Electronic and magnetic properties of phosphorene tuned by Cl and metallic atom co-doping *Phys Chem Chem Phys* **21** 18551–8

[64] Chen G-X, Li X-G, Wang Y-P, Fry J-N and Cheng H-P 2017 Two-dimensional lateral GaN/SiC heterostructures: First-principles studies of electronic and magnetic properties *Phys Rev B* **95** 045302

[65] De Parga A-V and Miranda R 2014 Scanning tunneling microscopy (STM) of graphene *In Graphene Woodhead Publishing* 124–55

[66] Coletti C et al. 2013 Revealing the electronic band structure of trilayer graphene on SiC: An angle-resolved photoemission study *Phys Rev B* **88** 155439

[67] Li G, Luican A and Andrei E Y 2009 Scanning tunneling spectroscopy of graphene on graphite *Phys Rev Lett* **102** 176804

6 Essential Properties of Metals/Transition Metals-Adsorbed Graphene Nanoribbons

Ngoc Thanh Thuy Tran, Shih-Yang Lin, and Ming Fa-Lin

CONTENTS

6.1 INTRODUCTION

The adatom-adsorbed graphene nanoribbon (GNR) has been the focus of a number of theoretical [1–11] and experimental [12–18] studies. This chapter will investigate two types of adatoms: metals (Al) and transition metals (Ti, Fe, Co, Ni). After adsorption on GNR surfaces, they are expected to induce more complicated multi-orbital hybridizations in the significant bonds with carbon atoms, especially the $2p_z$ orbital. These systems are very suitable for exploring the dramatic transformations induced by adsorption, e.g., the semiconductor-metal transition and the diverse spin distributions, which have potential applications in many areas [19–22].

For metal-doped systems, they are predicted to have very high conduction electron densities and thus high electrical conductivities, which is promising for anode materials in rechargeable batteries. Aluminum atoms have been identified to play a critical role in the great enhancement of current density in an aluminum-ion battery [23], where the predominant $AlCl_4$ anions are intercalated and de-intercalated between graphite layers during charge and discharge, respectively. As for transition metal adatoms (Ti, Fe, Co, Ni) doped systems, they are predicted to induce metallic band structures with free conduction electrons, in which the spin-split energy bands correspond to the ferromagnetic configuration [24, 25]. However, the geometrical relationships between the structures, the electronic properties, and the magnetic configuration for various adatom coverages have not been thoroughly reported for

DOI: 10.1201/9781003322573-6

GNRs. Orbital hybridizations between adatoms and GNRs, a key point in under-standing the modification of essential properties, still lack systematic investigation. Here, metals (Al) and transition metals (Ti, Fe, Co, Ni) -adsorbed GNRs will be able to create various orbital hybridizations and diversify the fundamental properties.

This study focuses on the geometric, electronic, and magnetic properties of metal/ transition metal-adsorbed GNR-related systems based on our proposed theoretical framework for quasiparticles. The essential properties arising from various types and concentrations of adatom dopings are studied in great detail by means of using the first principles. The principle focuses are the adatom-dependent binding ener-gies, the adatom-carbon lengths, the optimal position, the maximum adatom concen-trations, the free carrier density, the adatom-related valence and conduction bands, the various van Hove singularities in DOSs, the transition-metal-induced magnetic properties, and the significant competitions of the zigzag edge carbons and the metal/ transition metal adatoms in spin configurations of quasiparticles. The distinct chemi-cal bondings are clearly identified under the delicate physical quantities.

6.2   METAL ADSORPTION

The Al-adsorbed GNR remains the planar structure with a non-uniform honeycomb lattice, as revealed in Figure 6.1. This indicates that the Al adsorptions almost do not affect the well-behaved σ bondings of (2s, $2p_x$, $2p_y$) carbon atoms. The optimal posi-tions might correspond to the hollow sites if the adatoms are far away from the bound-aries, i.e., the xy-plane projections are the centers of hexagon lattice in the absence of y-shift (Table 6.1). However, when the Al adatoms are adsorbed near the armchair and zigzag boundaries, the obvious shifts are revealed along the transverse direction toward the edges. The height of Al adatoms are 2.07–2.20Å and 2.00–2.14 Å for arm-chair and zigzag systems, respectively. As to the highest concentrations, the stable structures are associated with the double-side adsorption of four Al adatoms in N_A = 10 armchair system and six Al adatoms in N_Z = 10 zigzag one. This result does not exceed the upper limit of 25% in Al-adsorbed monolayer graphene [26]. There exist the unusual geometric structures under the maximum concentration (Figure 6.1), in which the optimal position is dramatically transferred to the bridge side, clearly illustrating the complex competition/cooperation among the C-C, Al-C, Al-Al, and H-C bondings. This means that the two Al adatoms need to have a sufficiently long distance to achieve the stable structure. After Al adsorption, the similar 1D graphene plane suggests that only the $2p_z$ orbitals of carbons have significant chemical bonding with the half-occupied three kinds of orbitals (3s, $3p_x$, $3p_y$). High-resolution scanning tunneling microscopy (STM) and transmission electron microscopy (TEM) could be utilized to verify these predictions.

GNRs exhibit the special 1D band structures on account of the honeycomb lattice symmetry, finite-size quantum confinement, and edge structure. Pristine AGNRs (N_A = 10) possess a lot of 1D energy bands, as shown in Figure 6.2(a). The occu-pied valence bands are asymmetric to the unoccupied conduction bands around the

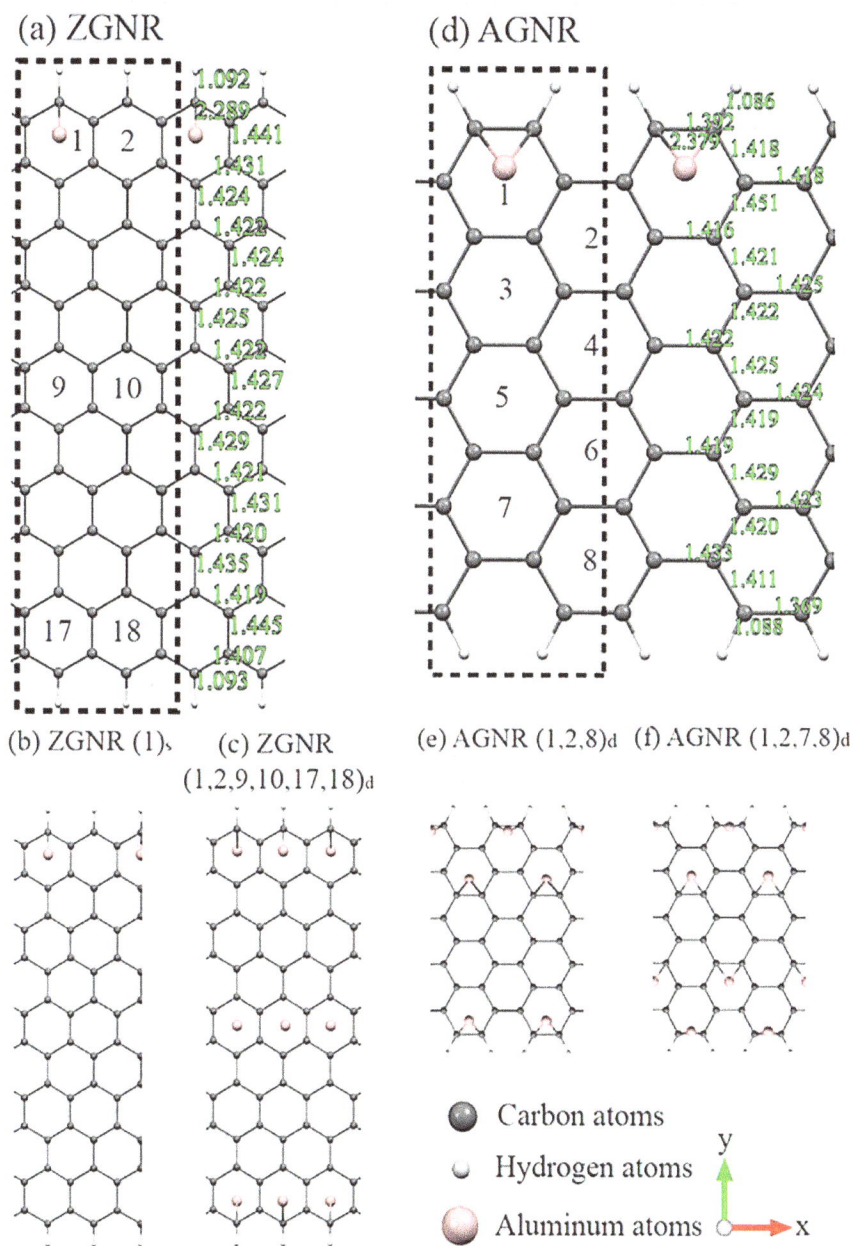

FIGURE 6.1   The optimal geometric structures for the Al-adsorbed $N_Z = 10$ ZGNRs (a)–(c) and $N_A = 10$ AGNRs (d)–(f), in which the adatom positions are marked in numbers.

TABLE 6.1

The Calculated Adatom Height, Adatom y-Shift, Binding Energies, Magnetic Moment/Magnetism for Al-adsorbed N_A = 10 Armchair and N_Z = 10 Zigzag GNRs under Various Distributions and Concentrations

Configurations		Height (Å)	Adatom y-shift (Å)	E_b (eV)	M (μ_B)
AGNR	$Al(5)_s$	2.207	X	−0.639	0/NM
	$Al(2,7)_d$	2.065	X	−1.969	0/NM
	$Al(1,2,7,8)_d$	2.153	0.144	−0.714	0/NM
ZGNR	$Al(1)_s$	2.049	0.172	−1.606	0.56/FM
	$Al(9)_s$	2.135	X	−1.142	0/AFM
	$Al(1,17)_s$	1.999	0.317	−1.789	0/NM
	$Al(1,2,9,10,17,18)_d$	2.142	0.167	−0.706	0/NM

Fermi level (E_F = 0), in which a direct energy gap of 1.1 eV arises from the finite-size confinement effect. Most of the energy dispersions are parabolic bands, while few of them are partially flat ones. All the energy dispersions depend on the wave vector monotonously except for the sub-band anti-crossings. The band-edge states, which occur at k_x = 0, 1, and others related to sub-band anti-crossings, will create van Hove singularities in DOSs (Figure 6.3(a). There are certain important differences between zigzag and armchair systems. Pristine ZGNRs (N_Z = 10), as shown in Figure 6.2(b), have a pair of partially flat at valence and conduction bands nearest to the Fermi level, corresponding to wave functions localized at the zigzag boundaries. Moreover, these energy bands also exhibit the double degeneracy for the spin degree of freedom. The band-edge states, which appear at k_x = 1/2, determine a direct gap of 0.4 eV as a result of the strong competition between quantum confinement and spin configuration. It is worth mentioning that the low-lying electronic states within $\pm$ 2 eV and the deeper ones are, respectively, contributed by the π bonds of parallel $2p_z$ orbitals and the σ bonds of (2s, $2p_x$, $2p_y$) orbitals, as indicated by the orbital-projected DOSs (Figure 6.3(a) & (b)).

The electronic structures exhibit the drastic changes in all GNRs, sensitive to the changes in edge structures, Al-adsorbed position, and concentration but not the single- or double-sided adsorption. In general, the Fermi level is shifted from the center of the energy gap to the conduction bands. Furthermore, the energy bands very close to E_F are mainly determined by carbon atoms, indicating the free electrons due to the distorted π bondings and the charge transfer from C atoms to Al adatoms. With higher Al concentrations, there are more carbon-dominated conduction bands intersecting with the Fermi level, leading to the increasing in 1D linear carrier density. However, the linear relation between conduction electron density and adatom concentration is not valid under the maximum adsorption concentration $(1,2,7,8)_d$ with a tiny indirect-gap Eg = 0.03 eV. For armchair systems, the energy spacing between the first pair of valence and conduction bands is related to $k_x \neq 0$ or k_x = 0, in which the former might depend on whether the Al adatoms are situated at armchair edges (the

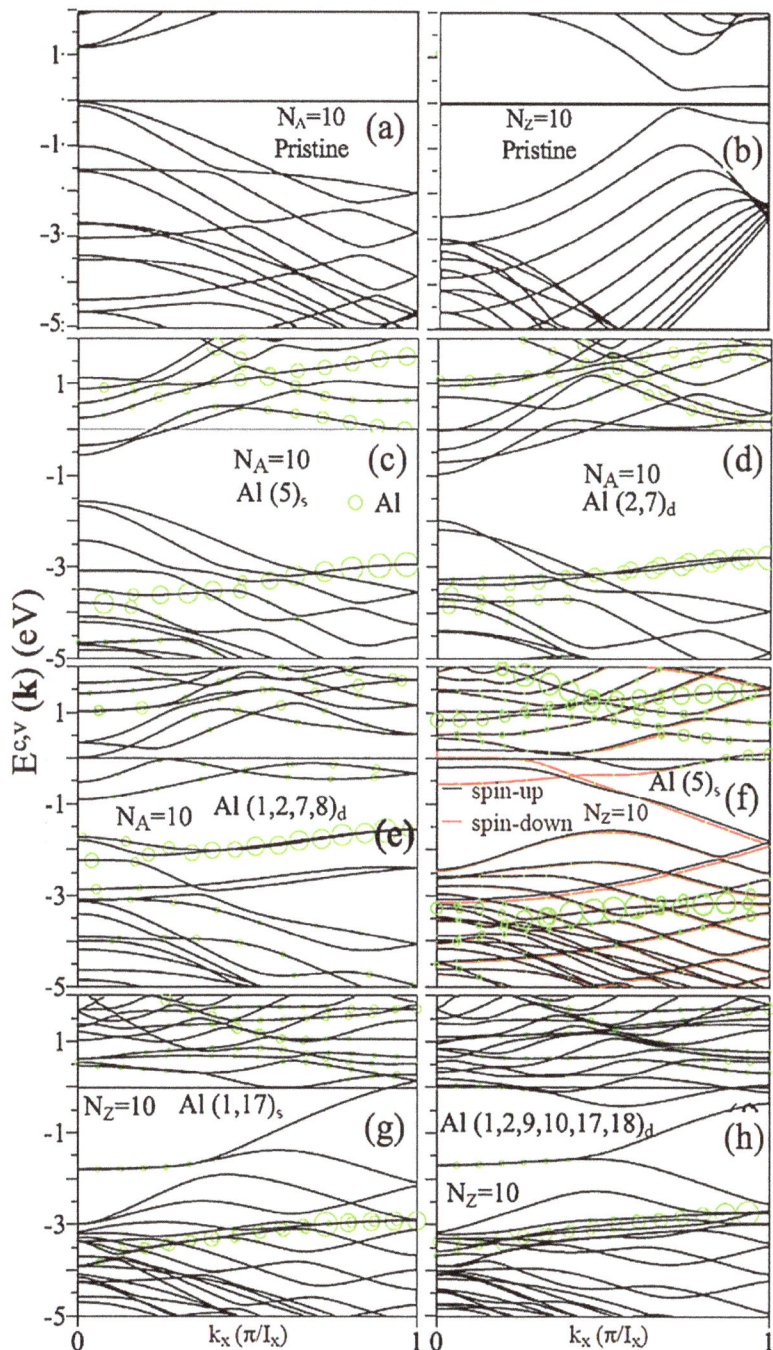

FIGURE 6.2 Band structures of (a) pristine AGNR, (b) pristine ZGRN, and Al-adsorbed GNR under various coverages: (c) (5)s, (d) (2,7)d, (e) (1,2,7,8)d for $N_A = 10$ armchair systems and (f) (1)s, (g) (1,17)s, (h) (1,2,9,10,17,18)s for $N_Z = 10$ zigzag systems.

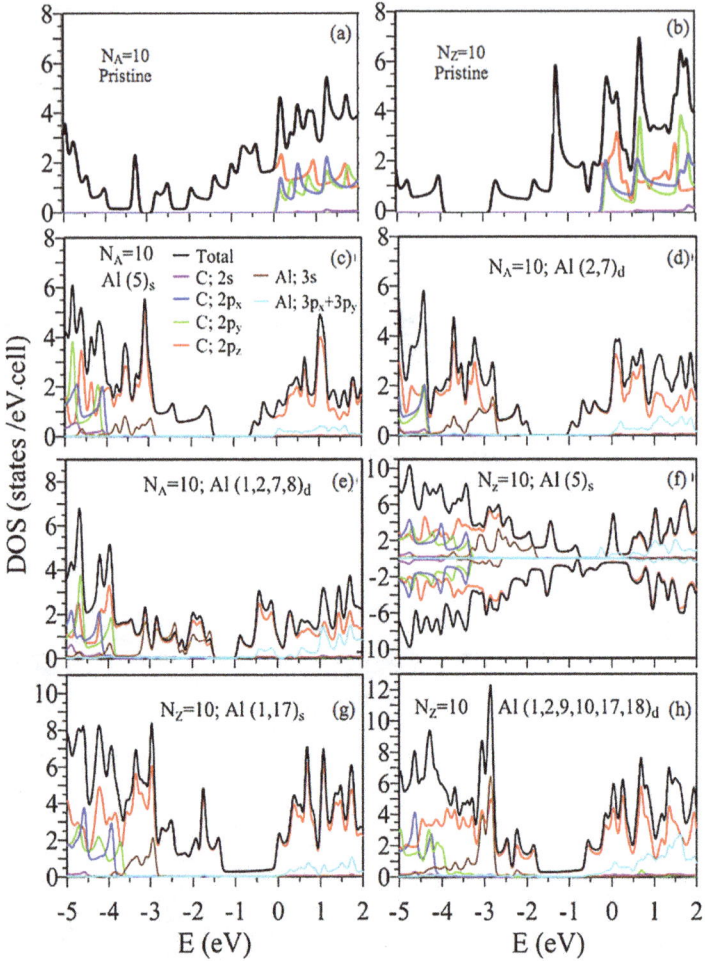

FIGURE 6.3 DOSs of (a) pristine AGNR, (b) pristine ZGRN, and Al-adsorbed GNR under various coverages: (c) (5)s, (d) (2,7)d, (e) (1,2,7,8)d for $N_A = 10$ armchair systems and (f) (1)s, (g) (1,17)s, (h) (1,2,9,10,17,18)s for $N_Z = 10$ zigzag systems.

(1)$_s$ case in Figure 6.2(c)). Specifically, the Al adatoms make important contributions to certain conduction and valence bands in the range of $0 < E_c < 2.0$ eV and -2.5 eV $> E_v > -4.5$ eV, in which the -2.5 eV $> E_v > -3.5$ eV valence states are weakly dispersive and almost doubly degenerate. The Al adsorption could create the extra band-edge states arising from the sub-band hybridizations. For zigzag systems, whether the low-lying energy bands exhibit the spin splitting is very sensitive to the position of Al. When the Al adatom is not located at the nanoribbon center, the FM metallic behavior comes to exist ((5)s case in Figure 6.2(f)). On the other hand, the central single-Al adsorption cannot create the spin splitting and thus preserve the original AFM spin configuration (the (9)s case in Table 6.1). Band structures become more complicated in the increase of adatom concentration. When there are two Al adatoms close to the distinct boundaries, the magnetic configuration is fully absent, e.g., (1,17) s adsorption in Figure 6.2(g).

The Al-adatom adsorptions can create the metallic DOSs in GNRs, while the main characteristics of van Hove singularities are in sharp contrast with those of pristine systems (Figures 6.3(a) and 6.3(b)). The low-energy DOSs are dominated by the C-2p_z orbitals, being consistent with the atom-dominated band structures (Figure 6.2). For armchair systems (Figures 6.3(c)–6.3(e)), there exist the specific zero DOSs below the Fermi level belonging to the initial valence and conduction bands. However, the similar DOSs have a finite value in zigzag systems (Figures 6.3(f)–6.3(h)). Concerning the clear evidences of the multi-orbital hybridizations in Al-C bonds, they show as the merged peaks in the range of $E_v < -1.5$ eV from the Al-3s and the C-2p_z orbitals and those for $E_c > 0.5$ eV due to the Al-(3p_x,3p_y) and C-2p_z orbitals. Energy bandwidths of the 3s and (3p_x + 3p_y) orbitals grows gradually in the increment of Al concentrations, indicating two kinds of orbital hybridizations in Al-Al bonds. The 3p_z orbitals of Al adatoms hardly contribute to the significant chemical adsorption, since each adatom only has three occupied orbitals in the outermost ones, in which it might possess two states in the 3s orbitals and one state in the (3p_x, 3p_y) orbitals.

All armchair Al-chemisorptions belong to non-magnetic configuration (Table 6.1). However, the Al-adsorbed zigzag systems exhibit the diverse magnetic configurations (Figure 6.4). A single Al close to the upper zigzag boundary will partially

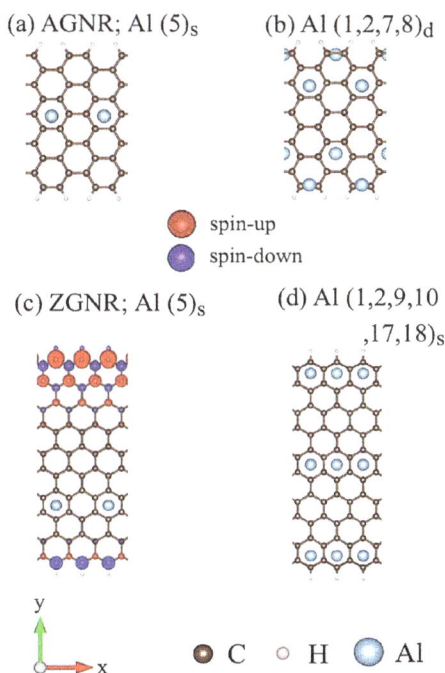

FIGURE 6.4 The spin-density distributions for Al-adsorbed armchair and zigzag GNR systems with (a) (5)s, (b) (1,2,7,8)d, (c) (5)s, (d) (1,2,9,10,17,18)s.

destroy the spin configuration due to the zigzag edge carbons, depending on the adatom positions. The lower zigzag edge preserves the spin-up-dominated configuration almost identical to the original case. However, the spin-down arrangement near the upper edge is fully annihilated, and the spin-up distribution might be drastically changed from edge to center. On the other hand, central Al adatom adsorptions hardly affect the original AFM spin distribution. The magnetic properties thoroughly vanish under symmetry Al adatoms across the zigzag edges (Figure 6.4(d)). These features of magnetic properties clearly show the very strong competition/cooperation between the multi-orbital hybridizations of Al-C bonds and the spin states due to zigzag edge carbons.

The group-V elements have three valence electrons in the outermost orbitals, so they can form the complex multi-orbital hybridizations after chemisorptions on GNR surfaces. The bonding strength of C-C, Al-C, and Al-Al bonds are revealed in Figures 6.5(a)–6.5(d). After the Al adatom adsorptions, all the C-C bonds possess the strong covalent σ bonds (black rectangle) and the somewhat weaker π bonds simultaneously (red rectangle). The former almost keep the same; furthermore, the π bonding also belongs to the extended state in a 1D system except that it is seriously distorted under the maximum-concentration case (Figure 6.5(f)). These are responsible for the carbon-dominated low-lying energy bands with the metallic or semiconducting behavior. In general, the 3s-orbital electrons are redistributed between Al and the six nearest C atoms, revealing a significant hybridization with the $2p_z$ orbitals. The multi-orbital hybridizations of 3s and $(3p_x, 3p_y)$ are, respectively, related to the Al-dominated valence and conduction bands (Figure 6.2).

6.3   TRANSITION-METALS ADSORPTION

Similar to Al adatoms, transition metals Ti/Fe/Co/Ni are also preferred to be adsorbed at the hollow-site (Figure 6.6), which is consistent with previous studies [1, 27, 28]. However, there are no shifts relative to the xy-projection centers under all the adsorption configurations. This is in sharp contrast with the significant shifts of the Al metal adsorptions. The transition-metal-adatom chemisorptions on 1D GNRs could create the optimal geometric structures similar to 2D graphene systems except for the non-uniform band lengths. The Ti heights keep in the short range of ~1.50 + –2.10 +, leading to the largest binding energies (−2.5 eV to −3.5 eV) among the absorbed adatoms (Table 6.2). This means it is relatively easy to induce the Ti chemisorptions, compared with Al, Fe, Co, and Ni adatoms. The main features of optimal structures strongly suggest that the $2p_z$ and $(2s, 2p_x, 2p_y)$ orbitals of carbon atoms, respectively, make important and minor contributions to the adatom-C bonds. Specifically, the buckling structures might be revealed for the highest-concentration adsorption in armchair/zigzag systems.

All the Ti/Fe/Co/Ni-adsorbed GNRs belong to the unusual metals, as clearly indicated in Figure 6.7. There are a lot of energy bands crossing the Fermi level, in which they mainly come from both adatoms and carbon atoms. Furthermore, the adatom-dominated energy bands might be thoroughly occupied or unoccupied. For armchair systems, all of them exhibit the spin-split energy bands, especially for those near Fermi level. Apparently, they correspond to the FM spin configurations, and the net

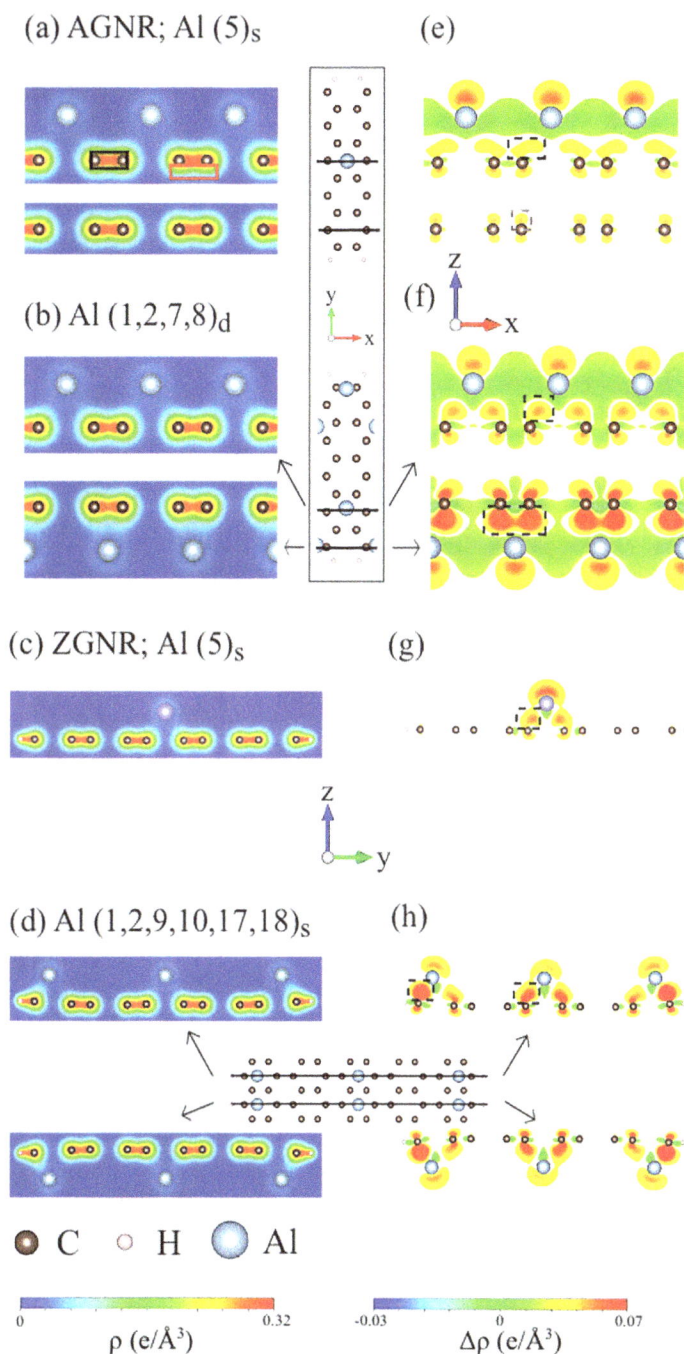

FIGURE 6.5 The spatial charge distribution and charge difference for Al-adsorbed armchair and zigzag GNR systems with (5)s, (1,2,7,8)d, (1)s, and (1,2,9,10,17,18)s, as shown in (a)–(d) and (e)–(h), respectively.

(a) ZGNR

(d) AGNR

(b) ZGNR (1)ₛ   (c) ZGNR; (1,2,5,6, 9,10,13,14,17,18)d

(e) AGNR (1)ₛ   (f) AGNR (1,2,3,4,5,6,7,8)d

⬤ Carbon atoms

◯ Hydrogen atoms

⬤ Titanium atoms

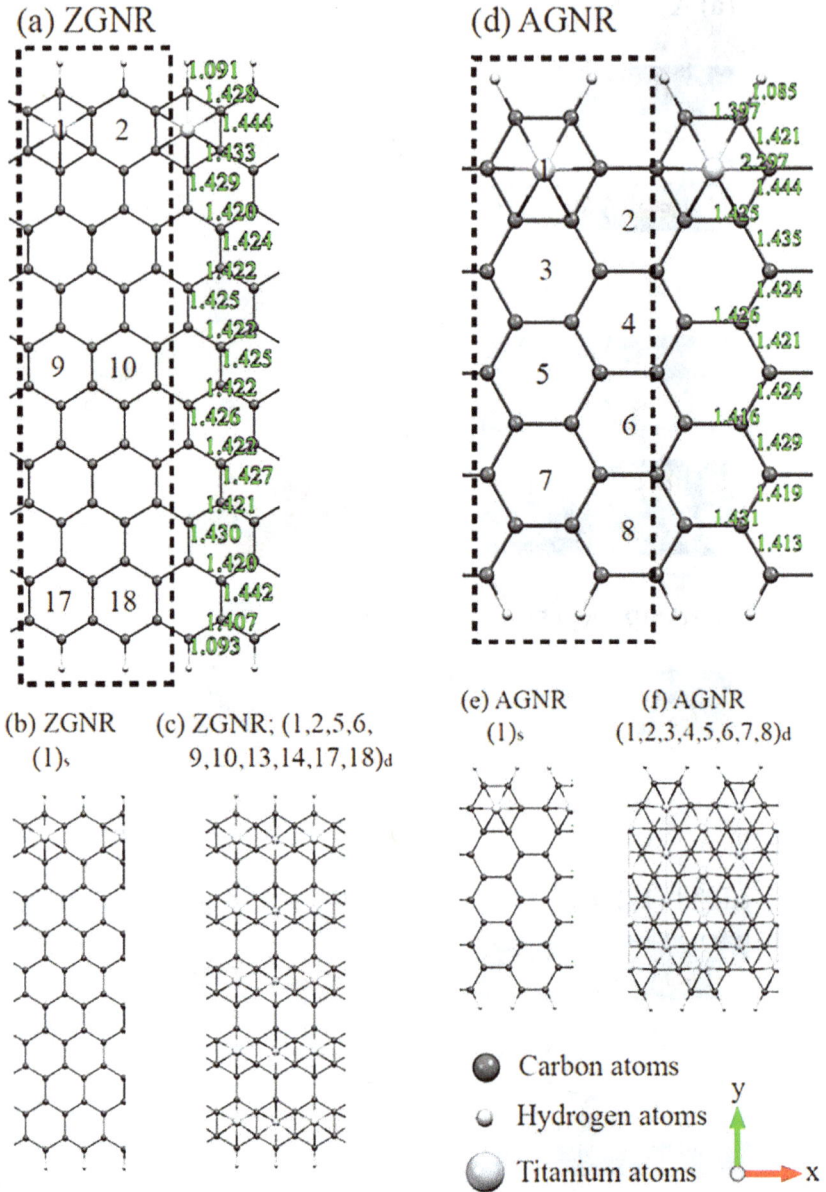

FIGURE 6.6 The optimal geometric structures for the Ti-adsorbed $N_Z = 10$ ZGNRs (a)–(c) and $N_A = 10$ AGNRs (d)–(f), in which the adatom positions are marked in numbers.

magnetic moments due to the Ti/Fe adsorbates are sensitive to the adatom distribution and concentration (Table 6.2).

The increasing adatom concentration, respectively, makes major and minor contributions to these two different ranges of energy bands. The Ti/Fe adatoms hardly

TABLE 6.2

The Calculated Adatom Height, Adatom y-Shift, Binding Energies, Magnetic Moment/Magnetism for Ti/Fe/Co/Ni-Adsorbed N_A = 10 Armchair and N_Z = 10 Zigzag GNRs under Various Distributions and Concentrations

Configurations		Height (Å)	Adatom y-shift (Å)	E_b (eV)	M (μ_B)
AGNR	Ti (1)$_s$	1.777	X	−2.828	1.28/FM
	Ti(1–8)$_d$	1.822	X	−3.454	5.93/FM
	Fe(1)$_s$	1.606	0.018	−1.125	2.37/FM
	Fe(2,7)$_d$	1.596	0.003	−2.529	4.78/FM
	Co(1) s	1.566	0.026	−1.291	1.39/FM
	Ni(1) s	1.602	0.013	−1.311	0/NM
ZGNR	Ti (1)$_s$	1.711	X	−2.774	1.46/FM
	Ti (9)$_s$	1.629	X	−1.991	0/AFM
	Ti (1,6,14,18)$_d$	1607	X	−2.354	0/AFM
	Ti (1,2,5,6,9,10,13,14,17,18)$_d$	1.665	X	−2.606	0/AFM
	Fe(1)$_s$	1.659	0.005	−1.227	2.4/FM

contribute to the deep valence states, indicating the weak chemical hybridizations from the (2s, $2p_x$, $2p_y$) orbitals in the adatom-C bonds. As for Co adatoms, they dominate the low-lying energy bands as well as Fe adatoms do, presenting similar spin-split spacing at low energy (Figure 6.7(d)). However, the Ni adatoms possess no magnetic moments, resulting in the absence of band split in low energy (Figure 6.7(e)). It should be noticed that the Ni adatoms also contribute to the energy band in the low energy range.

The Ti/Fe/Co/Ni chemisorptions could induce the diversified band structures because of the edge structures. There are certain important differences between the zigzag and armchair systems. As for the former, the partially flat bands across the Fermi level only survive under the single-adatom central adsorption (Figure 6.7(f)). However, most chemisorption cases, being shown in Figures 6.7(h)–6.7(j), lead to the dramatic changes in the low-lying energy bands because of the significant adatom-edge-C bonding; that is, the edge-C-dominated partially flat energy dispersions are thoroughly absent. The zigzag systems might be FM, AFM, and NM metals, in which the latter two need to be further examined from the spatial spin distributions. Apparently, this is purely due to the strong competition of edge carbon atoms and adatoms, similar to the Al case. Compared to others, Ni-adsorbed GNRs exhibit smaller spin split in low-lying bands, owing to the weak magnetic moments (Figure 6.7(j)).

The orbital- and spin-projected DOSs in Ti/Fe/Co/Ni-adsorbed GNRs, as clearly shown in Figure 6.8, could provide the very complicated van Hove singularities and thus identify the significant multi-orbital hybridizations in Ti-C and Ti-Ti bonds. There is an obvious DOS at the Fermi level, directly indicating the creation of the high free carrier density. The spin-split DOSs are revealed in most of absorption cases except for the symmetric adatom distributions in zigzag systems (Figure 6.8(f)). The

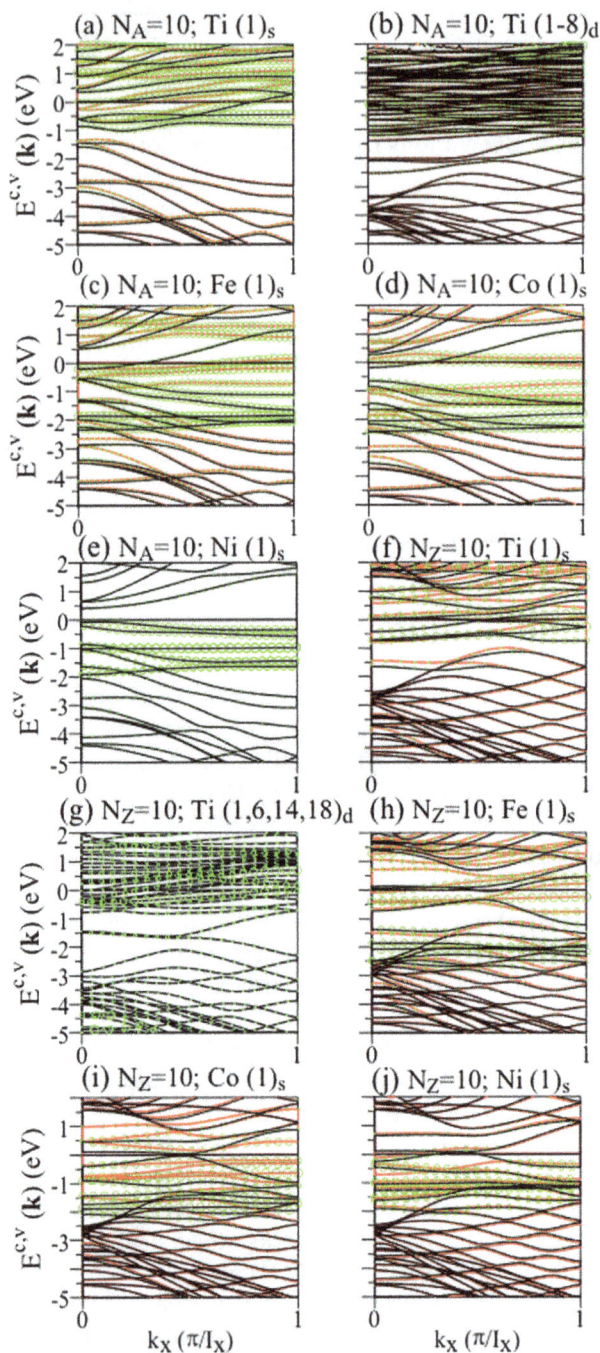

FIGURE 6.7 Band structures of Ti/Fe/Co/Ni-adsorbed GNR under various coverages: (a) Ti(1)s, (b) Ti(1–8)d, (c) Fe(1)s, (d) Co(1)s, (e) Ni(1)s for N_A = 10 armchair systems, and (f) Ti(9) s, (g) Ti(1,6,14,18)d, (h) Fe(1)s, (i) Co(1)s, (j) Ni(1)s for N_Z = 10 zigzag systems.

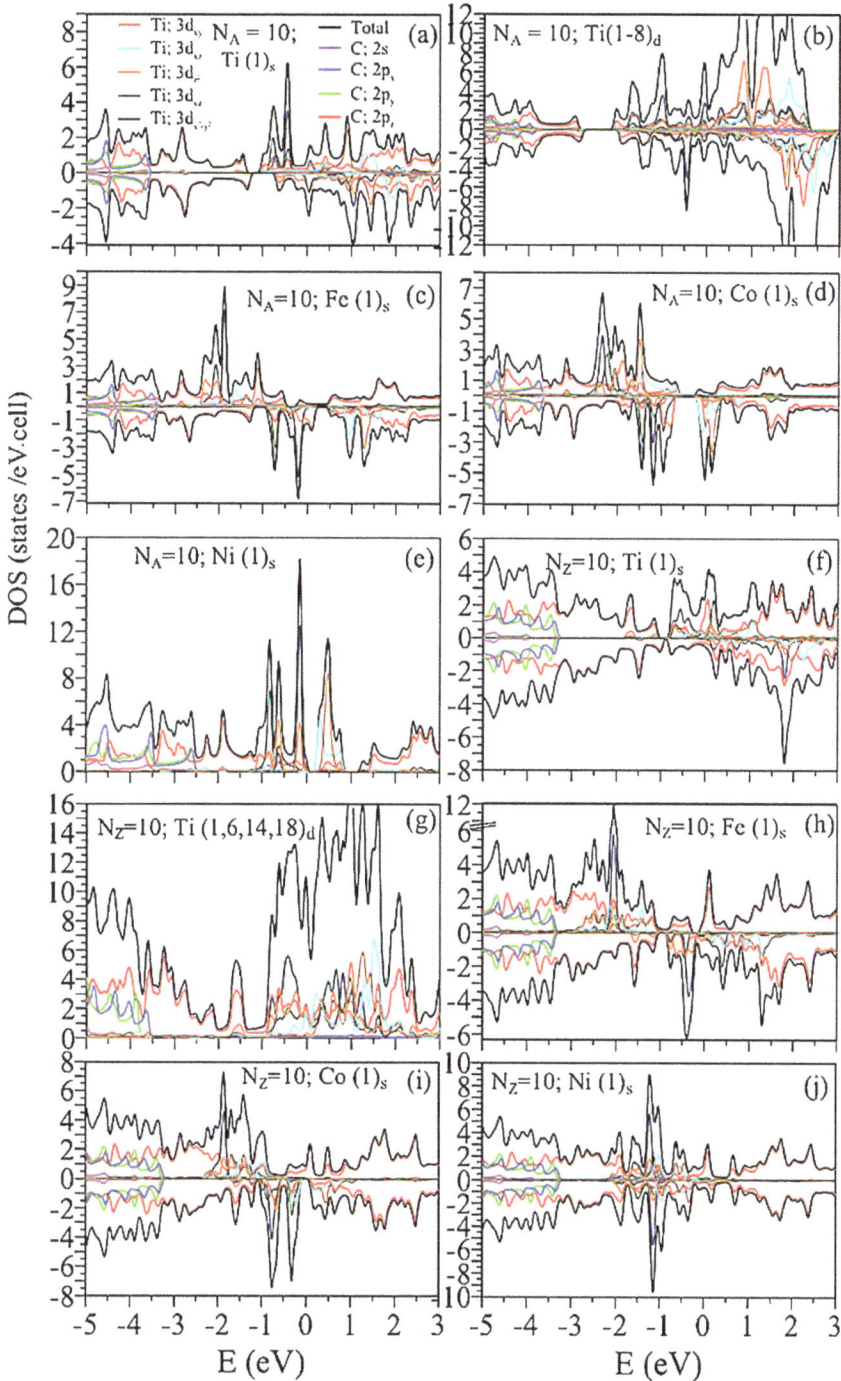

FIGURE 6.8 DOSs of Ti/Fe/Co/Ni-adsorbed GNR under various coverages: (a) Ti(1)s, (b) Ti(1–8)d, (c) Fe(1)s, (d) Co(1)s, (e) Ni(1)s for $N_A = 10$ armchair systems, and (f) Ti(9)s, (g) Ti(1,6,14,18)d, (h) Fe(1)s, (i) Co(1)s, (j) Ni(1)s for $N_Z = 10$ zigzag systems.

results further illustrate the adatom-induced spin states and their strong competitions with edge-carbon magnetic arrangement by the adatom-C chemical bonding. As to various orbital contributions to DOSs, 2s, $2p_x$, and $2p_y$ of carbons appear at E < −3.5 eV even for the highest adatom concentration ($(1-8)_d$ in Figure 6.8(b)). Such evidence strongly suggests their interactions with the outer five 3d orbitals of Ti adatoms should be weak. However, the $2p_z$ orbitals experience rather high hybridizations with the $3d_{xy}$, $3d_{xz}$, $3d_{yz}$, $3d_{z2}$, and $3d_{x2\text{-}y2}$, since the special structures of DOSs are seriously merged together. The Ti-Ti bonding is displayed in the enhancement of the metallic 3d-bandwidth. Similar behavior could be found in Fe- and Co-chemisorptions in both armchair and zigzag GNRs. However, the Ni-adsorbed armchair ribbons have no spin-split states (Figure 6.8(e)), and the corresponding zigzag systems possess rather symmetric peaks on spin-up and spin-down states (Figure 6.8(j)).

There exist the diverse various FM and AFM spin distributions. Any armchair systems have the distinct FM configurations, being sensitive to the single- and double-side absorptions and concentrations. The single-adatom chemisorptions, e.g., $Ti(1)_s$ (Figure 6.9(a)), could create the comparable magnetic moments. In general, the strength of magnetic response is proportional to the Ti concentration under the specific double-side adsorptions. For example, the spin-related volumes grow with the increasing concentrations, and so do the net magnetic moments. The unusual competitions between zigzag carbons and Ti adatoms could lead to the diverse spin distributions, being strongly associated with the partial or full spin suppressions of the former and the latter. The FM and AFM configurations, respectively, correspond to the asymmetric ($(1)_s$ in Figure 6.9(a)) and symmetric ($(9)_s$ in Figure 6.9(c)) adatom distributions in zigzag systems. Furthermore, there exists the spin-down dominant distribution on the other boundary. The absence of magnetism of Ti adatoms might arise from the symmetric chemical and magnetic environments. If Ti adatoms are evenly distributed along zigzag edges, the AFM configurations purely due to the adsorbates are created, and the spin states of the edge carbons are destroyed (Figure 6.9(d)). The strong magnetic moments of Ti- as well as Fe-adsorbed systems (Table 6.1) could lead to potential application in spintronic devices.

The transition metal adatoms have five kinds of d orbitals, so that the multiorbital hybridizations will be clearly shown on the xz-, yz-, and xy-planes. The chemical bonding between Ti and C is obvious and significant even for the single-adatom cases, regardless of the positions, such as $(1)_s$ for $N_A = 10$ and $N_Z = 10$, respectively (Figures 6.10(a) and 6.10(c)). Its strength grows with the increasing adatom concentration e.g., $(1-8)_d$ and $(1, 6, 14, 18)_d$ (Figures 6.10(b and 6.10(d)). The π bonding of carbon $2p_z$ orbitals is distorted after chemisorptions, while it is still extended along the longitudinal and transverse directions. The carbon-dominated conduction bands crossing the Fermi level also make certain important contributions to the high free carriers. For any absorption cases, the large charge transfers exist between Ti guest adatoms and carbon host atoms on the xz- and yz-planes, as clearly revealed in Figures 6.10(e)–6.10(h). It can only identify the significant orbital hybridizations of ($3d_{xz}$, $3d_{yz}$, $3d_{xy}$, $3d_{z2}$, $3d_{x2\text{-}y2}$) five orbitals and $2p_z$ orbital. As to the Ti-Ti chemical bonding, the charge distributions are easily observed only on the yz-plane at high concentration of armchair systems, e.g., $(1-8)_d$. The charge differences become obvious for any chemisorptions on the xz-plane and/or the

(a) AGNR; Ti $(1)_s$ (b) AGNR; Ti $(1\text{-}8)_d$

(c) ZGNR; Ti $(9)_s$ (d) ZGNR; $(1,6,14,17)_d$

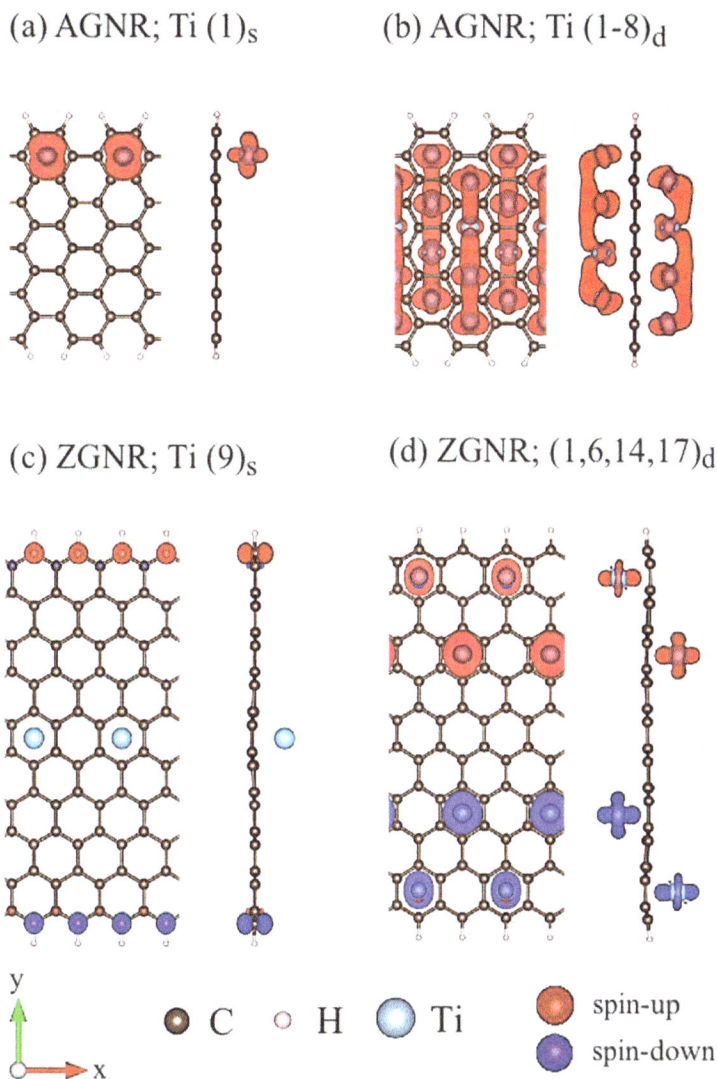

y

x

● C   ○ H   ◯ Ti   ● spin-up   ● spin-down

FIGURE 6.9   The spin-density distributions for Ti-adsorbed armchair and zigzag GNR systems with (a) (1)s, (b) (1–8)d, (c) (9)s, (d) (1,6,14,17)d.

yz-plane; furthermore, there are charge extensions and even overlaps on the xy-plane at the optimal heights Ti adatoms. The observable five-orbital hybridizations in Ti-Ti bonds are responsible for the low-lying Ti-dominated energy bands, being also one of the critical factors in the creation of the conduction electron density/ the metallic behavior.

The main features of band structures and DOSs could be examined by angle-resolved photoemission spectroscopy (ARPES) and scanning tunneling spectroscopy (STS) measurements. ARPES measurements are very successful in identifying the

Charge distribution

(a) AGRN; Ti (1)s

Charge difference

(e)

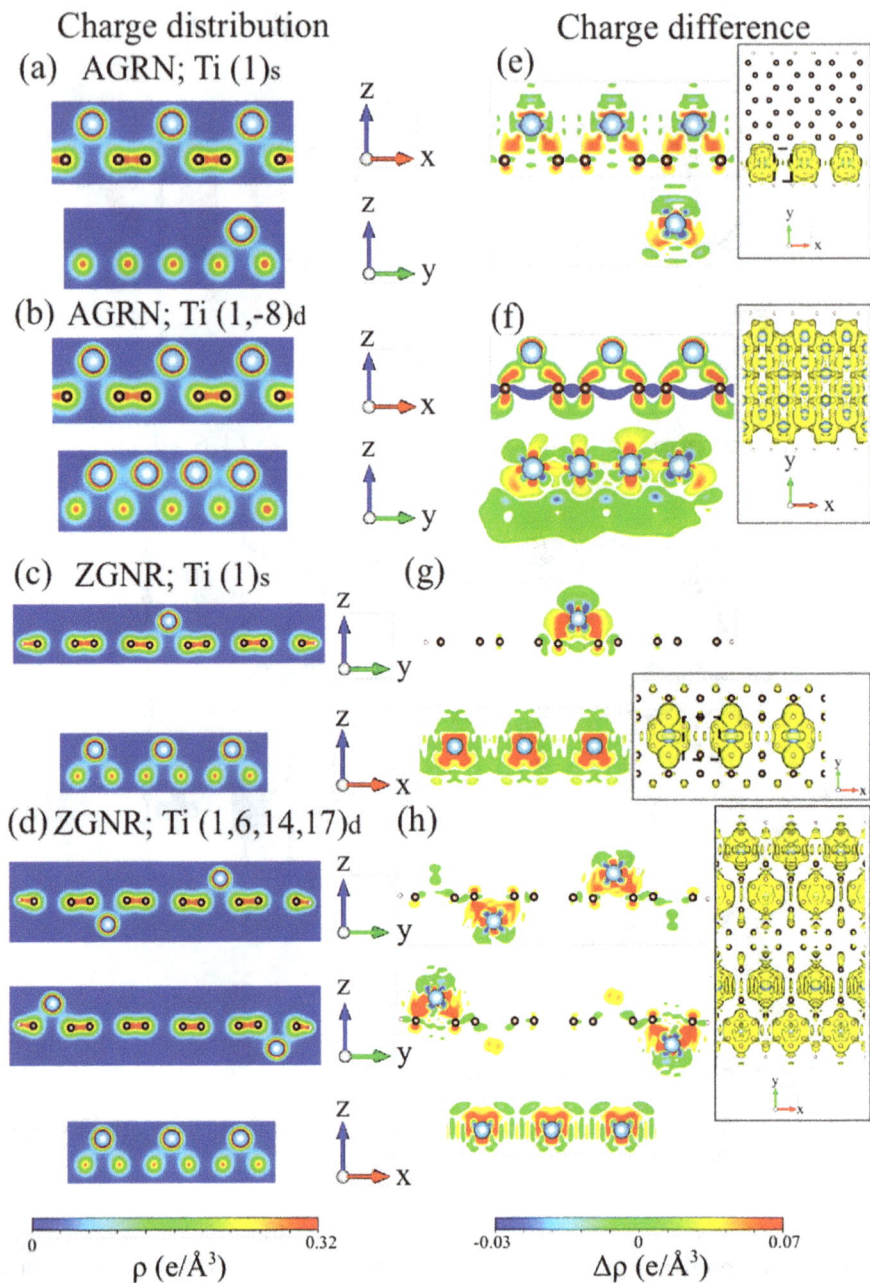

(b) AGRN; Ti (1,-8)d

(f)

(c) ZGNR; Ti (1)s

(g)

(d) ZGNR; Ti (1,6,14,17)d

(h)

ρ (e/Å^3)

$\Delta\rho$ (e/Å^3)

FIGURE 6.10 The spatial charge distribution and charge difference for Ti-adsorbed armchair and zigzag GNR systems with (1)s, (1–8)d, (1)s, and (1,6,14,17)d, as shown in (a)–(d) and (e)–(h), respectively.

diverse band structures of graphene-based systems, e.g., the gapless valence Dirac cone in monolayer graphene [29], the 1D parabolic valence bands near the Γ point accompanied with band gaps and distinct energy spacings of armchair GNRs in the presence/absence of hydrogen passivation [7, 30]. Furthermore, ARPES has been used to verify the effects arising from doping. The greatly modified band structures of the Ti-absorbed graphene [31] or the redshift (n-type doping) of 1–1.5 eV in the π bands has been confirmed by using the high-resolution ARPES measurements [32]. On the other hand, STS measurements of the dI/dV spectra has served to confirm the width- and edge-dominated energy gaps and the asymmetric peaks of 1D parabolic bands of GNRs [33–35]. Further ARPES and STS examinations are desirable for the aforementioned main structures of the electronic properties of metal/transition metals-adsorbed GNR systems.

6.4 CONCLUSIONS

The metal adatoms (Al) and transition metals (Ti, Fe, Co, Ni) adsorbed GNR systems have been investigated by means of the first-principles calculations. They clearly display the important differences in the essential properties. The Al and Ti/Fe/Co/Ni adatoms exhibit the hollow-site optimal positions, while the former one might have the y-direction shifts, especially for the non-symmetric distributions. The adatom chemisorptions are relatively easily observed in the Ti chemisorptions with the largest binding energies. Energy bands, which, crossing the Fermi level, mainly arise from carbon atoms or metal atoms for Al-adsorbed systems or Ti/Fe/Co/Ni ones, respectively. This feature is also supported by the spatial charge distributions and atom-projected DOSs. At higher adatom concentrations, the significant $3p_x$–$3p_x$ and $3p_y$–$3p_y$ hybridizations in Al-Al bonds are indicated by the enhanced Al-dependent conduction bands/DOSs with the wider energy widths. The multi-orbital hybridizations between Ti/Fe/Co/Ni-C bonds are very complicated, in which five orbitals of adatom, $(d_{xy}, d_{xz}, d_{yz}, d_z^2)$ and $d_{x^2-y^2}$ have the strong interactions with $2p_z$ orbitals of carbons. Such orbitals also take part in the adatom-adatom chemical bondings. As to the magnetic properties, the Al adatoms do not create the spin distributions, but their interactions with the zigzag carbon atoms can destroy the latter's ones and thus create the FM or NM configurations. Armchair and zigzag Al-adsorbed GNRs, respectively, belong to the NM and AFM/FM/NM metals. On the other side, most of Ti/Fe/Co-adsorbed asymmetric systems exhibit the FM configurations under the induced spin states themselves and the greatly reduced edge-carbon magnetic moments. The AFM spin arrangements is strongly associated with the edge carbons and adsorbates.

ACKNOWLEDGMENTS

This work was financially supported by the Hierarchical Green-Energy Materials (Hi-GEM) Research Center and the Ministry of Science and Technology (MOST 108–2112-M-006–022-MY3) in Taiwan.

REFERENCES

[1] Rigo V A, Martins T B, da Silva A J R, Fazzio A and Miwa R H 2009 Electronic, structural, and transport properties of Ni-doped graphene nanoribbons *Phys Rev B* **79**

[2] Yu S S, Zheng W T and Jiang Q 2010 Electronic properties of nitrogen-atom-adsorbed graphene nanoribbons with armchair edges *Ieee T Nanotechnol* **9** 243–7

[3] Srivastava P, Dhar S and Jaiswal N K 2015 Potential spin-polarized transport in gold-doped armchair graphene nanoribbons *Phys Lett A* **379** 835–42

[4] Martins T B, Miwa R H, da Silva A J R and Fazzio A 2007 Electronic and transport properties of boron-doped graphene nanoribbons *Phys Rev Lett* **98**

[5] Zhang X J, Zhang D, Xie F, Zheng X L, Wang H Y and Long M Q 2017 First-principles study on the magnetic and electronic properties of Al or P doped armchair silicene nanoribbons *Phys Lett A* **381** 2097–102

[6] Yang B, Li D B, Qi L, Li T B and Yang P 2019 Thermal properties of triangle nitrogen-doped graphene nanoribbons *Phys Lett A* **383** 1306–11

[7] Senkovskiy B V, Fedorov A V, Haberer D, Farjam M, Simonov K A, Preobrajenski A B, Martensson N, Atodiresei N, Caciuc V, Blugel S, Rosch A, Verbitskiy N I, Hell M, Evtushinsky D V, German R, Marangoni T, van Loosdrecht P H M, Fischer F R and Gruneis A 2017 Semiconductor-to-metal transition and quasiparticle renormalization in doped graphene nanoribbons *Adv Electron Mater* **3**

[8] Nguyen D K, Lin Y T, Lin S Y, Chiu Y H, Tran N T T and Fa-Lin M 2017 Fluorination-enriched electronic and magnetic properties in graphene nanoribbons *Phys Chem Chem Phys* **19** 20667–76

[9] Kaur L, Mahendia S, Saini S and Srivastava A 2021 Arsenic sensing using Al/Fe doped armchair graphene nanoribbons: Theoretical investigations *J Phys Chem Solids* **152**

[10] Abdelsalam H, Saroka V A, Atta M M, Osman W and Zhang Q F 2021 Tunable electro-optical properties of doped chiral graphene nanoribbons *Chem Phys* **544**

[11] Nguyen D K, Tran N T T, Nguyen T T and Lin M F 2018 Diverse electronic and magnetic properties of chlorination-related graphene nanoribbons *Sci Rep-Uk* **8**

[12] Johnson J L, Behnam A, Pearton S J and Ural A 2010 Hydrogen sensing using Pd-functionalized multi-layer graphene nanoribbon networks *Adv Mater* **22** 4877–80

[13] Lin J, Peng Z W, Xiang C S, Ruan G D, Yan Z, Natelson D and Tour J M 2013 Graphene nanoribbon and nanostructured SnO2 composite anodes for lithium ion batteries *Acs Nano* **7** 6001–6

[14] Pawlak R, Liu X S, Ninova S, D'Astolfo P, Drechsel C, Sangtarash S, Haner R, Decurtins S, Sadeghi H, Lambert C J, Aschauer U, Liu S X and Meyer E 2020 Bottom-up synthesis of nitrogen-doped porous graphene nanoribbons *J Am Chem Soc* **142** 12568–73

[15] Wang X Y, Urgel J I, Barin G B, Eimre K, Di Giovannantonio M, Milani A, Tommasini M, Pignedoli C A, Ruffieux P, Feng X L, Fasel R, Mullen K and Narite A 2018 Bottom-up synthesis of heteroatom-doped chiral graphene nanoribbons *J Am Chem Soc* **140** 9104–7

[16] Cortizo-Lacalle D, Mora-Fuentes J P, Strutynski K, Saeki A, Melle-Franco M and Mateo-Alonso A 2018 Monodisperse N-doped graphene nanoribbons reaching 7.7 nanometers in length *Angew Chem Int Edit* **57** 703–8

[17] Fu Y B, Yang H, Gao Y X, Huang L, Berger R, Liu J Z, Lu H L, Cheng Z H, Du S X, Gao H J and Feng X L 2020 On-surface synthesis of NBN-doped zigzag-edged graphene nanoribbons *Angew Chem Int Edit* **59** 8873–9

[18] Zhang Y F, Zhang Y, Li G, Lu J C, Que Y D, Chen H, Berger R, Feng X L, Mullen K, Lin X, Zhang Y Y, Du S X, Pantelides S T and Gao H J 2017 Sulfur-doped graphene nanoribbons with a sequence of distinct band gaps *Nano Res* **10** 3377–84

[19] Agrawal S, Srivastava A and Kaushal G 2021 Electron transport in boron function-alised armchair graphene nanoribbons: Potential interconnects *Solid State Commun* **327**

[20] Gopalsamy K, Balamurugan J, Thanh T D, Kim N H and Lee J H 2017 Fabrication of nitrogen and sulfur co-doped graphene nanoribbons with porous architecture for high-performance supercapacitors *Chem Eng J* **312** 180–90

[21] Zhang R Z, Zhang C M, Zheng F Q, Li X K, Sun C L and Chen W 2018 Nitrogen and sulfur co-doped graphene nanoribbons: A novel metal-free catalyst for high performance electrochemical detection of 2, 4, 6-trinitrotoluene (TNT) *Carbon* **126** 328–37

[22] Lee H W, Moon H S, Hur J, Kim I T, Park M S, Yun J M, Kim K H and Lee S G 2017 Mechanism of sodium adsorption on N-doped graphene nanoribbons for sodium ion battery applications: A density functional theory approach *Carbon* **119** 492–501

[23] Lin M C, Gong M, Lu B G, Wu Y P, Wang D Y, Guan M Y, Angell M, Chen C X, Yang J, Hwang B J and Dai H J 2015 An ultrafast rechargeable aluminium-ion battery *Nature* **520** 324–8

[24] Wang Z Y, Xiao J R and Li M 2013 Adsorption of transition metal atoms (Co and Ni) on zigzag graphene nanoribbon *Appl Phys a-Mater* **110** 235–9

[25] Sevincli H, Topsakal M, Durgun E and Ciraci S 2008 Electronic and magnetic properties of 3d transition-metal atom adsorbed graphene and graphene nanoribbons *Phys Rev B* **77**

[26] Lin S Y, Lin Y T, Tran N T T, Su W P and Lin M F 2017 Feature-rich electronic properties of aluminum-adsorbed graphenes *Carbon* **120** 209–18

[27] Lebon A, Carrete J, Longo R C, Vega A and Gallego L J 2013 Molecular hydrogen uptake by zigzag graphene nanoribbons doped with early 3d transition-metal atoms *Int J Hydrogen Energ* **38** 8872–80

[28] Cocchi C, Prezzi D, Calzolari A and Molinari E 2010 Spin-transport selectivity upon Co adsorption on antiferromagnetic graphene nanoribbons *J Chem Phys* **133**

[29] Ohta T, Bostwick A, McChesney J L, Seyller T, Horn K and Rotenberg E 2007 Interlayer interaction and electronic screening in multilayer graphene investigated with angle-resolved photoemission spectroscopy *Phys Rev Lett* **98**

[30] Ruffieux P, Cai J M, Plumb N C, Patthey L, Prezzi D, Ferretti A, Molinari E, Feng X L, Mullen K, Pignedoli C A and Fasel R 2012 Electronic structure of atomically precise graphene nanoribbons *Acs Nano* **6** 6930–5

[31] Chen J W, Huang H C, Convertino D, Coletti C, Chang L Y, Shiu H W, Cheng C M, Lin M F, Heun S, Chien F S S, Chen Y C, Chen C H and Wu C L 2016 Efficient n-type doping in epitaxial graphene through strong lateral orbital hybridization of Ti adsorbate *Carbon* **109** 300–5

[32] Virojanadara C, Watcharinyanon S, Zakharov A A and Johansson L I 2010 Epitaxial graphene on 6H-SiC and Li intercalation *Phys Rev B* **82**

[33] Sode H, Talirz L, Groning O, Pignedoli C A, Berger R, Feng X L, Mullen K, Fasel R and Ruffieux P 2015 Electronic band dispersion of graphene nanoribbons via Fourier-transformed scanning tunneling spectroscopy *Phys Rev B* **91**

[34] Chen Y C, de Oteyza D G, Pedramrazi Z, Chen C, Fischer F R and Crommie M F 2013 Tuning the band gap of graphene nanoribbons synthesized from molecular precursors *Acs Nano* **7** 6123–8

[35] Huang H, Wei D C, Sun J T, Wong S L, Feng Y P, Castro Neto A H and Wee A T S 2012 Spatially resolved electronic structures of atomically precise armchair graphene nanoribbons *Sci Rep-Uk* **2**

7 Essential Electronic Properties of Zigzag Carbon and Silicon Nanotubes

Hsin-Yi Liu and Ming-Fa Lin

CONTENTS

7.1 INTRODUCTION

An annular structure with respect to a planar structure material is more interesting and fascinating and arouses every one's attention. Carbon nanotubes (CNTs) were found by an arc-discharge evaporation method in 1991 [1]. Soon after, the experimental single-wall carbon nanotubes were discovered in 1993 [2], but the yield and purity were not very good. Those carbon nanotubes are usually found in large concentrations of carbon-based materials, and they contain a variety of radius and chirality distributions. Therefore, there are plenty of studies committed to research on the synthesis of carbon nanotubes [1, 3–8]. Relatively efficient methods of synthesis for single-wall carbon nanotubes are laser vaporization [4, 5] and carbon arc synthesis [6–8]. The tubular structure of carbon nanotubes has aroused a lot of interest due to their specific properties. They show superior electrical conductivities [9, 10] and have excellent thermal conductivity [11–14] and mechanical strength [15–18] due to the nanostructure and strength of the bonding between each carbon atom. Normally, carbon nanotubes often refer to single-wall carbon nanotubes (sCNTs) of various diameters in the range of nanometers. A single-wall carbon nanotube can be regarded as a quasi 1-D tubular structure, which is set up by honeycomb lattices; it presents either semiconducting or metallic behavior dependent on its diameter and chiral vectors. Silicon atoms belong to the group IV of the periodic table and have a similar electronic configuration to carbon atoms. Silicon nanotubes present a hollow

DOI: 10.1201/9781003322573-7

tube composed of silicon atoms. The geometry of silicon nanotubes (SiNTs) leads, due to the buckling angle, to deformations compared to carbon nanotubes. They present gear-like structures at cross-section view.

Silicon nanotubes (SiNTs) were initially successfully synthesized in 2000 via ozone to remove the tubular meso- and nanoporous silicate templates [19]. A lot of research teams have thereafter reported a different growth process [20–26]. Sha et al. used a nanochannel Al_2O_3 [21] and fabricated SiNTs by a method of chemical vapor deposition (CVD) [21, 27, 28]. After that, a variety of developing techniques were developed to create SiNTs, such as a method of molecular beam epitaxy (MBE) [22], a hydrothermal method [24], and a method of gas phase condensation [25, 29]. Ishai and Patolsky demonstrated that the diameter and wall thickness can be well controlled, ranging from 1.5 nm to 500 nm, which would be a great advantage in potential nanoscale electronics [30]. Yoo et al. used a surface sol-gel reaction to obtain optimal SiNTs for energy storage devices [31]. Other studies which report their fabricating SiNT techniques achieve better SiNT properties applied in many biological, chemical, and electrical fields, such as sensors, nanoscale electronics, and optoelectronics [32–34]. Especially for an application in lithium-ion batteries, silicon is one of the best candidates to replace current commercial materials of alloyed anodes, i.e. graphite. The nanoscale of SiNTs has overcome the disadvantages of volume expansion that lead to pulverization and rapid loss of reversible capacity [32]. Compared to the capacity of graphite (372 mAh g^{-1}) [35, 36], silicon nanotubes exhibit a high reversible charge capacity of 3247 mA h/g with a Coulombic efficiency of 89% [37].

Up to now, a lot of theoretical studies have provided important features of nanotubes dependent on their diameter and the chiral vectors [38–44], which have been verified experimentally [3–8, 20–31]. The predictions of theoretical calculations, such as conductivity properties of carbon nanotubes, lead to results consistent with experiments of observations [45]. Besides, chiral vectors and the curvature of carbon nanotubes affect the structure and energetic properties of carbon nanotubes [46]. A monolayer graphene is sp^2 hybridization and silicene, which contains sp^2 and sp^3 hybridization. The low-lying energy bands of graphene and silicene are dominated by π bonding contributed by the $2p_z$ and $3p_z$ orbitals, respectively [47]. The tubular structure of carbon and silicon nanotubes due to the band-folding effect has a σ^*-π^* hybridization [38]. There can be both the metallic and semiconducting characteristics due to the radii and chiralities, because the curvature effect leads to misorientation of p_z orbitals of both carbon and silicon nanotubes and the σ^*-π^* orbital hybridization on a cylindrical surface. Lin and Chuu report that the hopping integral on the dimer line is different from the other two, depending on the arrangement of three chemical bonds by detailed calculations [48].

The periodical boundary condition plays an important role in determining the metallic or semiconducting behavior of carbon nanotubes. When $2m + n = 3I$, carbon nanotubes are gapless metals. Linear bands of carbon nanotubes correspond to the Dirac-cone structure, or they can sample the Dirac point from graphene [49, 50]. Furthermore, $2m + n = 3I$, carbon nanotubes are moderate-gap semiconductors. On the other hand, the full orbital hybridizations cause the non-armchair $2m + n = 3I$ carbon nanotubes to become narrow-gap semiconductors, except for r < 2.5 Å, and they create metallic nanotubes with a very small radius. Briefly, the metallic, moderate-, and narrow-gap behavior of carbon nanotubes are characterized by the geometric structures: (I) m = n or r < 2.5 Å, (e.g., (5,0) and (6,0) nanotubes) [46, 38]; (II) 2m + n = 3/I; ((7,0) and (8,0) nanotubes); (III) 2m + n = 3I and m ≠ n ((9,0) and (12,0) nanotubes). The three different energy gap types of carbon nanotubes have been verified by accurate STS measurements [51–53]. The (m,m) armchair and (m,0) zigzag carbon nanotubes will be chosen for a model study to see the three types of electronic structures.

The electronic properties of silicon nanotubes are analyzed mainly by the tight-binding model [43, 44, 54–58]. However, the studies of the essential electronic properties of silicon nanotubes via the first-principle method are relatively rare. Fagan and Baierle et al. propose a hypothetical silicon nanotube to study its electronic and structural properties [59], since SiNTs had at the time not been created yet. The geometric appearance of Si nanotubes (SiNTs) takes Si hexagonal nanotubes (Si h-NTs) and Si gear-like nanotubes (Si g-NTs) into consideration. Both Si h-NTs and Si g-NTs differ in their hybridization of the Si atoms and are produced by rolling up a sp^2 and sp^3 Si atoms graphene sheet to form a nanotube. Besides, Si g-NTs are energetically more stable than Si h-NTs [54].

In this chapter, we focus on the diversified electronic properties of zigzag single-wall silicon and carbon nanotubes, investigated by first-principles calculations. A relativistic interaction of a carbon atom's spin with its motion inside a potential is normally weak, like in graphene. However, it is significant in a carbon nanotube due to the curvature of its surface [60]. The curve surface causes an asymmetric structure and leads to an intrinsic spin-orbit interaction in a carbon nanotube. The splitting estimated by theoretical calculations is around 100 μeV, and the spin-orbit splitting measured by experimental methods is reasonably consistent with theoretical predictions [61–64]. This work presents the bond lengths, ground state energy E_0, buckling distance Δh, energy bands, energy gap E_g, spatial charge density ρ, and orbital-projected density of states (PDOSs) and for both structures the chiral vectors and diameters are fully included in the calculations.

7.2 METHODS

First-principles calculations are based on density functional theory (DFT) carried out by the Vienna ab initio simulation package (VASP) [65, 66]. The exchange-correlation energy derived from many-particle Coulomb interactions is evaluated by use of the Perdew-Burke-Ernzerhof functional (PBE) [67] under generalized gradient approximation. Furthermore, the van der Waals (vdW) corrections are included via the semi-empirical DFTD2 correction of Grimme [68]. The one-dimensional periodic boundary condition is along z-direction and the vacuum distance at x- and y- space is set to 15 Å to make sure no resistance from neighboring SiNTs occurs. The maximal cutoff energy of the wave function expanded by the plane wave is 500 eV. The Brillouin zone and geometric optimization is implemented with 1x1x15 meshes and 1x1x500 for DOSs calculations via the Gamma scheme. The convergence of the Helmann-Feymann force is less than 0.01 eV Å^{-1} during ionic relaxation.

7.3   RESULTS AND DISCUSSION

7.3.1   CARBON NANOTUBES

The primary symmetry types of carbon nanotubes (CNTs) can be classified as chiral (non-symmorphic) and achiral (symmorphic). Chiral nanotubes show a spiral symmetry, which means they are without mirror images to the original one. The chiral vector is along the vector R_x shown in Figure 7.1(a). The equation is expressed as $R_x = ma_1 + na_2$, where a_1 and a_2 are primitive lattice vectors of a 2D sheet. The symbols m and n are integers, and m is not equal to n. The lattice vector R_y can be described as $R_y = pa_1 + qa_2$ shown as a rectangle in Figure 7.1(a), and the axial wave vectors in the 1D first Brillouin zone are $|k_y| \le \pi/|R_y|$. The radius r of a nanotube (m,n) behaves as $b\sqrt{3(m^2 + mn + n^2)} / 2\pi$.

The chiral angle θ, which is the angle between the vectors R_x and a_1, can characterize the non-chiral/chiral arrangements of the carbon hexagons on the cylindrical surface. It is confined in the range $0^o \le |\theta| \le 30^o$ for all carbon and silicon nanotubes. Hence, the chiral angle θ yields an expression for $\tan\theta$, which is $\tan\theta = \sqrt{3}/(2m + n)$. The other type of carbon nanotubes are armchair and zigzag ones. They are both classified as achiral nanotubes. (m,m) and (m,0), shown in Figure 7.1(a) as a green dashed line, are the symbols respectively for the achiral armchair and zigzag systems with $\theta = 30°$ and $\theta = 0°$. In Figure 7.1(b) and (c) show the rolled honeycomb lattice of zigzag (10,0) and armchair (10,10) carbon nanotubes, respectively. The number of carbon atoms N_u in a primitive unit cell is given by the equation of $N_u = 4\sqrt{(m^2 + mn + n^2)(p^2 + pq + q^2)/3}$, where (p,q) is determined by the primitive lattice vector R_y along the nanotube axis.

The schematic single wall zigzag carbon nanotubes are shown in Figure 7.2(a)–(p), which exhibit the cross-section and side view for zigzag C (4,0) to

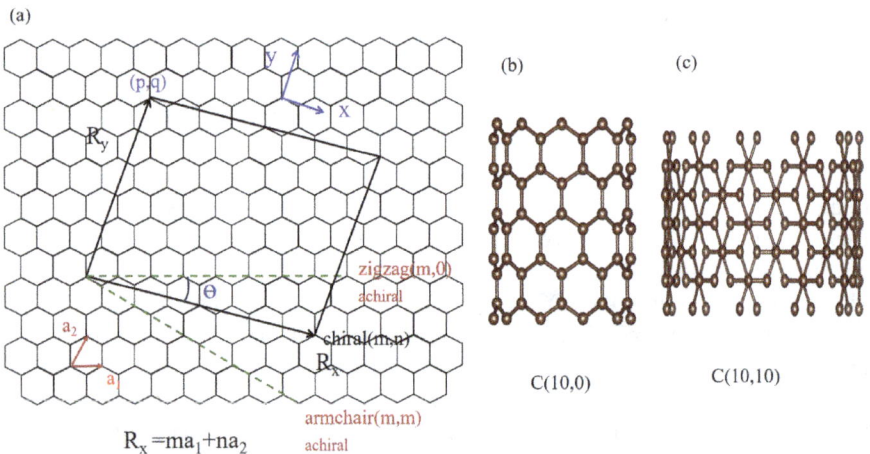

FIGURE 7.1   Unrolled honeycomb lattice of a nanotube (a). A single-wall zigzag C (10,0) and armchair C (10,10) carbon nanotube (b) and (c).

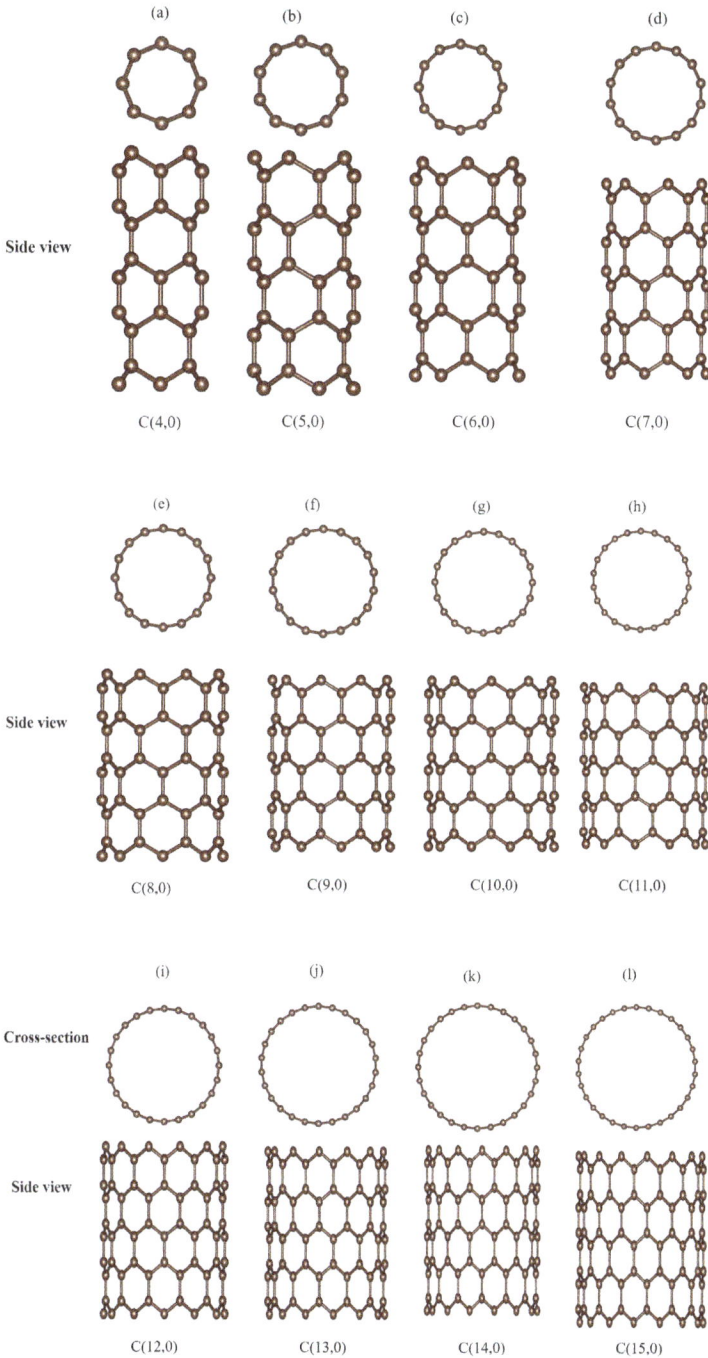

FIGURE 7.2 (a)–(l) Geometric structure of zigzag carbon nanotubes at side view and cross-section C (4,0), C (5,0), C (6,0), C (7,0), C (8,0), C (9,0), C (10,0), C (11,0), C (12,0), C (13,0), C (14,0), C (15,0). *(Continued)*

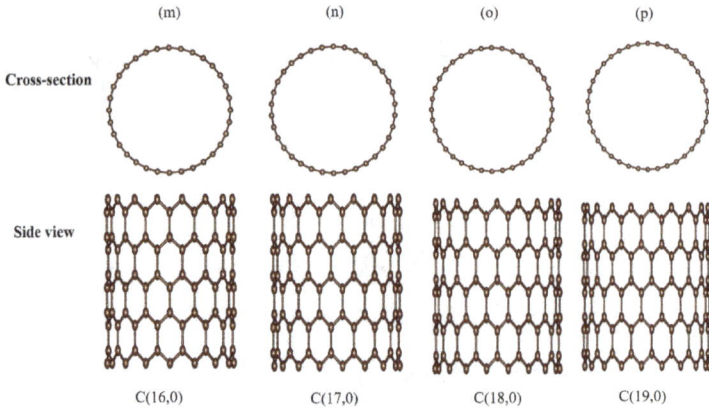

FIGURE 7.2 (*Continued*) (m)–(p) Geometric structure of zigzag carbon nanotubes at side view and cross-section C (16,0), C (17,0), C (18,0), and C (19,0).

C (19,0) nanotubes, respectively. Compared to silicon nanotubes, there are no gear-like shapes observed in this schema of the cross-section. The geometric parameters of zigzag carbon nanotubes (zCNTs) are summarized in Table 7.3.1.1. The chiralities (m,0) are in the range of 4 to 19. The smallest single-wall carbon nanotubes found experimentally by the HR-TEM method are around 4 Å in diameter, and the possible structures can be armchair (3,3), zigzag (5,0), and chiral (4,2) [69]. They are corresponding to our calculations for C (5,0) 4.144 Å. Besides, in a multiwall carbon nanotube, a stable 3 Å carbon nanotube can be grown [70]. The other diameter results of carbon nanotubes for C (4,0) to C (19,0) are 3.373 Å to 14.90 Å, respectively. There are two types of C-C bond lengths in the carbon tubular structure which are named b_1 and b_2. Compared with the standard carbon bond length b_0 (1.42 Å), b_1 and b_2 exhibit different results for small-sized tubes. When the tube size is as small as C (4,0), the bond length of b_1 is greater than b_0. On the contrary, the bond length of b_2 is shorter then b_0 for small carbon nanotubes. When the size of carbon nanotubes is increased, the values of b_1 and b_2 are close to b_0 and normalized equal to the one shown in Figure 7.3(b). The ground state energy E_0 is decreasing when the diameter of the carbon nanotubes increases, as illustrated in Figure 7.3(a). The ground state energy E_0 of small nanotubes is higher than that of bigger ones and can be related to the curvature effects. The low-energy essential properties are dominated by the π bonding of the parallel $2p_z$ orbitals. The disorientation of $2p_z$ orbitals and the sp^3 orbital hybridization on a cylindrical surface are not considered due to the curvature effects.

The band structures of zigzag carbon nanotubes (zCNTs) with chiral vectors of (4,0), (5,0), (6,0), (7,0), (8,0), (9,0), (10,0), (11,0), (12,0), (13,0), (14,0), (15,0), (16,0), (17,0), (18,0), and (19,0) are shown in Figure 7.4(a)–(p). They belong to 1D metals and are semiconductors with a zero energy gap E_g, even in the presence of significant curvature effects. They are initiated from the k_x-dependent

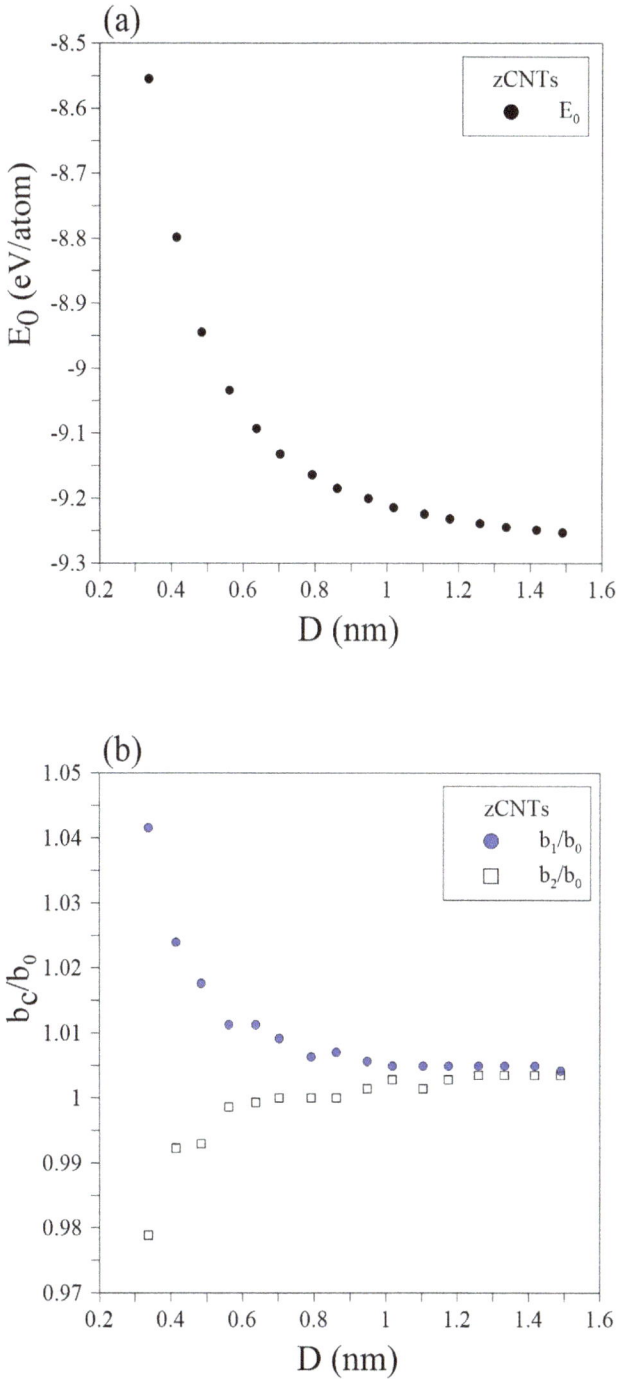

FIGURE 7.3   (a) Ground state energy E_0 are illustrated as function of diameter D. (b) C-C bond lengths are illustrated as function of diameter D.   *(Continued)*

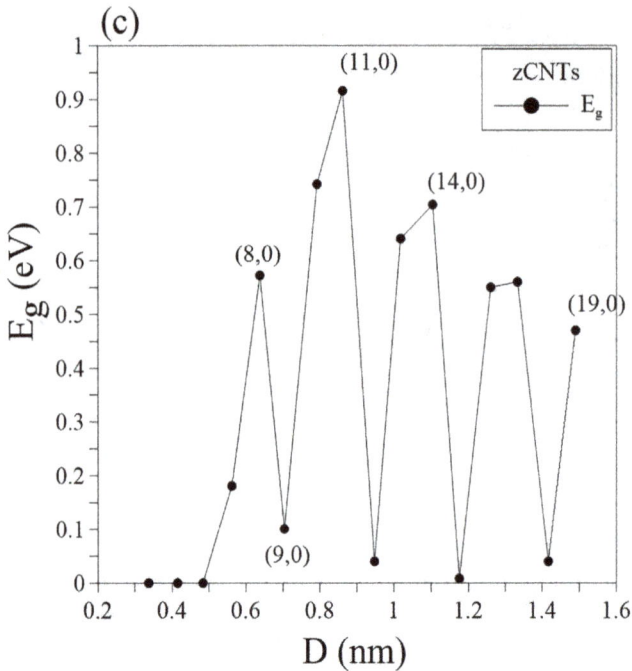

FIGURE 7.3   *(Continued)* (c) Energy gap E_g are illustrated as function of diameter D.

energy bands with significant dispersion relations, as shown in Figure 7.4 for the various zigzag systems. The energy band properties of carbon nanotubes are sensitive to the varied chiral angle and radius. There are certain important differences between zigzag and armchair carbon nanotubes. Due to the cylindrical boundary condition, the low-energy electronic states of zCNTs are doubly degenerate in determining the metallic property and the magnitude of E_g. The small-size zCNTs, such as (4,0), (5,0), and (6,0), exhibit a strong overlap of the valence and conduction bands, so they belong to the 1D metals. Normally, those particular metallic behaviors are not dominated by the periodical boundary condition and the π bonding. They mainly come from the significant sp^3 orbital hybridizations on the high-curvature cylindrical surface. In order to observe the quasiparticle decay rate and lifetimes of carbon nanotubes sp^2-bonding systems experimentally, the femtosecond pump-probe spectroscopies are very persuasive tools in exploring the ultrafast relation of photoexcited electrons. They contain the time-resolved photoemission spectroscopy [71–76], absorption/transmission/reflectivity [77–81], and flourence spectroscopies [82, 83], in which their measurements are very appreciable in understanding the quasiparticle behaviors of the excited conduction electrons. The results of carbon nanotubes exhibit the carrier relaxation of the excited electrons in metallic single-walled and multiwalled systems. The time-resolved photoemission spectroscopy shows the decay rate of the carbon nanotube is around 1 meV for the low-energy excited states [73,74].

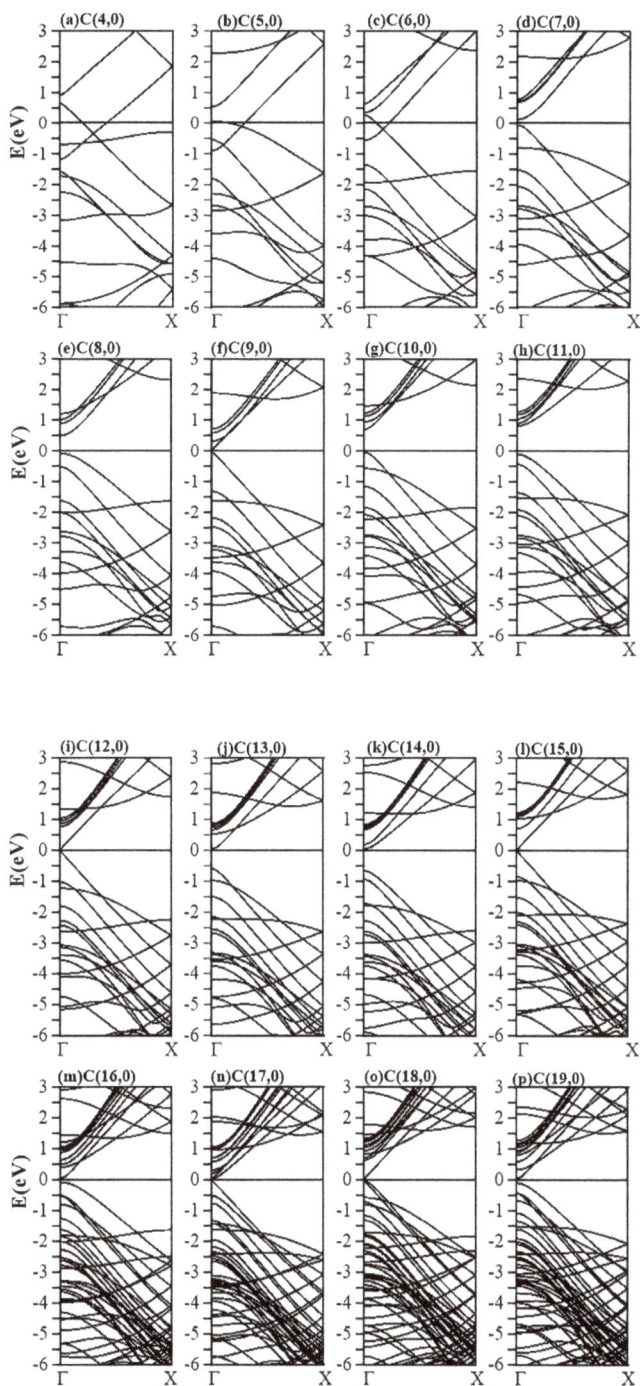

FIGURE 7.4   (a)–(p) Band structure of zigzag carbon nanotubes from C (4,0) to C (11,0) and C (12,0) to C (19,0).

The bundled and isolated single-walled carbon nanotubes exhibit the decay rate of the first, and the second conduction band is around 1–5 meV and 0.5–3 meV, respectively, analyzed by methods of the absorption [77, 84] and flourence [82, 83] spectroscopies.

Later, those features are confirmed and discussed by the spatial charge distributions and the orbital-projected density of states (PDOSs) (Figures 7.5 and 7.6). On the other hand, the (m ≥ 7, 0) of the zCNTs possess direct energy gaps at $k_x = 0$. The radius-dependent energy gaps E_g could be further divided into type-II for m = 3I + 1/3I + 2 and type-III for m = 3I, in which they are, respectively, proportional to the inverse of r and r^2, as shown in Figure 7.3(c). The narrow gaps of zCNTs are induced by the non-parallel and non-uniform π bonding [85, 86]. The edge-atom-dominated energy bands are unavailable in zCNTs, and each of the energy bands is equally contributed by carbon atoms on the cylindrical surface. Specifically, the metal-semiconductor transitions appear when a single-wall carbon nanotube is threaded by a uniform axial magnetic field (a magnetic flux ϕ_B). As a result of the cylindrical symmetry, the ϕ_B-dependent electronic structure exhibits the well-known periodical Aharonov-Bohm effect, with an oscillation period of flux quantum ($\phi_0 = hc/e$) [87–90]. Besides, the metal-semiconductor transitions are observable for carbon nanotubes in an external magnetic field; they occur at $\phi_B = I\phi_0$. Similar magneto electronic properties are disclosed in the type-II and type-III carbon nanotubes, in which the critical magnetic flux is close to $(I \pm 1/3)\phi_0$ and $I\phi_0$, respectively. The Aharonov-Bohm effect in carbon nanotubes has been verified by the experimental measurements on the magneto-optical [91, 92] and transport properties [93–95].

Spatial charge density ρ is determined by investigating the multi-/single-orbital hybridizations because of the various chemical bonds [96]. Carbon atoms contribute four valence electrons to create two specific chemical bondings, namely the strong σ bonding of (2s; $2p_x$; $2p_y$) orbitals and the π bonding of $2p_z$ orbitals. However, the cylinder carbon molecules exhibit another scenario which can be described by the charge density. The results of charge density ρ for zCNTs are illustrated in Figure 7.5 with the [x,y]-plane projection calculated from first-principle calculations. The charge density ρ of zCNTs clearly represents the chemical bonding as well as the charge transfer. In Figures 7.5(a)–(p), there are some red color shapes similar to rectangles that appear between each carbon atom. These are σ bonds. However, it is hard to define whether it is (2s, $2p_x$, $2p_y$) or (2s, $2p_x$, $2p_y$, $2p_z$) hybridization. In contrast to monolayer graphene, the results of the charge density ρ from zCNTs exhibits a weaker intensity of chemical bonding [47]. The zigzag tubular geometric structure seems not to effect the distribution of the charge density. The decreasing intensity could be due to the increasing bond length b_1. The results of the bond length b_1 from C (4,0) to C (19,0) are consistent with this phenomenon. The bond length b_1 of C (4,0) is 1.479 Å (see Table 7.1), which is 3.52% greater than the standard C-C bond length (1.42 Å). It then decreases to 1.426 Å when the bond length b_1 is C (19,0).

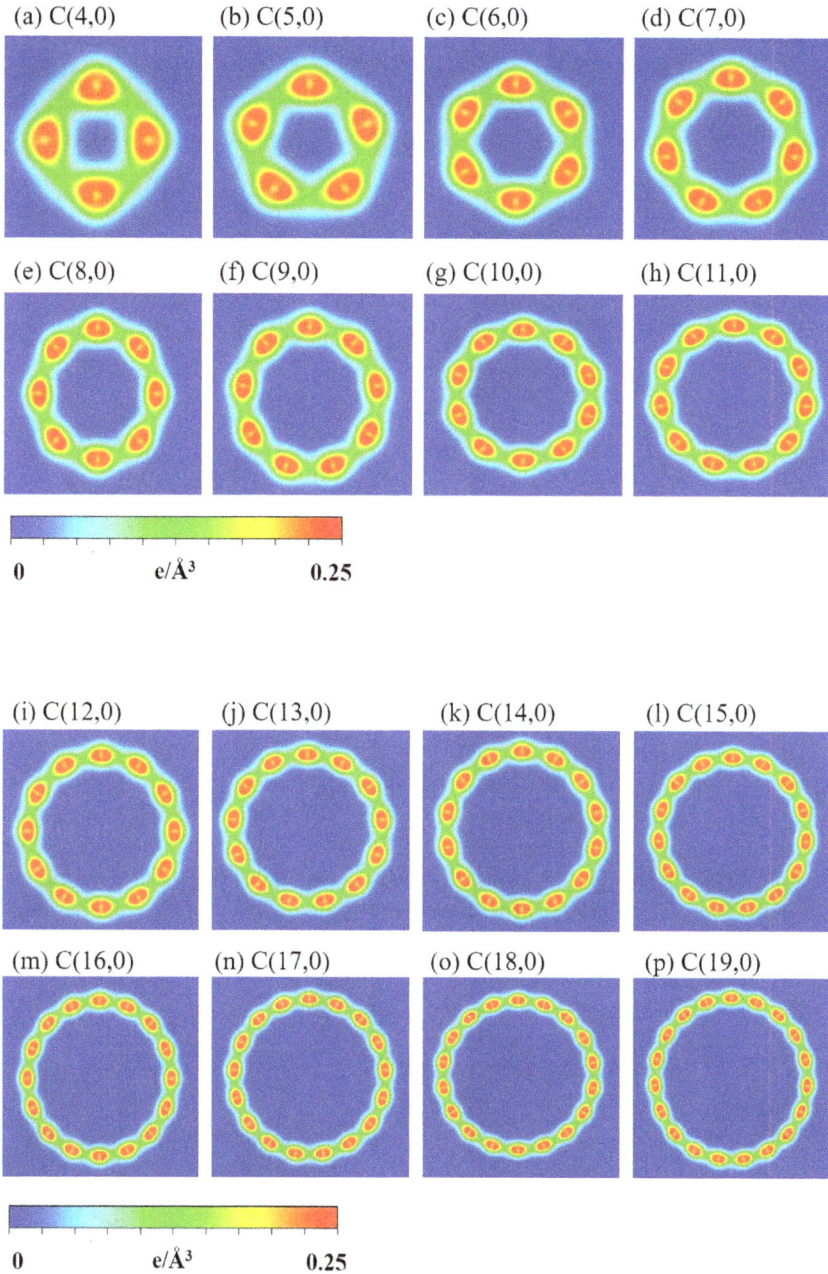

FIGURE 7.5 (a)–(p) Spatial charge density of zigzag carbon nanotubes from C (4,0) to C (11,0) and C (12,0) to C (19,0).

TABLE 7.1

Parameters of Zigzag Carbon Nanotubes Ground State Energy E_0, C-C Bond Length b_1 and b_2, Diameter D; b_0 =1.42Å

Sample	Lattice con. (Å)	E_0/unit cell (eV)	E_0/atom (eV)	b_1(Å)	B_2(Å)	b_1/b_0	b_2/b_0	D(Å)	E_g(eV)
C(4,0)	2.437	−136.87298	−8.55456	1.479	1.39	1.041549	0.978873	3.373	0
C(5,0)	2.462	−175.96264	−8.79813	1.454	1.409	1.023944	0.992254	4.144	0
C(6,0)	2.461	−214.66669	−8.94445	1.445	1.41	1.017606	0.992958	4.839	0
C(7,0)	2.461	−252.94793	−9.03385	1.436	1.418	1.011268	0.998592	5.622	0.1808
C(8,0)	2.468	−290.96927	−9.09279	1.436	1.419	1.011268	0.999296	6.373	0.572
C(9,0)	2.467	−328.75423	−9.13206	1.433	1.42	1.009155	1	7.036	0.1008
C(10,0)	2.463	−366.54792	−9.1637	1.429	1.42	1.006338	1	7.92	0.742
C(11,0)	2.466	−404.12975	−9.18477	1.43	1.42	1.007042	1	8.614	0.916
C(12,0)	2.467	−441.61468	−9.20031	1.428	1.422	1.005634	1.001408	9.477	0.04
C(13,0)	2.469	−479.13177	−9.21407	1.427	1.424	1.00493	1.002817	10.18	0.641
C(14,0)	2.467	−516.5532	−9.22416	1.427	1.422	1.00493	1.001408	11.041	0.704
C(15,0)	2.47	−553.87656	−9.23128	1.427	1.424	1.00493	1.002817	11.756	0.0085
C(16,0)	2.47	−591.27659	−9.2387	1.427	1.425	1.00493	1.003521	12.602	0.55
C(17,0)	2.47	−628.61862	−9.24439	1.427	1.425	1.00493	1.003521	13.329	0.56
C(18,0)	2.47	−665.90435	−9.24867	1.427	1.425	1.00493	1.003521	14.171	0.04
C(19,0)	2.47	−703.21061	−9.25277	1.426	1.425	1.004225	1.003521	14.9	0.47

The orbital-projected density of states (PDOSs) as well as the 1D van Hove singularities of the orbital dependence of zCNTs are clearly shown in Figure 7.6(a)–(p), directly reflecting the primary features of the electronic energy band structure. The partial orbital-project on density of states (PDOSs) can be investigated fully to realize the orbital-decomposition contributions. Besides, they can provide useful information about the cooperative/competitive relations among the chiral angle, the periodical boundary condition, and the multi-orbital hybridizations. The metallic zigzag nanotubes present several prominent asymmetric peaks near E_F, such as (4,0), (5,0), and (6,0) corresponding to Figures 7.6(a)–(c), being associated with the low-lying parabolic energy bands (Figures 7.4(a)–(c)). For small carbon nanotubes, the low-energy DOSs mainly come from the $2p_x$ and $2p_y$ orbitals; therefore, the strong multi-orbital hybridizations can create the metallic band structures. Unlike graphene [47], the PDOSs of carbon nanotube at low-lying energy range from −3 eV to 3 eV and are mostly contributed by $2p_x$ and $2p_y$ orbitals. In addition, the significant hybridization of these two orbitals also has a strong effect on the electronic properties of the highly curved nanoribbons. The contributions due to the $2p_x$ orbitals become weak for an increase of the radius; that is, the decrease of the curvature leads to the weakened sp^3 orbital hybridizations, or the low-energy essential properties are mainly determined by the neighboring $2p_y$ orbitals. The contributions of 2s and $2p_z$ orbitals are observed below −3 eV.

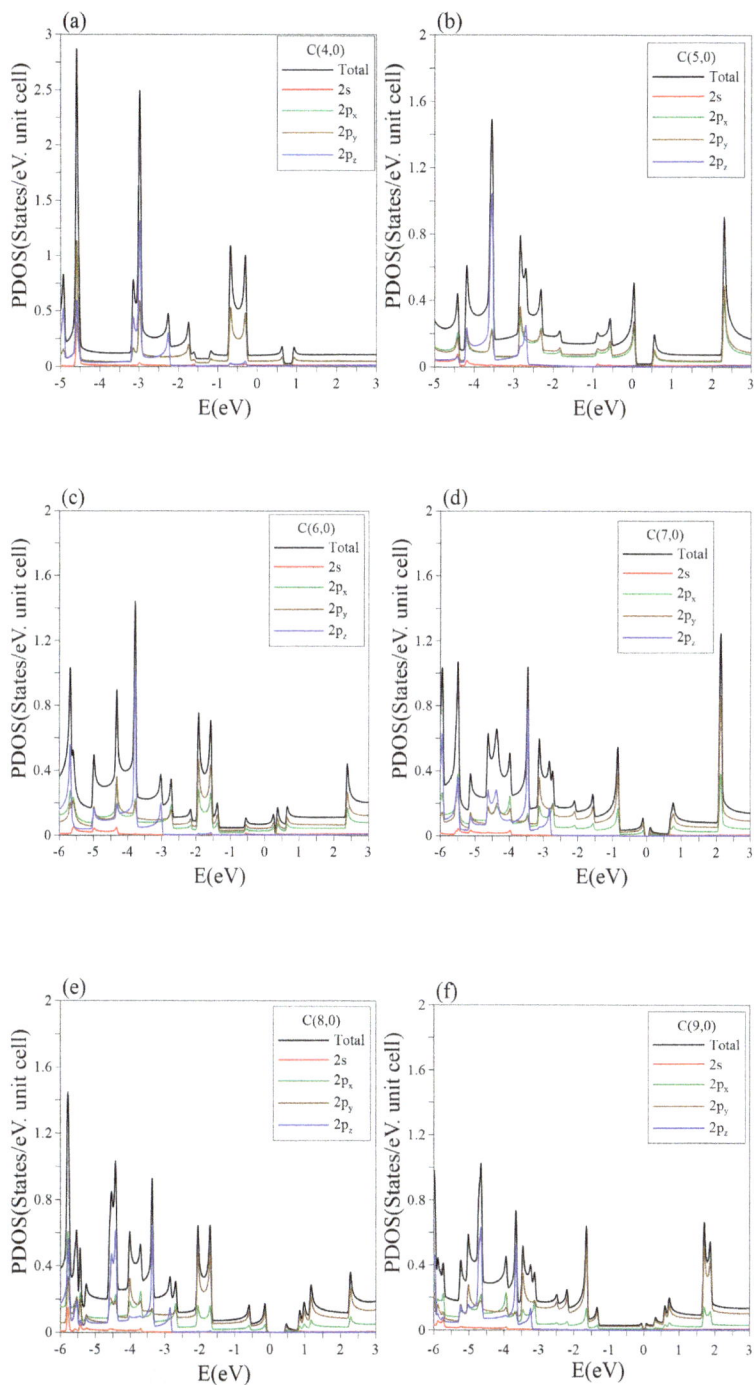

FIGURE 7.6 (a)–(f) Orbital-projected density of states of zigzag carbon nanotubes C (4,0), C (5,0), C (6,0), C (7,0), C (8,0), and C (9,0). *(Continued)*

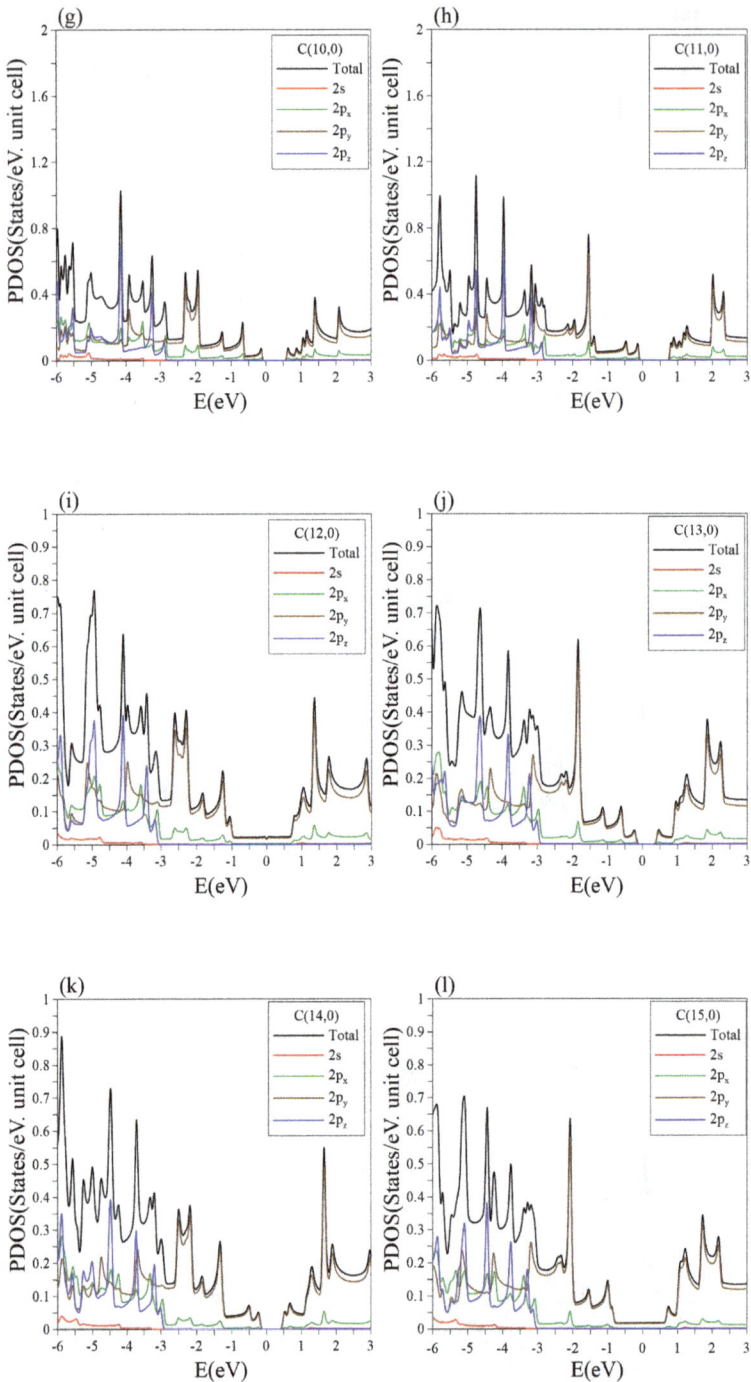

FIGURE 7.6   (*Continued*) (g)–(l) Orbital-projected density of states of zigzag carbon nano-tubes C (10,0), C (11,0), C (12,0), C (13,0), C (14,0), and C (15,0).

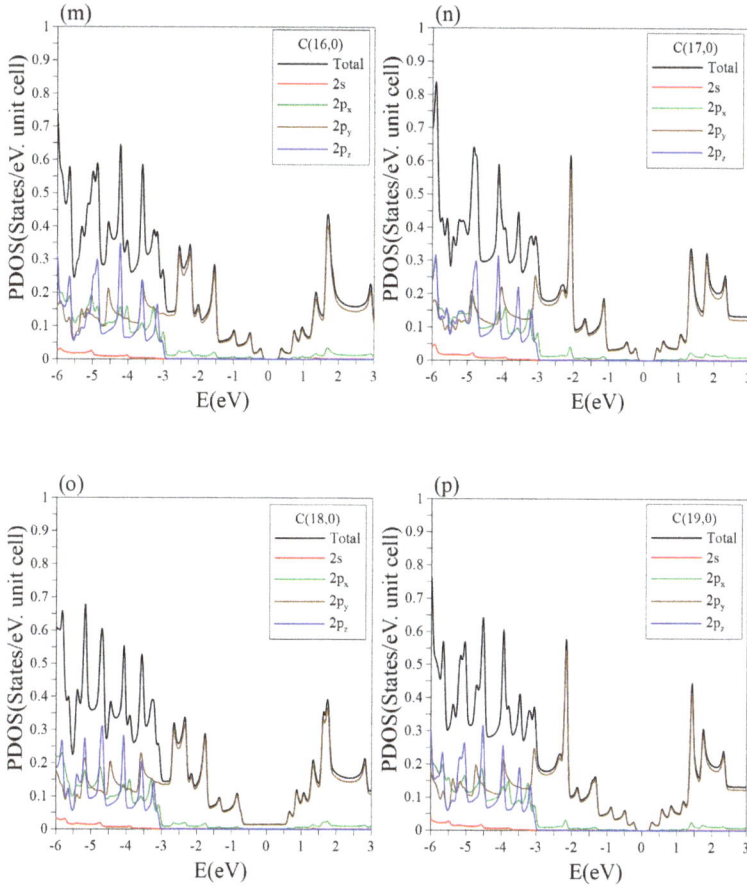

FIGURE 7.6 (*Continued*) (m)–(p) Orbital-projected density of states of zigzag carbon nanotubes C (16,0), C (17,0), (18,0), and C (19,0).

7.3.2 SILICON NANOTUBES

The schematic single-wall nanotubes of the zigzag structure are shown in Figure 7.7(a)–(r), which show the cross-section and side view for zigzag silicon nanotubes (zSiNTs) from (4,0) to (21,0), respectively. The chirality also follows the rule for CNTs that is expressed by $R_x = ma_1 + na_2 \equiv (m,n)$, where a_1 and a_2 are the real space unit vectors. In the case of the armchair configuration, it is m = n, i.e. (m,m), and in the case of zigzag configuration, it is n = 0, i.e., (m,0). The value m is proportional to the atomic number of the nanotubes. Compared with the zigzag carbon nanotubes (shown in Figure 7.2), the gear-like shapes can be observed in the cross-section schemata, especially for the small size of zSiNTs, because the sp^3 hybridization leads each neighboring Si atom to create a height difference. The zigzag structure with a periodical bond length in a unit cell is $\sqrt{3}b$ (b is a Si-Si bond length). After moderate adjustments of the zigzag structure, we obtained

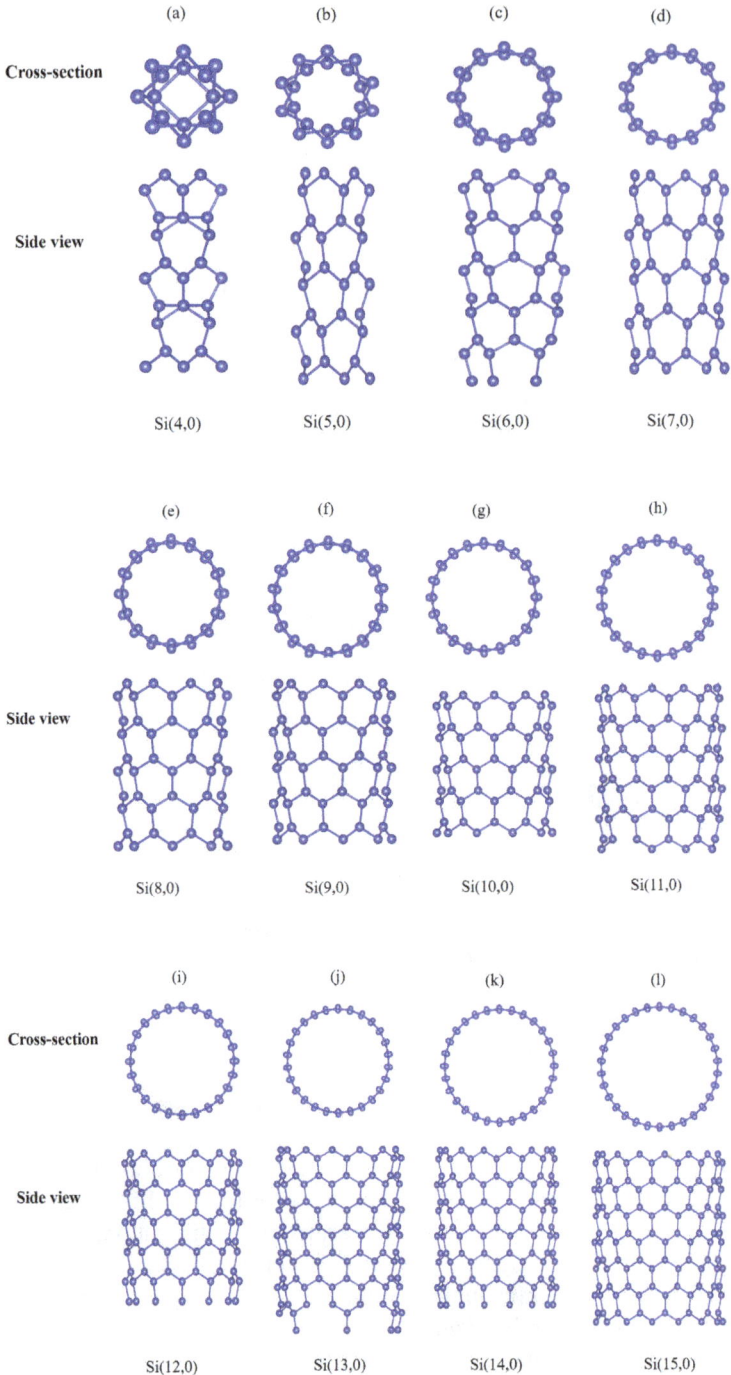

FIGURE 7.7 (a)–(l) Geometric structure of zigzag silicon nanotubes at side view and cross-section Si (4,0), Si (5,0), Si (6,0), Si (7,0), Si (8,0), Si (9,0), Si (10,0), Si (11,0), Si (12,0), Si (13,0), Si (14,0), and Si (15,0). *(Continued)*

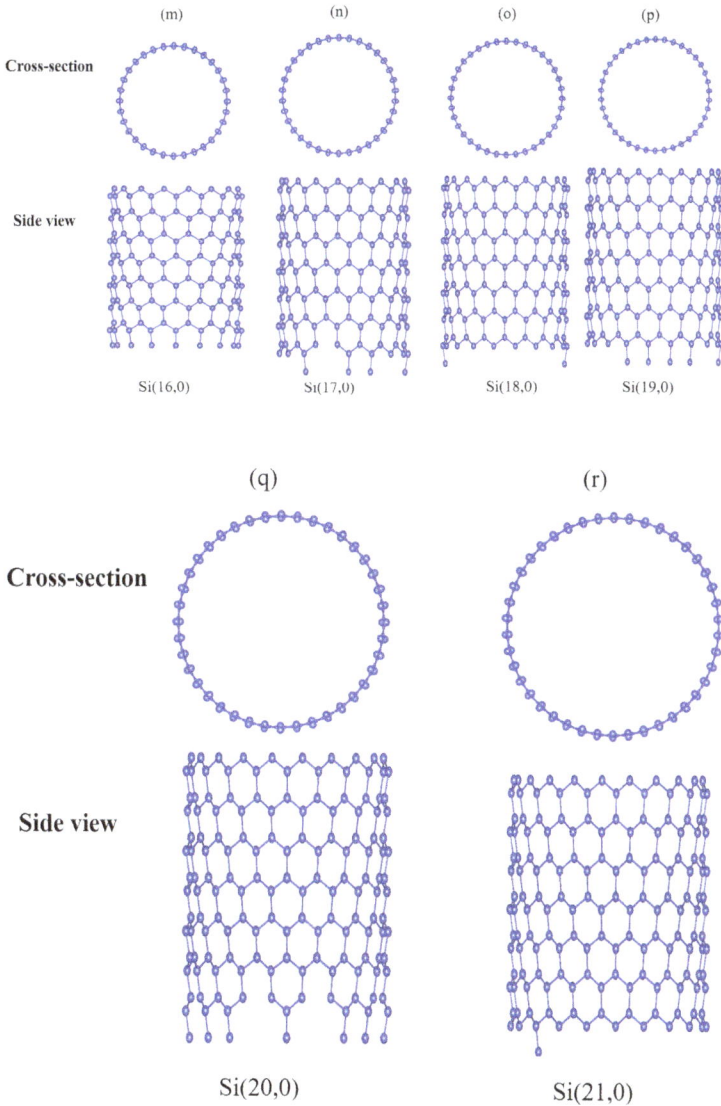

FIGURE 7.7 (*Continued*) (m)–(r) Geometric structure of zigzag silicon nanotubes at side view and crosssection Si (16,0), Si (17,0), Si (18,0), Si (19,0), Si (20,0), and Si (21,0).

the optimal lattice constant and relatively lowest ground state energy E_0, which are summarized in Table 7.2. Because of the buckling angle enhanced by the sp^3 hybridization, the silicon nanotubes have a gear-like structure unlike the carbon nanotubes.

Due to the gear-like structures, two adjacent atoms cause a distance difference to the center axis, which results in the two different diameters defined as D_1 and D_2 and shown in Figure 7.8. Meanwhile, the two types of Si-Si bonds are labeled as b_1 and b_2. b_1 is a horizontal Si-Si bond and b_2 is a vertical one. Both bond lengths are altered

TABLE 7.2

Parameters of zigzag silicon nanotubes ground state energy E_0, Si-Si bond length b_1 and b_2, diameter D_1 and D_2, buckling distance Δh; $b_0 = 2.28$ Å

Sample	Lattice con. (Å)	E_0/unit cell (eV)	E_0/atom (eV)	b_1(Å)	b_1/b_0	b_2/b_0	D_1(Å)	D_2(Å)	Buckling distance (Å)	E_g(Ev)
Si(4,0)	3.806	−75.246041	−4.70287756	2.581	2.355	1.132018	1.032895	3.651	5.761	0
Si(5,0)	3.814	−93.880358	−4.6940179	2.331	2.336	1.022368	1.024561	4.563	6.326	0
Si(6,0)	3.757	−112.98727	−4.70780292	2.298	2.304	1.007895	1.010526	6.813	8.15	0
Si(7,0)	3.76	−132.15185	−4.71970893	2.296	2.298	1.007018	1.007895	7.858	9.01	0
Si(8,0)	3.77	−151.40441	−4.73138781	2.294	2.296	1.00614	1.007018	9.322	10.445	0
Si(9,0)	3.773	−170.63903	−4.73997306	2.29	2.293	1.004386	1.005702	10.398	11.466	0
Si(10,0)	3.773	−189.864	−4.7466	2.288	2.291	1.003509	1.004825	11.811	12.868	0.186
Si(11,0)	3.773	−209.05319	−4.75120886	2.287	2.288	1.00307	1.003509	12.925	13.945	0.19
Si(12,0)	3.776	−228.23632	−4.75492333	2.2865	2.2885	1.002851	1.003728	14.289	15.306	0.2017
Si(13,0)	3.779	−247.42129	−4.75810173	2.286	2.287	1.002632	1.00307	15.4215	16.4115	0.32038
Si(14,0)	3.776	−266.57959	−4.76034982	2.285	2.285	1.002193	1.002193	16.782	17.765	0.3245
Si(15,0)	3.78	−285.73653	−4.7622755	2.283	2.284	1.001316	1.001754	17.89	18.856	0.248
Si(16,0)	3.777	−304.89153	−4.76393016	2.283	2.283	1.001316	1.001316	19.2415	20.208	0.3365
Si(17,0)	3.779	−324.0364	−4.76524118	2.283	2.283	1.001316	1.001316	20.3825	21.338	0.2987
Si(18,0)	3.782	−343.17398	−4.7663053	2.282	2.284	1.000877	1.001754	21.69	22.645	0.2022
Si(19,0)	3.781	−362.31814	−4.767344	2.282	2.283	1.000877	1.001316	22.8525	23.798	0.2821
Si(20,0)	3.78	−381.44967	−4.7681209	2.283	2.2825	1.001316	1.001096	24.176	25.127	0.2538
Si(21,0)	3.781	−400.57947	−4.7688032	2.281	2.282	1.000439	1.000877	25.317	26.257	0.16965

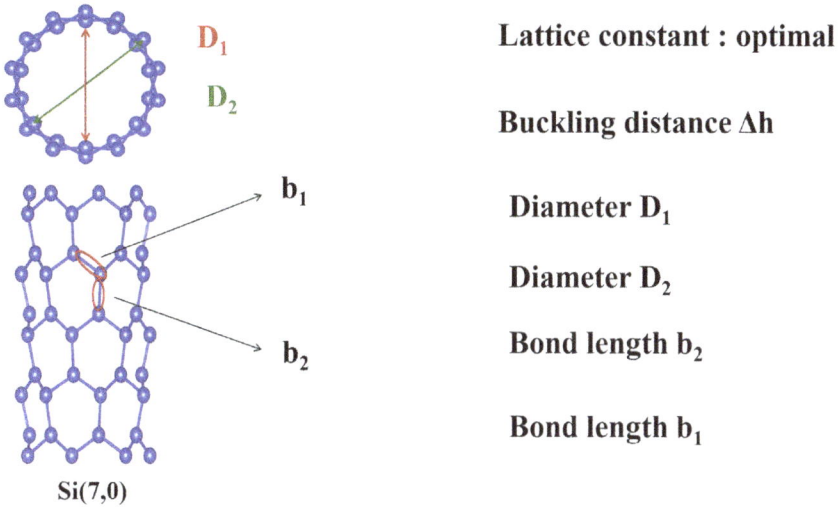

FIGURE 7.8 Illustration of silicon nanotubes parameters.

depending on the tube size, especially for small-sized nanotubes. Table 7.2 shows the geometric parameters of zigzag silicon nanotubes (zSiNTs), whose chiral vector (m,0) is in the range of 4 to 21. The smallest single wall silicon nanotubes are around 2 nm in diameter, as determined by the STM images. An atomic arrangement with a puckered structure and various chiralities are also observed [25]. The results of the zSiNTs diameters for D_1 and D_2 are 3.65 to 25.32 Å and 5.76 to 26.26 Å. The buckling distance Δh is determined by the distance difference of D_1 and D_2 and is dependent on the diameter and chirality of the silicon nanotubes (SiNTs). The buckling distance Δh is illustrated as a function of D_{mean}, which is the average value of D_1 and D_2 (in Figure 7.9(c)). The small silicon tubes lead to a large buckling distance. When the diameter and chirality of the silicon nanotubes (SiNTs) increases, the buckling distance of the SiNTs is reduced. The Si-Si bond lengths of the silicon nanotubes (SiNTs) are different at the position of the horizontal and vertical atoms; they are named in this chapter as b_1 and b_2. The bond lengths b_1 and b_2 are highly dependent on the diameter and chirality, especially in the case of b_1. When the symbol m for the chirality is small, the Si-Si bond length b_1 is extended. When the symbol m is large, b_2 returns close to the standard value.

The Si-Si bond length of b_0 (2.28 Å) is shown in Figure 7.9(b). Small-size silicon nanotubes (SiNTs) contain only a few Si atoms. When they roll up to form a tube, the distance between each Si atom is lengthened. This phenomenon can be observed in b_2 of the Si-Si bonds as well. The normalized values of the bond length b_1/b_0 and b_2/b_0 are summarized in Table 7.2. Both results are close to 1 when the diameter and chirality m increase. The ground state energy E_0 obviously decreases as the diameters and chirality m increase as shown in Figure 7.9(a). The decreasing ground state energy E_0 reflects the stable geometric structure of the silicon nanotubes (SiNTs) and could be a result of the reduced curvature energy [46].

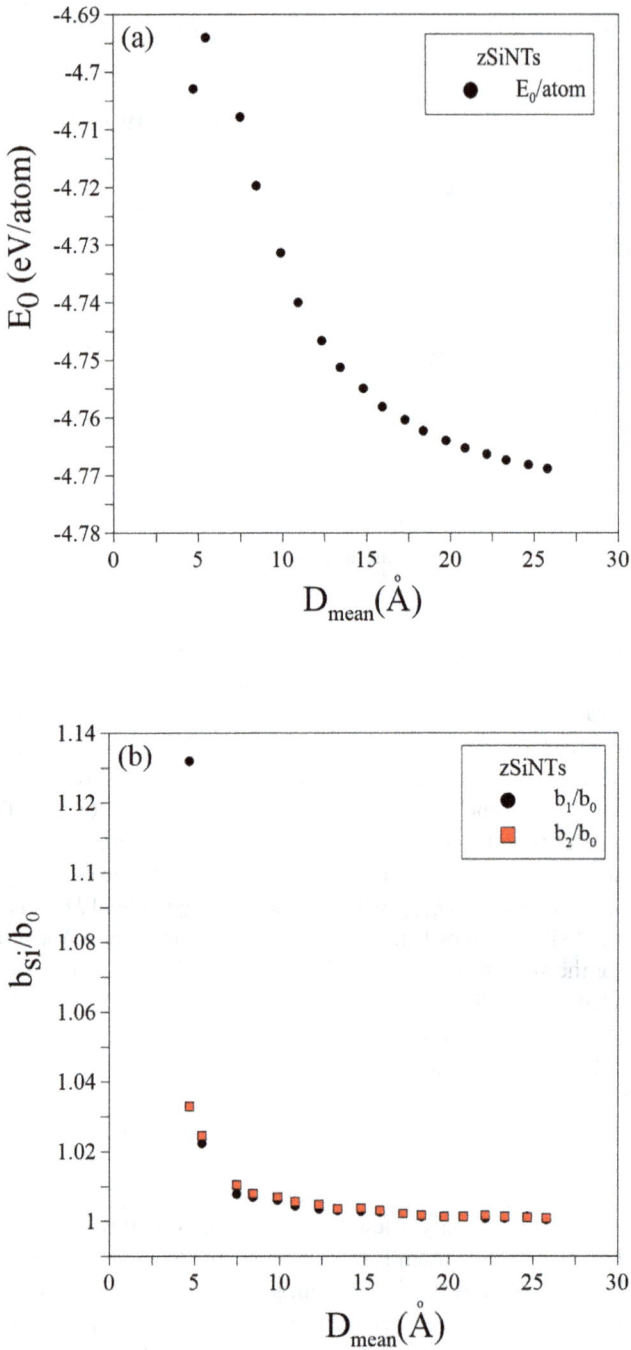

FIGURE 7.9 (a)–(b) (a) Ground state energy E_0 is illustrated as function of average diameter D_{mean}. (b) Si-Si bond lengths are illustrated as function of average diameter D_{mean}. *(Continued)*

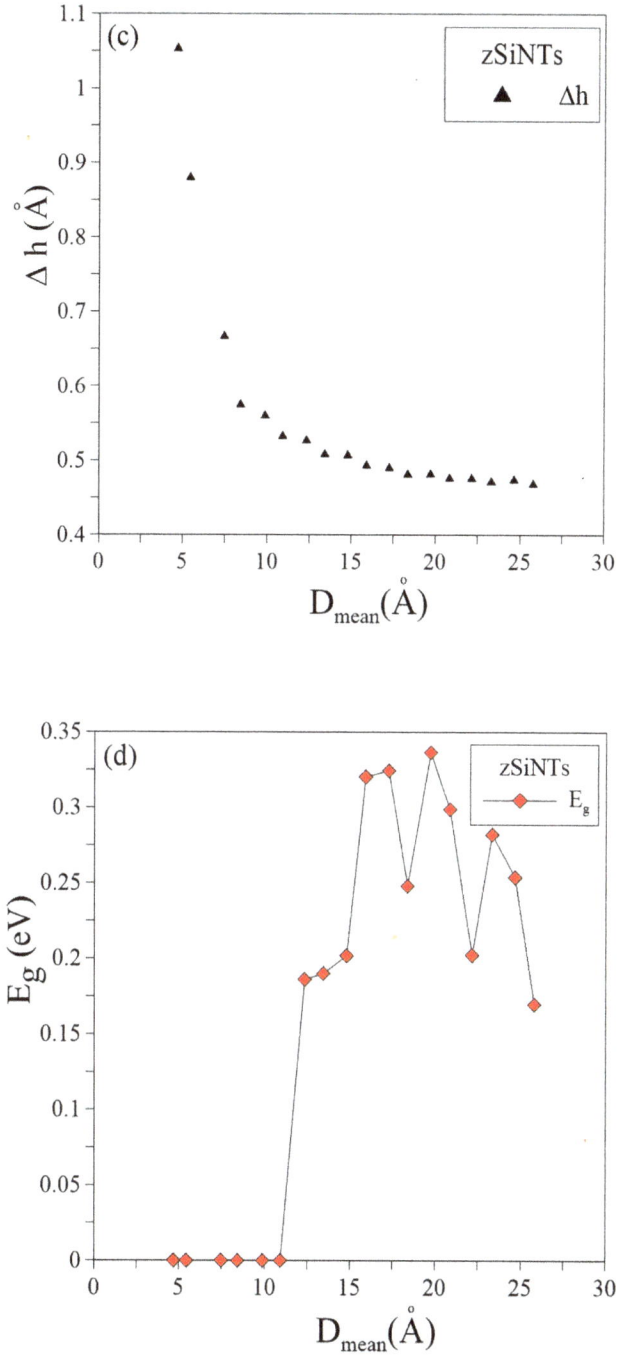

FIGURE 7.9 (*Continued*) (c)–(d) (c) Buckling distance Δh is illustrated as function of average diameter D_{mean}. (d) Energy gap E_g is illustrated as function of average diameter D_{mean}.

The band structures of zigzag silicon nanotubes (zSiNTs) with chiral vectors of (4,0), (5,0), (6,0), (7,0), (8,0), (9,0), (10,0), (11,0), (12,0), (13,0), (14,0), (15,0), (16,0), (17,0), (18,0), (19,0), (20,0), and (21,0) are shown in Figure 7.10(a)–(r). There are certain important differences between the armchair and zigzag systems in their energy band structure. The chiral index (m,0) n from 4 to 9 exhibits metallic features, while the others (m = 10–21) are semiconductors. The metallic characteristics alter the electronic structure of the silicon nanotubes (SiNTs); they could be due to the mixing of σ^* and π^* bands [38, 43, 51, 85]. For the band structure of Si (4,0), there are two flat sub-bands parallel to the Fermi level (E^F) between 0 eV and −0.5 eV. There are two other parabolic dispersions located starting at the Γ point around 0.7 and 1.0 eV and ending at the X point around 0 and 0.25 eV. Other zSiNTs with metallic behavior present crossing and partial overlapping flat sub-bands at E_F. The appearance of a crossing energy dispersion is located near the X point with a small radius of zSiNTs Si (5,0). When the radius of zSiNTs increases, the position of the crossing energy dispersion shifts and approaches the Γ points. For the case of Si (9,0), there are no crossing sub-bands at the Fermi level E_F. In the case of metallic nanotubes, the fundamental differences of electronic properties are observable by the curvature effect [46]. In the curved system, the σ^* sub-band shifts upward and the π^* sub-band move downward. Both sub-bands result in the mixed states shown in the energy dispersion. The rehybridization of the σ and π states reflects the consequence of the curvature. In contrast to non-curved systems, the σ and π orbitals appear orthogonal to each other and therefore cannot be a mixed state [97]. The alternation of the electronic structure is dramatically different, especially for silicon nanotubes with small radii and chiral vectors. In the case of carbon nanotubes, those with small diameters for semiconducting behaviors, their π^* bands slide into the band gap. This phenomenon leads to an overlap with the valence band and results in appearance of the metallic feature [51]. The large zSiNTs of Si(m,0) m ≥ 10 show a direct energy band gap E_g at the Fermi level E_F. The values of band gap E_g from Si (4,0) to Si (21,0) are illustrated as a function of D_{mean} in Figure 7.9(d). They are located at the Γ point, at which the nearest sub-bands of occupied valance and unoccupied conduction parabolic dispersions around the Fermi level E_F form band edge states. The shortest energy difference leads to semiconducting behaviors.

In the case of zigzag (m,0) carbon nanotubes, the metallic character presents itself when m = 3q (q is an integer). For zSiNTs (m,0) the semiconducting character is retained when m = 3q, which m is in the range of 10–24 [43]. There are quite a few sub-band anti-crossings observable in the results for the band structures of zSiNTs; only Si (4,0) in Figure 7.10(a) presents an anti-crossing sub-band around −1.1 eV. Most of them are crossing energy dispersions distributed among valence and conduction sub-bands.

The spatial charge density ρ of zSiNTs from (4,0) to (21,0) are available for investigating the multi-/single-orbital hybridizations because of the various chemical bonds [96]. They are illustrated in Figure 7.11(a)–(r) with the [x,y]-plane projection calculated from first-principle calculations. The results of ρ clearly represent the chemical bonding as well as the charge transfer. In Figure 7.11, there are some red color

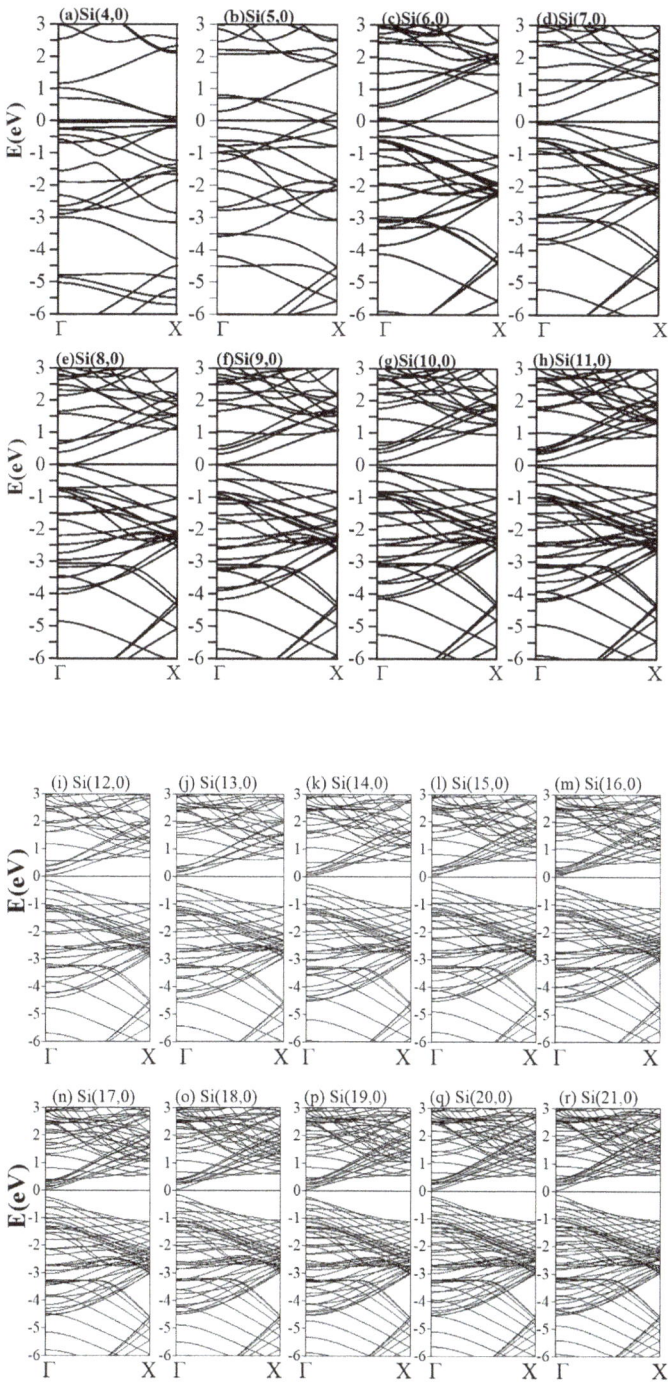

FIGURE 7.10 (a)–(r) Band structure of zigzag silicon nanotubes from Si (4,0) to Si (11,0) and Si (12,0) to Si (21,0).

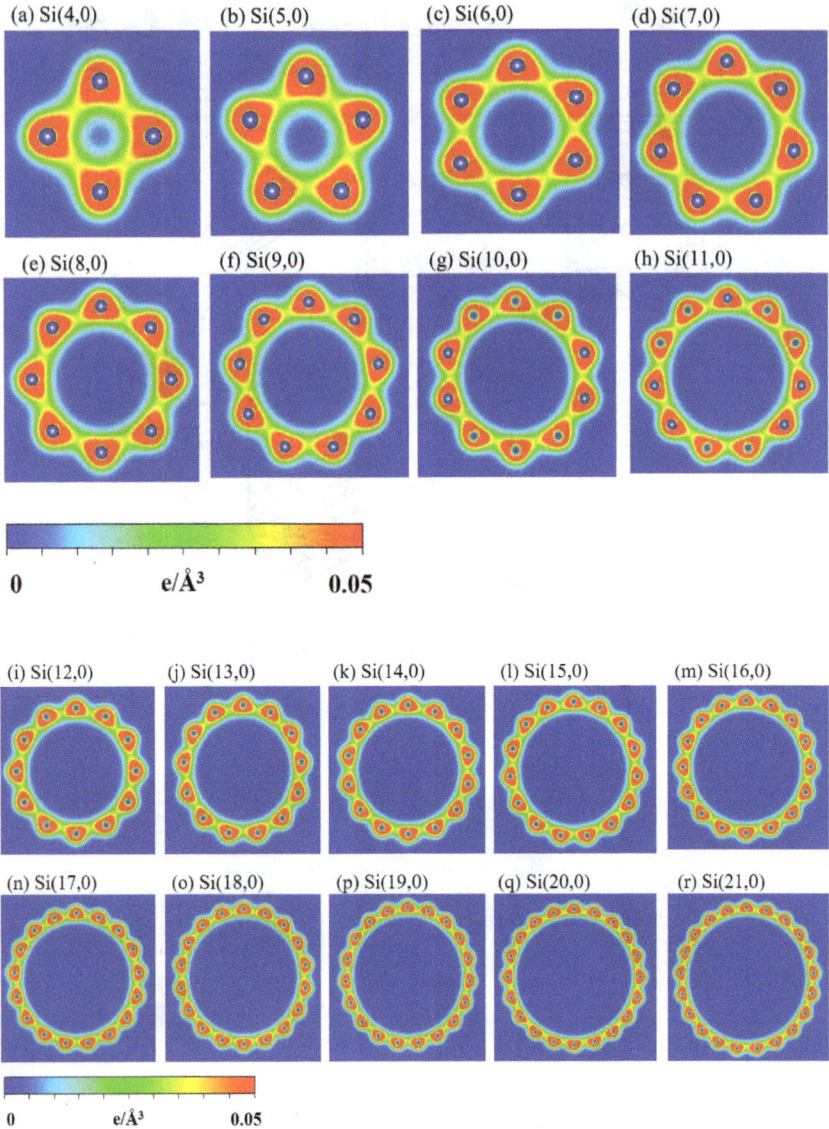

FIGURE 7.11   (a)–(h) Spatial charge density of zigzag silicon nanotubes from Si (4,0) to Si (11,0) and Si (12,0) to Si (21,0).

shapes around each silicon atom. These form σ bonds. However, it is hard to define whether they are the $(3s, 3p_x, 3p_y)$ or $(3s, 3p_x, 3p_y, 3p_z)$ hybridization. In contrast to the monolayer silicene, the results of the charge density ρ from zSiNTs have a chemical bonding of weaker intensity. The zigzag tubular geometric structure seems to affect the distribution of the charge density. The results of the bond length b_1 from Si (4,0) to

Si (21,0) are consistent with this phenomenon. The bond length b_1 of Si (4,0) is 2.355 Å, which is only 3.29% greater than the standard Si-Si bond length (2.28 Å).

It then decreases to 2.282 Å when the bond length b_1 is Si (21,0). For the Si (4,0), a σ bonding is hard to connect between the tubular surrounding silicon atoms. There is a yellow region clearly shown in the middle of each atom that indicates the weak σ bonding. The yellow region around the silicon atom is distinctly present for a small radius of zSiNTs. When the radii of zSiNTs increase, the strength of the yellow regions weakens. Compared to aSiNTs, the spatial charge density of zSiNTs, which is distributed between two neighboring silicon atoms on the same layer (the red regions), is obviously reduced and transferred.

The orbital-projected density of states (PDOSs) as well as the van Hove singularities of the energy dependence of zSiNTs are clearly shown in Figure 7.12, directly reflecting the primary features of the electronic energy band structure. The partial orbital-projected PDOSs can be investigated fully to realize the orbital-decomposed contributions. For example, the results of 2D materials of monolayer silicene appear to be zero at $E_F = 0$. The contributions of the $3p_z$-orbital present a linear energy dependence at low energy, originating from an almost disappeared energy gap 0 eV and two symmetric logarithmic peaks at 1 and −1 eV, respectively, and the isotropic linear bands and the saddle points of the π bands. Other features are a shoulder structure and a prominent peak corresponding to the saddle point of σ subbands at the M point. They appear in the middle energy region of the DOSs and are mainly contributed by $3p_x$ and $3p_y$ orbitals. In addition, $3p_z$-orbitals contribute partial $(3s, 3p_x, 3p_y)$-orbitals and yield a distinct peak contributed by the hybridization of $(3s, 3p_x, 3p_y, 3p_z)$-orbitals.

Figures 7.12(a)–(r) show the multi-orbital projection of DOSs of zSiNTs Si (4,0)-Si (21,0), respectively. The PDOSs of Si (4,0)-Si (9,0) have a non-zero value at the Fermi level (E_F) and are metallic silicon nanotubes due to the mixing of σ^* and π^*. However, the PDOSs of Si (10,0)—Si (21,0) possess a zero value at (E_F) and exhibit semiconducting features. Each peak is contributed by the 3s-, $3p_x$-, $3p_y$-, and $3p_z$-orbitals; because of curved structure, most of them generate either sp^2 or sp^3 σ bonding. Independent $3p_z$ π bonding is not observable in the results of zigzag silicon nanotubes by the first-principles method. The strength of the curvature effect induced by the hybridization of the $3p_z$-orbitals is noticeable in the results of the PDOSs of zSiNTs. The first peak of $3p_z$ for Si (4,0) is located at ca. −0.1 eV in the valence regions. This peak is then shifted to −0.6 eV for Si (5,0) and remains in the same location for Si (6,0) and Si (7,0). Finally, the peak shifts slowly but distinctly to between 1.0 eV and 1.5 eV when the size of tubes increases to Si (21,0), as shown in Figure 7.12(r). It is located at ca. −1.1 eV, as shown by the Si (11,0) nanotubes in Figure 7.12(h). It is overall very difficult to distinguish the initial, middle saddle-point structure and final π/σ bands; thus, both zigzag carbon and silicon nanotubes (Figure 7.6(a)–(p) and Figure 7.12(a)–(r)) are significantly different from each other. There are a lot of symmetric peaks, antisymmetric ones, and shoulders over the entire energy range, mainly owing to the multi-orbital hybridizations, the oscillatory energy dispersions, and the frequent anti-crossings.

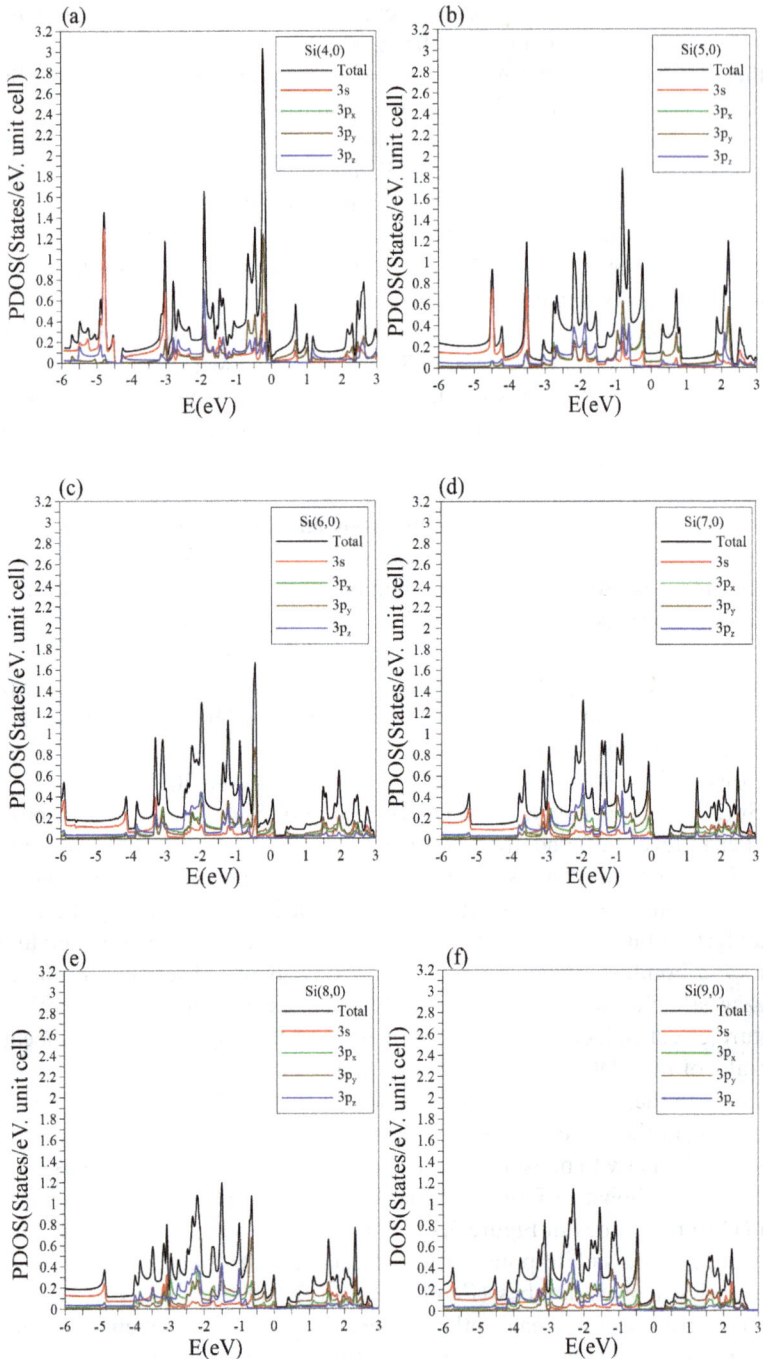

FIGURE 7.12 (a)–(f) Orbital-projected density of states of zigzag silicon nanotubes Si (4,0), Si (5,0), Si (6,0), Si (7,0), Si (8,0), and Si (9,0). *(Continued)*

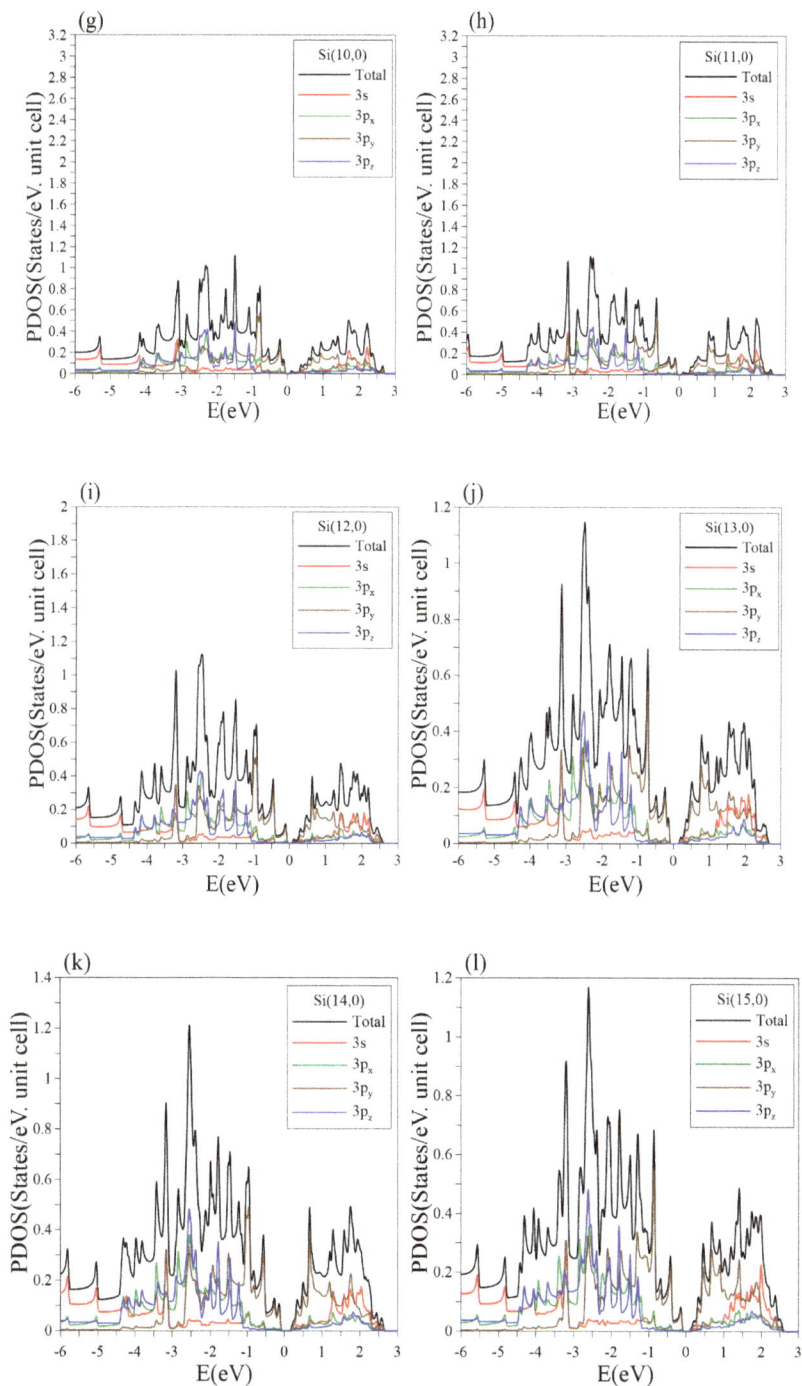

FIGURE 7.12 (*Continued*) (g)–(l) Orbital-projected density of states of zigzag silicon nanotubes Si (10,0), Si (11,0), S (12,0), Si (13,0), Si (14,0), and Si (15,0).

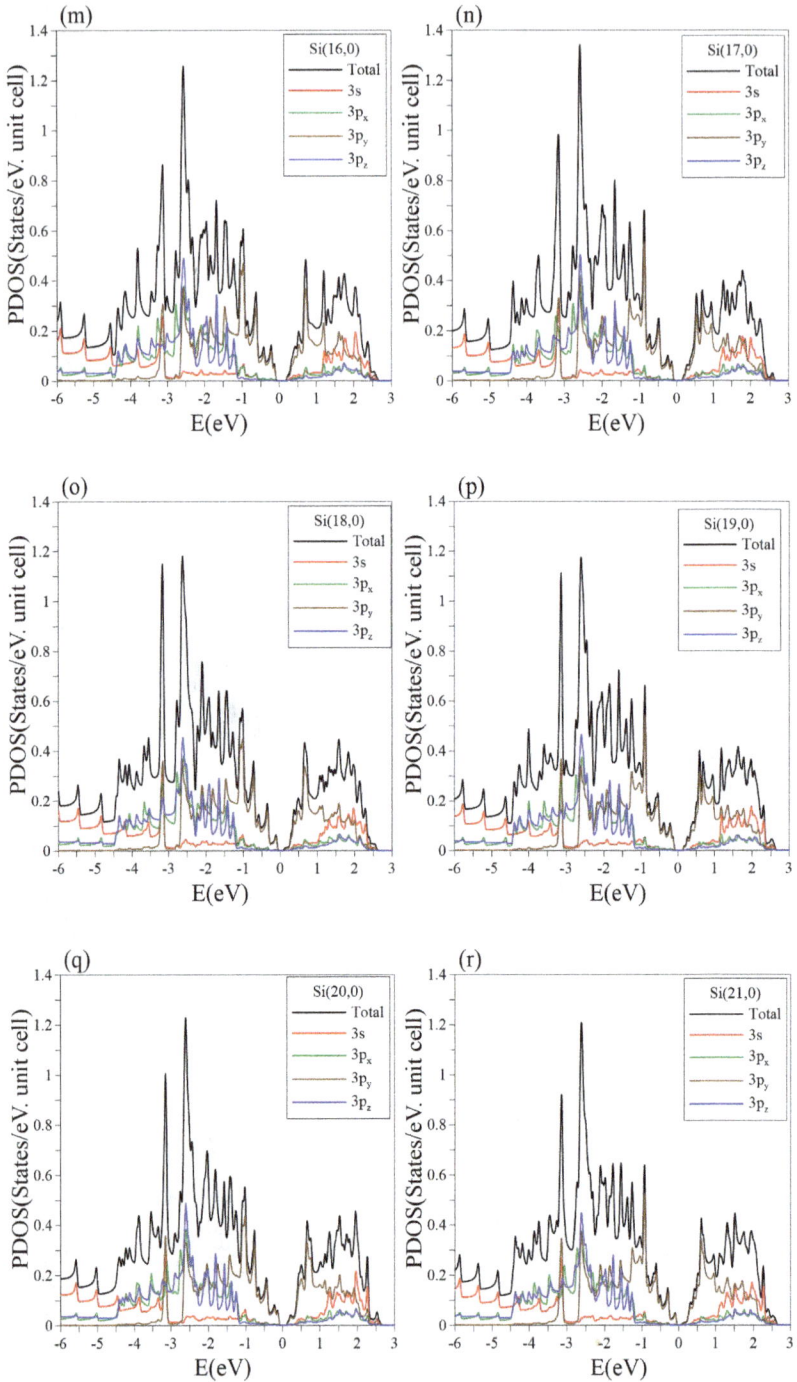

FIGURE 7.12 (*Continued*) (m)–(r) Orbital-projected density of states of zigzag silicon nanotubes Si (16,0) and Si (17,0), Si (18,0), Si (19,0), Si (20,0), and Si (21,0).

REFERENCES

[1] Iijima S 1991 Helical microtubules of graphitic carbon *Nature* **363**(6430) 1993
[2] Iijima S and Ichihashi T 1993 Single-shell carbon nanotubes of 1-nm diameter *Nature* **354**(6348) 56
[3] Journet C, Maser W K, Bernier P, Loiseau A, delaChapelle M L, Lefrant S, Deniard P, Lee R and Fischer J E 1997 Large-scale production of single-walled carbon nanotubes by the electric-arc technique *Nature* **388**(6644) 756
[4] Thess A, Lee R, Nikolaev P, Dai H, Petit P, Robert J, Xu C, Lee Y H, Kim S G, Rinzler A G, Colbert D T, Scuseria G E, Tomanek D, Fischer J E and Smalley R E 1996 Crystalline ropes of metallic carbon nanotubes *Science* **273**(5274) 483
[5] Yudasaka M, Komatsu T, Ichihashi T and Iijima S 1997 Single-wall carbon nanotube formation by laser ablation using double-targets of carbon and metal *Chem Phys Lett* **278**(1–3) 102
[6] Ebbesen T W and Ajayan P M 1992 Large-scale synthesis of carbon nanotubes *Nature* **358**(6383) 220
[7] Ebbesen T W, Hiura H, Fujita J, Ochiai Y, Matsui S and Tanigaki K 1993 Patterns in the bulk growth of carbon nanotubes *Chem Phys Lett* **209**(1–2) 83
[8] Seraphin S, Zhou D, Jiao J, Withers J C and Loutfy R 1993 Effect of processing conditions on the morphology and yield of carbon nanotubes *Carbon* **31**(5) 685
[9] Tans S J, Devoret M H, Dai H, Thess A, Smalley R E, Geerligs L J and Dekker C 1997 Individual single-wall carbon nanotubes as quantum wires *Nature* **386**(6624) 474
[10] Mintmire J W, Dunlap B I and White C T 1992 Are fullerene tubules metallic? *Phys Rev Lett* **68**(5) 631
[11] Pop E, Mann D, Wang Q, Goodson K and Dai H 2005 Thermal conductance of an individual single-wall carbon nanotube above room temperature *Nano Lett* **6**(1) 96
[12] Sinha S, Barjami S, Iannacchione G, Schwab H and Muench G 2005 Off-axis thermal properties of carbon nanotube films *J Nanoparticle Res* **7**(6) 651
[13] Kumanek B and Janas D 2019 Thermal conductivity of carbon nanotube networks: A review *J Mater Sci* **54**(10) 7397
[14] Mingo N, Stewart D A, Broido D A and Srivastava D 2008 Phonon transmission through defects in carbon nanotubes from first principles *Phys Rev B* **77**(3) 033418
[15] Yu M, Lourie O, Dyer M J, Moloni K, Kelly T F and Ruoff R S 2000 Strength and breaking mechanism of multiwalled carbon nanotubes under tensile load *Science* **287**(5453) 637
[16] Peng B, Locascio M, Zapol P, Li S, Mielke S L, Schatz G C and Espinosa H D 2008 Measurements of near-ultimate strength for multiwalled carbon nanotubes and irradiation-induced crosslinking improvements *Nat Nanotechnol* **3**(10) 626
[17] Filleter T, Bernal R, Li S and Espinosa H D 2011 Ultrahigh strength and stiffness in cross-linked hierarchical carbon nanotube bundles *Adv Mater* **23**(25) 2855
[18] Jensen K, Mickelson W, Kis A and Zettl A 2007 Buckling and kinking force measurements on individual multiwalled carbon nanotubes *Phys Rev B* **76**(19) 195436
[19] Kiricsi I, Fudala A, Konya Z, Hernadi K, Lentz P and Nagy J B 2000 The advantages of ozone treatment in the preparation of tubular silica structures *Applied Catalysis A: General* **203**(1) L1
[20] Schmidt O G and Eberl K 2001 Nanotechnology—Thin solid films roll up into nanotubes *Nature* **410**(6825) 168
[21] Sha J, Niu J, Ma X, Xu J, Zhang X, Yang Q and Yang D 2002 Silicon nanotubes *Adv Mater* **14**(17) 1219
[22] Jeong S Y, Kim J Y, Yang H D, Yoon B N, Choi S H, Kang H K, Yang C W and Lee Y H 2003 Synthesis of silicon nanotubes on porous alumina using molecular beam epitaxy *Adv Mater* **15**(14) 1172

[23] Chen Y W, Tang Y H, Pei L Z and Guo C 2005 Self-assembled silicon nanotubes grown from silicon monoxide *Adv Mater* **17**(5) 564

[24] Tang Y H, Pei L Z, Chen Y M and Guo C 2005 Self-assembled silicon nanotubes under supercritically hydrothermal conditions *Phys Rev Lett* **95**(11) 116102

[25] De Crescenzi M, Castrucci P, Scarcelli M, Diociauti M, Chaudhari P S, Balasubramanian C, Bhave T M and Bhoraskar S V 2005 Experimental imaging of silicon nanotubes *Appl Phys Lett* **86**(23) 231901

[26] Xie M, Wang J S, Fan Z Y, Lu J G and Yap Y K 2008 Growth of p-type Si nanotubes by catalytic plasma treatments *Nanotechnology* **19**(36) 1

[27] Taghinejad M, Taghinejad H, Abdolahad M and Mohajerzadeh S 2013 A nickel-gold bilayer catalyst engineering technique for self-assembled growth of highly ordered Silicon Nanotubes (SiNT) *Nano Lett* **13**(3) 889

[28] Morata A, Pacios M, Gadea G, Flox C, Cadavid D, Cabot A and Tarancon A 2018 Large-area and adaptable electrospun silicon-based thermoelectric nanomaterials with high energy conversion efficiencies *Nat Commun* **9** 1

[29] Castrucci P, Scarselli M, De Crescenzi M, Diociaiuti M, Chaudhari P S, Balasubramanian C, Bhave T M and Bhoraskar S V 2006 Silicon nanotubes: Synthesis and characterization *Thin Solid Films* **508**(1–2) 226

[30] Ishai M B and Patolsk F 2009 Shape- and dimension-controlled single-crystalline silicon and SiGe nanotubes: Toward nanofluidic FET devices *J Am Chem Soc* **131**(10) 3679

[31] Yoo H, Park E, Bae J, Lee J, Chung D J, Jo Y N, Park M S, Kim J H, Dou S X, Kim Y J and Kim H 2018 Si nanocrystal-embedded SiOx nanofoils: Two-dimensional nanotechnology-enabled high performance li storage materials *Sci Rep* **8**

[32] Kasavajjula U, Wang C S and Appleby A J 2007 Nano- and bulk-silicon-based insertion anodes for lithium-ion secondary cells *J Power Sources* **163**(2) 1003

[33] Chan C K, Peng H, Liu G, McIlwrath K, Zhang X F, Huggins R A and Cui Y 2008 High-performance lithium battery anodes using silicon nanowires *Nat Nanotechnol* **3**(1) 31

[34] Zhu J, Yu Z, Burkhard G F, Hsu C M, Connor S T, Xu Y, Wang Q, McGehee M, Fan S and Cui Y 2009 Optical absorption enhancement in amorphous silicon nanowire and nanocone arrays *Nano Lett* **9**(1) 279

[35] Wakihara M and Yamamoto O 1998 Lithium ion batteries: Fundamentals and performance *Wiley-VCH* Germany

[36] Boukamp B A, Lesh G C and Huggins R A 1981 All-solid lithium electrodes with mixed-conductor matrix *Nano Lett* **128**(4) 725

[37] Park M H, Kim M G, Joo J, Kim K, Kim J, Ahn S, Cui Y and Cho J 2009 Silicon nanotube battery anodes *Nano Lett* **9**(11) 3844

[38] Blase X, Benedict L X, Shirley E L and Louie S G 1994 Hybridization effects and metallicity in small radius carbon nanotubes *Phys Rev Lett* **72**(12) 1878

[39] Kleiner A and Eggert S 2001 Band gaps of primary metallic carbon nanotubes *Phys Rev B* **63**(7) 073408

[40] Zolyomi V and Kurti J 2004 First-principles calculations for the electronic band structures of small diameter single-wall carbon nanotubes *Phys Rev B* **70**(8) 085403

[41] Solange B, Fagan R, Baierle J and Mota R 2000 Ab initio calculations for a hypothetical material: Silicon nanotubes *Phys Rev B* **61**(15) 9994

[42] Barnard A S and Russo S P 2003 Structure and energetics of single-walled armchair and zigzag silicon nanotubes *J Phys Chem B* **107**(31) 7577

[43] Yang X and Ni J 2005 Electronic properties of single-walled silicon nanotubes compared to carbon nanotubes *Phys Rev B* **72**(19) 195426

[44] Seifert G, Koehler Th, Urbassek H M, Hernandez E and Frauenheim Th 2001 Tubular structures of silicon *Phys Rev B* **63**(19) 193409

[45] Ebbesen T W, Lezect H J, Hiura H, Bennett J W, Ghaemi H F and Thio T 1996 Electrical conductivity of individual carbon nanotubes *Nature* **382** 54

[46] Guelseren O, Yildirim T and Ciraci S 2002 Systematic ab initio study of curvature effects in carbon nanotubes *Phys Rev B* **65**(15) 153405

[47] Liu H, Lin S Y and Wu J 2020 Stacking-configuration-enriched essential properties of bilayer graphenes and silicenes *J Chem Phys* **153** 154707

[48] Lin M F and Chuu D S 1998 Persistent currents in toroidal carbon nanotubes *Phys Rev B* **57**(11) 6731

[49] Saito R, Fujita M, Dresselhaus G and Dresselhaus M S 1992 Electronic structure of chiral graphene tubules *Appl Phys Lett* **60**(18) 2204

[50] Saito R, Fujita M, Dresselhaus G and Dresselhaus M S 1992 Electronic structure of graphene tubules based on C_{60} *Phys Rev B* **46**(3) 1804

[51] Ouyang M, Huang J L, Cheung C L and Lieber C M 2001 Energy gaps in "metallic" single-walled carbon nanotubes *Science* **292** 702

[52] Wilder J W, Venema L C, Rinzler A G, Smalley R E and Dekker C 1998 Electronic structure of atomically resolved carbon nanotubes *Nature* **391** 59

[53] Odom T W, Huang J L, Kim P and Lieber C M 1998 Atomic structure and electronic properties of single-walled carbon nanotubes *Nature* **391** 62

[54] Chegel R and Behzad S 2003 Bandstructure modulation for Si-h and Si-g nanotubes in a transverse electric field: Tight binding approach *Superlattices Microstruct* **63** 79

[55] Ahmadi N, Shokri A A and Elahi S M 2016 Optical transition of zigzag silicon nanotubes under intrinsic curvature effect *SILICON* **8**(2) 217

[56] Behzad S and Chegel R 2016 Magnetic field-induced splitting of optical spectra in silicon nanotubes: Tight binding calculations *SILICON* **8**(1) 43

[57] Chegel R and Behzad S 2014 Electronic properties of SiNTs under external electric and magnetic fields using the tight-binding method *J Electron Mater* **43**(2) 329

[58] Zhang R Q, Lee H L, Li W K and Teo B K 2005 Investigation of possible structures of silicon nanotubes via density-functional tight-binding molecular dynamics simulations and ab initio calculations *J Phys Chem B* **109**(18) 8605

[59] Fagan S B, Baierle R J, Mota R, da Silva A J R and Fazzio A 2000 Ab initio calculations for a hypothetical material: Silicon nanotubes *Phys Rev B* **61**(15) 9994

[60] Steele G A, Pei F, Laird E A, Jol J M, Meerwaldt H B and Kouwenhoven L P 2013 Large spin-orbit coupling in carbon nanotubes *Nat Commun* **4**

[61] Kuemmeth F, Ilani S, Ralph D and McEuen P 2008 Coupling of spin and orbital motion of electrons in carbon nanotubes *Nature* **452**(7186) 448

[62] Churchill H O H, Kuemmeth F, Harlow J W, Bestwick A J, Rashba E I, Flensberg K, Stwertka C H, Taychatanapat T, Watson S K and Marcus C M 2009 Relaxation and dephasing in a two-electron C-13 nanotube double quantum dot *Phys Rev Lett* **102**(16) 166802

[63] Jhang S H, Marganska M, Skourski Y, Preusche D, Witkamp B, Grifoni M, van der Zant H, Wosnitza J and Strunk C 2010 Spin-orbit interaction in chiral carbon nanotubes probed in pulsed magnetic fields *Phys Rev B* **82**(4) 041404

[64] Jespersen T S, Grove-Rasmussen K, Paaske J, Muraki K, Fujisawa T, Nygard J and Flensberg K 2011 Gate-dependent spin-orbit coupling in multielectron carbon nanotubes *Nat Phys* **7**(4) 348

[65] Kresse G and Furthmueller J 1996 Efficient iterative schemes for ab initio total-energy calculations using a plane-wave basis set *Phys Rev B* **54**(16) 11169

[66] Kresse G and Joubert D 1999 From ultrasoft pseudopotentials to the projector augment-edwave method *Phys Rev B* **59**(3) 1758

[67] Perdew J P, Burke K and Ernzerhof M 1996 Generalized gradient approximation made simple *Phys Rev Lett* **77**(18) 3865

[68] Grimme S 2006 Semiempirical GGA-type density functional constructed with a long-range dispersion correction *J Comput Chem* **27**(15) 1787

[69] Guan L, Suenaga K and Iijima S 2008 Smallest carbon nanotube assigned with atomic resolution accuracy *Nano Lett* **8**(2) 459

[70] Zhao X, Liu Y, Inoue S, Suzuki T, Jones R O and Ando Y 2004 Smallest carbon nanotube is 3 Å in diameter *Phys Rev Lett* **92**(12) 125502

[71] Valla T, Camacho J, Pan Z-H, Fedorov A V, Walters A C, Howard C A and Ellerby M 2009 Anisotropic electron-phonon coupling and dynamical nesting on the graphene sheets in superconducting CaC_6 using angle-resolved photoemission spectroscopy *Phys Rev Lett* **102** 107007

[72] Bostwick A, Ohta T, Seyller T, Horn K, Rotenberg E 2007 Quasiparticle dynamics in graphene *Nat Phys* **3** 36

[73] Hertel T and Moos G 2000 Electron-phonon interaction in single-wall carbon nanotubes: A time-domain study *Phys Rev Lett* **84** 5002

[74] Ichida M, Hamanaka Y, Kataura H, Achiba Y and Nakamura A 2004 Ultrafast relaxation dynamics of photoexcited carriers in metallic and semiconducting single-walled carbon nanotubes *J Phys Soc Jpn* **73** 3479

[75] Xu S, Cao J, Miller C C, Mantell D A, Miller R J D and Gao Y 1996 Energy dependence of electron lifetime in graphite observed with femtosecond photoemission spectroscopy *Phys Rev Lett* **76** 483

[76] Moos G, Gahl C, Fasel R, Wolf M and Hertel T 2001 Anisotropy of quasiparticle life-times and the role of disorder in graphite from ultrafast time-resolved photoemission spectroscopy *Phys Rev Lett* **87** 267402

[77] Lauret J-S, Voisin C, Cassabois G, Delalande C, Roussignol Ph, Jost O and Capes L 2003 Ultrafast carrier dynamics in single-wall carbon nanotubes *Phys Rev Lett* 90 057404

[78] Bai Y, Olivier J-H, Bullard G, Liu C and Therien M J 2018 Dynamics of charged exci-tons in electronically and morphologically homogeneous single-walled carbon nano-tubes *Proc Natl Acad Sci USA* **115** 674

[79] Maekawa K, Yanagi K, Minami Y, Kitajima M, Katayama I and Takeda J 2018 Bias-induced modulation of ultrafast carrier dynamics in metallic single-walled carbon nanotubes *Phys Rev B* **97** 075435

[80] Tang X-P, Kleinhammes A, Shimoda H, Fleming L, Bennoune K Y, Sinha S, Bower C, Zhou O, Wu Y 2000 Electronic structures of single-walled carbon nanotubes deter-mined by NMR *Science* **288** 492

[81] Seibert K, Cho G C, Kutt W, Kurz H, Reitze D H, Dadap J I, Ahn H, Downer M C and Malvezzi A M 1990 Femtosecond carrier dynamics in graphite *Phys Rev B* **42** 2842

[82] Hagen A, Steiner M, Raschke M B, Lienau C, Hertel T, Qian H, Meixner A J and Hartschuh A 2005 Exponential decay lifetimes of excitons in individual single-walled carbon nanotubes *Phys Rev Lett* **95** 197401

[83] Wang F, Dukovic G, Brus L E and Heinz T F 2004 Time-resolved fluorescence of car-bon nanotubes and its implication for radiative lifetimes *Phys Rev Lett* **92** 177401

[84] Ostojic G N, Zaric S, Kono J, Strano M S, Moore V C, Hauge R H and Smalley R E 2004 Interband recombination dynamics in resonantly excited single-walled carbon nanotubes *Phys Rev Lett* **92** 117402

[85] Kane C L and Mele E 1997 Size, shape, and low energy electronic structure of carbon nanotubes *Phys Rev Lett* **78**(10) 1932

[86] Son Y W, Cohen M L and Louie S G 2006 Energy gaps in graphene nanoribbons *Phys Rev Lett* **97**(21) 216803

[87] Lin M F and Shung K W K 1995 Magnetization of graphene tubules *Phys Rev B* **52**(11) 8423

[88] Lin M F and Shung K W K 1995 Magnetoconductance of carbon nanotubes *Phys Rev B* **51**(12) 7592

[89] Lu J P 1995 Novel magnetic properties of carbon nanotubes *Phys Rev Lett* **74**(7) 1123

[90] Ajiki H and Ando T 1995 Magnetic properties of ensembles of carbon nanotubes *J Phys Soc Jpn* **64**(11) 4382

[91] Zaric S, Ostojic G N, Kono J, Shaver J, Moore V C, Strano M S, Hauge R H, Smalley R E and Wei X 2004 Optical signatures of the Aharonov-Bohm phase in singlewalled carbon nanotubes *Science* **304**(5674) 1129

[92] Akima N, Iwasa Y, Brown S, Barbour A M, Cao J, Musfeldt J L, Matsui H, Toyota N, Shiraishi M and Shimoda H 2006 Strong anisotropy in the far-infrared absorption spectra of stretch-aligned single-walled carbon nanotubes *Adv Mater* **18**(9) 1166

[93] Bachtold A, Strunk C, Salvetat J P, Bonard J M, Forro L, Nussbaumer T and Schoenenberger C 1999 Aharonov-Bohm oscillations in carbon nanotubes *Nature* **397**(6721) 673

[94] Cao J, Wang Q, Rolandi M and Dai H 2004 Aharonov-Bohm interference and beating in single-walled carbon-nanotube interferometers *Phys Rev Lett* **93**(21) 216803

[95] Cao J, Wang Q and Dai H 2005 Electron transport in very clean, as-grown suspended carbon nanotubes *Nat Mater* **4**(10) 745

[96] Jacobsen H 2009 Chemical bonding in view of electron charge density and kinetic energy density descriptors *J Comput Chem* **30**(7) 1903

[97] Lin S Y, Chang S L, Thuy Tran N T, Yang P H and Lin M F 2015 H-Si bonding induced unusual electronic properties of silicene: A method to identify hydrogen concentration *Phys Chem Chem Phys* **17**(39) 26443

8 Electronic and Optical Properties of Boron-/Carbon- and Nitrogen-Substituted Silicene Systems
A DFT Study

Jheng-Hong Shih, Jhao-Ying Wu, and Ming-Fa Lin

CONTENTS

8.1 INTRODUCTION

Silicenes, the new 2D materials, are composed of silicon atoms and have attracted many considerable investigations in the field of chemistry, physics, and material sciences. Because they belong to the same group IV, silicene and graphene have several similar geometric and electronic properties, such as they are both hexagonal honeycomb structures that make them have the Dirac cone near the Fermi level [1]. Silicene and graphene exhibit the well-behavior of π and σ electronic band structure. However, compared with graphene, silicene has many prominent advantages like a narrow band gap that can easily conduct quantum Hall spin effects at low temperatures due to possessing intrinsic spin-orbit coupling [2]. Most importantly, the band gap is turnable by modulating critical factors of the system. This is an outstanding feature, since the capability of adjusting semiconducting materials' band gap is

DOI: 10.1201/9781003322573-8

extremely necessary for many applications, e.g., manufacturing effective field-effect transistors [3] and photovoltaic [4] and other electronic devices [5].

Several experimental methods have been performed to enhance the potential of materials [6–8], one of which, chemical substitution [9], is especially promising. Recently, various kinds of guest-atoms doped graphene systems have been successfully synthesized [10, 11]. By chemical vapor deposition, the nitrogen-doped graphene presents an enhancement in optical transmittance and high electron mobility [12]. This study is expected to facilitate direct synthesis, since this process does not affect the underlying equipment. Furthermore, many concentrations of boron-substituted graphene compounds were synthesized and verified by high-resolution measurements using scanning tunneling microscopy (STM), transmission electron microscope (TEM), and Raman spectroscopy [13]. The boron-graphene compounds could be used to produce an active material for flexible in-plane micro-supercapacitors, since they exhibit the facile and robust laser induction process [14]. Interestingly, boron-doped laser-induced graphene presented that the electrochemical performance is enriched with 3 times larger areal capacitance and 5–10 times larger volumetric energy density at many power densities [15]. Such studies have motivated the authors to accomplish a thorough investigation of the electronic and optical properties of guest-atom substituted silicene.

Chemisorption and substitution of guest atoms on silicene have stirred a lot of experimental and theoretical investigations, since they are efficient methods in greatly diversifying the essential properties and enhancing potential applications. In this study, B-/C-/N-substituted silicene systems are comprehensively explored the quasiparticle characteristics such as the electronic, magnetic, and optical properties. These guest atoms possessing the three, four, and five outer electrons, respectively, would be well suited for a complete systematic study, while a theoretical framework based on the quasiparticle framework as well as the multi-orbital hybridizations could be used to explain such properties. For example, electron-related quantum particles carry charges and spins, in which the former and the latter, respectively, induce the orbital hybridizations (the hopping integrals) and the spin polarization, which can induce the magnetic configuration. The fundamental properties of materials include that the finite/zero-gap semiconductor, the semi-metallic/metallic behaviors, the detailed identifications of atom dominance through band structure, the spatial charge densities, and the various 2D van Hove singularities could be understood from the relations among the hybridized π, σ, and sp^3 chemical bondings. More especially, in this study, we are able to determine the orbital hybridization for each excitation channel. Finally, a detailed comparison of quasiparticle properties in boron-/carbon-/ and nitrogen-substituted silicene systems was also performed.

8.2 COMPUTATIONAL DETAILS

The optimal geometries and quasiparticle properties such as the electronic band structure, the atom-projected density of state, spin configuration, and optical spectra of boron-/carbon-/nitrogen-substituted silicene systems are investigated by the density functional theory using the Vienna Ab Initio simulation Package (VASP) [16, 17]. The Perdew-Burke-Ernzerhof functional [18] (PBE) with the generalized gradient approximation are used to simulate the exchange and correlation energies, while the projector augmented wave (PAW) pseudopotentials [19] can be described as the electron-ion interactions. To optimize and calculate the optical and electronic properties, the Monkhorst-Pack scheme along the two-dimensional periodic direction is sampled as 9x9x1, 20x20x1, and 100x100x1 k-points, respectively, in the first Brillouin zones (BZ). A vacuum distance along the z-axis is set to be 20 Å to avoid the interaction between the adjacent layers. To expand the wave function, the plane wave basis set with a maximum kinetic energy cutoff of 500 eV is chosen. Moreover, during the ionic relaxations, the maximum Hellmann-Feynman force acting on each atom is less than 0.01 eVÅ^{-1}, while the energy convergence is set to be 10^{-8} eV between two consecutive steps.

8.3 RESULTS AND DISCUSSION

8.3.1 GEOMETRIC STRUCTURES

The main characteristics of geometric structures of B-/C-/N-substituted silicene, including the bond lengths and bucklings, are calculated in the delicate first-principles calculations. The uniform stable honeycomb presents the bond length of 2.02 Å, 1.86 Å, and 1.93 Å for B-, C-, and N-Si systems, respectively. Compared with the pristine silicene (Figure 8.1(a)), all guest-host bond lengths get shorter, implying the strong

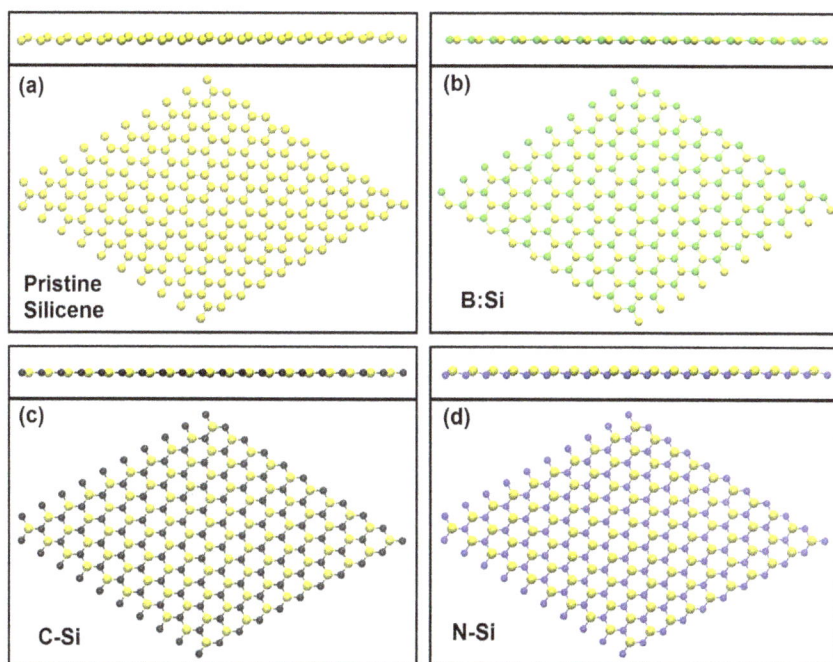

FIGURE 8.1 The side and top views of geometric structures for (a) pristine silicene, (b) boron-, (c) carbon-, and (d) nitrogen-substituted silicene.

TABLE 8.1

The Optimal Geometric Structures of B-/C-/N-Substituted Silicene Systems with Bond Lengths, Magnetic Moments, Band Gaps, and Ground State Energies Per Unit Cell and the Buckling

Systems	Bond lengths (Å)	Magnetic moment (μ_B)	E_g(eV) /Metal	Ground energy (eV)	Buckling (Å)
Pristine Silicene	2.550	0	0.001	−9.533	0.48
B-Si	1.953	0	Metal	−9.570	0
C-Si	1.785	0	2.56	−14.052	0
N-Si	1.783	0.25	0.27	−13.783	0.57

covalent bonds. Interestingly, the buckling becomes planar/buckled depending on each kind of guest atoms (Table 8.1). While B and C cases (Figure 8.1(b, c)) abruptly transform to planar, exhibiting the dominance of sp^2 bondings, the high difference between two sublattices of the N-Si system (Figure 8.1(d)) becomes more serious and might relate to the sp^3 hybridization of π and σ chemical bondings (discussed in Section 3.3). However, the B-Si and C-Si dependent spin interactions are totally suppressed by very powerful and complicated chemical bondings. Consequently, these systems do not contain magnetic configuration.

8.3.2 BAND STRUCTURES

Boron, carbon, and nitrogen possess three, four, and five valence electrons in the 2s, $2p_x$, $2p_y$, and $2p_z$ orbitals, which can create unique electronic band structures and are totally different from the pristine one. Figure 8.2 presents all the electronic states in the Brillouin zone scheme, which are plotted along Γ-K-M-Γ paths. For boron-substituted silicene, the asymmetric energy spectrum of valence and conduction bands due to the Fermi level dramatically shifts down. Obviously, the Si atom [the green circle] is almost dominant over the B atoms [the blue circle] in all electronic states. Since there are only three valence orbitals in the boron, the B-Si system becomes p-type doping with lots of free holes at K and Γ valleys near the Fermi level. At K valley, a gap of approximately 0.30 eV makes the emanation of π/π^* at $E_v = 1.70$ eV and $E_c = 2.00$ eV. The π/π^* and the first σ bands could be defined along the K-M-Γ path according to the concave-downward parabolic dispersions. Along the K and Γ valleys, the crossing between π and σ bands demonstrates the dominance of sp^2 chemical bondings. In other words, the sp^3 orbital hybridizations are negligible in the B-Si system.

Interestingly, the almost perfect Dirac cone from pristine silicene will be destroyed in carbon-substituted silicene. The system belongs to the semiconductor with a direct band gap of 2.56 eV at K valley. This gap might be due to the ionization energy difference between C-$2p_z$ and Si-$3p_z$. In both KMΓ and KΓ paths, the dissimilar group velocities present the highly anisotropic of the π/π^* states in the valence and conduction band. Similar to the boron case, the crossing feature between π and σ bands

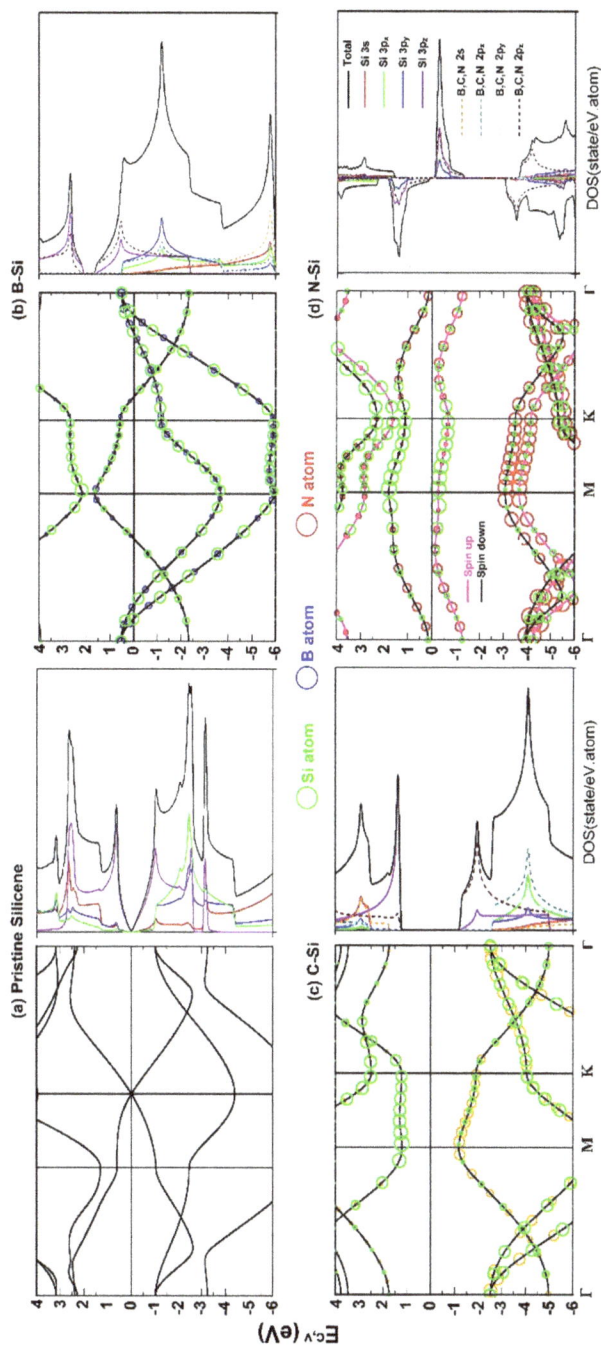

FIGURE 8.2 Band structures with the atom-dominances and the orbital-project density of states of various cases: (a) silicene, (b) boron-, (c) carbon-, and (d) nitrogen-substituted silicene.

in carbon-substituted silicene indicates the dominance of sp^2. The π and the first-σ bandwidth can be measured in both B- and C-Si systems. That means we could simulate the first-principle results using the tight-binding method [20] and study further properties such as magnetic-quantization [21], Coulomb excitation [22], and quantum spin Hall effect [23].

The buckling structure and four/five valence electrons in silicon/nitrogen atoms cause the complicated orbital-hybridization of [2s, 2px, 2py]-[3s, 3px, 3py] in N-Si bonds, the ferromagnetic spin configuration, and create an unusual N-Si-based electronic states. Obviously, the disappearance of the Dirac cone along with the unwell behavior of μ and σ bands can be realized. The system becomes an indirect-gap semiconductor of 0.21 eV with the lowest unoccupied-state at the K point, and the highest one is located between G and K points. The band-edge states appear between high-symmetry points K, M, and G, presenting the non-monotonic wavevector dependence that can be related to the van Hove singularities and optical excitation spectra. Most importantly, the ferromagnetic spin states are combined from both silicon and nitrogen atoms with the spin-up and spin-down energy split of Es ~ 0.1–2.0 eV.

8.3.3 ORBITAL-PROJECTED DENSITY OF STATE

The atom-, orbital, and spin-decomposed density of states (DOS) of pristine silicene and B-, C-, N-substituted silicene, as clearly illustrated in Figures 8.2(a) through 8.2(d), can provide further information about the orbital-hybridization, p-type doping phenomena, and magnetic configuration. The band-edge states of parabolic and oscillatory energy dispersions mainly create many special structures in the density of states. For example, the Dirac cone structure, local minima or maxima of parabolic electronic band structure, saddle points, constant energy loops, and the flat or partial flat band are closely related to the various kinds of van Hove singularities like the V-shape form, discontinuous shoulders, logarithmically divergent peaks, asymmetric peaks in the square root divergence, and delta function–like peaks, respectively.

Similar to the pristine case, the well-behavior of π and σ bands in B-, C-substituted silicene can be observed in the way we can trace their beginning, intermediate, and terminal total paths. They exhibit the same characteristic that the electronic, magnetic, and optical properties of these compounds are governed by the μ bonds of B-, C-$2p_z$, and Si-$3p_z$ at low energy. However, for a boron-silicene compound, the modified Dirac cone structure generates the asymmetric V-shape DOS with an energy gap of about 0.30 eV, which is mainly determined by the Si-$3p_z$ and B-$2p_z$ orbitals. In the carbon-silicene system, the structure which no density of state demonstrates the band gap of 2.56 eV. The widths of the π and first σ bands of B-/C-substituted silicene are, respectively, 3.80/3.74 and 4.80/4.26 eV. Apparently, there was no evidence of π, σ bands' miscibility in these two substitutional systems. The reason is mainly due to the absence of the four orbitals structures simultaneously. Nevertheless, the opposite is true for the nitrogen-silicene-based system. The complicated orbital hybridization makes the identification of the π and σ bands impossible. We cannot verify the band-widths or the initial and final state of π and σ bands. As a result, the N-Si system presents a semiconductor phenomenon with an indirect band gap of 0.24 eV. Interestingly, compared with B- and C-Si cases, only the N-Si substitutional system possesses the magnetic property. The spin up and spin down are illustrated in E > 0 and E < 0 regions. Obviously, the ferromagnetic configuration is primarily governed by the valence subbands near the Fermi level.

8.3.4 OPTICAL PROPERTIES

In this study, the optical properties of pristine and substituted silicene by three kinds of guest atoms B, C, and N are investigated. All the features such as dielectric function, electron loss function, absorption, refractive index, and reflection will provide more information about the optical properties. Obviously, the main part of the optical properties is the dielectric function calculated by the formula $\varepsilon(\omega) = \varepsilon_1(\omega) + i\varepsilon_2(\omega)$. ε_1 is the real part that exhibits the scattering of a photon by medium—in other words, the ability to store the energy of materials. It is calculated by the Kramers-Kronig equation [24]

$$\varepsilon_1 = 1 + \frac{2}{\pi} P \int_{\infty}^{0} \frac{\omega' \varepsilon_2(\omega') d\omega'}{\omega'^2 - \omega^2}$$

where P is the principal value and the imaginary part, ε_2, presents the energy absorption capacity of the material determined by

$$\varepsilon_2(\omega) = \left(\frac{4\pi^2 e^2}{m^2 \omega^2} \right) \sum \int \left(\langle i|M|f \rangle^2 \right) f_i (1 - f_i) \delta (E_f - E_i - \omega) d^3 k$$

where ω is the frequency of the electromagnetic. M is the dipole matrix and E_f and E_i refer to the energy of conduction and valence band, respectively. In this work, the difference of direction of electric polarization is taken into account. The optical properties including dielectric function, electron loss function, absorption coefficient, refractivity, and reflectivity are calculated in both directions parallel $(\perp)$ and perpendicular $(\parallel)$ the axis normal to the silicene nanosheets. These optical constants can be deduced using the real and imaginary parts of the dielectric function.

The electron energy-loss spectrum (EEL) function, being defined by $\text{Im}\left[\frac{-1}{\varepsilon(\omega)} \right]$, is given by [25]

$$L(\omega) = \frac{\varepsilon_2(\omega)}{\varepsilon_1^2 + \varepsilon_2^2}$$

The reflectivity and absorption coefficient of the electromagnetic wave can be calculated using the expression [26]

$$R(\omega) = \frac{(n-1)^2 + k^2}{(n+1)^2 + k^2} \quad \text{and} \quad \alpha(\omega) = \sqrt{2}\omega \left[\sqrt{\varepsilon_1^2 + \varepsilon_2^2} - \varepsilon_1(\omega) \right]$$

where n, k are the real and imaginary parts of the refractive index $\left[\tilde{N} = n(\omega) + ik(\omega) \right]$ and can be defined by [27]

$$n(\omega) = \sqrt{\frac{\sqrt{\varepsilon_1^2 + \varepsilon_2^2} + \varepsilon_1(\omega)}{2}} \text{ and } k(\omega) = \sqrt{\frac{\sqrt{\varepsilon_1^2 + \varepsilon_2^2} - \varepsilon_1(\omega)}{2}}$$

In this study, the independent-particle approximation (IPA) has been used. The real and imaginary parts of the dielectric function in all systems, as clearly shown in Figure 8.3, are anisotropic below 10 eV and become isotropic above 10 eV. The static value of the real part of the dielectric function ($\varepsilon_1(0)$) for the pristine, boron-, carbon-, and nitrogen-substituted silicene are 10.8, 4.8, 2.8, and 2.3 eV, respectively. The static dielectric constant is a dielectric constant related to a material's properties in a constant electric field or a low frequency. Apparently, the static dielectric constant in substitutional systems is found to be less than the pristine case.

The zero value of the real part of the dielectric function ($\varepsilon_1(\omega_p) = 0$), which is called the plasmon frequency, also plays an important role in understanding a material's characteristics as a dielectric or metal. At these points, the real part of the dielectric function will change from positive to negative and vice versa, which is closely related to the collection of excitation electrons. As a result, pristine silicene and substitution systems demonstrate the plasma frequency at different energy. For example, pristine silicene, B-Si, and C-Si have plasma frequencies at 4.0, 1.8, and 6.9 eV, respectively. Interestingly, the absence of plasma frequencies in the N-Si compound represents an excitation that is always available for the entire optical range of the electromagnetic spectrum. Furthermore, based on the plasma frequency, we can get the extreme values of the reflectivity and refractive index. In particular, the reflectivity will be minimal at plasma frequencies. Below the plasma frequency, the reflectivity will be maximum, while the refractive index will be minimal.

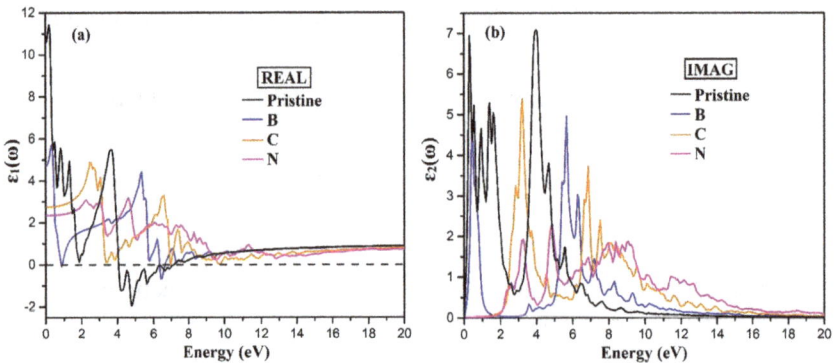

FIGURE 8.3 (a) The real part and (b) imaginary part of dielectric function of pristine silicene and substitution silicene systems.

The imaginary part of the dielectric function that has a physical meaning demonstrates the excitation feature of available channels and is closely related to the dispersal of energy within the medium. While carbon- and nitrogen-silicene compounds present an optical gap E_g of 1.8 eV, this does not exist in boron-silicene and pristine cases due to the fact that these two systems are metal and very narrow gap semiconductor behavior. Comparison between the optical gap and band gap, the carbon-silicene systems exhibit a red-shift behavior that implies the strong Coulomb interaction between the excited holes and electrons. However, the N-Si system exhibits a blue-shift phenomenon. This might come from the complicated hybridization in chemical bondings. The imaginary part of the dielectric function of all systems drops to zero at energies below 16 eV.

The electron energy loss function of pristine, B-, C-, and N-substituted silicene are calculated in both parallel and perpendicular polarization direction, as clearly illustrated in Figure 8.4. The electron energy loss function is defined as the probability that an electron passing through materials loses its energy. The observed peaks describe the parameters related to the screened response function due to the significant valence charges. The most prominent peaks in the loss function are called the collective excitations (a plasmon mode). The variation of the EEL peaks in perpendicular polarization of pristine and substitutional systems. The first prominent peaks are 2.0, 1.0, 4.0, and 3.5 for the pristine, boron, carbon, and nitrogen systems, respectively. However, in parallel polarization, the EEL peak is absent within low energy (< 4.5 eV).

Figure 8.5 shows the absorption coefficient of all systems in both in-plane and out-of-plane polarized direction. Under the single-particle picture, the optical absorption spectrum is mainly determined by the density of the states relative to the covalence and conduction bands and the electric dipole moment. After substitution, the intensity of the peaks is almost unchanged for in-plane light polarization. However, for out-of-plane light polarization, the absorption coefficient peak is higher than pristine silicene, especially for the carbon case. The intensity of the absorption peaks was almost unchanged, demonstrating that substitution of

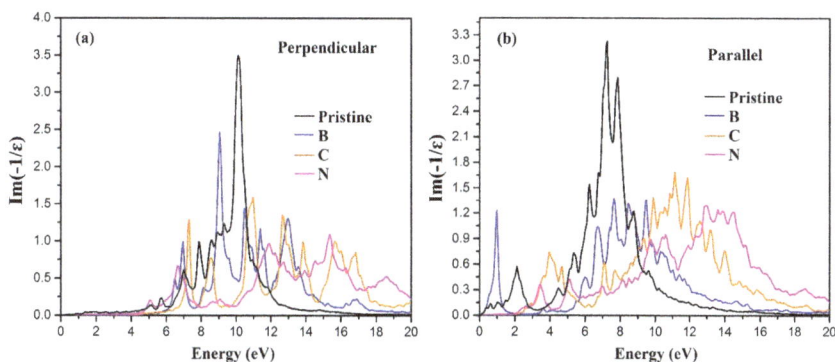

FIGURE 8.4 Electron loss function diagrams of pristine silicene and substitution silicene compounds of (a) in-plane ($\perp$) polarization and (b) out-of-plane ($\parallel$) polarization.

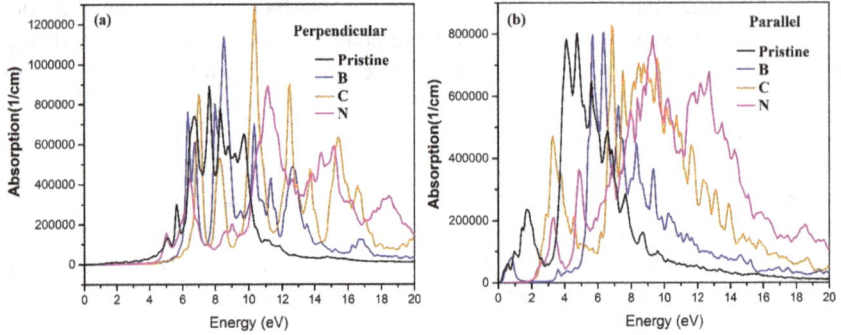

FIGURE 8.5 The absorption coefficients of pristine silicene and guest-substituted silicene (a) in-plane ($\perp$) polarization and (b) out-of-plane ($\parallel$) polarization.

B and N is not very beneficial if we want to adjust the optical properties of the silicene.

The high-resolution STM/TEM/LEED measurements [28–30] could be examined by the theoretical predictions on the boron-/carbon-/nitrogen-substituted silicene bond lengths, while the angle-resolved photoemission spectroscopy (ARPES) [31] can examine the wave vector–dependent energy spectra. Besides, the high-resolution SP-STM [32] and SQUID [33] measurements could be used to verify the spin magnetic moments. Furthermore, the verification of frequency-dependent optical properties can be reliably determined by spectroscopic reflectance, absorption, and transmission methods [34–36].

8.4 CONCLUSION

The geometric, electronic, and optical properties of boron-, carbon-, and nitrogen-substituted silicene are studied using DFT calculations. The delicate calculations are conducted on the buckling/planar honeycomb lattices, atom-dominated energy spectra, the atom-, orbital-, and spin-projected density of state, which can be used to get a deep understanding of the orbital hybridizations in chemical bonds. All substitution results present diverse properties that are very different from the pristine case. Buckling geometric structures become less/more dependent on the guest atom that are closely related to the complicated relations among the π, sp^2, and sp^3 bondings and the spin configurations. Possessing three, four, and five electrons in the outermost cells of B, C, and N atoms plays a crucial role to creating a variety of different anomalous band structures. The modified/destroyed Dirac cone accompanies the well-/ill-defined behavior of μ and σ bands. Moreover, these optical properties such as dielectric function, electron energy loss function, and absorption coefficient also are investigated. The imaginary part of dielectric function presents the red shift and blue shift that can be observed in carbon- and nitrogen-silicene compounds. Such optical properties may be useful for interesting applications such as designing optoelectronic industries.

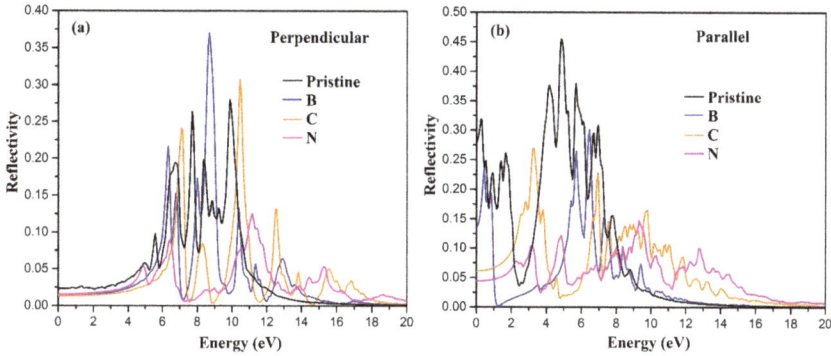

FIGURE 8.6 The reflectivity of pristine silicene and substitution silicene compounds of (a) in-plane ($\perp$) polarization and (b) out-of-plane (∥) polarization.

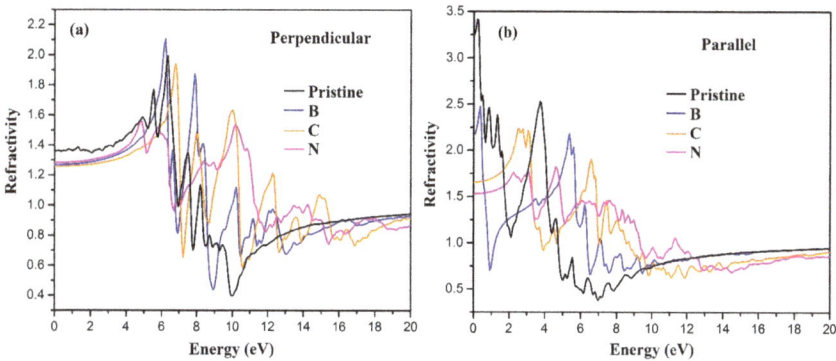

FIGURE 8.7 The refractive index of pristine silicene and substitution silicene compounds of (a) in-plane ($\perp$) polarization and (b) out-of-plane (∥) polarization.

REFERENCES

[1] Castro Neto A H, Guinea F, Peres N M R, Novoselov K S and Geim A K 2009 The electronic properties of graphene *Rev Mod Phys* **81** 109–62

[2] Mao Y L, Xu H Q, Yuan J M and Zhong J X 2018 Functionalization of the electronic and magnetic properties of silicene by halogen atoms unilateral adsorption: A first-principles study *J Phys-Condens Mat* **30**

[3] Chuan M W, Wong K L, Hamzah A, Rusli S, Alias N E, Lim C S and Tan M L P 2020 2D honeycomb silicon: A review on theoretical advances for silicene field-effect transistors *Curr Nanosci* **16** 595–607

[4] Afzal A M, Javed Y, Hussain S, Ali A, Yaqoob M Z and Mumtaz S 2020 Enhancement in photovoltaic properties of bismuth ferrite/zinc oxide heterostructure solar cell device with graphene/indium tin oxide hybrid electrodes *Ceram Int* **46** 9161–9

[5] Won U Y, Lee B H, Kim Y R, Kang W T, Lee I, Kim J E, Lee Y H and Yu W J 2020 Efficient photovoltaic effect in graphene/h-BN/silicon heterostructure self-powered photodetector *Nano Res* **14** 1967–72

[6] Chuan M W, Wong K L, Hamzah A, Rusli S, Alias N E, Lim C S and Tan M L P 2020 Electronic properties and carrier transport properties of low-dimensional aluminium doped silicene nanostructure *Physica E* **116**

[7] Galashev A Y and Vorob'ev A S 2020 Electronic and mechanical properties of silicene after nuclear transmutation doping with phosphorus *J Mater Sci* **55** 11367–81

[8] Bafekry A and Neek-Amal M 2020 Tuning the electronic properties of graphene-graphitic carbon nitride heterostructures and heterojunctions by using an electric field *Phys Rev B* **101**

[9] Chen L, Yu J Y, Zhang X F, Guan P F and Su R 2020 Theoretical Modeling of Site Selectivity and Chemical Substitution Effect of H(2)O(2)Production Efficiency on Modified Graphene *Catal Lett* **151** 390–7

[10] Ampadu E K, Kim J, Oh E, Lee D Y and Kim K S 2020 Direct chemical synthesis of PbS on large-area CVD-graphene for high-performance photovoltaic infrared photodetectors *Mater Lett* **277**

[11] Keshvardoostchokami M, Bigverdi P, Zamani A, Parizanganeh A and Piri F 2018 Silver@ graphene oxide nanocomposite: Synthesize and application in removal of imidacloprid from contaminated waters *Environ Sci Pollut R* **25** 6751–61

[12] Garino N, Zeng J Q, Castellino M, Sacco A, Risplendi F, Fiorentin M R, Bejtka K, Chiodoni A, Salomon D, Segura-Ruiz J, Pirri C F and Cicero G 2021 Facilely synthesized nitrogen-doped reduced graphene oxide functionalized with copper ions as electrocatalyst for oxygen reduction *Npj 2d Mater Appl* **5**

[13] Bleu Y, Bourquard F, Barnier V, Lefkir Y, Reynaud S, Loir A S, Garrelie F and Donnet C 2020 Boron-doped graphene synthesis by pulsed laser co-deposition of carbon and boron *Appl Surf Sci* **513**

[14] Murugan K, Nainamalai D, Kanagaraj P, Nagappan S G and Palaniswamy S 2020 Green-synthesized nickel nanoparticles on reduced graphene oxide as an active and selective catalyst for Suzuki and Glaser-Hay coupling reactions *Appl Organomet Chem* **34**

[15] Li J W, Li X F, Xiong D B, Wang L Z and Li D J 2019 Enhanced capacitance of boron-doped graphene aerogels for aqueous symmetric supercapacitors *Appl Surf Sci* **475** 285–93

[16] Kresse G and Furthmuller J 1996 Efficient iterative schemes for ab initio total-energy calculations using a plane-wave basis set *Phys Rev B* **54** 11169–86

[17] Kresse G and Joubert D 1999 From ultrasoft pseudopotentials to the projector augmented-wave method *Phys Rev B* **59** 1758–75

[18] Perdew J P, Burke K and Ernzerhof M 1996 Generalized gradient approximation made simple *Phys Rev Lett* **77** 3865–8

[19] Blochl P E 1994 Projector augmented-wave method *Phys Rev B* **50** 17953–79

[20] Shih P H, Do T N, Gumbs G, Pham H D and Lin M F 2019 Electric-field-diversified optical properties of bilayer silicene *Opt Lett* **44** 4721–4

[21] Shih P H, Do T N, Gumbs G, Huang D, Pham H D and Lin M F 2019 Rich magnetic quantization phenomena in AA bilayer silicene *Sci Rep-Uk* **9**

[22] Wu J Y, Chen S C and Lin M F 2014 Temperature-dependent Coulomb excitations in silicene *New J Phys* **16**

[23] Lee K W and Lee C E 2019 Quantum spin-valley hall effect in AB-stacked bilayer silicene *Sci Rep-Uk* **9**

[24] Jezierski K 1984 A linear-equations algorithm for reflectivity extrapolation determination in Kramers-Kronig analysis *J Phys C Solid State* **17** 475–82

[25] Wang S, Cai J, Xu H D, Tao H L, Cui Y, Zhang Z H, Song B, Liu S M and He M 2019 Investigations on electronic structure of YMnO3 by electron energy loss spectra and first-principle calculations *Powder Diffr* **34** 339–44

[26] Ech-chamikh E, Aboudihab I, Azizan M, Essafti A and Ijdiyaou Y 2004 Determination of the absorption coefficient of X-rays from X-ray reflectivity measurements *Can J Phys* **82** 75–9

[27] Rasheed M N, Maryam A, Fatima K, Iqbal F, Afzal M, Syvajarvi M, Murtaza H, Zhu M and Asghar M 2020 Enhanced electrical properties of nonstructural cubic silicon carbide with graphene contact for photovoltaic applications *Dig J Nanomater Bios* **15** 963–72

[28] Dwyer K J, Dreyer M and Butera R E 2019 STM-induced desorption and lithographic patterning of Cl-Si(100)-(2 x 1) *J Phys Chem A* **123** 10793–803

[29] Kim K M, Lee J, Choi S I, Ahn G H, Paik J G, Ryu B T, Kim Y H and Won Y S 2019 A combined study of TEM-EDS/XPS and molecular modeling on the aging of THPP, ZPP, and BKNO3 explosive charges in PMDs under accelerated aging conditions *Energies* **12**

[30] Yue C G, Ying P, Xu B and Tian Y J 2019 Surface reconstructions of SrTiO3(110) calibrated with STM and LEED *Phys Status Solidi B* **256**

[31] Tan S N, Mou Y P, Liu Y Q and Feng S P 2020 ARPES autocorrelation in electron-doped cuprate superconductors *J Supercond Nov Magn* **33** 2305–11

[32] Ara F, Oka H, Sainoo Y, Katoh K, Yamashita M and Komeda T 2019 Spin properties of single-molecule magnet of double-decker Tb(III)-phthalocyanine (TbPc2) on ferromagnetic Co film characterized by spin polarized STM (SP-STM) *J Appl Phys* **125**

[33] Yang K, Chen H, Lu L, Kong X Y, Yang R H and Wang J L 2019 SQUID array with optimal compensating configuration for magnetocardiography measurement in different environments *Ieee T Appl Supercon* **29**

[34] Wang Y N, Zhang L, Yang W, Lv S S, Su C H, Xiao H, Zhang F Y, Sui Q M, Jia L and Jiang M S 2020 An in situ reflectance spectroscopic investigation to monitor two-dimensional MoS2 flakes on a sapphire substrate *Materials* **13**

[35] Penzkofer A, Silapetere A and Hegemann P 2020 Absorption and emission spectroscopic investigation of the thermal dynamics of the archaerhodopsin 3 based fluorescent voltage sensor Archon2 *Int J Mol Sci* **21**

[36] Kamnev A A, Tugarova A V, Shchelochkov A G, Kovacs K and Kuzmann E 2020 Diffuse Reflectance Infrared Fourier Transform (DRIFT) and Mossbauer spectroscopic study of Azospirillum brasilense Sp7: Evidence for intracellular iron(II) oxidation in bacterial biomass upon lyophilisation *Spectrochim Acta A* **229**

9 Adatom-Enriched Essential Quasiparticle Properties of Germanene
A DFT Study

Yu-Ming Wang, Jhao-Ying Wu, Thi Dieu Hien Nguyen, Vo Khuong Dien, Thi My Duyen Huynh, and Ming-Fa Lin

CONTENTS

9.1 INTRODUCTION

The group-IV layered systems, which comprise C, Si, Ge, Sn, and Pb atoms [1–5], have been successfully synthesized in experimental laboratories since the first discovery of monolayer graphene systems by mechanical exfoliation on a graphite surface in 2004. Besides graphene, other group-IV two-dimensional systems have attracted a lot of interest from scientists in both theoretical and experimental studies [6–9]. Among like-graphene systems, silicene and germanene are of great interest because of their compatibility with the current equipment or devices. Recently, few-layer $GeCH_3$ nanosheets were synthesized by liquid-phase exfoliation and demonstrated as a promising high-energy density anode material, which exhibited a remaining capacity of 1058 mAh g^{-1} after 100 cycles at 0.2 A g^{-1} [10]. This encouraging result may shed light on the application of germanene in Li-ion batteries. In another study, the pure GeH compounds could be produced in a large quantity since the first experimental synthesis by Günther Vogg [11]. They belonged to the semiconductors with a band gap of 1.57 eV, which demonstrated strong optical photoluminescence in the near-infrared and revealed potential applications in optoelectronic devices. These studies show that the diverse properties of functionalized forms of germanene can be achieved by modulating the various critical factors.

DOI: 10.1201/9781003322573-9

Chemical modification including absorption and substitution could provide outstanding chemical environments in mainstream materials, especially for the planar/buckled/curved/folded honeycomb lattices, since they possess a very active surface, create a variety of novel quasiparticle properties, and extensively boost the potential applications [12]. For example, the dilute impurities could be added/substituted into the diamond structures associated with C, Si, and Ge elements, greatly enriching the featured band structures [13, 14]. Moreover, the previous predictions show that V, Cr, Mn, Fe, and Co-doped monolayer germanene presented magnetic metallic behavior. Especially, Mn-substituted germanene displayed a massive magnetic moment ∼ 3.5 μ_B even though at a very low guest concentration [15]. Their discovery paves the way for more extensive research into exploring the novel properties of germanene through chemical modification methods.

In this chapter, germanene with a buckled structure plays as a host material, and three kinds of guest adatoms, boron, carbon, and nitrogen, can greatly enrich the quasiparticle characteristics such as the modulation of band gaps, the semiconductor/metal and transitions. The theoretical framework, which is built from the first-principles simulations, is successfully developed for the chemical substitutions in emergent silicene-related emergent systems [16]. In addition to the development of the quasiparticle framework, the theoretical viewpoints clearly show much progress, especially the relations among the π, σ, and sp³ chemical bondings, and spin-dependent interactions are also main concerns.

Boron, carbon, and nitrogen, respectively, possess three, four, and five valence electrons in terms of an isolated atomic configuration, being suitable for full understanding of the diversified essential quasiparticle properties. The band properties near the Fermi level (the modified Dirac-cone structures or the multi-orbital-induced low-lying energy bands; carrier dopings or finite-gap semiconductors), the well-behaved/undefined π and σ valence bands, the mixings and crossings of π- and σ-electronic energy spectra, their whole bandwidths, the various 2D van Hove singularities due to the substitution-induced band-edge states, and the presence/absence of spin configurations are the focus of this research. A detailed comparison between boron, carbon, and nitrogen substitutions is also made. Moreover, the relation between VASP and the tight-binding model [17] in band structures is worthy of detailed discussion.

9.2 COMPUTATIONAL DETAILS

The geometric, electronic, and magnetic properties of B/C/N-substituted germanene are investigated by the density functional theory (DFT) in the generalized gradient approximation (GGA) implemented in the Vienna Ab-Initio Simulation Package (VASP) [18–20]. The projected augmented wave method and Perdew, Burke, and Ernzerhof functional [19] are used to calculate the exchange-correlation energies. Furthermore, the projector augmented wave (PAW) [21] pseudopotentials can characterize electron-ion interactions. In our calculations, all works were computed in the spin-unrestricted manner [22] with the first Brillouin zones sampled by 9x9x1 and 100x100x1 k-points, respectively, for optimal geometric structures and calculations of the electronic and magnetic properties later via the Monkhorst-Pack scheme. In the z-direction, a vacuum region of 15 Å is added to avoid the interaction between adjacent cells. During the ionic relaxations process, the convergence precision of total energy is chosen to be $10-_6$ eV, and the maximum Hellmann-Feynman force is less than 0.01 eV/Å.

9.3   RESULTS AND DISCUSSION

Geometric structures of B-/C-/N-substituted germanenes present the stable honeycomb lattices under any substitution cases. The uniform chemical bondings in a Moiré superlattice are 2.02 Å, 1.86 Å, and 1.93 Å for B, C, and N cases, respectively. While the B-, C-substituted systems, as shown in Figure 9.2(b, c), suddenly change to planar implying the dominance of sp² bonding, the opposite is true for N case (Figure 9.2(d)); the height difference between the A and B sublattices is bigger than that in the absence of guest atoms (Figure 9.2(a)). This feature might be closely related to the multi-orbital hybridizations of the ill-defined π and σ chemical bondings. However, due to the strong and complicated chemical bondings, the B-, C-, and Ge-dependent

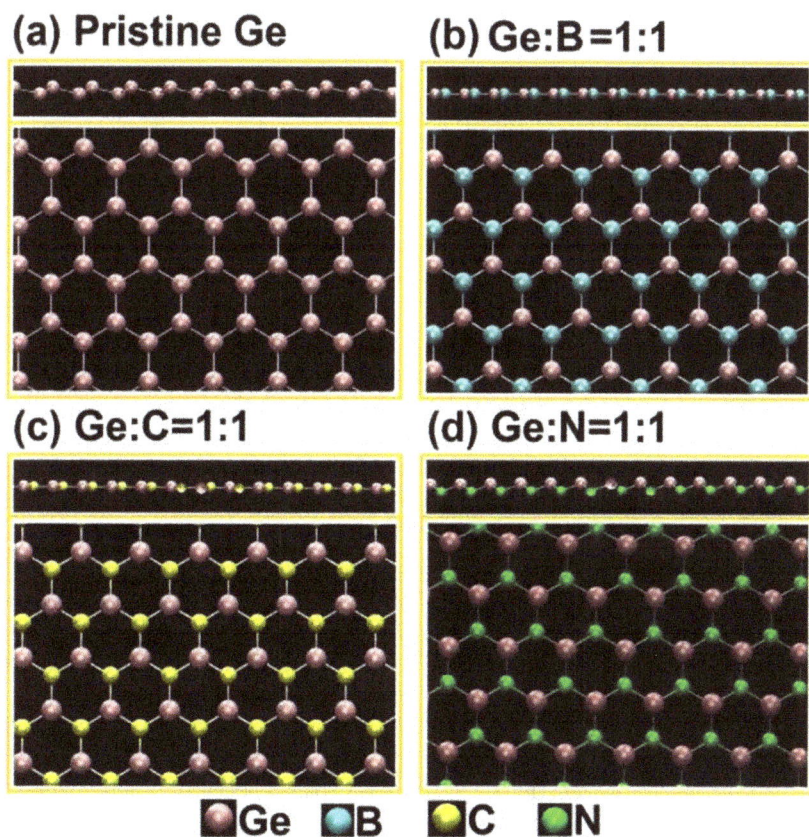

FIGURE 9.1   The side and top views of optimal geometric structures for (a) pristine germanene, (b) boron, (c) carbon, and (d) nitrogen substitutions.

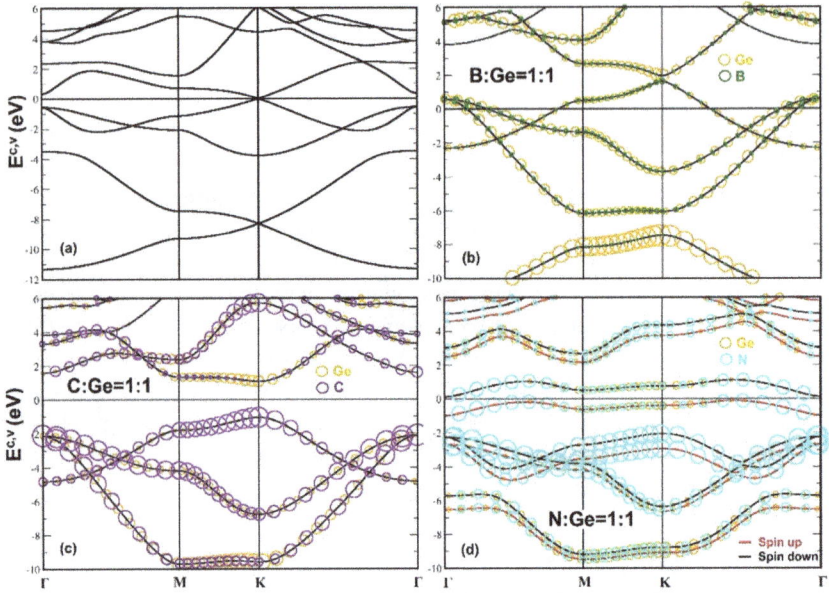

FIGURE 9.2 Band structures with the atom-dominances [the yellow, green, purple, and aqua-marine circles corresponding to germanium, boron, carbon, and nitrogen atoms, respectively] under various cases: (a) pristine germanene, (b) boron, (c) carbon, and (d) nitrogen substitution.

spin configurations/interactions are fully suppressed, that is to say, the spin-induced magnetic configuration is absent in any B- and C-substituted germanene.

The specific relations between the π, σ, and sp^3 bondings in pristine monolayer germanene describe the main features of the band structure. The electron-hole symmetry near the Fermi level is only weakly modified at low energy, as shown in Figure 9.2(a). A narrow band gap, $E_g \sim 22.0$ meV, is attributed to the enhanced spin-orbit couplings in the buckled honeycomb lattice. The first pair of valence and conduction bands nearest to $E_F = 0$ is initiated from the stable K valley. Moreover, they show the valence and conduction saddle M-point structure at -1.15 eV/0.68 eV. Finally, the μ band is finished at the Γ valley, in which the π-bandwidth for valence states is about 3.48 eV. On the other hand, [$4p_x$, $4p_y$]-orbital bondings, which mainly create the four-fold degenerate σ bands, exhibit the parabolic energy dispersion from the stable Γ valley at $E_v \sim -0.56$ eV. The anti-crossing behavior between σ and π valence bands along the ΓK and ΓM directions implies there exists a weak but significant sp^3 bonding. Generally, in monolayer germanene, all π-bandwidth could be well defined. However, it is very difficult to measure the σ-bandwidth.

The full boron-/carbon-/nitrogen-substituted germanene systems exhibit unusual electronic energy spectrums that are totally different from the pristine case. For the boron-substituted germanene system (Figure 9.2(b)), with a planar honeycomb lattice, presents a p-type metal. The three valence electrons of B-atom in the [$2s$, $2p_x$, $2p_y$, $2p_z$] orbitals create the asymmetric valence and conduction bands about the Fermi level. Near the K and Γ valleys, plenty of free holes are created simultaneously. The Dirac-cone band structure is greatly changed by the impure hybridizations of $2p_z$–$4p_z$ orbitals compared with the pristine germanene (Figure 9.2(a)). Furthermore, the hybridized σ bondings of [$2s$, $2p_x$,

$2p_y$]-[$4s$, $4p_x$, $4p_y$] orbitals have an influence on the energy dispersions and the number of band-edge states. The double degenerate of σ-band state at the Γ valley that is located upon the Fermi level can make a lot of free valence holes. Most importantly, both valence and conduction bands are dominated by germanium atom [yellow circle], and the π and σ energy bands are totally well defined, as similar to the monolayer graphene.

Very interestingly, the carbon-substituted germanene system (Figure 9.2(c)) is classified as a direct-gap semiconductor with a band gap of 2.12 eV at the K valley. The band gap might depend on the different ionization energies of the C-$2p_z$ and Ge-$4p_z$ orbitals. The valence and conduction of π/π^* states are highly anisotropic due to showing different group velocities along the KM and KΓ directions. Similar to boron substitution, the anti-crossing behaviors of π and σ bands in the C-Ge honeycomb lattice indicate that their chemical bondings are orthogonal to each other. However, at low-energy, the first valence and conduction structure are dominated by C-$2p_z$ and Ge-$4p_z$ orbitals, respectively.

The characterization of the impure/pure μ and σ chemical bondings, as well as their non-orthogonality or orthogonality, are able to reveal in the spatial charge density distributions. Both μ and σ chemical bondings that survive in pristine germanene become non-orthogonal to each other, as clearly observed in Figure 9.3(a).

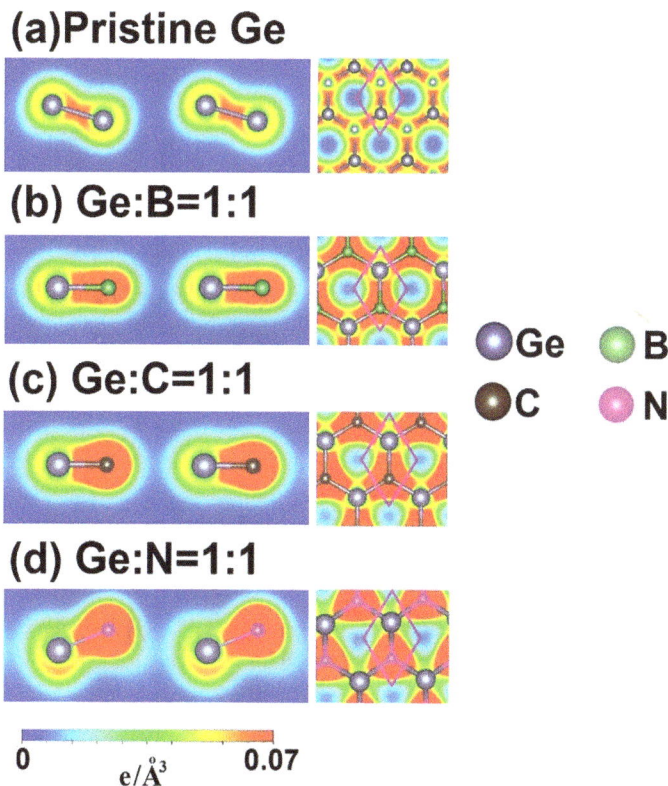

FIGURE 9.3 The spatial charge distributions in (a) the pristine germanene, (b) boron, (c) carbon, and (d) nitrogen substitution.

This feature once again confirmed the existence of a weak and significant sp³ bonding. Apparently, most of the charge is clustered in the area between two germanium atoms [the red region] in proportion to the very strong σ bonding of three orbitals [4s, 4p$_x$, 4p$_y$]. Moreover, perpendicular to the [x; y] plane, the parallel 4p$_z$ orbitals that create the μ bonding are also observed by the light green in the outer region.

Apparently, the charge density distribution is quite similar for boron- and carbon-substituted germanene (Figures 9.3(b, c)). The higher carrier density is accumulated around guest atoms, revealing the charge transfer from germanium to boron/carbon atoms. The charge density is highly asymmetric with respect to the B/C-Ge bond center, mainly owing to the different electron affinities of the guest and host atoms. The chemical bonding of C-Ge is stronger than B-Ge because the red region around carbon is heavier than boron. This result is also consistent with the bond lengths as shown in Table 9.1. Most importantly, due to the planar honeycomb lattice, the σ and μ bondings which roughly established by 2p$_z$-4p$_z$ and [2s, 2p$_x$, 2p$_y$]-[4s, 4p$_x$, 4p$_y$] orbital hybridizations are orthogonal to each other. Therefore, in B-/C-substituted germanenes, the sp³ orbital hybridizations might be negligible. Each B/C-Ge bond has the same environment, with only one chemical bond in a unit cell. This phenomenon indicates that the tight-binding model might be reliable in simulating first-principles band structure results.

The buckling structure in the nitrogen-substituted germanene compound displays a unique charge density distribution, as illustrated in Figure 9.3(d). The formation of the μ bonds is affected by the charge distribution. It is shown that almost all carriers are located around the nitrogen atom. Consequently, the charge density becomes highly asymmetrical about the center of the N-Ge bonds, as shown in both [x; z]- and [x; y]-plane projections. The N-Ge bonds are estimated to have significant and complex [2s, 2p$_x$, 2p$_y$, 2p$_z$]-[4s, 4p$_x$, 4p$_y$, 4p$_z$] sp³-orbital hybridizations. That is to say, it is almost impossible to clearly characterize the μ and σ chemical bondings. The presence of mixed four-orbital hybridizations implies that it would be difficult to simulate the VASP band structures by employing a tight-binding

TABLE 9.1

The Optimal Geometric Structures of Pristine Germanene, B-Ge, C-Ge, and N-Ge with Bond Length, Lattice Constant, Magnetic Moment, Band Gaps, Total Energy, and Buckling

Configuration	X-Ge bond length (Å)	Lattice constant	Magnetic moment (μ$_B$)/ magnetism	E_g^d(eV)/ Metal	Optimization E$_0$ (eV)	Δ (Å)
Pristine Germanene	2.428	4.050	X	0.022	−8.0471	0.65
B-Ge	2.021	3.500	X	Metal	−9.7367	0
C-Ge	1.860	3.222	X	2.12	−11.980	0
N-Ge	1.922	3.100	0.45	0.243	−11.571	0.7

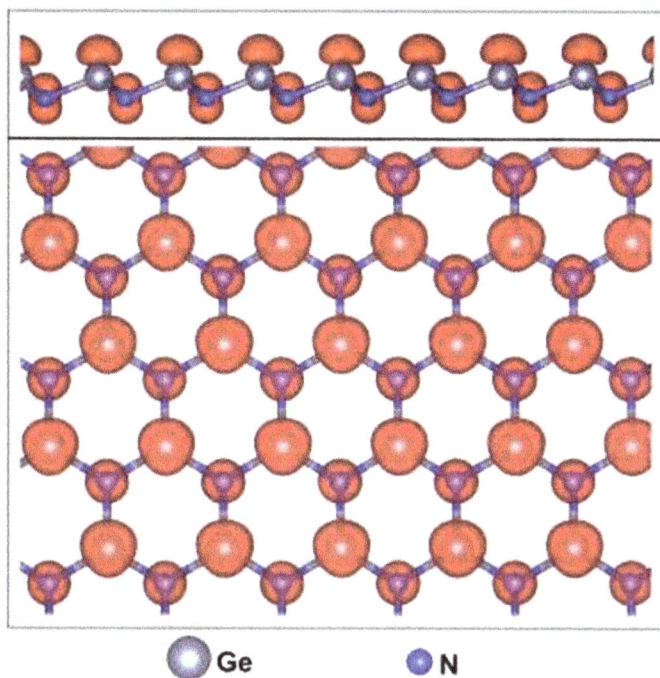

Ge   N

FIGURE 9.4 The side and top view of spin-charge distributions in a nitrogen-substituted germanene system.

model with orbital-induced hopping integrals, site energies, and the spin-dependent interactions.

Very interestingly, only nitrogen-substituted germanene present the ferromagnetic spin configuration, as shown in Figure 9.4, compared with the carbon- and boron-related ones. In our notation, the spatial distribution of the spin density is extremely nonuniform and anisotropic. In the stable honeycomb lattice of a uniform 2D nitrogen- germanium compound, only spin-up density (illustrated by the red region) manifests. Furthermore, the spin-up density seems symmetrical about the center of the N atom (blue spheres), while its distribution is only displayed in the upper planar of the host atom (gray spheres), as demonstrated in Figure 9.4. Apparently, the magnetic properties are largely caused by this asymmetry. In our calculations, the p-orbital of nitrogen atom plays a large part in the magnitude of the magnetic moment (in supplement files). This result is consistent with the N and Ge-dominances of the spin-split valence band below the Fermi level (Figure 9.2(d)) or the comparable spin-up densities of states due to the $2p_z$ and $4p_z$ orbitals (discussed in Figure 9.5(d)).

The complicated orbital hybridizations in B-/C-/N-substituted germanenes can be clarified by the atom-, orbital-, and spin-decomposed density of states, as clearly illustrated in Figures 9.5(a) through 9.5(d). The existence of special structures in the density of state mainly comes from band-edge states of parabolic and oscillatory

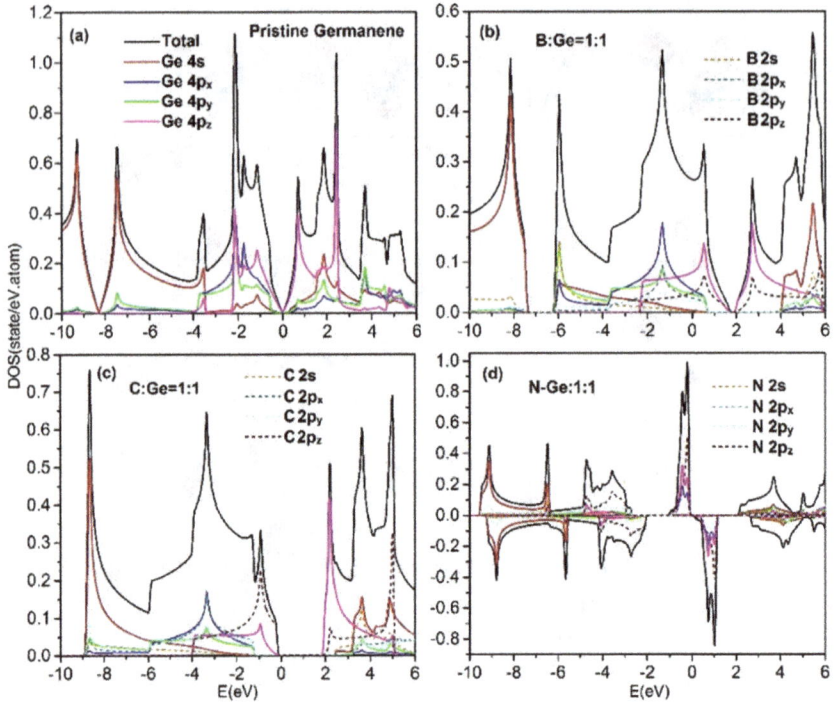

FIGURE 9.5   The atom- and orbital-decomposed density of states in various cases: (a) pristine germanene, (b) boron, (c) carbon, (d) and nitrogen substitution.

energy dispersions. Apparently, in all the substitution cases, the π bondings due to B-/C-/N-2p$_z$ and Ge-4p$_z$ are dominant at low energy. The modified Dirac-cone structure across the Fermi level in B-substituted germanene (Figure 9.5(b)) is mainly determined by the Ge-4p$_z$ and B-2p$_z$ orbitals corresponding to the asymmetric V-shapes with an observable energy spacing. Obviously, Ge-[4s, 4p$_x$, 4p$_y$, 4p$_z$] orbitals make larger contributions in every region. As a result, the π- and σ-bandwidths are ~ 3.80 eV and 4.80 eV, respectively. This feature of van Hove singularities further illustrates the well-behaved π and σ chemical bondings as discussed in previous sections.

The dominance of π bondings is also observed at low energy in C-substitution (Figure 9.5(c)). However, the first valence and conduction bands are dominated by the C-2p$_z$ orbitals and Ge-4p$_z$ ones, respectively. Compared with the pristine germanene, the almost perfect Dirac cone is totally destroyed. The special structures cover vanishing density- of-states within a band gap of E$_g$ ~ 2.12 eV at the K valley, the initial μ shoulder/delta function-like π^* peak at −0.25 eV/1.93 eV, the −0.97 eV symmetric μ peak in the logarithmic divergence, the final μ shoulder at −3.90 cV, the first σ shoulder at −1.25 eV, the −3.37 eV logarithmic peak, and several delta function-like peaks at −8.6 eV. Like the B case, C-substitution clearly reveals no evidence of μ-σ band mixing. This is mainly due to the absence of the simultaneous four-orbital structures. The fact that the sp^3 bonding is absent, which agrees with the direct μ-σ sub-band crossings (Figures 9.2(b)–9.2(c)). The not too complex chemical bonds give an opportunity for the simulation of these results by phenomenological methods. When the first-principles electronic energy spectra along the high-symmetry paths are successfully simulated by the tight-binding model with the uniform and multi-/single-orbital hopping integrals, the diversified essential properties could be fully explored in the near-future studies, e.g., the rich and unique magnetic quantization phenomena, the quantum spin Hall effect, Coulomb excitation, and optical properties.

The atom-, orbital-, and spin-dependent interactions play critical roles in the creation of various van Hove singularities in the N-Ge compound. According to the density of states near the Fermi level, the system is a semiconductor with a narrow gap of 0.24 eV. Roughly speaking, within the energy range of −1.0 eV < E < 1.2 eV, the density of states is dominated by the Ge-$4p_z$ and N-$2p_z$ orbitals [the dashed purple and wine solid curves], in which the [4s, $4p_x$, $4p_y$, 2s, $2p_x$, $2p_y$] orbitals also make significant contributions. Concerning the spin-split density of states, the ferromagnetic configuration is mainly determined by the sub-bands near the Fermi level. The spin-up and spin-down ones appear at E < 0 and E > 0, respectively. Apparently, in this case, the valence electronic states determine the net magnetic moment. However, the buckling honeycomb lattice might create the complicated [2s, $2p_x$, $2p_y$, $2p_z$]-[4s, $4p_x$, $4p_y$, $4p_z$] sp^3 -orbital hybridizations and thus make it difficult to trace the itinerary as well as to measure the π and σ bandwidths.

The theoretical predictions on the B-/C-/N-Ge bond lengths could be examined by the high-resolution STM/TEM/LEED measurements [23–25], while the ARPES measurements [26] on the main features of occupied states are very useful in examining the guest-atom substitution effect such as the π and σ valence bands, the modified Dirac-cone structures with anisotropic dispersions and Fermi momenta, and the unoccupied hole states near the stable K or Γ valleys. Besides, the high-resolution STS/SP-STS measurements [27, 28] are the most efficient and accurate way of exploring the van Hove singularities of layered materials. Very interestingly, direct combinations of SP-STM and SP-STS are capable of thoroughly examining the spin-induced magnetic properties. Furthermore, the net magnetic moments could be verified through the superconductor quantum interference device (SQUID) [29].

9.4   CONCLUSIONS

The first-principles method is available in thoroughly exploring the diversified fundamental properties in binary compounds, fully B-/C-/N-substituted germanenes. The delicate calculations and analyses, which are conducted on the buckling/planar honeycomb lattices, the atom-dominated energy spectra, the top-/side-view charge densities, the spin arrangements, and the atom-, orbital-, and spin-projected van Hove singularities, can determine the orbital hybridizations in chemical bonds and the atom-induced magnetic configurations. The concise pictures cover the pure sp^2 and dominating sp^3 bondings, respectively, in [B, C]- and N-guest systems and the N-created spin configurations. As for B-/C-/N-related compounds, they can account for the planar/planar/buckled honeycomb lattices, the metallic/wide-/narrow-gap band properties, the slight modification drastic change/destruction of Dirac-cone structures, the magnetic/nonmagnetic/ferromagnetic properties, the crossings/anti-crossings of μ and σ bands, and the pure sp^3 energy bands.

REFERENCES

[1]   Geim A K and Novoselov K S 2007 The rise of graphene *Nat Mater* **6** 183–91
[2]   Vogt P, Capiod P, Berthe M, Resta A, De Padova P, Bruhn T, Le Lay G and Grandidier B 2014 Synthesis and electrical conductivity of multilayer silicene *Appl Phys Lett* **104**
[3]   Derivaz M, Dentel D, Stephan R, Hanf M C, Mehdaoui A, Sonnet P and Pirri C 2015 Continuous Germanene layer on A(111) *Nano Lett* **15** 2510–16
[4]   Chen R B, Chen S C, Chiu C W and Lin M F 2017 Optical properties of monolayer tinene in electric fields *Sci Rep-Uk* **7**
[5]   Yu X L, Huang L and Wu J S 2017 From a normal insulator to a topological insulator in plumbene *Phys Rev B* **95**
[6]   Hanh T T T, Phi N M and Hoa N V 2020 Hydrogen adsorption on two-dimensional germanene and its structural defects: An ab initio investigation *Phys Chem Chem Phys* **22** 7210–17
[7]   Abdullah N R, Kareem M T, Rashid H O, Manolescu A and Gudmundsson V 2021 Spin-polarised DFT modeling of electronic, magnetic, thermal and optical properties of silicene doped with transition metals *Physica E* **129**
[8]   Zhuang J C, Xu X, Peleckis G, Hao W C, Dou S X and Du Y 2017 Silicene: A promising anode for lithium-ion batteries *Adv Mater* **29**
[9]   Tan X, Cabrera C R and Chen Z F 2014 Metallic BSi3 silicene: A promising high capacity anode material for lithium-ion batteries *J Phys Chem C* **118** 25836–43
[10] Zhao F L, Wang Y, Zhang X, Liang X J, Zhang F, Wang L, Li Y, Feng Y Y and Feng W 2020 Few-layer methyl-terminated germanene-graphene nanocomposite with high capacity for stable lithium storage *Carbon* **161** 287–98
[11] Vogg G, Brandt M S and Stutzmann M 2000 Polygermyne: A prototype system for layered germanium polymers *Adv Mater* **12** 1278–81
[12] Ye M, Quhe R, Zheng J X, Ni Z Y, Wang Y Y, Yuan Y K, Tse G, Shi J J, Gao Z X and Lu J 2014 Tunable band gap in germanene by surface adsorption *Physica E* **59** 60–5
[13] Shen S N, Shen W, Liu S, Li H, Chen Y H and Qi H Q 2020 First-principles calculations of co-doping impurities in diamond *Mater Today Commun* **23**
[14] Amato M, Kaewmaraya T and Zobelli A 2020 Extrinsic doping in group IV hexagonal-diamond-type crystals *J Phys Chem C* **124** 17290–8
[15] Sun M L, Ren Q Q, Wang S, Zhang Y J, Du Y H, Yu J and Tang W C 2016 Magnetism in transition-metal-doped germanene: A first-principles study *Comp Mater Sci* **118** 112–16
[16] Lin S Y L H Y, Nguyen D K, Tran N T T, Pham H D, Chang S L, Lin C Y, Lin M F 2020 *Silicene-Based Layered Materials* (Bristol: IOP Publishing)
[17] Chegel R and Behzad S 2020 Tunable electronic, optical, and thermal properties of two-dimensional Germanene via an external electric field *Sci Rep-Uk* **10**
[18] Kresse G and Joubert D 1999 From ultrasoft pseudopotentials to the projector augmented-wave method *Phys Rev B* **59** 1758–75
[19] Perdew J P, Burke K and Ernzerhof M 1996 Generalized gradient approximation made simple *Phys Rev Lett* **77** 3865–8
[20] Kresse G and Furthmuller J 1996 Efficient iterative schemes for ab initio total-energy calculations using a plane-wave basis set *Phys Rev B* **54** 11169–86
[21] Blochl P E 1994 Projector augmented-wave method *Phys Rev B* **50** 17953–79
[22] Pople J A, Gill P M W and Handy N C 1995 Spin-unrestricted character of Kohn-Sham orbitals for open-shell systems *Int J Quantum Chem* **56** 303–5

[23] Tapaszto L, Dobrik G, Nemes-Incze P, Vertesy G, Lambin P and Biro L P 2008 Tuning the electronic structure of graphene by ion irradiation *Phys Rev B* **78**

[24] Stobinski L, Lesiak B, Malolepszy A, Mazurkiewicz M, Mierzwa B, Zemek J, Jiricek P and Bieloshapka I 2014 Graphene oxide and reduced graphene oxide studied by the XRD, TEM and electron spectroscopy methods *J Electron Spectrosc* **195** 145–54

[25] Hamalainen S K, Boneschanscher M P, Jacobse P H, Swart I, Pussi K, Moritz W, Lahtinen J, Liljeroth P and Sainio J 2013 Structure and local variations of the graphene Moiré on Ir(111) *Phys Rev B* **88**

[26] Hwang E H and Das Sarma S 2008 Quasiparticle spectral function in doped graphene: Electron-electron interaction effects in ARPES *Phys Rev B* **77**

[27] Ridene M, Girard J C, Travers L, David C and Ouerghi A 2012 STM/STS investigation of edge structure in epitaxial graphene *Surf Sci* **606** 1289–92

[28] Yamada T K, Bischoff M M J, Heijnen G M M, Mizoguchi T and van Kempen H 2003 Observation of spin-polarized surface states on ultrathin bct Mn(001) films by spin-polarized scanning tunneling spectroscopy *Phys Rev Lett* **90**

[29] Indolese D I, Karnatak P, Kononov A, Delagrange R, Haller R, Wang L J, Makk P, Watanabe K, Taniguchi T and Schonenberger C 2020 Compact SQUID realized in a double-layer graphene heterostructure *Nano Lett* **20** 7129–35

10 Excitonic and Spin-Orbit Coupling Effects on Optical Properties of Plumbene Adsorption Hydrogen

Vo Khuong Dien, Nguyen Thi Han, and Ming-Fa Lin

CONTENTS

10.1 INTRODUCTION

The new era of two-dimension (2D) material sciences has arisen since graphene, the monolayer of carbon atoms with a hexagonal lattice, was discovered. Such materials have attracted a lot of attention [1–5] owing to the exceptional properties that absent in the bulk structure [6–8], the potential applications for nano-electronic devices, and the possibility to downscale the thickness [9, 10] of the channel at the atomic level. Beyond the graphene, other counterpart structures also have been successfully fabricated [11, 12], especially for other group-IV elements like silicon (Si), germanium (Ge), or tin (Sn) [13–18]. Differences from the planar hexagonal structure of graphene, monolayer silicene, germanene, or stanene are energetically favorable in the bulked structure. These corrugated structures of Si, Ge, and Sn are promising materials in the design of field-effect transistors, as an application of vertical electric fields can open and control the energy band gaps. Because of the low buckled structure and greater spin-orbit coupling, silicene/germanene/stanene can also become

DOI: 10.1201/9781003322573-10

important materials for spintronic applications. Furthermore, hydrogen-passivated graphene (graphane), silicene (silicane), germanene (germanane), and stanene (stanane) have also attracted much attention in both theoretical and experimental investigations because of the drastic changes in chemical bonding that occur upon hydrogenation.

Plumbene, the latest cousin of graphene, has been mentioned as a material for topological insulators (TI) [19, 20], which perform as a new state of quantum matter with a non-trivial energy gap in the bulk while in a conducting state at the edges protected by time-reversal symmetry. Recently, Yuhara et al. reported the successful fabrication of 2D plumbene by molecular beam epitaxy (MBE) [21], which has prompted the development of related research, e.g., chemical decoration and/or hydrogenated plumbene. Although the high potentials of plumbene in nanoelectronics are expected [22], the fundamental electronic and optical properties of plumbene and related nanostructures are still unclear due to the following reasons: (i) Pb, as one of the heavy elements of group-IV atoms, needs to be considered with the spin-orbit coupling (SOC) effect. (ii) The low-dimensional systems have strong many-body interactions due to geometrical confinement and reduced screening. Previous studies on the fundamental electronic structure of plumbene are usually using the standard density functional theory (DFT) level [23–25], and therefore underestimate its band gap value. (iii) For optical properties of plumbene, such as absorption, reflectance, or photoluminescence, measurements have not been reported yet. Further studies need to investigate the optical properties and explore their potential applications. (iv) The physical/chemical pictures dominated by the orbital hybridization between H-1s with Pb orbitals in the hydrogenated plumbene systems have not been achieved up to now.

Fortunately, previous theoretical simulations based on the quasi-particle viewpoints can clearly depict the essential rich and unique features of emergence materials, especially for emergent 2D layered [26–28] and 3D ternary anode/cathode/electrolyte materials [29–31]. For example, systematic studies have been conducted on the outstanding features of ternary $Li_4Ti_5O_{12}$ anode, Li_2SiO_3, and Li_2GeO_3 electrolyte material [32–34], the essential properties of graphene/silicone-related materials [35–37]. Such investigations clearly illustrate that the quasiparticle charges, orbitals, and spin govern all the fundamental quasiparticle properties. The high accurate simulation results and delicate analyses are capable of proposing significant pictures/mechanisms to fully understand the geometric, electronic, and magnetic properties and optical excitations. The important single-/multi-orbital hybridizations in various chemical bonds are obtained from the geometric, the atom-dominated quasiparticle band structures, the spatial charge densities, and their variations after chemical modifications. The magnetic configurations in the host/guest material could be comprehended through spin-split/spin-degenerate energy bands, the spin density distributions, the net magnetic moments, and the spin-projected van Hove singularities. Furthermore, the energy-dependent optical excitations and the influence of electron-hole interactions on optical properties are thorough investigations by the dielectric function, absorption, reflectance coefficients, and energy loss function [38–40]. This developed framework theory, which is successfully conducted on silicene/graphene-related systems [41, 42] and the anode/cathode/electrolyte compounds and could be generalized to other emergent materials, needs to be thoroughly tested in further investigations. It is thus expected to be very suitable for studying the geometric, electronic, magnetic, and optical properties of hydrogenated systems.

In this chapter, the presented theoretical framework, which is focused on the orbital hybridizations in chemical bonding, was used to determine the essential properties in monolayer plumbene and its hydrogenated systems. This strategy is based on the density functional theory calculation (DFT method) and many-body perturbation theory (GW method) on the position-dependent chemical bonding of optimized structure, the atom/orbital quasiparticle band structures, the charge density distribution due to different orbitals, and the atom-/orbital-projected density of states associated with the overlap of orbitals. The specific orbital hybridizations will be used to explain the main features of optical properties, such as the optical gap, the excitonic effect, various prominent absorption structures, and a strong collective excitation in terms of the energy loss function. The predictions in this work require highly resolved experimental measurements.

10.2 COMPUTATIONAL DETAILS

In this study, we used the first-principle calculations [43] that integrated into the Vienna Ab-Initio Simulation Package (VASP) [44] to optimize the geometric structure and calculation of the electronic and optical excitations. The Perdew-Burke-Ernzerhof (PBE) [45] generalized gradient approximation was used for the exchange-correlation functional. The interaction between the ions and valence charges was described by the projector augmented wave (PAW) method [45]. The energy cutoff for the plane wave expansion was set to 400 eV. The Brillouin zone was integrated with a special k-point mesh of 35x35x1 in the Monkhorst-Pack sampling technique for geometric optimization. To avoid the interaction of the monolayer with its images, the horizontal vacuum was set equal to 25 Å. The convergence condition of the ground-state is set to be $10-^8$ eV between two consecutive simulation steps, and all atoms could be allowed to fully relax during geometric optimization until the Hellmann-Feynman force acting on each atom was smaller than 0.001 eV.

When perturbed by a photon of suitable energy, the electron will move vertically from the filled state to the empty state. To evaluate the optical response of plumbene and its hydrogenate systems, we adopted the strategy of Michael Rohlfing and Steven G. Louie [46], in which the quasi-particle Green's function and the screened Coulomb interactions (G_0W_0) approach using 200 eV energy cutoff for the response functions. The Brillouin zone was integrated with a special k-points mesh of 30x30x1 in the Gamma sampling technique. The corrected density of states and quasiparticle electronic band structure with and without SOC was achieved under WANNIER90 codes. The picture for single-particle excitation can be described by the Kubo formula [47]:

$$\epsilon_2\left(\omega\right)=\frac{8\pi^2e^2}{\omega^2}\sum_{vck}\left|e.vk|v|ck\right|^2\delta\left(\omega-E_{ck}-E_{vk}\right),$$

Where the intensity of each excitation peak and the available transition channels is directly related to the velocity matrix element, $\left|e.vk|v|ck\right|^2$, and joined with the density of states $\delta\left(\omega-E_{ck}-E_{vk}\right)$, respectively.

In addition to the independent optical excitations, the excited electrons and holes could be strongly combined through a suitable condition, such as a large band gap or suppression of temperature broadening. Using the k-point sampling, energy cutoff, and number of bands setting the same as in the GW calculation, the electron-hole states can be determined by solving the standard Bethe-Salpeter equation (BSE) [48].

$$\epsilon_2(\omega) = \frac{8\pi^2 e^2}{\omega^2} \sum_{vck} |e.\langle 0 | v | S \rangle|^2 \delta(\omega - \Omega_s).$$

The presence of SOC and stable excitons may significantly affect the main features of optical spectra. The close connection of the quasi-particle charges, the orbital hybridization, the effect of the coupled quasi-particle, and the impact of spin-orbital interactions on optical excitations will be discussed in detail in the current work.

10.3 RESULTS AND DISCUSSIONS

10.3.1 GEOMETRIC STRUCTURE

Figure 10.1(a) presents the geometric structure of a free-standing Pb monolayer (Plumbene). Different from sp^2 hybridization of graphene with perfect hexagon planar sheet [49], the buckled configuration is more energetically favorable in plumbene due to the contribution of sp^3 hybridization [50], in which the height difference of two sublattices and the optimization Pb bond length are 0.92 Å and 1.42 Å, respectively. In this study, we used the first-principles calculations to systematically investigate the optimized geometric structure of plumbane, double-side hydrogenations of plumbene, as has been done for graphene [51], in which 2 H atoms are alternately located on the top and the bottom of the monolayer plumbene plane, as shown in Figure 10.1(b). After optimization, the corresponding Pb-Pb chemical bond/the bulking height of plumbane is slightly increased/decreased to 3.01 Å and 0.81 Å as a consequence of the transformation from the mixed sp^2-sp^3 to sp^3 configuration in the neighbor Pb atoms.

(a) Plumbene   (b) Plumbane

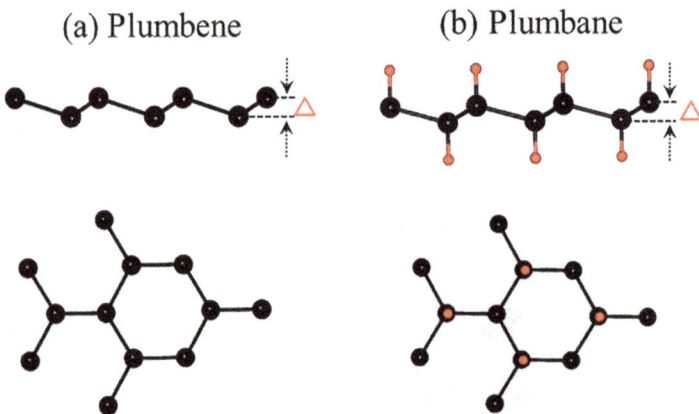

FIGURE 10.1   Side and top views of the (a) plumbene and (b) plumbane monolayers.

10.3.2 ELECTRONIC PROPERTIES

The energy band structure of plumbene is obviously presented in Figure 10.2(a). The Fermi level was set at the zero-energy and it also could serve as the reference energy. There are four valence energy sub-bands related to the [6s, $6p_x$, $6p_y$, $6p_z$] orbitals, and their energy dispersions are strongly anisotropic and mainly present the parabolic, dispersionless relations. The crossing, non-crossing, or anti-crossing phenomena exist frequently due to the composite effects. Various band-edge states, such as extreme points, survive at the high symmetry points, and these critical points will be responsible for strong optical excitations. Very interestingly, the valence and conduction bands are linear crossing at the Fermi level, showing the modified Dirac cone at the K point due to the contribution of p_z orbital (π band) similar to that of graphene or silicene [52, 53]. The spin-orbit coupling (SOC), the splitting of energy states caused by the interaction of spin magnetic momentum with the internal magnetic

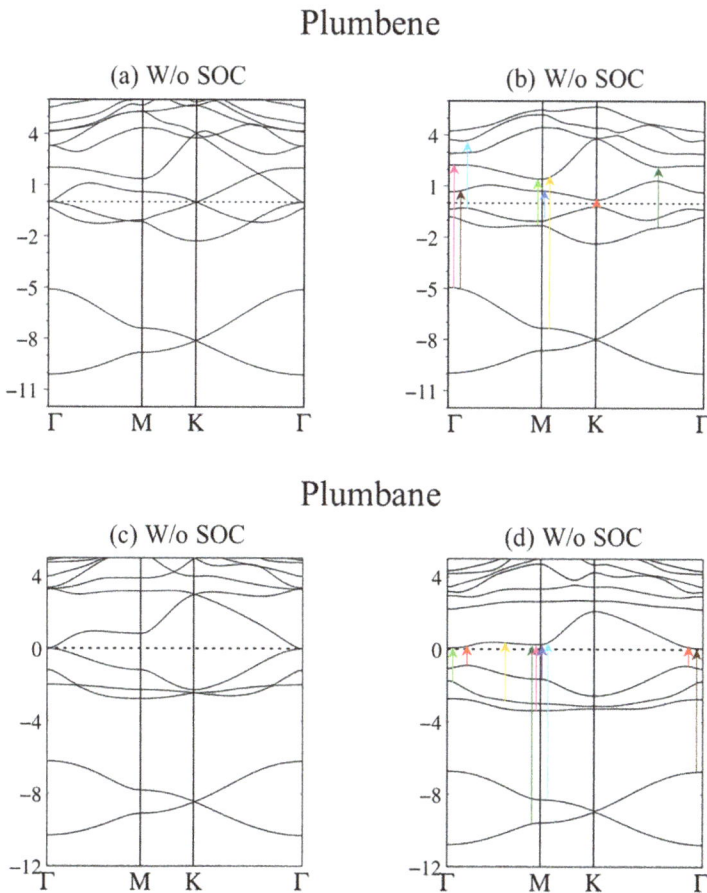

FIGURE 10.2 The quasi-particle band structures of (a)/(b) plumbene and (c)/(d) plumbane without/with spin-orbit couplings.

field established by the orbital motions, extremely affects the electronic properties of un-plana 2D materials. Since the Pb atom processes large intrinsic SOC, the energy spacing at both symmetry points is significantly enhanced when this interaction is taken into account. For example, the low-energy electronic properties, which come to exist near the stable K and Γ valleys ([very strong sp^3 bondings in [17]), are drastically changed by the significant SOCs. A pair of parabolic valence and conduction sub-bands, with a small energy spacing, occur at the Γ point, and a separated and slightly modified Dirac-cone structure appear at the K point; therefore, a direct gap of ~ 0.6 eV is generated at the K/K' valleys.

After hydrogenations, the coupling between two Pb atoms may change from sp^2-sp^3 mixing to sp^3 hybridization. This process is able to dramatically modify the energy band structure, especially for the low-energy states. For example, after hydrogenations, the π bondings at the K point are totally destroyed, but σ bonding remain at the Γ high-symmetry point in the monolayer plumbane and therefore maintain the gapless nature. The presence of five valence energy sub-bands clearly evidences the s-sp^3 orbital hybridizations. In general, the low-lying energy bands still present high anisotropic behaviors with the energy states dominant at the same energy range. The H dominance occurs in the deep energy region because the orbital of H-1s strongly hybridizes with the Pb orbitals. As the spin-orbit coupling is applied, the energy degeneracies of σ orbitals are lifted away from the Γ point and the conduction bands are upshifted, whereas the valence bands are downshifted, which produces a remarkable indirect-gap of 1.3 eV, ample for room-temperature applications. These features suggest that plumbane could be TIs upon opening a band gap at the touching point by SOC.

In order to elucidate the chemical bonding in hydrogenated plumbene, we consider the spatial charge density and charge density difference. The former reveals the serious orbital hybridizations in Pb-Pb and Pb-H bonds, while the latter could give the fully chemical reaction pictures in the real space. As for plumbene, the charge density ρ between two Pb atoms is very dense due to the presence of very strong covalence σ bonds (the red rectangle). Such bondings have slightly changed under double-side hydrogenations (Figures 10.3(b)–(c)). The spatial charge density clearly reveals that the 1s orbital of the H atom is strongly hybridized with the 6p$_z$ orbital of the Pb atom (the blue rectangle). It explains why π bonds are eliminated after the hydrogenation (Figure 10.4(b)), and thus, accounting for the absence of Dirac cone in the low-lying energy band structures (Figure 10.2(b)). In addition to the single orbital couplings, the heterogeneous charge density difference reveals that there are serious orbital hybridizations between H-1s and Pb-(6p$_x$, 6p$_y$, 6p$_z$, 6s) orbitals. The mentioned results combine with further analysis on the projected density of states. The fully chemical pictures of the adsorption configuration could be achieved.

The atom and orbital projected density of states could give the full information to comprehend the orbital hybridization in the hydrogenation systems. For 2D plumbene (Figure 10.4(a)), the density of state presents various van Hove singularities, which originated from the extreme/critical points in the energy band structure—for example, the V-shape, the shoulder, and the asymmetric peaks at the vicinity of the Fermi level, which are mainly raised from the linear Dirac cone of the π band, the parabolic or almost dispersionless relation energy of the σ bands. The electron and hole states are asymmetrical about the Fermi level. The co-existence of various orbitals is evident in

Charge density distribution

(a) Plumbene

Charge density difference

(b) Plumbane

(c) Plumbane

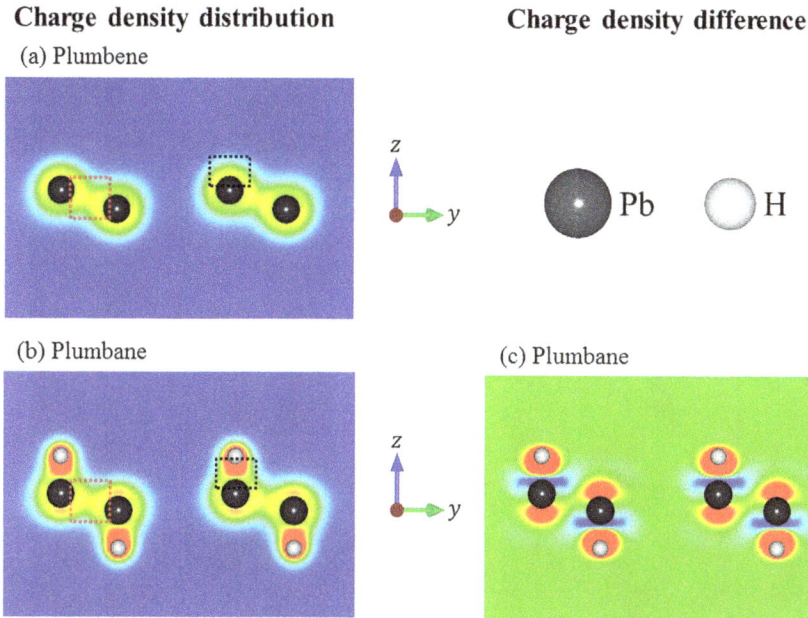

FIGURE 10.3 The charge density of (a) plumbene and (b) plumbane and (c) the charge density difference of plumbane.

the presence of sp^2-sp^3 mixing hybridizations. When applying the spin-orbit effects, the occupied states are all downward shifting, while the unoccupied states shift upward. Very interestingly, the electronic states about the Fermi level disappear and open a sizeable electronic band gap (Figure 10.4(a)). Although the π bonding is destroyed after hydrogenation, the two-dimensional plumbane without spin-orbit coupling is still a zero-gap semiconductor, since it remains the touching point of the σ bonding. The V-shape belonging to the p_z orbital disappears, while the shoulder structures and the asymmetric peaks, which arise from the px and py orbitals, arise from the parabolic and oscillatory band structure. The strong hybridizations of the 1s orbital of hydrogen with Pb orbitals are observed at the deep energy level. The spin-orbit coupling induces the plumbane to open an electronic gap of 1.3 eV (Figure 10.4(b)).

10.3.3 OPTICAL PROPERTIES

When the 2D pristine/hydrogenated plumbene are perturbed by an electromagnetic wave, all the charge carriers are capable of dynamically screening this external field and thus generate the induced current density. From the quantum mechanics point of view, this coupling could be considered the scattering between the photon quanta of electromagnetic waves and electrons in the systems. Apparently, due to the conservation of momentum and energy, the electrons could be referred to as vertically excited from the occupied states to the unoccupied ones under absorption of photons with suitable energy. This scattering mechanism could be well characterized by the

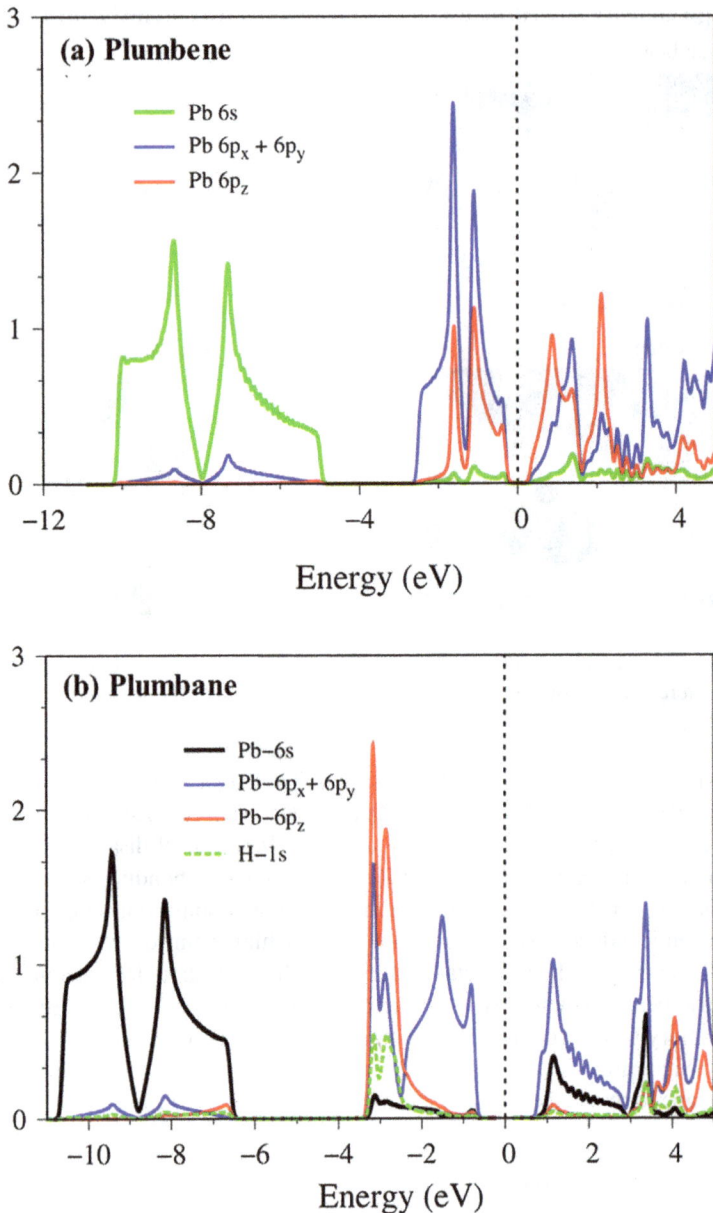

FIGURE 10.4   The orbital-projected density of states of (a) plumbene and (b) plumbane with spin-orbit coupling (SOC).

imaginary part $[\epsilon_2(\omega)]$ and the real part $[\epsilon_1(\omega)]$ of dielectric functions under long wavelength limits.

A pristine plumbene clearly exhibits the unusual optical excitation phenomena in the absence/presence of excitonic effects, as indicated in Figures 10.5(a)/(b) and

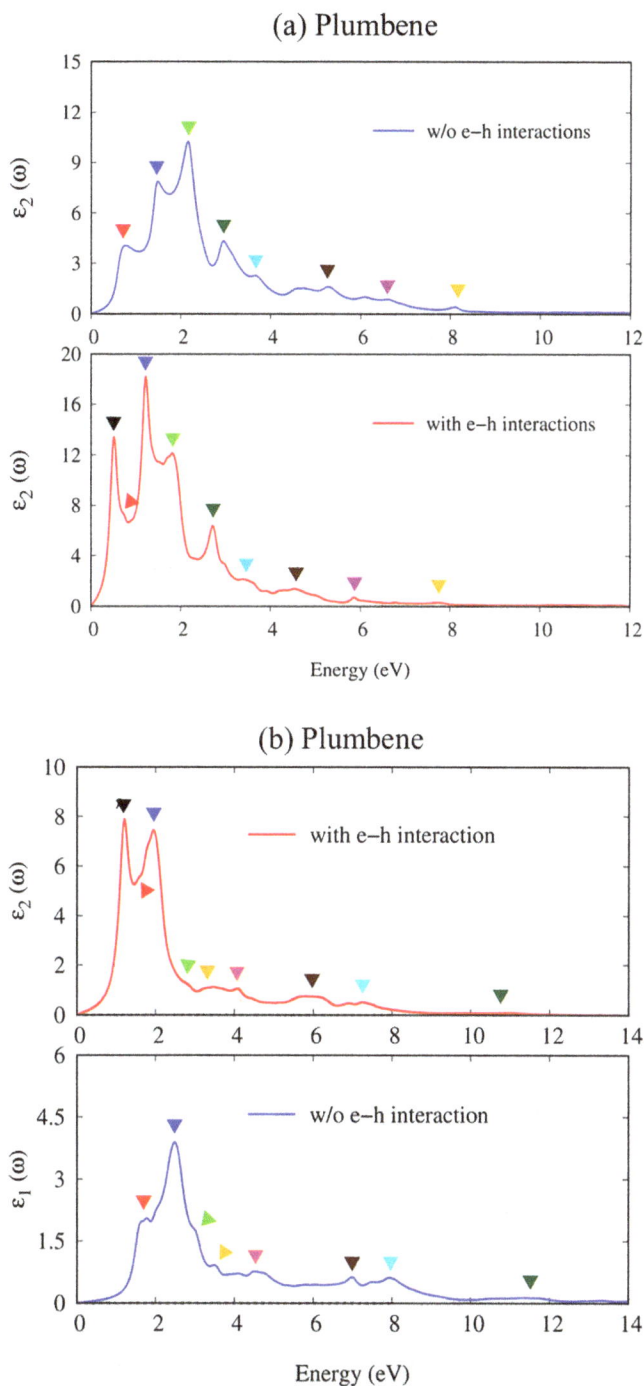

FIGURE 10.5   Comparison of the imaginary parts of the dielectric function of (a) plumbene and (b) plumbane with and without excitonic effects.

Table 10.2(a): for example, the prominent excitation structures at 0.60 eV/0.55 eV (the inter-band transitions due to Pb-$6p_z$ orbitals near the K valleys), 1.73 eV/1.56 eV (the blue arrow related to the inter-π-band transitions of the saddle M point), 2.20 eV/1.88 eV [the purple arrow due to the ($6p_x$, $6p_y$, $6p_z$) → $6p_z$ from the second valence and first conduction subbands at the M point], 2.95 eV/2.74 eV (the green arrow arising from the hybridized channels of ($6p_x$, $6p_y$, $6p_z$) → ($6p_x$, $6p_y$) in the second pair of valence and conduction subbands along the KT path), 3.55 eV/3.50 eV [the cyan arrow corresponding to ($6p_x$, $6p_y$, $6p_z$) → ($6p_x$, $6p_y$, $6p_z$) through the first valence and third conduction subbands under the "M" path], 5.31 eV/4.75 eV (the brown arrow indicating the dominant 6s → ($6p_x$, $6p_y$) channel of the third valence and first conduction subbands near the Γ point), 6.15 eV/5.90 eV (the pink arrow illustrating the available 6s → $6p_z$ transition of the third valence and second conduction subbands at the valley), and 8.14 eV/7.90 eV (the yellow arrow showing the dominating 6s → $6p_z$ channel through the third valence and second conduction sub-bands at the M point). Very interestingly, the monolayer plumbene also shows a stable exciton peak at the specific threshold frequency of 0.49 eV (the black arrow), being stronger than the band-gap structure initiated from the K valley (the red arrow). This middle-gap system of E_g^d, which has the weakest charge screening ability, cannot quickly reduce the long-range attractive Coulomb potential and thus built the stable/quasi-stable excitonic bound states.

The chemical, physical pictures as well as optical excitations of plumbene will be changed after hydrogenation. The onset energy of plumbane is expanded to 1.4 eV. The number of channels, and their energy is slightly reduced/modified due to the termination of π bands/change of σ bands. Meanwhile, the electron-hole interaction is also largely enhanced with the binding energy of about 0.4 eV. The excitonic effects in plumbane are smaller than in other hydrogenated systems, such as graphane (1.6 eV) [54], silicane (1.07 eV) [55], and germanane (0.92 eV) [55] due to the different X-H chemical bonding (X = C, Si, Ge, Sn, Pb). In addition to the threshold frequency, the single-particle excitation spectrum of plumbene also presents a lot of prominent peaks with different intermediate orbital-hybridizations, such as the 1.60 eV-/1.60 eV singularities due to the multi-orbital mixings about Pb-($6p_x$, $6p_y$) → Pb-(6s, $6p_x$, $6p_y$) (the red arrow in Figure 10.2(b)) are the consequence of the vertical transition from the first pair of valence and conduction energy subbands at the Γ valley through, the strongest 2.30-eV/1.90-eV absorption peak with a similar intermediate chemical bonding of the former, but related to the saddle M point [the blue arrow], the observable 2.90-/2.70-eV excitation structure associated with the second valence and first conduction sub-bands at the Γ valley through (6s, $6p_x$, $6p_y$, $6p_z$) → (6s, $6p_x$, $6p_y$) transitions (the light-green arrow), the weak, but significant 3.70-/3.50-eV absorption structure related to the second valence and the first conduction sub-bands along the ΓM path by the multi-orbital hybridizations of 1s-(6s, $6p_x$, $6p_y$, $6p_z$) → (6s, $6p_x$, $6p_y$) [the yellow arrow], the obviously 4.60 eV-/4.10 eV responses are related to the third occupied and first unoccupied states from the M valley using the 1s-(6s, $6p_x$, $6p_y$, $6p_z$) → (6s, $6p_x$, $6p_y$) chemical bondings (the pink arrow), the non-negligible 7.00 eV-/5.90 eV-absorption peak from the fourth valence and first conduction subbands at the stable Γ valley, the singularities of 8.00-/7.30-eV are due to the fourth valence and first conduction subbands from the M point using the 1s-(6s, $6p_x$, $6p_y$, $6p_z$) → (6s,

TABLE 10.1

Structure and Electronic Parameters of Plumbene and Plumbane

Structure	Pb-Pb	Pb-H	Buckle	Band gap (eV)	
				W/o SOC	With SOC
Plumbene	3.0		0.92	0	0.6
Plumbane	3.01	1.81	0.81	0	1.4

TABLE 10.2

Calculated Prominent Absorption Structures and the Leading Transition of Each Peak

System	Colors	Energy (eV)		Specific orbital hybridizations
		w/o excitonic effects	with excitonic effects	
Plumbene	Black	–	0.49	–
	Red	0.6	0.55	Pb-(6pz) $\rightarrow$ Pb-(6pz)
	Blue	1.73	1.56	Pb-(6pz) $\rightarrow$ Pb-(6pz)
	Light Green	2.20	1.88	Pb-(6px, 6py, 6pz) $\rightarrow$ Pb-(6pz)
	Green	2.95	2.74	Pb-(6px, 6py, 6pz) $\rightarrow$. Pb-(6px, 6py)
	Cyan	3.55	3.50	Pb-(6px, 6py, 6pz) $\rightarrow$ Pb-(6px, 6py, 6pz)
	Brown	5.31	4.75	Pb-(6s) $\rightarrow$ Pb-(6px, 6py)
	Pink	6.15	5.90	Pb-(6s) $\rightarrow$ Pb-(6pz)
	Yellow	8.14	7.90	Pb-(6s) $\rightarrow$ Pb-(6pz)
Plumbane	Black	–	1.1	–
	Red	1.6	1.6	Pb-(6px, 6py) $\rightarrow$ Pb-(6s, 6px, 6py)
	Blue	2.3	1.9	Pb-(6px, 6py) $\rightarrow$ Pb-(6s, 6px, 6py)
	Light Green	2.9	2	Pb-(6s, 6px, 6py, 6pz) $\rightarrow$ Pb-(6s, 6px, 6py)
	Yellow	3.7	3.5	Pb-(6s, 6px, 6py, 6pz) + H-1s $\rightarrow$ Pb-(6s, 6px, 6py)
	Pink	4.6	4.1	Pb-(6s, 6px, 6py, 6pz) + H-1s $\rightarrow$ Pb-(6s, 6px, 6py)
	Brown	7	5.9	Pb-(6s, 6px, 6py) + H-1s $\rightarrow$ Pb-(6s, 6px, 6py)
	Cyan	8.0	7.3	Pb-(6s, 6px, 6py) + H-1s $\rightarrow$ Pb-(6s, 6px, 6py)
	Green	11.7	10.8	Pb-(6s, 6px, 6py) + H-1s $\rightarrow$ Pb-(6s, 6px, 6py)

$6p_x$, $6p_y$) intermediate states (the cyan arrow), the highest-frequency 11.70-/10.80-eV absorption peak between the fifth valence and first conduction band under the similar mechanisms of the previous (the green arrow).

The energy loss function (ELF), which is defined as $Im\left[1/\epsilon\left(\omega\right)\right]$ in Figure 10.6, can be used to describe the collective excitation of certain valence charges. Each prominent peak in ELF-plasmon mode corresponds to the zero point in the real part of dielectric functions $[\epsilon_1(\omega)]$ accompanied with a weak Landau damping ($\epsilon_2(\omega)$ vanishes). For the monolayer plumbene, there are four observable plasmon modes located at 6.0 eV/5.6 eV, 7.0 eV/6.60 eV, 7.60 eV/7.20 eV, and 8.30/8.00 eV. It is very important to notice that the well-known π plasmon modes in graphene/silicene/germanene systems characteristic of the dominating sp^2 bondings are almost absent in monolayer plumbene because of too-strong sp^3 hybridizations. The chemical/physical pictures of plumbene will be changed after being hydrogenated, and this behavior could be depicted on the single-particle excitation as we mentioned in Figure 10.5. For the energy loss function, its spectra are moderately reshaped. For example, there exist four prominent plasmon modes. Their characteristic frequencies come to exist at 3.30 eV/3.00 eV, 5.00 eV/4.50 eV, 7.00 eV/6.59 eV, and 8.50 eV/7.50 eV. Moreover, the coherent charge oscillations are identified

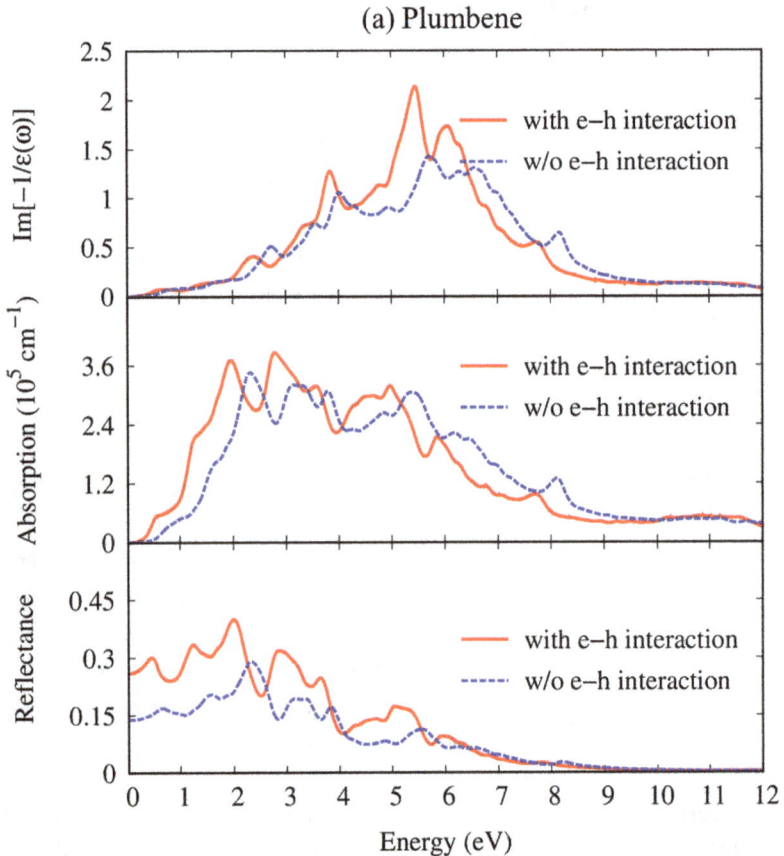

FIGURE 10.6 Various optical properties: energy loss functions, optical absorption, and reflection coefficients of (a) pristine plumbene. *(Continued)*

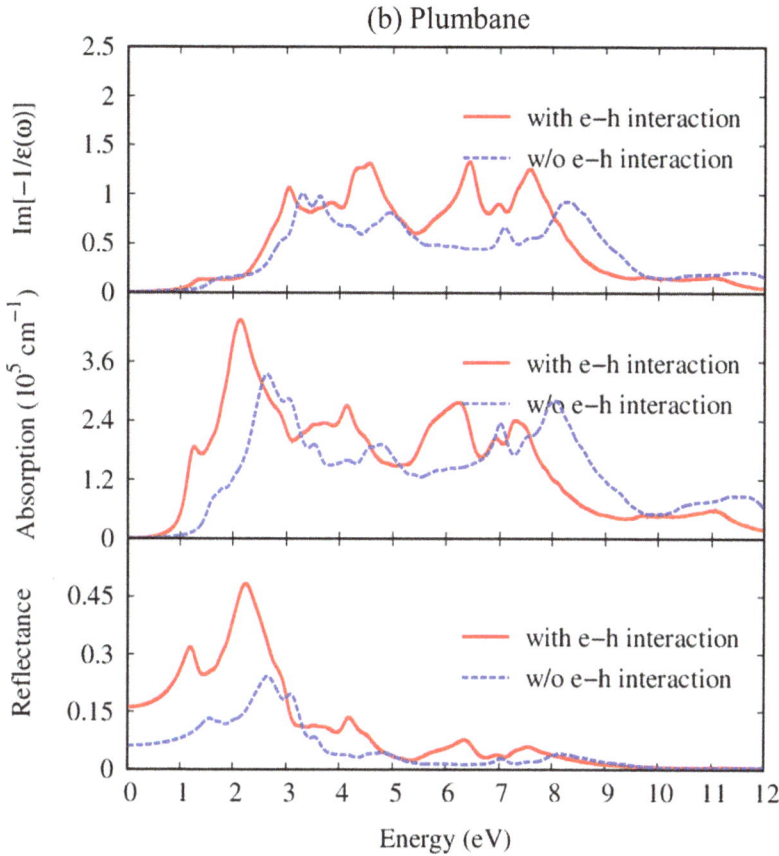

FIGURE 10.6 (*Continued*) (b) plumbane.

to mainly come from the 1s-(6s, $6p_x$, $6p_y$) multi-orbital hybridizations in Pb-Pb and Pb-H bonds.

The absorption and reflectance coefficients are two main characteristics of optical properties and reflect the main features of collective and single optical excitations. The absorptions and reflection-related energy spectra are shown in Figures 10.6(b)–(c). Below the optical gap, the absorption coefficients vanish while the static reflection coefficients are about 2.7 and 1.6 for plumbene and plumbane, respectively. Beyond the threshold frequency, the coefficients of both reflection and absorption drastically fluctuate with the increase of photon energy. There also exists a dramatic drop-out near the plasmon frequencies for the frequency-related absorption and reflection spectra. Very interestingly, $R(\omega)$ totally vanishes within the specific range $\omega > 4$ eV for both systems. It indicates that only a partial perturbation is absorbed by the monolayer plumbene/plumbane, while most of them will be transmitted.

10.4   CONCLUSIONS

In summary, monolayer plumbene and its hydrogenated systems exhibit diverse quasiparticle behaviors under distinct orbital hybridizations of the chemical bonds by using first-principles calculations. The effects, which are due to the significant overlap of different constitutions, the strong spin-orbit coupling, and the weak but important electron-hole coupling, are responsible for outstanding features. As for plumbene, the consequence of sp^2-sp^3 orbital mixing, its geometric structure exhibits a significant buckle in the freestanding configuration. The strong σ band and rather weak π band reveal in the parabolic, the oscillatory, and the modified Dirac cone/the touching of valence-conduction bands at the K/Γ high symmetry points of the energy band structure. Similarly, the projected density of states exhibits the shoulder, asymmetric, and V-shaped van Hove singularities. The presence of spin-orbit coupling induced expansion of the conduction and valence bands, thus leading to a sizable band gap. The single-particle excitation spectrum exhibits the strong like-step profile peaks, the robust exciton states, and the most prominent collective excitation peak depicted by the electron loss functions, reflection, and absorption coefficients.

The hydrogenated systems show a drastic change in their geometric structures, thus leading to change in the electronic and optical properties. The weakening of the Pb-Pb chemical bond and termination of p_z orbital leading energy band structures exhibit the absence of the direct cone, the slightly modified the σ bands after double-side hydrogenations. Furthermore, the extremely strong Pb-H chemical bond contributes to the flat bands at deep energy for all systems, creating similar van Hove singularities. Under the spin and orbital interactions, the energy states are significantly splitting, and thus, the energy gap is opened. The optical transition features cover the expansion of the optical spectrum, numbers of the single excitation channels, and the plasmon modes. The optical excitation events of all systems mainly occur in the visible range, and thus, the monolayer plumbene and plumbane can be used for electro-optic devices. Under current investigations, the state-of-the-art theoretical framework could be generated for other emergent 2D material systems.

ACKNOWLEDGMENTS

This work is supported by the Hi-GEM Research Center and the Taiwan Ministry of Science and Technology under grant number MOST 108–2212-M-006–022-MY3, MOST 109–2811-M-006–505 and MOST 108–3017-F-006–003.

REFERENCES

[1]   Castro Neto A H, Guinea F, Peres N M R, Novoselov K S and Geim A K 2009 The electronic properties of graphene *RvMP* **81** 109–62
[2]   Seabra A B, Paula A J, de Lima R, Alves O L and Durán N 2014 Nanotoxicity of graphene and graphene oxide *Chemical Research in Toxicology* **27** 159–68
[3]   Geim A K 2009 graphene: Status and prospects *Science* **324** 1530–4

[4] Banhart F, Kotakoski J and Krasheninnikov A V 2011 Structural defects in graphene *ACS Nano* **5** 26–41

[5] Zhu Y, Murali S, Cai W, Li X, Suk J W, Potts J R et al. 2010 graphene and graphene oxide: Synthesis, properties, and applications *Advanced Materials* **22** 3906–24

[6] Li X, Tao L, Chen Z, Fang H, Li X, Wang X et al. 2017 graphene and related two-dimensional materials: Structure-property relationships for electronics and optoelectronics *Applied Physics Reviews* **4** 021306

[7] Gupta A, Sakthivel T and Seal S 2015 Recent development in 2D materials beyond graphene *Progress in Materials Science* **73** 44–126

[8] Ionita M, Pandele A M, Crica L and Pilan L 2014 Improving the thermal and mechanical properties of polysulfone by incorporation of graphene oxide *Composites Part B: Engineering* **59** 133–9

[9] Ghosh D, Calizo I, Teweldebrhan D, Pokatilov E P, Nika D L, Balandin A et al. 2008 Extremely high thermal conductivity of graphene: Prospects for thermal management applications in nanoelectronic circuits *Applied Physics Letters* **92** 151911

[10] Marmolejo-Tejada J M and Velasco-Medina J 2016 Review on graphene nanoribbon devices for logic applications *Microelectronics Journal* **48** 18–38

[11] Lim W S, Kim Y Y, Kim H, Jang S, Kwon N, Park B J et al. 2012 Atomic layer etching of graphene for full graphene device fabrication *Carbon* **50** 429–35

[12] Zhuang M, Ou X, Dou Y, Zhang L, Zhang Q, Wu R et al. 2016 Polymer-embedded fabrication of Co2P nanoparticles encapsulated in N, P-doped graphene for hydrogen generation *Nano Letters* **16** 4691–8

[13] Kouvetakis J, Menendez J and Chizmeshya A 2006 Tin-based group IV semiconductors: New platforms for opto-and microelectronics on silicon *Annu Rev Mater Res* **36** 497–554

[14] Reboud V, Gassenq A, Hartmann J, Widiez J, Virot L, Aubin J et al. 2017 Germanium based photonic components toward a full silicon/germanium photonic platform *Progress in Crystal Growth and Characterization of Materials* **63** 1–24

[15] Bradac C, Gao W, Forneris J, Trusheim M E and Aharonovich I 2019 Quantum nanophotonics with group IV defects in diamond *Nature Communications* **10** 1–13

[16] Doherty J, Biswas S, Galluccio E, Broderick C A, Garcia-Gil A, Duffy R et al. 2020 Progress on Germanium—Tin nanoscale alloys *Chemistry of Materials* **32** 4383–408

[17] Doh S-J, Rhee S-W, Kim J-P, Lee J-H, Lee J-H and Kim Y-S 2006 Method of fabricating silicon-doped metal oxide layer using atomic layer deposition technique *Google Patents*

[18] Tang W, Liu Y, Peng C, Hu M Y, Deng X, Lin M et al. 2015 Probing lithium germanide phase evolution and structural change in a germanium-in-carbon nanotube energy storage system *Journal of the American Chemical Society* **137** 2600–7

[19] Pang Z-X, Wang Y, Ji W-X, Zhang C-W, Wang P-J and Li P 2020 Two-dimensional ligand-functionalized plumbene: A promising candidate for ferroelectric and topological order with a large bulk band gap *Physica E: Low-Dimensional Systems and Nanostructures* 114095

[20] Yu X-L, Huang L and Wu J 2017 From a normal insulator to a topological insulator in plumbene *Physical Review B* **95** 125113

[21] Yuhara J, He B, Matsunami N, Nakatake M and Le Lay G 2019 graphene's latest cousin: plumbene epitaxial growth on a "nano watercube" *Advanced Materials* **31** 1901017

[22] Grazianetti C, Martella C and Molle A 2020 Two-dimensional Xenes and their device concepts for future micro-and nanoelectronics and energy applications in *Emerging 2D Materials and Devices for the Internet of Things* (Elsevier) pp 181–219

[23] Bihlmayer G, Sassmannshausen J, Kubetzka A, Blügel S, von Bergmann K and Wiesendanger R 2020 Plumbene on a magnetic substrate: A combined scanning tunneling microscopy and density functional theory study *Physical Review Letters* **124** 126401

[24] Katoch N, Jamdagni P, Ahluwalia P and Kumar J 2020 Optical properties of mono and bilayer plumbene: A DFT study in *AIP Conference Proceedings* p 030704

[25] Zhang L, Zhao H, Ji W-X, Zhang C-W, Li P and Wang P-J 2018 Discovery of a new quantum spin hall phase in bilayer plumbene *Chemical Physics Letters* **712** 78–82

[26] Tran N T T, Nguyen D K, Glukhova O E and Lin M-F 2017 Coverage-dependent essential properties of halogenated graphene: A DFT study *Scientific Reports* **7** 1–13

[27] Nguyen D K, Tran N T T, Chiu Y-H and Lin M-F 2019 Concentration-diversified magnetic and electronic properties of halogen-adsorbed silicene *Scientific Reports* **9** 1–15

[28] Lin S-Y, Tran N T T, Chang S-L, Su W-P and Lin M-F 2018 *Structure-and Adatom-Enriched Essential Properties of Graphene Nanoribbons* (Boca Raton: CRC Press)

[29] Giorgi G, Fujisawa J-I, Segawa H and Yamashita K 2014 Cation role in structural and electronic properties of 3D organic—inorganic halide perovskites: A DFT analysis *The Journal of Physical Chemistry C* **118** 12176–83

[30] da Silveira Lacerda L H and de Lazaro S R 2018 Multiferroism and magnetic ordering in new NiBO3 (B= Ti, Ge, Zr, Sn, Hf and Pb) materials: A DFT study *Journal of Magnetism and Magnetic Materials* **465** 412–20

[31] Mirtamizdoust B, Ghaedi M, Hanifehpour Y, Mague J T and Joo S W 2016 Synthesis, structural characterization, thermal analysis, and DFT calculation of a novel zinc (II)-trifluoro-β-diketonate 3D supramolecular nano organic-inorganic compound with 1, 3, 5-triazine derivative *Materials Chemistry and Physics* **182** 101–9

[32] Nguyen T D H, Pham H D, Lin S-Y and Lin M-F 2020 Featured properties of Li+-based battery anode: Li 4 Ti 5 O 12 *RSC Advances* **10** 14071–9

[33] Han N T, Dien V K, Thuy Tran N T, Nguyen D K, Su W-P and Lin M-F 2020 First-principles studies of electronic properties in lithium metasilicate (Li₂SiO₃) RSC Advances **10** 24721-9

[34] Khuong Dien V, Thi Han N, Nguyen T D H, Huynh T M D, Pham H D and Lin M-F 2020 Geometric and electronic properties of Li2GeO3 *Frontiers in Materials* **7**

[35] Tran N T T, Lin S-Y, Glukhova O E and Lin M-F 2015 Configuration-induced rich electronic properties of bilayer graphene *The Journal of Physical Chemistry C* **119** 10623–30

[36] Tran N T T, Lin S-Y, Lin C-Y and Lin M-F 2017 *Geometric and Electronic Properties of Graphene-Related Systems: Chemical Bonding Schemes* (Boca Raton: CRC Press)

[37] Tran N T T, Gumbs G, Nguyen D K and Lin M-F 2020 Fundamental properties of metal-adsorbed silicene: A DFT study ACS Omega **5** 13760–9

[38] Chantler C and Bourke J 2019 Low-energy electron properties: Electron inelastic mean free path, energy loss function and the dielectric function: Recent measurements, applications, and the plasmon-coupling theory *Ultramicroscopy* **201** 38–48

[39] Onida G, Reining L and Rubio A 2002 Electronic excitations: Density-functional versus many-body Green's-function approaches *Reviews of Modern Physics* **74** 601

[40] Senthilkumar P, Dhanuskodi S, Thomas A R and Philip R 2017 Enhancement of non-linear optical and temperature dependent dielectric properties of Ce: BaTiO3 nano and submicron particles *Materials Research Express* **4** 085027

[41] Nguyen D K, Tran N T T, Nguyen T T and Lin M-F 2018 Diverse electronic and magnetic properties of chlorination-related graphene nanoribbons *Scientific Reports* **8** 1–12

[42] Lin S-Y, Chang S-L, Shyu F-L, Lu J-M and Lin M-F 2015 Feature-rich electronic properties in graphene ripples *Carbon* **86** 207–16

[43] Parr R G 1983 Density functional theory *Annual Review of Physical Chemistry* **34** 631–56

[44] Hafner J 2008 Ab-initio simulations of materials using VASP: Density-functional theory and beyond *Journal of Computational Chemistry* **29** 2044–78

[45] Hammer B, Hansen L B and Nørskov J K 1999 Improved adsorption energetics within density-functional theory using revised Perdew-Burke-Ernzerhof functionals *Physical review B* **59** 7413

[46] Rohlfing M and Louie S G 1999 Optical excitations in conjugated polymers *Physical Review Letters* **82** 1959

[47] Bischoff J-M and Jeckelmann E 2017 Density-matrix renormalization group method for the conductance of one-dimensional correlated systems using the Kubo formula *Physical Review B* **96** 195111

[48] Rohlfing M and Louie S G 2000 Electron-hole excitations and optical spectra from first principles *Physical Review B* **62** 4927

[49] Duplock E J, Scheffler M and Lindan P J 2004 Hallmark of perfect graphene *Physical Review Letters* **92** 225502

[50] Li Y, Zhang J, Zhao B, Xue Y and Yang Z 2019 Constructive coupling effect of topological states and topological phase transitions in plumbene *Physical Review B* **99** 195402

[51] Pujari B S, Gusarov S, Brett M and Kovalenko A 2011 Single-side-hydrogenated graphene: Density functional theory predictions *Physical Review B* **84** 041402

[52] Quhe R, Yuan Y, Zheng J, Wang Y, Ni Z, Shi J et al. 2014 Does the Dirac cone exist in silicene on metal substrates? *Scientific Reports* **4** 1–8

[53] Wang Y-P and Cheng H-P 2013 Absence of a Dirac cone in silicene on Ag (111): First-principles density functional calculations with a modified effective band structure technique *Physical Review B* **87** 245430

[54] Cudazzo P, Attaccalite C, Tokatly I V and Rubio A 2010 Strong charge-transfer excitonic effects and the Bose-Einstein exciton condensate in graphane *Physical Review Letters* **104** 226804

[55] Wei W, Dai Y, Huang B and Jacob T 2013 Many-body effects in silicene, silicane, germanene and germanane *Physical Chemistry Chemical Physics* **15** 8789–94

11 Diverse Phenomena in Stage 2/3/4 in AlCl$_4$/Al$_2$Cl$_7$ Graphite Intercalation Compounds of Aluminum-Ion-Based Battery Cathodes

Wei-Bang Li, Ming-Hsiu Tsai, and Ming-Fa Lin

CONTENTS

11.1 INTRODUCTION

Bulk graphite has attracted a lot of studies for more than one hundred years, both experimentally and theoretically [1], mainly owing to its unusual features. This layered system consists of the dominating AB stacking (the Bernal structure in Figure 1.2. 1(a) [2]) and a few ABC ones (the rhombohedral structures in Figure 11.1(b) [3]), in which the intralayer and interlayer atomic interactions are, respectively, dominated by the C-[2s, 2p$_x$, 2p$_y$] and C-2p$_z$ orbitals. Interestingly, the well-characterized π and σ chemical bondings have created a lot of unusual quasiparticle behavior, such as the planar honeycomb lattices (without buckling in [4]), Dirac-like or non-Dirac band structures [5], the coupling of collective charge/atom oscillations [plasmons/phonons in [6]], and the diverse magnetic quantization phenomena (the quantized Landau sub-bands and magneto optical selection rules [7]). Moreover, chemical modifications can greatly diversify these essential properties. For example, both n- and p-type

DOI: 10.1201/9781003322573-11

graphite intercalation compounds have shown dramatic transformations in the geometric symmetries [8], band structures [9], the density of states [9], optical absorption spectra [9], electron-hole excitations and plasmons, electrical conductivity [10], and superconducting temperatures [11]. There also exist a lot of phenomenological models [12] and numerical simulations [13]. The theoretical framework is useful in providing concise physical/chemical/material viewpoints. This chapter focuses on the rich properties of stage-n $AlCl_4$-graphite intercalation compounds through delicate calculations and analyses of the VASP results.

Many theoretical predictions have been made based on phenomenological models and first-principles simulations [14] conducted on bulk graphites (AA/AB/ABC/random stackings [15–18]), graphite intercalation compounds, their composite systems with polymers, and nanotubes [19]. For example, the tight-binding model/generalized tight-binding model [20] are available for exploring fully the essential π-electronic properties in the absence/presence of magnetic fields, as the systematic investigations of [21] show. The superlattice model, without the interlayer atomic interactions, is combined with the random-phase approximation to comprehend the layer-dependent middle-energy π plasmon modes of layered graphite systems [22], as well as the quasiparticle energy spectra and the Coulomb decay rates (the quasiparticle self-energies in [23]). Moreover, the first-principles calculations are reliable in thoroughly exploring the geometric, electronic, optical, and phonon properties, before and after the intercalation/de-intercalation processes, such as the diversified phenomena of fundamental properties in graphite-related compounds. This method is available for conducting systematic investigations about the various intermediate configurations during the charging and discharging processes of lithium-ion-based batteries [24]. Specifically, the complicated excitonic effects of semiconducting systems could only be solved by it, being useful in determining the optical threshold absorption frequency and the exciton-bound states [25].

Graphite and graphite intercalation compounds exhibit rich and unique properties, as revealed in the experimental measurements for the essential properties [26]. The AA stacking is observed in the lithium graphite intercalation compounds [27], while the AB stacking is confirmed in most of the other ones (e.g., K-, Rb-, and Cs-related

(a) AB stackings   (b) ABC stackings

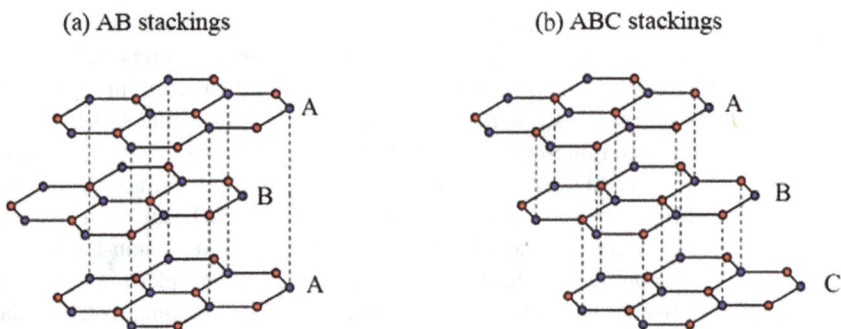

FIGURE 11.1 Geometric structures of pristine graphites: (a) AB stackings and (b) ABC stackings.

compounds in [28]). The diverse crystal symmetries, the various stage-n cases, have been successfully synthesized in experimental laboratories, being responsible for the quasiparticle behavior. Pristine graphite is verified to be a band-overlapping semi-metal by measuring the low-lying band structure (ARPES examinations in [29]) and density of states (STS observations in [30]). In general, the high free electron/hole densities are mainly generated by the n-/p-type dopings, such as alkali-/$FeCl_3$- and H_2SO_4-graphite intercalation compounds [28, 31]. Such π-electronic states [28] are able to achieve very high electrical conductivity values [28], as shown in copper. Furthermore, they belong to the traditional group of superconductors with transition temperatures below 10 K [32]. The low-frequency Coulomb excitations become unusual, e.g., the stage-n-dependent plasmon modes in the presence of hole carriers. Apparently, the optical threshold absorption frequency is enlarged under an obvious blue or red shift [33]. Most importantly, the uniform/non-uniform graphitic spacing, which is caused by the weak but significant van der Waals interactions [34], is available for the intercalations/de-intercalations of guest atoms/molecules/ions, especially for the charging and discharging processes of batteries [35]. There are many examinations of their capacities, providing one of the critical factors in determining the high performance of these batteries. How to obtain the best graphite-related materials is currently under investigation.

$AlCl_4$ intercalations into graphite will be fully explored by the VASP calculations, being greatly enriched by its large structure [36]. Many kinds of chemical bonds and orbital hybridizations in determining the quasiparticle charge properties are achieved from the consistent first-principles simulations of the optimal crystal symmetry, the atom-dominated band structure, the spatial charge density, and its variation after intercalation, and the atom- and orbital-decomposed van Hove singularities. Their direct connection with the tight-binding model [37] under a Moiré superlattice is discussed in detail. The highly anisotropic chemical environment is expected to be reflected in the featured quasiparticle properties, such as a blue shift of the Fermi level [28], a strong asymmetry of hole and electron energy spectra, zone-folding effects [38], well-behaved or undefined π and σ [39], the non-crossing/crossing/anti-crossing energy sub-bands, the high/significant charge density/variation of the carrier distribution, the atom dominances within the specific energy ranges, and the various orbital hybridizations related to the emerging special structures of the van Hove singularities. The delicate analyses are performed on the accurate VASP results. The theoretical predictions require near-future experimental examinations [40]. The similarities and differences among the various graphite compounds are also stated, being useful in basic sciences and potential applications.

11.2 RESULTS AND DISCUSSIONS

11.2.1 THE STAGE-N CRYSTAL STRUCTURES

According to previous experimental and theoretical calculations [41, 42], the guest atom or molecule intercalations/de-intercalations into the graphite spacing would generate on stage-n stacking configuration with a specific intercalant concentration. Their crystal symmetries are optimized by the current VASP simulations (details in

Chapter 2). The high-precision results, which cover the total ground state energy per unit cell, the lengths of the C-C/Al-Cl/C-Cl/Cl-Cl bonds, the specific angle in the Al-Cl bond, and the sensitive stage-n dependences, are listed in Table 11.1. Their features correspond to the highly non-uniform environments, being useful in creating the chemical reactions during the charging and discharging processes of the ion transport [36]. The predicted stacking configurations and intercalant distributions could be examined by the X-ray diffraction patterns [44].

The crystal structures of stage-n $AlCl_4$ graphite intercalation compounds possess periodical arrangements on the xy-plane as well as along the z-direction [36]. Figures 11.2(a)–11.2(d) clearly show the stage-1, stage-2, stage-3, and stage-4 cases, in which such layered superlattices are well characterized by the number of graphitic layers between two periodical intercalants. There exist 2D honeycomb lattices and intercalant layers in the presence of the highest intercalation concentration. Stable geometric structures could be achieved during the VASP optimal processes. The used parameters are listed in Table 11.1. Moiré superlattices of stage-1-stage-4 systems, respectively, have 23, 41, 59, and 77 atoms in an enlarged unit cell. Specifically, only the stage-1 system exhibits the AA stacking, while the other compounds show AB stackings. This behavior has been revealed in the stage-n lithium graphite intercalation compounds [28]. More atoms appear under a diluted chemical environment, which leads to simulation difficulties in the first-principles calculations. It might be

TABLE 11.1

The Optimal Geometric Parameters of Stage-n $AlCl_4$ Graphite Intercalation Compounds; the Ratio between Molecule and Carbon, the Total Ground State Energy Per Unit Cell, the C-C/Al-Cl/C-Cl/Cl-Cl Bond Lengths, and a Specific Angle of Al-Cl Bond

	Molecule Layer distance (Å)	C-C Layer distance (with $AlCl_4$) (Å)	C-C Layer distance (without $AlCl_4$) (Å)	C-C Bond length (Å)	Al-Cl Bond length (Å)	Al-Cl Bond angel (Å)
Primitive	–	–	3.35	1.420	2.159	109.00
Stage 1	8.35	8.35	–	1.429	2.034	118.43
Stage 2	12.04	8.71	3.31	1.421	2.152	115.92
Stage 3	16.44	8.73	3.82	1.422	2.153	113.01
Stage 4	20.91	9.02	3.88	1.422	2.157	110.09
	Primitive AA (2atom /cell)	Primitive AB (4atom /cell)	Stage1 (23atom /cell)	Stage2 (41atom /cell)	Stage3 (59atom /cell)	Stage4 (77atom /cell)
Ground state energy per unit cell (eV)	−18.418	−36.839	−184.65	−353.40	−515.66	−681.81

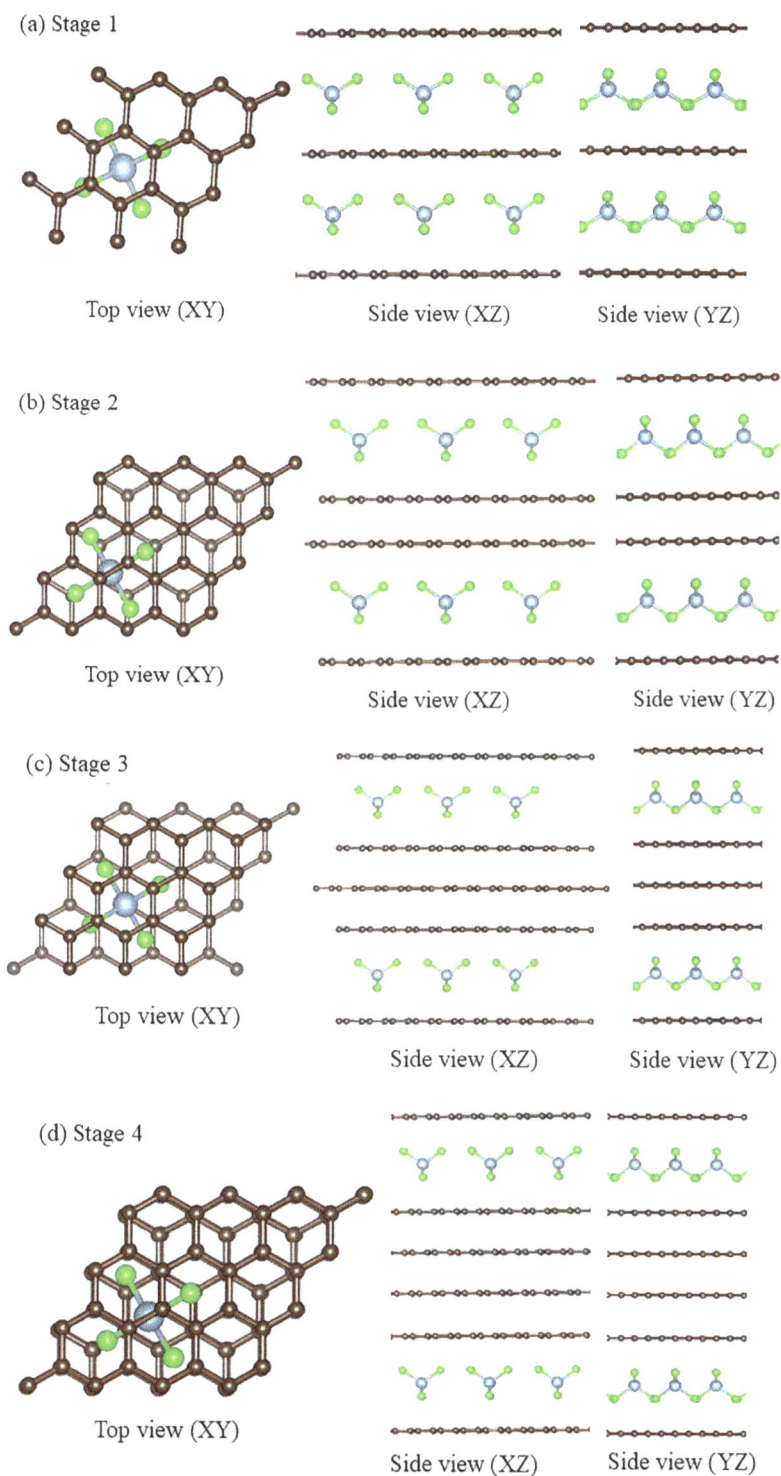

(a) Stage 1

Top view (XY)   Side view (XZ)   Side view (YZ)

(b) Stage 2

Top view (XY)   Side view (XZ)   Side view (YZ)

(c) Stage 3

Top view (XY)   Side view (XZ)   Side view (YZ)

(d) Stage 4

Top view (XY)   Side view (XZ)   Side view (YZ)

FIGURE 11.2 AlCl$_4$-graphite intercalation compounds: (a) stage 1, (b) stage 2, (c) stage 3, (d) stage 4.

mission impossible to explore the dynamic chemical reactions of these intercalation/ de-intercalation reactions [45]. Very interestingly, the chemical/physical/material environments are highly non-uniform, as obviously indicated by the various bond lengths/bond angles and the longitudinal and transverse periods. That is to say, intra-layer/interlayer C-C bondings, interlayer C-intercalant interactions, and intra-/inter-molecule ones will be the focus of studies, being responsible for the other essential quasiparticle properties [46].

Layered graphite consists of layered honeycomb lattices, with very strong σ chemical bondings, even in the presence of large-molecule/-ion intercalations and de-intercalations. Up-to-date green energy applications clearly show that they are frequently utilized as the cathode/anode of aluminum-related batteries [47], mainly owing to the high-performance charging and discharging processes. The critical role is the graphitic spacing that can be modulated within a large fluctuation. The weak but significant van der Waals forces are the featured interlayer interactions [48]. The outstanding physical/chemical/material environment of pristine graphites is capable of generating a lot of graphite intercalation compounds, such as acceptor- and donor-type systems (free valence holes and conduction electrons in [49]). They display many merits, providing the full information on basic researches and potential applications [50]. Systematic investigations will be necessary in near-future studies.

The theoretical predictions on the crystal structures of stage-n $AlCl_4$ graphite intercalation compounds could be directly examined by the X-ray diffraction patterns, as has been very successfully done on alkali-adatom ones [51]. For example, lithium and other alkali atoms exhibit similar stacking configurations but rather different concentrations and distributions [28]. That is to say, all the alkali-metal graphite-related compounds present planar structures of a carbon-honeycomb and intercalant lattices. Two periodical intercalant layers cover n-layer graphenes. These high-symmetric layered structures present the planar distributions for the smallest lithium ions and others [52], respectively, leading to the well-known chemical structures of $LiC_{\{6n\}}$ and $MC_{\{8n\}}$ (M = Na, K, Rb; Cs). Obviously, both kinds possess apparent differences in terms of interlayer distances and bond lengths [52]. Accurate X-ray examinations can clarify the quasiparticle properties and the highly non-uniform chemical environment.

11.2.2 BAND STRUCTURES

The featured band structures and wave functions are analyzed in detail under the VASP simulations, in which they are clearly shown only along the high-symmetry points within the first Brillouin zone (Figure 11.3(e)).

The doping-enriched phenomena, which cover the blue shift of the Fermi level, the asymmetry of the valence hole and conduction electron energy spectra [29], the modified Dirac-cone structures [53], the coexistence of non-crossings/crossings/anti-crossings [9], the characterizations of the π, σ, and intercalant-related energy bands [54], and the specific energy ranges corresponding to the different atom dominances/ co-dominances, will be investigated thoroughly. How to examine the main features via the high-precision ARPES measurements [29] and develop the tight-binding model [55] is also discussed.

Before and after chemical intercalations, graphite-related systems exhibit unusual energy spectra and wave functions, as clearly indicated in Figure 11.3(a) and 11.3(c)–(f). A pristine Bernal possesses two pairs of valence and conduction bands near the K and H valleys (the π and π^* bands in 1). This infinitely layered system belongs to a

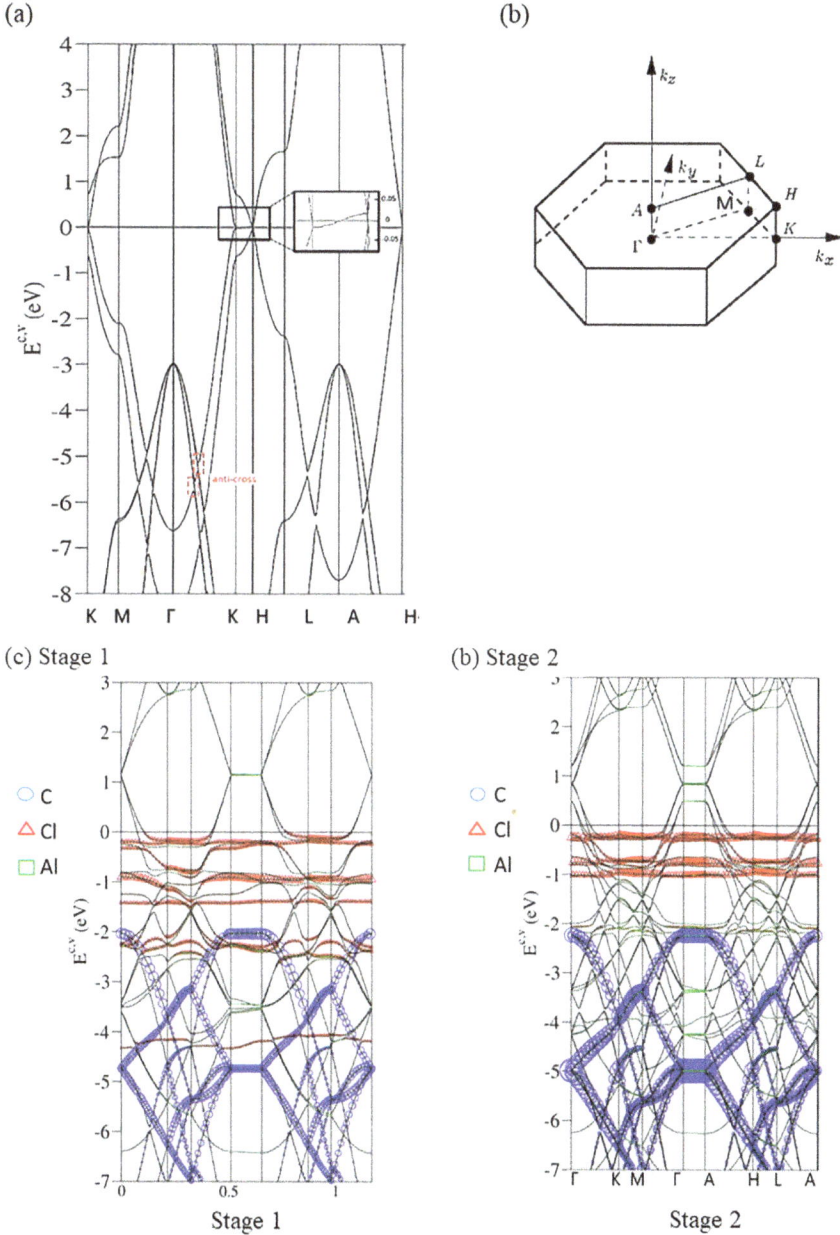

(a)

(b)

(c) Stage 1

(b) Stage 2

FIGURE 11.3 Band structures for (a) a pristine AB-stacked graphite within (b) the first Brillouin zone, (c) stage 1, (d) stage 2, (e) stage 3, (f) stage 4. *(Continued)*

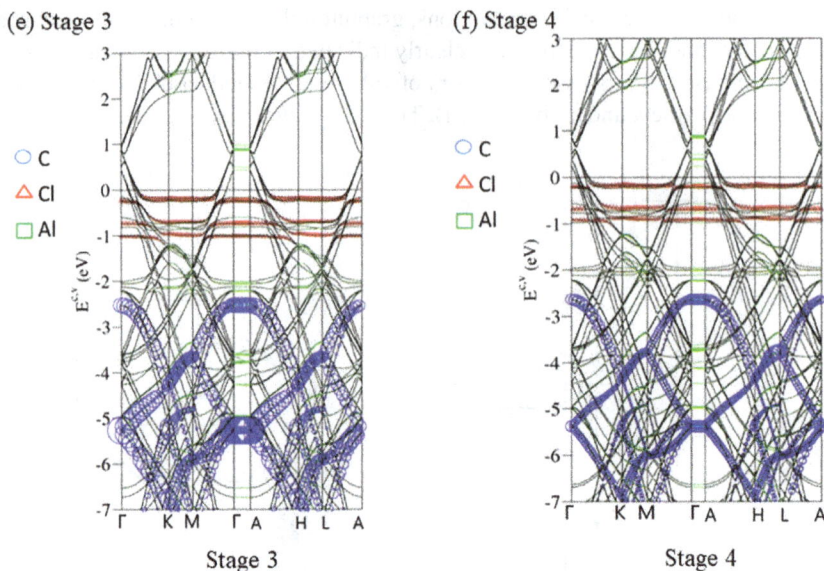

FIGURE 11.3   *(Continued)* (e)–(f) stage 3 and stage 4.

standard semi-metal since it simultaneously has low 3D free valence holes and conduction electrons. Only the weak but significant band overlaps are revealed along the K-H line, such as the co-existence of free electron and hole pockets near the stable K and H valleys. These positive and negative charges make the same contributions to all the essential properties, e.g., the thermal capacity [56], optical absorption spectra [9], Coulomb excitations [2], and transport properties [7]. Apparently, the van der Waals interactions lead to strong modifications of the low-lying band properties, in which the parabolic energy dispersions are quite different from the isotropic and gapless Dirac cone of monolayer graphene. The occupied valence energy spectrum is highly asymmetric to the unoccupied conduction one, mainly owing to the interlayer atomic interactions and the multi-orbital hybridizations [57]. The entire π and σ bands, which, respectively, arise from the π and σ bondings of $2p_z$ and (2s, $2p_x$, $2p_y$) orbitals, are well-defined by their energy spectra along the 3D high-symmetry points within the first Brillouin zone (Figure 11.3(b)), The former and the latter are associated with KMΓHLA and ΓMK/AHL, where the exact paths can determine the entire band widths. In addition, a few anti-crossings are revealed in the red rectangles. This unusual phenomenon clears a rather weak non-orthogonality of the π and σ bondings [58] and is characteristically absent in single-layer graphene [59].

    Chemical intercalations can greatly diversify the main features of electronic energy spectra and wave functions, as shown in the stage-n cases (Figures 11.3(e)–11.3(f)). The featured quasiparticle behavior covers the initial π-electronic states of the σ valley but not the K one, the various energy dispersions of the linear, parabolic, and linear wave-vector dependences [60], a lot of band-edge states (many critical points in the energy-wave-vector space, inducing the van Hove singularities in [61]), the enhanced asymmetric hole and electron bands, the red shift of the Fermi levels (p-type doping behaviors in [62]), more valence and conduction bands due to the Moiré superlattices [63], the normal/undefined π and σ energy bands, and the C-/Al-/Cl-dominated or co-dominated energy sub-bands (blue circles, red triangles, and green squares, respectively), as well as their sensitive dependence on the stacking configuration or the intercalant concentration. Specifically, the free valence holes, which survive near the stable Γ valley, are closely related to the planar $AlCl_4$-molecule distributions, as revealed in the lithium cases [28]. Obviously, their density declines during the decrease of the intercalant concentration (in increments of the stage-n). Generally speaking, the intercalation-induced carrier density is higher than that in pristine graphite. That is, the stage-n $AlCl_4$ graphite compounds and Bernal graphite, respectively, belong to metals and semi-metal. The chemical intercalations are able to induce the unusual semi-metal-metal transition [41].

The VASP band structures can further provide the atom-decomposed contributions, being useful in determining the critical orbital hybridizations of quasiparticle charge. Figures 11.3(c)–(f) clearly show the C-, Cl-, and Al-atom dominances (by the various magnitudes of the colored symbols). The first ones make the most important contributions within a whole energy range, while the other atoms have a partial flat band, especially for the conduction sub-bands. Furthermore, their energy spectra have strong wave-vector dependence so that the localized states hardly survive, even under chemical intercalation (the propagating features of π- and σ-electronic states [64]). These results are consistent with a lot of carbon host atoms within a Moiré superlattice (details in Table 11.1 and the energy-dependent density of states in Figures 11.5(a)–(p)). On the other hand, the intercalant-dominated energy bands, which come into existence below the Fermi level, might display weak energy dispersions. Interestingly, the Cl- and C-atoms co-dominate the low-lying occupied and unoccupied valence sub-band close to E_F [$E^V > -1.5$ eV]. This clearly indicates that their intercalation has created a p-type charge transfer (quasiparticles from the latter to the former due to the relative affinity [62]) and thus the red shift of E_F (the semi-metal-metal transition [41]). As for the deeper-energy states, there exist Cl- and Al-related valence sub-bands simultaneously. For example, the stage-1 $AlCl_4$ graphite intercalation compound reveals the partial flat valence bands at $E^V \sim -2.5$ eV, $E^V \sim -4.4$ eV. Similar localized behavior is revealed in all the stage-n systems, in which they directly reflect the very strong intra-molecule interactions, whereas the opposite is true for the inter-molecule cases. The predicted results are expected to greatly diversify the other quasiparticle phenomena, e.g., the diverse dependences of magnetic [65], optical, Coulomb-excitation, and transport properties.

The high-precision ARPES measurements, as clearly illustrated in Chapter 3.3, can be utilized to detect the significant intercalation effects in the current predictions. Very interestingly, the featured band structure of a pristine Bernal graphite along the k-direction is available for overcoming the momentum non-conservation issues due to the surface boundary. Its semi-metal behavior, which is due to the van der Waals interactions [48], is directly verified by the ARPES examinations. In addition, the band-overlap free electrons and holes are very sensitive to thermal excitations; therefore, they exhibit very rich Coulomb excitations, single-particle and collective excitations with unusual temperature- and momentum-dependences under a frequency of ~10–100 meVs [66]. The ARPES observations are rather useful in verifying the blue shift of the Fermi level, the enhanced asymmetry spectra of the electron and hole energy spectra, the non-crossing/crossing/mixing behavior, the well-defined or undefined π and σ bands, and intercalant-induced occupied energy bands, as well as their strong dependence on the intercalant concentrations and distributions. Such clarifications could provide a partial support of the quasiparticle properties, i.e., the important multi-orbital hybridizations [57].

11.2.3 ACTIVE ORBITAL HYBRIDIZATIONS

The spatial charge density distributions ($\rho(r)s$ in Figure 11.4), their variations after $AlCl_4$-molecule intercalations ($\Delta\rho(r)s$ in Figure 11.4), and the van Hove singularities in density of states (Figures 11.5(a)–(p)), can clearly identify the critical single-multi-orbital hybridizations in the active chemical bonds under the partial support of the atom-dominated energy bands (Figures 11.3(a)–(e) [57]), Only the first ones are not sufficient in verifying all the significant orbital mixings associated with the non-negligible chemical bonds. The merged special structures in the density of states, which arise from specific atom orbitals simultaneously, indicate their non-negligible hybridization. These need to be delicately examined step by step. They are consistent with the specific energy ranges of atom-dominated energy spectra (Figures 11.3(a)–(e)). The quasiparticle properties will be determined [5] and thus very useful in developing a theoretical framework.

Both $\rho(r)s$ and $\Delta\rho(r)s$ are available for identifying the concise quasiparticle picture related to the chemical bonds with active orbital hybridizations. By the delicate calculations, the [x, y]-top, [x, z]-side, and [y, z]-side views clearly illustrate the intralayer, interlayer, intramolecular, and intermolecular orbital mixings. First, the most prominent σ bonding of the C-[2s, $2p_x$, $2p_y$] orbitals, which appears in a pristine honeycomb lattice, clearly shows very strong covalent bonds between two neighboring carbon atoms (a rather high carrier density indicated by the red color on the (x,y)-plane in Figure 11.4(a)). Their charge variations are only slightly modified after the $AlCl_4$-molecule intercalations, as indicated in Figures 11.4(d)/11.4(g), 11.4(j)/11.4(m), 11.4(p)/11.4(s), 11.4(v), and 11.4(y), respectively, for the stage-1, stage-2, stage-3 and stage-4 cases. Moreover, the π bonding, which is due to the parallel $2p_z$-orbital hybridizations, is revealed by the wave-like charge distributions of both the (x,z) and the (y,z) planes (Figures 11.4(b) and 11.4(c), respectively). Its charge density distribution along the z-direction is somewhat extended by the significant van der Waals interactions [34]. Secondly, the significant interlayer carbon-chloride couplings are directly reflected in the obvious variations of $\Delta\rho(r)s$ near the C- and Cl-atoms (respectively, the blue and red regions in the side views of Figures 11.4(h), 11.4(i), 11.4(n), 11.4(o), 11.4(t), 11.4(u), 11.4(z), and 11.4 (α)), where there exist reduced and enhanced charge densities for the former and the latter. This charge transfer phenomenon is mainly determined by the electron affinities [68]. However, it is very difficult to observe the obvious evidence of charge variations about the carbon-aluminum interactions.

As for both, the Al-Cl and Cl-Cl bonds, their features are clearly characterized by the strongly anisotropic charge density distributions near them (Figures 11.4(e)/11.4(f), 11.4(k)/11.4(l), 11.4(q)/11.4(r), 11.4(w), and 11.4(x)). Evidence of the Al-Al bonds are absent in the VASP predictions. Accurate analyses are expected to be very useful in understanding the intercalation behavior of the various molecules and ions, such as the diverse chemical bonds in H_2SO_4- [68], HNO_3- [69], and $FeCl_3$-related [70] graphite intercalation compounds.

The strong evidence, being thoroughly identified from $\rho(r)s$ and $\Delta\rho(r)$, indicates these significant chemical bondings: C-C bonds of a honeycomb lattice, significant

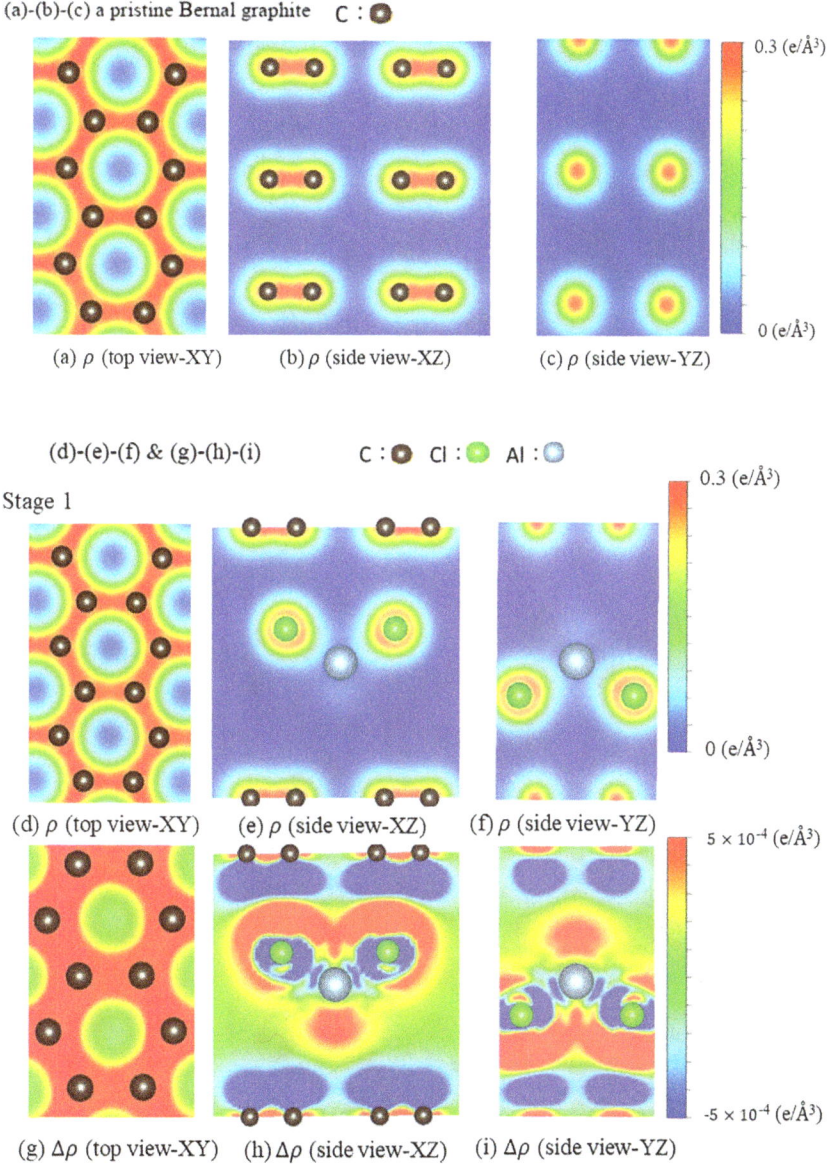

FIGURE 11.4   The spatial charge densities for top view: (a) and (d), and for side view: (b), (c) (e), and (f). The charge variations after intercalations for top view: (g), and for side view: (h) and (i). *(Continued)*

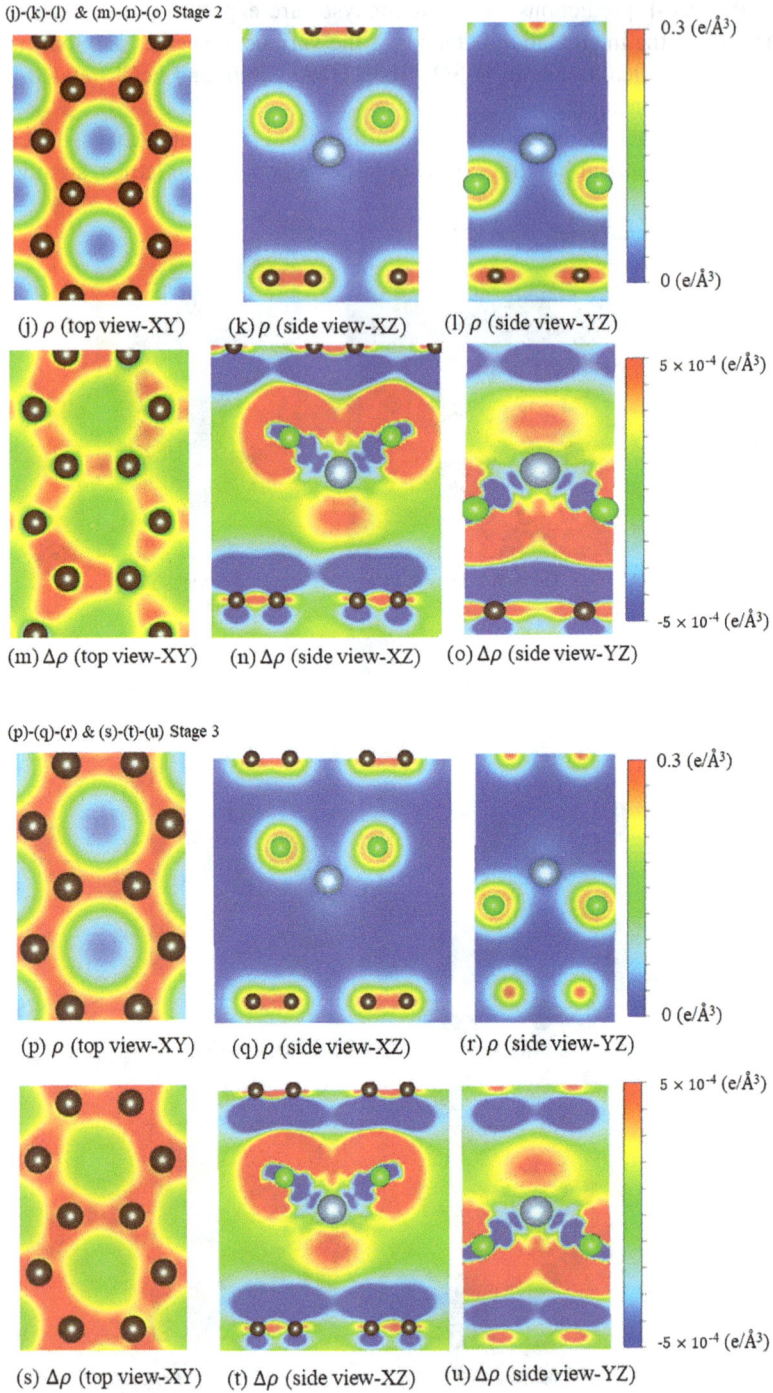

(j)-(k)-(l) & (m)-(n)-(o) Stage 2

(j) ρ (top view-XY)   (k) ρ (side view-XZ)   (l) ρ (side view-YZ)

(m) $\Delta\rho$ (top view-XY)   (n) $\Delta\rho$ (side view-XZ)   (o) $\Delta\rho$ (side view-YZ)

(p)-(q)-(r) & (s)-(t)-(u) Stage 3

(p) ρ (top view-XY)   (q) ρ (side view-XZ)   (r) ρ (side view-YZ)

(s) $\Delta\rho$ (top view-XY)   (t) $\Delta\rho$ (side view-XZ)   (u) $\Delta\rho$ (side view-YZ)

FIGURE 11.4   (*Continued*) The spatial charge densities for top view: (a) and (d), and for side view: (b), (c) (e), and (f). The charge variations after intercalations for top view: (g), and for side view: (h) and (i).

(v)-(w)-(x) & (y)-(z)-(α) Stage 4

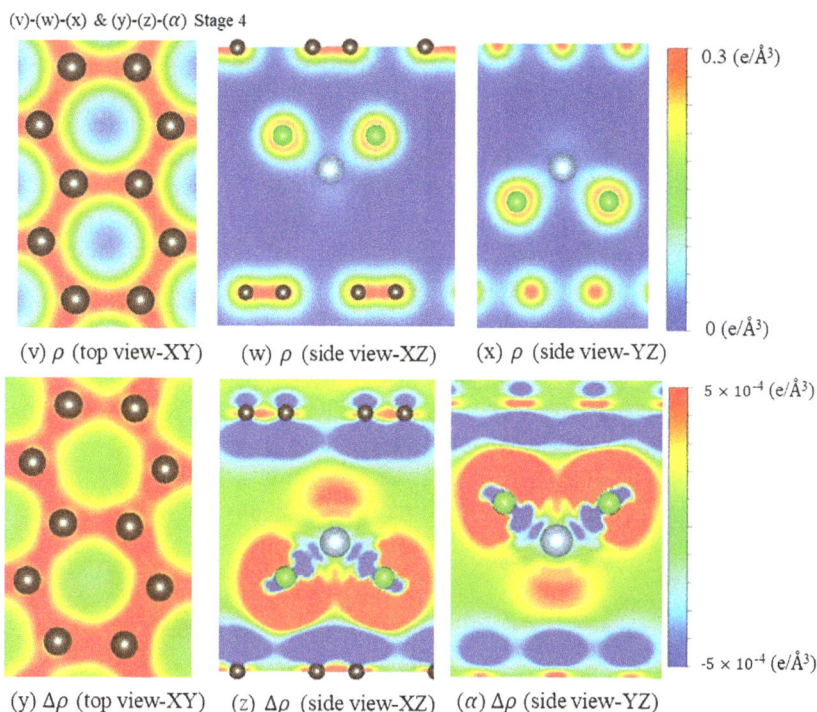

(v) ρ (top view-XY)   (w) ρ (side view-XZ)   (x) ρ (side view-YZ)

0.3 (e/Å³)

0 (e/Å³)

5 × 10⁻⁴ (e/Å³)

-5 × 10⁻⁴ (e/Å³)

(y) $\Delta\rho$ (top view-XY)   (z) $\Delta\rho$ (side view-XZ)   (α) $\Delta\rho$ (side view-YZ)

FIGURE 11.4 (*Continued*) The spatial charge densities for top view: (v), and for side view: (w), (x). The charge variations after intercalations for top view: (y), and for side view: (z), (/alpha/).

interlayer C-Cl bonds, and intra-molecule/inter-molecule Al-Cl and Cl-Cl/Cl-Cl bonds. Also, they are consistent with the featured density of states (the merged van Hove singularities in Figures 11.5(a)–(p)) and the electronic energy spectra (atom dominances in Figures 11.3(c)–(f)). However, both Al-Al and Al-C bonds hardly appear in the charge density distributions. The predicted charge densities are directly reflected in the elastic X-ray diffraction patterns. How to achieve their close relations is worthy of a thorough investigation.

The density of states is characterized by the integration of the inverse of the group velocity on the constant-energy surface for 3D graphite-related materials. The critical points [71], which are present in the energy-wave-vector space, are responsible for the unusual van Hove singularities. Their special structures are diversified by the different band-edge states of the parabolic, linear, and partially flat energy dispersions (extreme and saddle points [71]). In addition, they are sensitive to distinct dimensionalities. The calculated results are delicately decomposed into the specific contributions of the active atoms and orbitals. As a result, significant orbital hybridizations are achieved from the merged unusual structures at specific energies. These detailed analyses could be generalized to other complex condensed-matter systems [41]. The current predictions strongly depend on the intercalant concentrations and distributions, such as the different redshifts of the Fermi level in the stage-n AlCl$_4$ graphite intercalation compounds. How to directly examine them by high-resolution STS, as done for a Bernal graphite [72], requires near-future studies.

Pristine graphite only exhibits the semi-metallic property near the Fermi level, as clearly shown by the red dashed curve. The density of states is very low compared with that of the intercalation cases (stage 1 in Figure 11.5(a), stage 2 in 11.5(e), stage 3 in 11.5(i), and stage 4 in 11.5(m)). This large-molecule intercalation has successfully resulted in the semi-metal-metal transition. There are two/one prominent van Hove singularities for stage-1/other systems. Both Cl and C atoms [red and blue colors] make the most important contributions, while Al ones are almost negligible. Obviously, the p-type dopings mainly come from the interlayer C-Cl chemical bondings, where they are characterized by the active $2p_z$-($3p_x$, $3p_y$, $3p_z$) orbital hybridizations (the merged van Hove singularities in Figures 11.5(b), 11.5(c), 11.5(f), 11.5(g), 11.5(j), 11.5(k), 11.5(n), and 11.5(o)). Interestingly, the conduction density of states, with a range of $E^C < 6$ eV, is dominated by the C-$2p_z$ orbitals (the green curves in Figure 11.5(b), 11.5(f), 11.5(j), and 11.5(n)). That is, the unoccupied higher-energy electronic states hardly depend on the intercalant molecules. On the other side, the deeper valence energy spectra display more complicated structures, revealing the merged van Hove singularities due to the distinct atoms and orbitals. They are

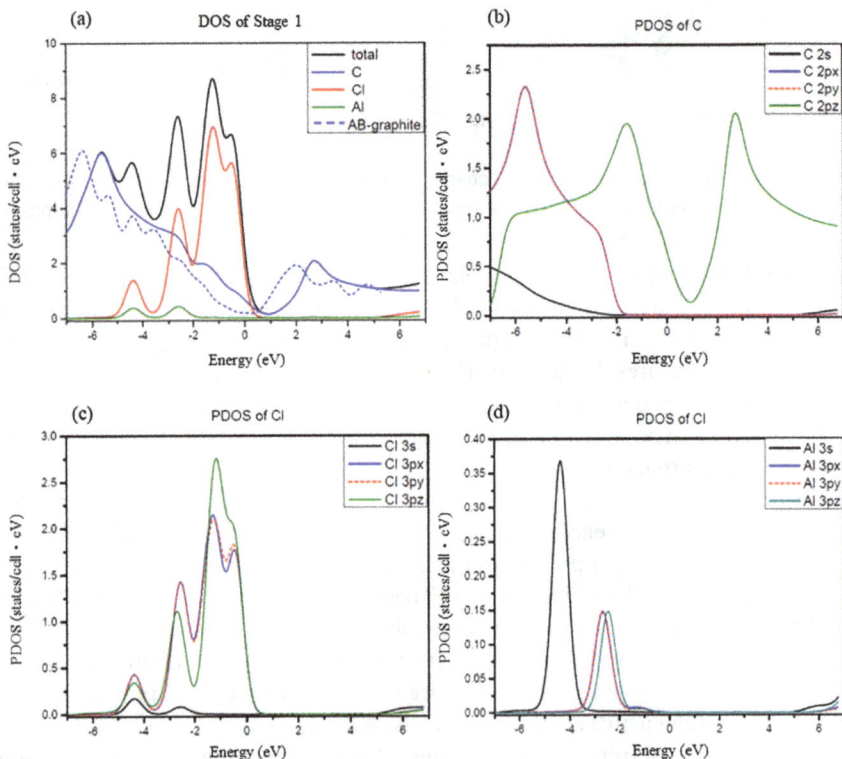

FIGURE 11.5 The atom- and orbital-projected density of states of stage 1 AlCl4-graphite intercalation compounds for (a) all atoms, (b) C atoms, (c) Cl atoms, and (d) Al atoms, mainly arising from the different contributions of atoms and active orbitals. Those of graphite [the dashed color curers] are also displayed in the first case for a clear comparison. *(Continued)*

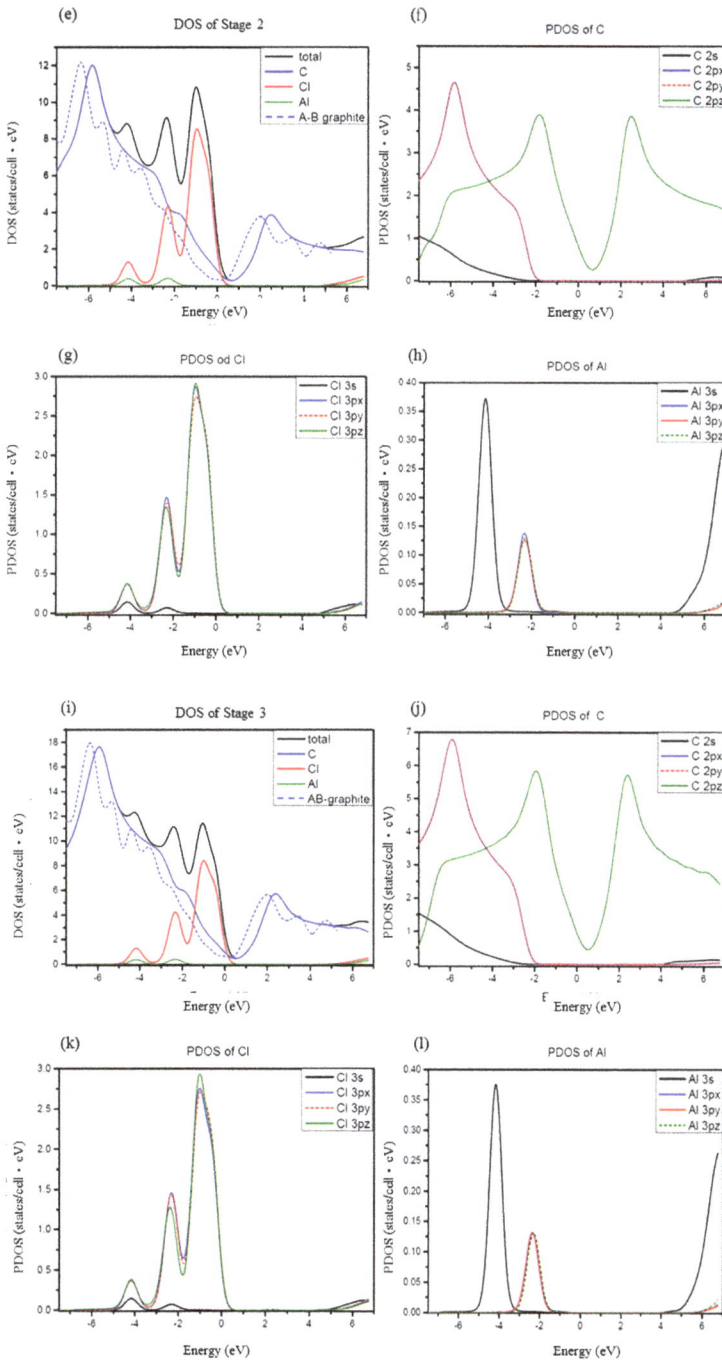

FIGURE 11.5 (*Continued*) The atom- and orbital-projected density of states of stage 2 AlCl4-graphite intercalation compounds for (e) all atoms, (f) C atoms, (g) Cl atoms, and (h) Al atoms; and of stage 3 for (i) all atoms, (j) C atoms, (k) Cl atoms, and (l) Al atoms.

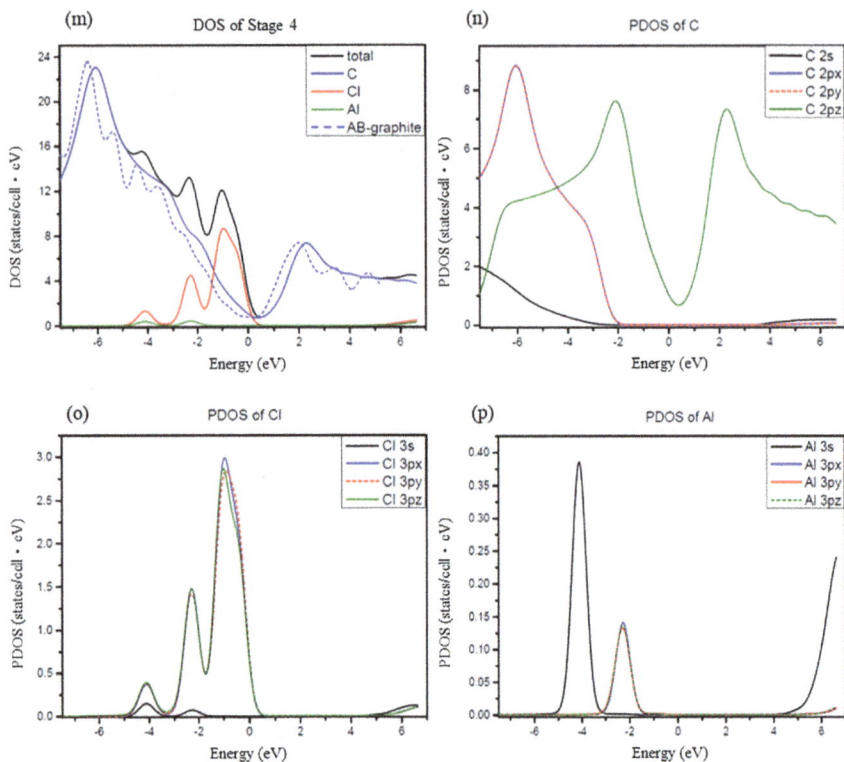

FIGURE 11.5 (*Continued*) The atom- and orbital-projected density of states of stage 4 $AlCl_{j15}$-graphite intercalation compounds for (m) all atoms, (n) C atoms, (o) Cl atoms, and (p) Al atoms.

capable of providing very useful information about the specific orbital mixings. First, the carbon-related orbital contributions clearly display separate characteristics. The π and σ bondings, which are due to $2p_z$ and $(2s, 2p_x, 2p_y)$ orbitals, respectively, are well distinguished from each other within the ranges of $E > -2$ eV and others (the distinct colors in Figures 11.5(b), 11.5(f), 11.5(j), and 11.5(n)). This result suggests the almost orthogonal features of these two kinds of chemical bondings; therefore, the honeycomb lattice keeps the planar geometry during the intercalation and de-intercalation processes [73]. In addition, it represents a geometric merit in the charging and discharging processes of the lithium-ion-based batteries [74]. As for the $(3s, 3p_x, 3p_y, 3p_z)$ orbitals of Cl-atoms (Figures 11.5(c), 11.5(g), 11.5(k), and 11.5(o)), their merged van Hove singularities come into existence at $E = -2.4$ eV and -4.5 eV, revealing the sp^3 bondings in the intra-/inter-molecule Cl-Cl bonds. However, the $(3p_x, 3p_y, 3p_z)$ and 3s orbitals of Al atoms, respectively, dominate the special structures at $E = -2.4$ eV and -4.5 eV (Figures 11.5(d), 11.5(h), 11.5(l), and 11.5(p)). Such unusual features mainly arise from the significant intra-molecule Al-Cl bonds and the almost vanishing Al-Al bonds, the former being due to the sp^3-bondings.

Figure 11.5 shows the atom- and orbital-projected density of states of $AlCl_4$-graphite intercalation compounds, mainly arising from the different contributions of atoms and active orbitals. Those of graphite [the dashed color curers] are also displayed in the first case for a clear comparison.

In short, the most important quasiparticle property, the charge interactions, could be revealed by the delicate analyses of the van Hove singularities (Figure 11.5), spatial charge density distributions (Figure 11.4) and the atom-dominated band structures [29]. The significant orbital hybridizations in the C-C/C-Cl/Al-Cl/Cl-Cl chemical bonds, respectively correspond to $(2s, 2p_x, 2p_y)$-$(2s, 2p_x, 2p_y)$ and $(2p_z$–$2p_z/2p_z)$-$(3p_x, 3p_y, 3p_z)/(3s, 3p_x, 3p_y, 3p_z)$-$(3s, 3p_x, 3p_y, 3p_z)/(3s, 3p_x, 3p_y, 3p_z)$-$(3s, 3p_x, 3p_y, 3p_z)$. Furthermore, both C-Al and Al-Al bonds are almost absent and do not play a significant role in the physical/chemical/material properties. Other intercalants are expected to exhibit different phenomena. Systematic investigations of the various p- and n-type graphite intercalation compounds [62] are required in further studies.

High-resolution STS measurements (details in Chapter 3) can directly examine the energy-dependent van Hove singularities only under their concise structures. For example, such examinations are very successful in exploring the low-lying features of the density of states for graphene-related systems [75], such as few-layer graphenes [76], carbon nanotubes [77], and graphene nanoribbons [78], respectively, with their semi-metallic behavior, metallic or semiconducting properties, and semiconducting one. In addition, the measured special structures strongly depend on the dimensionalities and energy dispersions [71]. As for electron doping of stage-n $AlCl_4$ graphite intercalation compounds, the blue shift of the Fermi level [28], the Dirac-cone minimum [53], and the strong peak due to the π-electronic saddle-point structure [71] are expected to be observable. However, the deeper- and higher-energy van Hove singularities might be very difficult to clearly identify, mainly owing to the zone-folding effects [38] and the multi-merged structures.

11.2.4 OTHER QUASIPARTICLE PROPERTIES

The electronic energy spectra and wave functions could be further used to investigate other fundamental properties, such as vertically optical transitions [79] and dynamic Coulomb excitations [2], especially for the former through the accurate simulations. While an incident electromagnetic wave is normally incident on a semi-infinite sample, the frequency-dependent reflectance spectra are measured within a sufficiently wide ω-range beyond 30 eV. The measured results are supported by the Kramer-Kronig relations [the principle-value integration relations between the real and imaginary parts of the dielectric function in Ref [80], and then Re $[\varepsilon(\omega)]$ and Im $[\varepsilon(\omega)]$ are estimated under the long wavelength limit [81]. The single-particle excitations and plasmon modes are, respectively, characterized by $Im[\varepsilon(\omega)]$ and $Im[-1/\varepsilon(\omega)]$ (energy loss function in [82]). Obviously, they represent the bare and screened charge response abilities due to the external perturbations. The rich and unique excitations are expected to be closely related to the chemical bonds in the carbon honeycomb lattices and molecular intercalants, as well as between them. For example, the various electron-hole excitation ranges (boundaries in [83]) and plasmon frequencies would be dominated by the specific energy ranges of atom- and orbital-decomposed van Hove singularities [84]. Furthermore, the collective excitations are revealed as prominent peaks in the energy-loss spectra, in which they mainly arise from the specific conduction bands and valence ones. According to previous theoretical and experimental studies [85], carbon honeycomb lattices can present the low-frequency optical plasmons [~ 1 eV], the middle-energy π-plasmon modes [~ 5–7 eV], and the π and σ ones [> 15 eV] [86, 87]. Carbon-intercalant and intra-/inter-intercalant bonds might create extra collective excitations within the specific frequency ranges, being dominated by the joint van Hove singularities [84]. On the side of reflectance and the absorption coefficient [88], the special transition structures principally come from the joint van Hove singularities, in which they are affected by the active orbital hybridizations and excitonic effects [81]. Most importantly, the stable or quasi-stable excitons at low temperatures might induce the extra absorption peaks below or close to the threshold frequency. Their existence would result in a dramatic change of the initial optical excitations. This is responsible for further de-excitation processes (the photoluminescence spectra in [89]) and thus potential applications.

There are certain significant differences among the various graphite intercalation compounds, being revealed in the featured crystal symmetries, band structures, orbital hybridizations (chemical bonds), optical absorption spectra, Coulomb excitations, and electrical conductivities. According to previous experimental measurements and theoretical predictions [90], the diverse quasiparticle phenomena cover the following: (i) different AA or AB stackings/interlayer distances/stage-n dependences [2], (ii) the blue or red shift, respectively, due to metal atoms and molecules, as well as a lot of the dramatic changes in the electronic structures (e.g., Figure 11.3 [28]), (iii) the distinct threshold absorption frequencies [91], quasi-stable excitonic bound states [25] and extra prominent peaks beyond those of a pristine case, and (iv) the diversified momentum, frequency-excitation phase diagrams arising from the zone-folding [38] and doping effects, and the great enhancement of the transport properties [92]. For example, $AlCl_4$ molecules are quite different from alkali atoms in terms of orbital hybridizations after the chemical intercalations, mainly owing to their complicated multi-orbital mixings and the interlayer and intralayer atomic interactions (e.g., charge density distributions in Figure 11.4 and density of states in Figure 11.5); that is, the essential properties would become more complex under a highly non-uniform chemical environment, such as the bond-generated single-particle and collective excitations. Even for alkali-guest graphite intercalation compounds [28], lithium exhibits the highest performance in serving as transport ions of batteries because of its many merits [93]. How to achieve the optimal current density and energy capacity remains a profound challenge; the different critical mechanisms, being unified under a great theoretical framework, need to be included in further predictions, such as the asymmetry-initiated chemical reactions and ion currents.

Specifically, the intercalations of neutral and ionic molecules into graphite will be quite different from each other in terms of chemical bondings and low-energy fundamental properties. For example, the stage-1 $AlCl_4$ and $AlCl_4^{-1}$ graphite intercalation compounds (the stage-1 Li-atom/Li$^+$-ion intercalation compounds in [94]), respectively, belong to an n-type metal and a zero-gap semiconductor. That is, the chemical intercalations/de-intercalations are capable of creating the semi-metal-metal and semi-metal-semiconductor transitions. Very interesting, a pure Dirac-cone structure, being initiated from the K/K' valleys, is recovered under the saturated cases of intercalant ions. This clearly illustrates the negligible, but significant interlayer atomic interactions between the carbon-honeycomb lattices and the intercalant layers, mainly owing to the closed shell configuration of ionic molecules. The unusual intercalation effects of the latter are expected to cover the absences of the blue-shift Fermi level, the optical threshold frequency, the low-frequency intra-band electron-hole Coulomb excitations and optical plasmon modes, and the greatly reduced electrical conductivities [95]. These diverse phenomena are very useful in providing the full information for ionic transports of batteries [96]. Most importantly, how to establish the tight-binding/phenomenological models with the critical differences between them are worthy of near-future investigations.

How to modulate the specific concentrations and stacking configurations of intercalants into graphite might have two approaches, according to the first-step theoretical viewpoint. In the previous work [21], the intercalant concentrations are directly reduced under the specific stage-1 case. For example, the plane-density ratios of $AlCl_4$/C are 1:18, 1:24, 1:32, and 1:54, being chosen for a model study. The first planar distribution corresponds to all the stage-n cases, while the current systems are also characterized by the number of graphitic layers and interlayer distances. That is, this work includes the interlayer van der Waals interaction for $n > 2$. The intercalation and de-intercalations processes strongly depend on the chemical and physical mechanism, e.g., catalysts [97], vacancies/impurities/deformations pressures [98], and temperatures [99]. Such complicated growth behaviors could be solved by the combination of molecular dynamics [100] and first-principles simulations, in which the delicate calculations and analyses are able to clarify the optimal crystal symmetry. The theoretical calculations clearly show that these two kinds of stacking configurations have significant differences in terms of physical/chemical/material properties: the ground state energies, the interlayer distances, the intercalant concentrations, the non-uniform chemical environments, the red shifts of the Fermi levels, the creation/absence of linear Dirac-cone structures, the band-edge states, the non-crossing/crossing/anti-crossing behaviors, the characterizations of π- and σ-electronic energy bands, the spatial charge densities and their variations due to chemical modifications, the intralayer/interlayer C-C bonds, the molecule-related Al-Cl and Cl-Cl bonds, the interlayer C-Cl bonds, and atom-orbital-decomposed van Hove singularities.

How to directly link the first-principles method and the tight-binding model is very important in fully comprehending the essential properties under the concise physical/chemical/material pictures. For example, the magnetic quantization phenomena cannot be solved by the former up to now. Based on the important conclusions of the VASP simulations, the significant orbital hybridizations are obtained from the consistent predictions (e.g., Figures 11.3–11.5). The various atomic interactions would become the reliable hopping integrals and site energies [101], such as a lot of parameters due to the multi-orbital mixings in C-C, Cl-C, Al-Cl, and Cl-Cl bonds for $AlCl_4$ graphite intercalation compounds. The reliable intrinsic interactions are associated with a good fitting with the VASP simulations in the low-lying valence and conduction bands. Obviously, this is a tough and complicated technique. The generalized tight-bind model [55], which is developed for the magneto-electronic properties [102], magneto-optical selection rules [103], and quantum Hall conductance [104], will be available in examining the diversified features after the chemical intercalations/de-intercalations. Such phenomenological models could further combine with the modified random-phase approximation [105] and the self-energy method. Furthermore, the (momentum, frequency)-dependent Coulomb excitation/de-excitation phase diagrams are achieved from the delicate calculations and analyses, as done for the systematic investigations about graphene-related systems [106]. Both dimensionalities and chemical modifications [107] are expected to play a critical role in driving the diverse quasiparticle behaviors of graphene-related layered structures [46]. A similar strategy could be generalized to other condensed-matter systems, such as the lithium oxides of cathode/electrolyte/anode materials in ion-based batteries.

11.3 SUMMARY

Very interestingly, pristine graphite and stage-1-stage-4 $AlCl_4$-molecule graphite intercalation compounds show the diverse quasiparticle behaviors, according to the VASP simulations [108] and the delicate analyses. This work clearly illustrates the theoretical development of the quasiparticle framework, being consistent with the previous systematic investigations [21]. The active orbital hybridizations, which possess the intralayer and interlayer chemical bonds, are thoroughly clarified from the highly non-uniform chemical environments of Moiré superlattices, the atom-dominated band structures at specific energy ranges, the spatial charge densities and their variations after intercalations, and atom- and orbital-merged van Hove singularities at the various energies. The important chemical bondings could survive in the intralayer/interlayer C-C bonds, the interlayer C-intercalant bonds, and the intra-molecule/inter-molecule bonds. The observable multi-/single-orbital hybridizations of C-C/C-Cl/Al-Cl/Cl-Cl chemical bonds cover $(2s, 2p_x, 2p_y)$-$(2s, 2p_x, 2p_y)$ and $(2p_z$–$2p_z/2p_z)$-$(3p_x, 3p_y, 3p_z)/(3s, 3p_x, 3p_y, 3p_z)$-$(3s, 3p_x, 3p_y, 3p_z)/(3s, 3p_x, 3p_y, 3p_z)$-$(3s, 3p_x, 3p_y, 3p_z)$. However, the evidence almost disappears for the C-Al and Al-Al bonds. Moreover, the strong charge transfers between graphitic layers and molecular intercalants, the zone-folding effects due to the intercalant stackings and arrangements, and the van der Waals interactions are responsible for the featured quasiparticle behaviors. The main features of quasiparticle properties include the semi-metal transitions, the enhanced asymmetry of hole and electron energy spectra, the large modifications about the low-lying energy dispersions/band-edge states, the well-defined or undefined π- and σ-electronic bands, the intercalant-induced/C- and intercalant co-dominated energy sub-bands, and the greatly diversified van Hove singularities associated with sp^3 four orbitals of C, Al, and Cl atoms.

   The first-principles energy spectra and wave functions could be further utilized to explore the perturbations of electromagnetic and Coulomb fields (or the screening abilities of quasiparticle charges). There are a lot of significant differences among the various graphite intercalation compounds in terms of the featured n- or p-type dopings [62], optical absorption features [9], and electron-hole excitations and plasmon modes [6]. Specifically, the molecule- and ion-intercalations, respectively, lead to the semi-metal-metal and semi-metal-semiconductor transitions. The essential quasiparticle properties are very sensitive to the stacking configurations and concentrations of intercalants. Very interestingly, how to link the VASP simulations on band structures and the tight-binding model is expected to be available for planar graphite intercalation compounds, since the intralayer/interlayer C-C orbital hybridizations, the interlayer C-intercalant interactions, and intra- and inter-intercalant ones are able to provide the full information in establishing the suitable hopping integrals of the latter [101].

Maybe the diverse magnetic quantization could be clarified in electron- or hole-doped graphite intercalation compounds, as systematically done for layered graphene systems [Refs]. Most of theoretical predictions require high-resolution X-ray [44, 109], ARPES [110], and STS examinations [47]. The calculated results clearly illustrate intercalant-induced quasiparticle phenomena closely related to the theoretical development of quasiparticle frameworks.

REFERENCES

[1] Lin C Y, Wu J Y, Chiu Y H and Lin M F 2014 Stacking-dependent magneto-electronic properties in multilayer graphenes *Physical Review B* **90** 205434

[2] Ho J H, Lu C L, Hwang C C, Chang C P and Lin M F 2006 Coulomb excitations in AA- and AB-stacked bilayer graphites *Physical Review B* **74** 85406

[3] Lin C Y, Lee M H and Lin M F 2018 Coulomb excitations in trilayer ABC-stacked graphene *Physical Review B Rapid communication* **98** 41408

[4] Lin C Y, Wu J Y, Ou Y J, Chiu Y H and Lin M F 2015 Magneto-electronic properties of multilayer graphenes *Physical Chemistry Chemical Physics* **17** 26008

[5] Lin C Y, Wu J Y, Chiu Y H and Lin M F 2014 Stacking-dependent magneto-electronic properties in multilayer graphenes *Physical Review B* **90** 205434

[6] Borghi G, Polini M, Asgari R and MacDonald A H 2009 Dynamical response functions and collective modes of bilayer graphene *Physical Review B* **80** 241402

[7] Do T N, Chang C P, Shih P H, Wu J Y and Lin M F 2017 Stacking enriched magneto transport properties of few-layer graphenes *Physical Chemistry Chemical Physics* **19** 29525–33

[8] Lipson H and Stokes A R 1942 The structure of graphite *Proceedings of the Royal Society of London Series A-Mathematical and Physical Sciences* **181** 101–5

[9] Lin C Y, Chen R B, Ho Y H and Lin M F 2018 Electronic and optical properties of graphite-related systems *CRC Press, Boca Raton, Florida*

[10] Hwang E H and Das Sarma S 2007 Dielectric function, screening, and plasmons in two-dimensional graphene *Physical Review B* **75**(20) 205418

[11] Wu J Y, Chen S C, Roslyak O, Gumbs G and Lin M F 2011 Plasma excitations in graphene: Their spectral intensity and temperature dependence in magnetic field *ACS Nano* **5** 1026

[12] Lin C Y, Wu J Y, Ou Y J, Chiu Y H and Lin M F 2015 Magneto-electronic properties of multilayer graphenes *Physical Chemistry Chemical Physics* **17**(39) 26008

[13] Charlier J C, Gonze X and Michenaud J P 1991 1st-principles study of the electronic-properties of graphite *Physical Review B* **43** 4579–89

[14] Charlier J C, Michenaud J P and Gonze X 1992 First-principles study of the electronic properties of simple hexagonal graphite *Physical Review B* **46** 4531

[15] Tsai S J, Chiu Y H, Ho Y H and Lin M F 2012 Gate-voltage-dependent Landau levels in AA-stacked bilayer graphene *Chemical Physics Letters* **550** 104

[16] Do T N, Lin C Y, Lin Y P, Shih P H and Lin M F 2015 Configuration enriched magneto-electronic spectra of AAB-stacked trilayer graphene *Carbon* **94** 619

[17] Lin C Y, Lee M H and Lin M F 2018 Coulomb excitations in trilayer ABC-stacked graphene *Physical Review B Rapid Communication* **98**(4) 41408

[18] Ho J H, Lu C L, Hwang C C, Chang C P and Lin M F 2006 Coulomb excitations in AA- and AB-stacked bilayer graphites *Physical Review B* **74**(8) 85406

[19] Lin C Y, Yang C H, Chiu C W, Chung H C, Lin S Y and Lin M F 2021 Many-particle interactions in carbon nanotubes *IOP Concise Physics, San Rafel, CA, USA: Morgan & Claypool Publishers. in print*

[20] Charlier J C, Michenaud J P and Gonze X 1991 Tight-binding model for the electronic properties of simple hexagonal graphite *Physical Review B* **44** 13237

[21] Nguyen T D H, Lin S Y, Chung H C, Tran N T T and Lin M F 2021 First-Principles Calculations for Cathode, Electrolyte and Anode Battery Materials (IOP publishing: Bristol)

[22] Shin S Y, Hwang C G, Sung S J, Kim N D, Kim H S and Chung J W 2011 Observation of intrinsic intraband pi-plasmon excitation of a single-layer graphene *Physical Review B* **83** 161403

[23] Lin C Y, Wu J Y, Chiu C W and Lin M F 2019 Coulomb excitations and decays in graphene-related systems *CRC Press Boca Raton Florida* **83** 161403

[24] Ambroz F, Macdonald T J and Nann T 2017 Trends in aluminium-based intercalation batteries *Advanced Energy Materials* **7** 1602093

[25] Shung K W K 1986 Dielectric function and plasmon structure of stage-1 intercalated graphite *Physical Review B* **34** 979

[26] Gould T, Liu E, Liu J Z, Dobson J F, Zheng Q and Lebègue S 2013 Binding and inter-layer force in the near-contact region of two graphite slabs: Experiment and theory *J Chem Phys* **139** 224704

[27] Ding Y and Yu G H 2020 When graphite meets Li metal *National Science Review* **7** 1521–22

[28] Li W B, Lin S Y, Tran N T T, Lin M F and Lin K I 2020 Essential geometric and electronic properties in stage-n graphite alkali-metal-intercalation compounds *RSC Advances* **10** 23573–81

[29] Hwang E H and Das Sarma S 2008 Quasiparticle spectral function in doped graphene: Electron-electron interaction effects in ARPES *Physical Review B* **77** 81412

[30] Matsui T, Kambara H, Niimi Y, Tagami K, Tsukada M and Fukuyama H 2005 STS observations of landau levels at graphite surfaces *Physical Review Letters* **94** 226403

[31] Li Y and Yue Q 2013 First-principles study of electronic and magnetic properties of FeCl3-based graphite intercalation compounds *Physical B-Condensed Matter* **425** 72–7

[32] Kubo T, Ohashi Y F and Kinoshita T 2004 Superconducting transition temperature of niobium/graphite bilayers *Physical C-Superconductivity and ITS Applications* **417** 58–62

[33] Marinopoulos A G, Reining L, Rubio A and Olevano V 2004 Ab initio study of the optical absorption and wave-vector-dependent dielectric response of graphite *Physical Review B* **69** 245419

[34] Crowell A D 1965 Van der Waals interactions of simple molecules with graphite *Abstracts of Papers of the American Chemical Society* APR C005

[35] Flandrois L S 1982 Graphite-intercalation compounds as electrode materials in batteries *Synthetic Metals* **4** 255–66

[36] MS Wua, B Xua, LQ Chenb and CY Ouyanga 2016 Geometry and fast diffusion of AlCl4 cluster intercalated in graphite *Electrochimica Acta* **195** 158–65

[37] Konschuh S, Gmitra M and Fabian J 2010 Tight-binding theory of the spin-orbit coupling in graphene *Physical Review B* **82** 245412

[38] Lin Y, Chen G, Sadowski J T, Li Y Z, Tenney S A, Dadap J I, Hybertsen M and Osgood R M 2019 Observation of intercalation-driven zone folding in quasi-free-standing graphene energy bands *Physical Review B* **99** 35428

[39] Rozplocha F, Patyk J and Stankowski J 2007 Graphenes bonding forces in graphite institute of physics *Nicolaus Copernicus University Grudzi adzka* **5**(7) 87–100

[40] Grüneis A et al. 2008 Electron-electron correlation in graphite: A combined angle resolved photoemission and first-principles study *Phys Rev Lett* **100** 37601

[41] Dresselhaus M S and Dresselhaus G 2002 Intercalation compounds of graphite *Advances in Physics* **51** 1–186

[42] Li N and Su D 2019 In-situ structural characterizations of electrochemical intercalation of graphite compounds *Carbon Energy* **1** 2

[44] Chun J P et al. 2018 An operando X-ray diffraction study of chloroaluminate anion-graphite intercalation in aluminum batteries *PNAS* **115** 22

[45] Liu H T, Liu Y Q and Zhu D B 2011 Chemical doping of graphene *Journal of Materials Chemistry* **21** 3335–45

[46] Bostwick A, Ohta T, Seyller T, Horn K and Rotenberg E 2007 Quasiparticle dynamics in graphene *Nature Physics* **3** 36–40

[47] Jung S C, Kang Y J, Yoo D J, Choi J W and Han Y K 2016 Flexible few-layered graphene for the ultrafast rechargeable aluminum-ion battery *Journal of Physical Chemistry C* **120** 13384–9

[48] Duong D L F, Yun S J and Lee Y H 2017 Van der Waals layered materials: Opportunities and challenges *ACS Nano* **11** 11803–30

[49] Huang J R, Lin J Y, Chen B H and Tsai M H 2008 Structural and electronic properties of few-layer graphenes from first-principles *Physical Status Solid B-Basic Solid State Physics* **245** 136–41

[50] Brownson D A C and Banks C E 2010 Graphene electrochemistry: An overview of potential applications *Analyst* **135** 2768–78

[51] Mansour A, Schnatterly S E and Ritsko J J 1987 Electronic structure of alkali-intercalated graphite studied by soft-x-ray emission spectroscopy *Phys Rev Lett* **58** 614

[52] Kaneko T and Saito R 2017 First-principles study on interlayer state in alkali and alkaline earth metal atoms intercalated bilayer graphene *Surface Science* **665** 1–9

[53] Malko D, Neiss C, Vines F and Gorling A 2012 Competition for graphene: Graphynes with direction-dependent Dirac cones *Physical Review Letteres* **108** 86804

[54] Tasaki K 2014 Density functional theory study on structural and energetic characteristics of graphite intercalation compounds *Journal of Physical Chemistry C* **118** 1443–50

[55] Reich S, Maultzsch J, Thomsen C and Ordejon P 2002 Tight-binding description of graphene *Physical Review B* **66** 35412

[56] Abergel D S L, Apalkov V, Berashevich J, Ziegler K and Chakraborty T 2010 Properties of graphene: A theoretical perspective *Advances in Physics* **59** 261–482

[57] Dresselhaus M S and Dresselhaus G 2002 Intercalation compounds of graphite *Advances in Physics* **51** 1–186

[58] Pham H D, Lin S Y, Gumbs G, Khanh N D and Lin M F 2019 Diversified properties of carbon substitutions in silicene *Front Phys* **8** 561350

[59] Late D J, Ghosh A, Chakraborty B, Sood A K, Waghmare U V and Rao C N R 2011 Molecular charge-transfer interaction with single-layer graphene *Journal of Experimental Nanoscience* **28** 641–51

[60] Venghaus H 1974 Wave vector dependence of the electron energy loss functions of graphite *IPSS* **66** 145–50

[61] Rosenzweig P, Karakachian H, Marchenko D, Kuster K and Starke U 2020 Over doping graphene beyond the van hove singularity *Physical Review Letters* **125** 176403

[62] Xiuqing X M, Tongay S and Kang J 2013 Stable p- and n-type doping of few-layer graphene/graphite *Carbon* **57** 507–14

[63] Patil S, Kolekar S and Deshpande A 2017 Revisiting HOPG superlattices: Structure and conductance properties *Surface Science* **658** 55–60

[64] Wallace P R 1947 The band theory of graphite *Phys Rev* **71** 622

[65] Goerbig M O 2011 Electronic properties of graphene in a strong magnetic field *Reviews of Modern Physics* **83** 1193–243

[66] Nozieres P and Pines D 1959 Electron Interaction in solids-characteristic energy loss spectrum *Physical Review* **113** 1254–67

[67] Cole L A and Perdew J P 1982 Calculated electron affinities of the elements *Phys Rev A* **25** 1265

[68] Shioyama H and Fujii R 1987 Electrochemical reactions of stage 1 sulfuric acid-graphite intercalation compound *Carbon* **25** 771–4

[69] Sorokina N E, Maksimova N V, Nikitin A V, Shornikova O N and Avdeev V V 2001 Synthesis of intercalation compounds in the graphite-HNO_3-H_3PO_4 system *Inorganic Materials* **37** 584–90

[70] Zhan D, Sun L, Ni Z H, Liu L, Fan X F, Wang Y Y, Ting Yu T, Lam Y M, Huang W, Shen Z X 2010 $FeCl_3$-based few-layer graphene intercalation compounds: single linear dispersion electronic band structure and strong charge transfer doping *Advanced Functional Materials* **20** 3504–9

[71] Lin S Y, Chang S L, Shyu F L, Lu J M and Lin M F 2015 Feature-rich electronic properties in graphene ripples *Carbon* **86** 207–16

[72] Matsui T, Kambara H, Niimi Y, Tagami K, Tsukada M and Fukuyama H 2005 STS observations of landau levels at graphite surfaces *Phys Rev Lett* **94** 226403

[73] Gao Y, Zhu C, Chen Z Z and Lu G 2017 Understanding ultrafast rechargeable aluminum-ion battery from first-principles *J Phys Chem* **121** 7131–8

[74] Wang Q, Zheng D, He L and Ren X 2019 Cooperative effect in a graphite intercalation compound: Enhanced mobility of alcl4 in the graphite cathode of aluminum-ion batteries *Physical Review Applied* **12** 44060

[75] Hu M, Dong X, Wu Y J, Liu L Y and Zhao Z S 2018 Low-energy 3D sp^2 carbons with versatile properties beyond graphite and graphene *Dalton Trans-Actions* **47** 6233–9

[76] Jung N, Kim N, Jockusch S, Turro N J, Kim P and Brus L 2009 Charge transfer chemical doping of few layer graphenes: Charge distribution and band gap formation *Nano Lett* **9** 4133–7

[77] Lien J Y and Lin M F 2007 Low-energy electronic properties of a pair of carbon nanotubes *IPSS* **4** 512–4

[78] Li T S, Huang Y C, Chang S C, Chuang Y C and Lin M F 2008 Transport properties of AB-stacked bilayer graphene nanoribbons in an electric field *The European Physical Journal B* **64** 73–80

[79] Klucker R, Skibowski M and Steinmann W 1974 Anisotropy in the optical transitions from the π and σ valence bands of graphite *IPSS* **65** 703–10

[80] Grosse P and Offermann V 1991 Analysis of reflectance data using the Kramers-Kronig relations *Applied Physics A* **52** 138–44

[81] Grigorenko A N, Polini M and Novoselov K S 2012 Graphene plasmonics *Nature Photonics* **6** 749–58

[82] Nagatomi T, Shimizu R, Ritchie R H 1999 Energy loss functions for electron energy loss spectroscopy *Surface Science* **419** 158–73

[83] Rez P and Muller D A 2008 The theory and interpretation of electron energy loss near-edge fine structure *Annual Review of Materials Research* **38** 535–58

[84] Kravets V G, Grigorenko A N, Nair R R, Blake P, Anissimova S, Novoselov K S and Geim A K 2010 Spectroscopic ellipsometry of graphene and an exciton-shifted van Hove peak in absorption *Physical Review B* **81** 155413

[85] Venghaus H 1975 Redetermination of the dielectric function of graphite *IPSS* **71** 609–14

[86] Jablan M, Soljačić M, Buljan H 2014 Plasmons in graphene: Fundamental properties and potential applications *IEEE* **101** 1689–704

[87] Lin M F, Huang C S and Chuu D S 1997 Plasmons in graphite and stage-1 graphite intercalation compounds *Phys Rev B* **55** 13961

[88] Aloia A G D, Marra F, Tamburrano A, Bellis G D and Sarto M S 2014 Electromagnetic absorbing properties of graphene-polymer composite shields *Carbon* **73** 175–84

[89] Cao L, Meziani M J, Sahu S and Sun Y P 2013 Photoluminescence properties of graphene versus other carbon nanomaterials *Acc Chem Res* **46** 171–80

[90] Marinopoulos AG, Reining L, Rubio A and Olevano V 2004 Ab initio study of the optical absorption and wave-vector-dependent dielectric response of graphite *Phys Rev B* **69** 245419

[91] Sodemann I and Fogler M M 2012 Interaction corrections to the polarization function of graphene *Phys Rev B* **86** 115408

[92] Mahata P B and Pathak B 2017 The staging mechanism of AlCl4 intercalation in a graphite electrode for an aluminium-ion battery *Phys Chem Chem Phys* **19** 7980

[93] Gao L, Liu S and Dougal R A 2002 Dynamic lithium-ion battery model for system simulation *IEEE Transactions on Components and Packaging Technologies* **25** 495–505

[94] Basu S, Zeller C, Flanders P J, Fuerst C D, Johnson W D and Fischer J E 1979 Synthesis and properties of lithium-graphite intercalation compounds *Materials Science and Engineering* **38** 275–83

[95] Inagaki M 1989 Applications of graphite intercalation compounds *Journal of Materials Research* **4** 1560–68

[96] Jung S C, Kang Y J, Yoo D J, Choi J W and Han Y K 2016 Flexible few-layered graphene for the ultrafast rechargeable aluminum-ion battery *J Phys Chem C* **120** 13384–89

[97] Wang W, Xu S, Wang K, Liang J and Wei Zhang W 2019 De-intercalation of the intercalated potassium in the preparation of activated carbons by KOH activation *Fuel Processing Technology* **189** 74–79

[98] Clarke R and Uher C 2006 High pressure properties of graphite and its intercalation compounds *Advances in Physics* **33** 469–566

[99] Solin S A 1982 The nature and structural properties of graphite intercalation compounds *Advances in Chemical Physics* **13** 456–528

[100] Alder B J and Wainwright T E 1959 Studies in molecular dynamics. 1. General method *J Chem Phys* **31** 459

[101] Nobuhara K, Nakayama H, Nose M, Nakanishi S and Iba H 2013 First-principles study of alkali metal-graphite intercalation compounds *Journal of Power Sources* **243** 585–7

[102] Safran S A and DiSalvo F J 1979 Theory of magnetic susceptibility of graphite intercalation compounds *Phys Rev B* **20** 4889

[103] Chunga D D L and Dresselhausa M S 1977 Magneto-optical studies of graphite intercalation compounds *Physica B+* **89** 131–8

[104] Bernevig B A, Hughes T L, Raghu S and Arovas D P 2007 Theory of the three-dimensional quantum Hall effect in graphite *Phys Rev Lett* **99** 146804

[105] Leconte N, Jung J, Lebègue S and Gould T 2017 Moiré-pattern interlayer potentials in van der Waals materials in the random-phase approximation *Phys Rev B* **96** 195431

[106] Ho J H, Chang C P and Lin M F 2006 Electronic excitations of the multilayered graphite *Physics Letters A* **352** 446–50

[107] Noel M and Santhanam R 1998 Electrochemistry of graphite intercalation compounds *Journal of Power Sources* **72** 53–65

[108] Hafner J 2008 Ab-initio simulations of materials using VASP: Density-functional theory and beyond *Journal of Computational Chemistry* **29** 2044–78

[109] Wang D Y, Huang S K, Liao H J, Chen Y M, Wang S W, Kao Y T and An J Y 2019 Insights into dynamic molecular intercalation mechanism for Al-C battery by operando synchrotron X-ray techniques *Carbon* **146** 528–34

[110] Childress A S, Parajuli P, Zhu J Y, Podila R and Rao AM 2017 A Raman spectroscopic study of graphene cathodes in high-performance aluminum ion batteries *Biotechnology Advances* **29** 189–98

12 Geometric and Electronic Properties of LiFeO$_2$

Vo Khuong Dien, Nguyen Thi Han, and Ming-Fa Lin

CONTENTS

12.1 INTRODUCTION

Increasing demands for storing energy from wind and solar energy, mobile electronic equipment, and electronic transportation promote the development of reliable and cost-effective batteries [1–4]. Compared with other energy storage systems, lithium-ion batteries (LIBs) have received a great deal of attention since they possess desirable features, such as lightweight, long life cycle, fast charging time, and ability to provide a sizable electronic current for electronic devices [5–8]. The typical LIB is a combination of the electrolyte sandwiched between the negative (cathode) and the positive (anode) electrodes. During the charging/discharging process, the Li$^+$ will continuously transport from the cathode $\rightarrow$ electrolyte $\rightarrow$ anode/anode $\rightarrow$ electrolyte $\rightarrow$ cathode materials [9] (Figure 12.1). The energy will be stored/released in/from this system in terms of chemical energy. Apparently, the physical/chemical properties and material environments of these components are very complicated and strongly related to the efficiency of an energy-stored system [10–12].

Generally, the commercial LIBs use solid or liquid electrolytes, such as Li$_2$SiO$_3$ or Li$_3$OCl ternary compounds which can provide a wide energy window [13, 14], while the graphite layers or the ternary Li$_2$GeO$_3$ and Li$_4$Ti$_5$O$_{12}$ compounds were used for the anode ones [15, 16]. On the other hand, the traditional cathode usually uses ternary compounds, such as LiCoO$_2$, LiNiO$_2$, and LiMn$_2$O$_4$ [17, 18]. However, these cathode materials have economic and environmental problems that limit their use in large-scale Li-ion batteries. Recently, a series of investments in research have been done to search for new cathode materials that satisfy the market demand and then

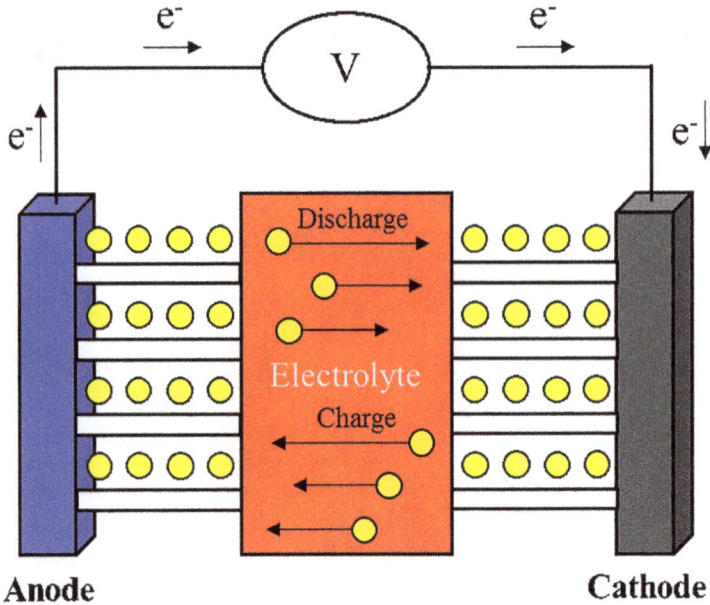

FIGURE 12.1   The charging and discharging processes in Li^+-based batteries.

reduce the increase in pollution. Among them, Li-Fe-O compound has been paid more attention due to greater abundance and non-toxicity of Fe. As a matter of fact, various types of lithium iron oxides have been investigated—for example, inverse spinel-type Fe_3O_4 [19], corundum-type α-Fe_2O_3 [20], β-$NaMnO_2$-type $LiFeO_2$ [21], β-FeOOH-type $LiFeO_2$ [22], β-$NaFeO_2$-type $LiFeO_2$ [23], and layered α-$NaFeO_2$ type of $LiFeO_2$ [24]. Among these materials, α-$NaFeO_2$ type of $LiFeO_2$ has a relatively high theoretical capacity of 282 mAhg^{-1} in a one-electron reaction [25] and simple and eco-friendly synthesis [26].

On the theoretical aspects, the various molecule dynamic (MD) and density functional theory (DFT) calculations were carried out to investigate the stability and the electronic and magnetic properties and the Li^+ transport mechanism of electrode and electrolyte materials. Although the fundamental properties of the geometric, electronic, and magnetic properties of these condensed materials have been successfully investigated by both these methods [27, 28], a systematic strategy to study essential properties of these kinds of materials is absent. For example, the critical chemical/physical/material pictures are very important, since they relate to the ion transport mechanism and magnetic configuration of material, but they are lacking up to now. Furthermore, the close connection of the quasiparticle charge, spin, and the specific orbital hybridization have also not been achieved so far.

Previous theoretical simulations based on first-principle calculations can depict clearly the essential rich and unique quasi-particle features of emergent materials, especially for 2D layered [29–32] and 3D ternary anode/cathode/electrolyte materials [33–35]. For example, systematic studies have been conducted on the essential properties of silicene-/graphene-related materials, the outstanding features of ternary

$Li_4Ti_5O_{12}$ anode, Li_2SiO_3, and Li_2GeO_3 electrolyte materials [36–38]. Such investigations clearly illustrate that the diversified phenomena in materials are governed by quasiparticle charges, orbitals, and spins. The highly accurate results and delicate analyses are capable of proposing significant mechanisms/pictures to fully comprehend the geometric, electronic, magnetic properties. The important multi-/single-orbital hybridizations in various chemical bonds are obtained from the geometric optimizations, the atom-dominated band structures, the spatial charge densities and their variations after chemical modifications, and the atom- and orbital-decomposed density of states. The magnetic properties are identified from the spin-split/spin-degenerate energy bands, the spin density distributions, the net magnetic moments, and the spin-projected van Hove singularities. This framework is successfully conducted on silicene-/graphene-related systems [39, 40] and the anode/cathode/electrolyte compounds [41–43] and could be applied to other emergent materials. It is thus expected to be very appropriate for investigating the complex geometric, electronic, and magnetic properties of mainstream Li^+-based batteries.

In this chapter, the theoretical framework is developed to comprehend the quasiparticle charges, spin orientations, and orbital hybridizations in the chemical bonding of the ternary $LiFeO_2$ compound. This strategy is based on the first-principles calculations on an optimized structure with position-dependent chemical bonding, the spin-dependent energy band structure with atom-dominance at different ranges, the spatial spin and charge densities due to various orbitals, and the atom- and orbital-projected density of state related to spin directions and orbital overlaps. The current study is of paramount importance not only for fundamental physics but also for technical applications. The predictions on the optimization geometric, the spin-split occupied states and energy gap, and the spin-polarization van Hove singularities can be respectively examined via powder x-ray diffraction (PXRD) [44]/tunneling electron microscopy (TEM) [45]/scanning electron microscopy (SEM) [46], optical measurements/spin-polarized angle-resolved photo-emission spectroscopy (ARPES) [47], and spin-polarized scanning tunneling spectroscopy (STS) [48]. Furthermore, to get a deeper understand of the magneto-electronic and optical properties of materials, the close relationship between phenomenological models and the current numerical simulations will be discussed in detail. The current work provides more insights into the understanding of the physical/chemical pictures in the ternary $LiFeO_2$ cathode for future applications.

12.2 DELICATELY NUMERICAL VASP CALCULATIONS

The theoretical investigation could be divided into two groups, the phenomenological models and the numerical simulations, in which the complex and simple geometric structures, respectively, can be investigated by the former and the latter. For example, the single or monolayer graphene systems, which only have the interlayer atomic interactions due to the $C-2p_z$ single-orbital hybridizations, are well characterized by the first-principles calculations and the tight-binding model simultaneously. Specifically, the generalized tight-binding model, being developed for external magnetic and electric fields, is successful in predicting the unique and rich quantization phenomena, e.g., the diversified Landau-level energy spectra in monolayer graphene [49], bilayer tri-layer ABA-, AAA-, AAB-, ABC-stacked ones [50], and AB- and AB-stacked stackings [51], and apparently, the first-principles calculations are capable of fully exploring the geometric and electronic and magnetic features in the 3D ternary $LiFeO_2$ cathode materials. The close connections between them are very interesting topics under the current investigations.

The Vienna Ab Initio Simulation Package, VASP [52], which is built on the density functional theory (DFT) [53], was adopted to investigate the geometric relaxation, electronic band structure, charge and spin density distributions, and van Hove singularities in density of states. The interactions between electron and ion core are described by the projector-augmented wave (PAW) [54] pseudopotentials. While the many-particle correlation and exchange energies, the electron-electron Coulomb interactions beyond the classical electrodynamics, are based on the Perdew-Burke-Ernzerhof of functional (PBE) [55] under the generalized gradient approximation (GGA). In general, the plane waves, with the kinetic energy cutoff of 500 eV, are very reliable in serving as a complete set. That is, their linear superposition is rather suitable in characterizing Bloch wave functions and band structures. For the 3D ternary $LiFeO_2$ compounds, the first Brillouin zone is sampled by $20 \times 20 \times 20$ and $15 \times 15 \times 15$ k-point meshes within the Γ-center scheme [56] for the geometric optimizations and electronic energy spectra, respectively. Most importantly, the convergence condition of the ground-state energy is $\sim 10^{-5}$ eV between two consecutive evaluation steps, in which the maximum Hellmann-Feynman force acting on each ion is less than 0.01 eV/Å during the atom relaxations.

Most importantly, how the phenomenological models and the numerical simulations can be combined with each other becomes the challenge, in which the former can provide more physical/chemical pictures on the critical mechanisms and thus can be used to investigate other vital features (optical [57], transport [58], and magneto-electronic properties [59] and excitonic effects [60]), while the energy dispersion of conduction and valence bands can be well established by VASP calculation. These approaches were successful in investigating the diversified phenomena of layered graphene and other group-IV and group-V layered systems, e.g., the diversified Landau-level energy spectra of graphene [61], silicene [62], germanene [63], tinene [64], bismuthene [65], and phosphorene [66].

This chapter will point out that the 3D ternary $LiFeO_2$ material has a very extremely non-uniform chemical environment in a unit cell (later discussions in Figure 12.3) and thus complicated band structures (Figure 12.4). As a result, the tight-binding models, with the various hopping integrals due to the different chemical bonding strengths, would be rather difficult in simulating the first-principles electronic structures.

12.3   UNUSUAL CRYSTAL STRUCTURES OF 3D TERNARY LIFEO₂ MATERIAL

Based on the delicate first-principle calculations for the optimal geometric structure, the 3D ternary $LiFeO_2$ cathode material presents the unique crystal lattice symmetries. In this work, one certain meta-stable of $LiFeO_2$ was chosen to clearly illustrate the complex physical and chemical environments; it crystallizes in the trigonal structure with R-3m space group (Figures 12.2(a)–(b)). The calculated lattice constants of this meta-stable structure are 2.88 Å, 2.88 Å, and 14.31 Å for x, y, and z directions, respectively, which are very good in agreement with previous theoretical [67] and experimental values [68]. The basic structure of ternary LiFeO$_2$ is the corner shearing of the FeO$_2$ layer with the LiO$_2$ layer along the z-axis, in which each Fe/Li atom occupies the center of an octahedron of O atoms (Figure 12.2(c)). As a result, there is a total of 18 equivalent Fe-O bonds (1.97 Å) and 18 identical Li-O bonds (2.13 Å). The highly ordered arrangement of the atoms and the anisotropy of the geometric structure, which is originated from the complex orbital hybridization, will support for Li$^+$ migration. In addition to the lattice constant of the 3D ternary compound, other quantities, such as the morphology or the size of the material particle, could be measured by SEM, and the top view of nano-scale materials could be clarified by TEM, while STM measurement is usually used to provide the side-view information. Such experiments have been successfully applied for multi-wall carbon nanotubes [69], the stacking configuration in multi-layer graphene [70], and the geometric profile of graphene nanoribbons [71] and therefore are very suitable for ternary LiFeO$_2$ compound. Apparently, the high ordering of the geometric structure will induce extra challenges in exploring phenomenological models.

12.4   RICH AND UNIQUE ELECTRONIC PROPERTIES

The 3D ternary LiFeO$_2$, a strong candidate for the cathode material, presents unusual geometric structures and thus created rich and unique electronic quasiparticle properties. The energy band structure along the high symmetry point of the LiFeO$_2$

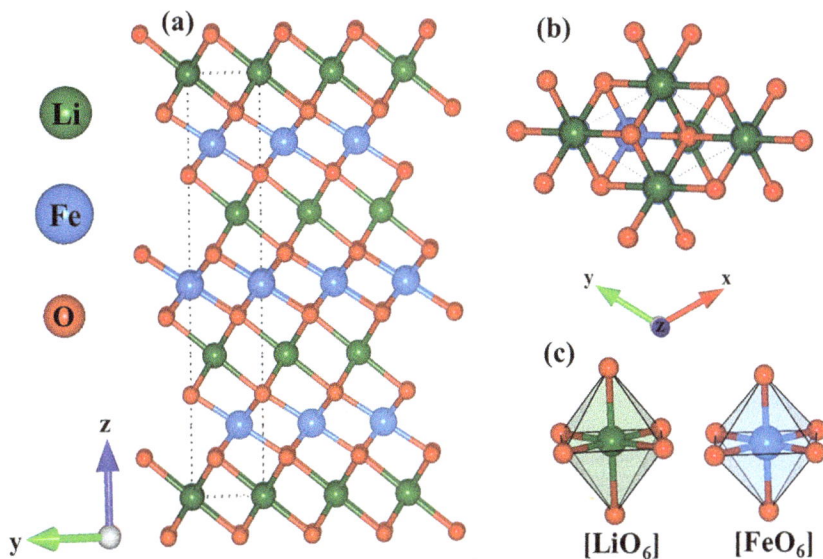

FIGURE 12.2   The optimal geometric structure of LiFeO$_2$ compound with (a) side view and (b) top view, respectively; and (c) the octahedron structure of [LiO$_6$] and [FeO$_6$].

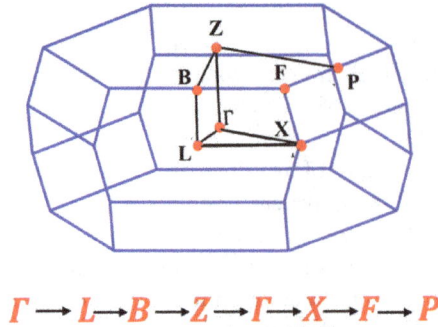

$$\Gamma \rightarrow L \rightarrow B \rightarrow Z \rightarrow \Gamma \rightarrow X \rightarrow F \rightarrow P$$

FIGURE 12.3 The first Brillouin zone with the high-symmetry points within the three orthogonal axes of $LiFeO_2$.

compound was calculated and shown in Figure 12.3. The Fermi level understood as the reference point was set at the middle of the valence and conduction bands. As a result of the trigonal symmetry and many valence electrons for each atom, the electronic band structure of $LiFeO_2$ exhibits various valence and conduction bands with complex dispersion characteristics such as parabolic, oscillatory, or dispersionless. A lot of crossing, non-crossing, or anti-crossing phenomena are frequently present at both valence and conduction bands. Furthermore, the occupied states are highly asymmetrical to the unoccupied states about zero energy. These evidence the complicated of electronic band structure and therefore the orbital hybridizations. Very interestingly, the spin-dependent energy band structure of the $LiFeO_2$ compound is obviously expressed, especially for the remarkable spin-splitting near the Fermi level. For the spin-up, spin-down states, the energy bands nearest to the E_F are fully occupied and unoccupied, respectively, which leads to an indirect gap of 1.9 eV along the Γ-X path. The large spin-splitting with complicated energy band structure reflected the ferromagnetic configuration in the $LiFeO_2$ compound.

In addition to the electronic band structure, the atom-decomposed energy spectrum for the valence and the conduction states can provide the briefly orbital hybridizations in Li-O and Fe-O chemical bonds. As present in Figures 12.4, the dominations of the Li, Fe, and O atoms are represented by the green, red, and blue circles. The effective energy range related to the Li-O and Fe-O chemical bonds is in the energy range from −20 eV to 15 eV. For all energy spectra, it is very hard for eyes to detect the dominations of the Li atom since it contributes only one valence electron. However, we cannot ignore this contribution, because some outstanding features will be absent with the disappearance of Li atoms. According to the atom-dominated band structure (Figures 12.4(b)−(d)), the energy spectrum of $LiFeO_2$ can be sub-divided into four specific regions: (i) from 6.5 eV to 14.7 eV, which is dominated by the Fe atom; (ii) from −1 eV

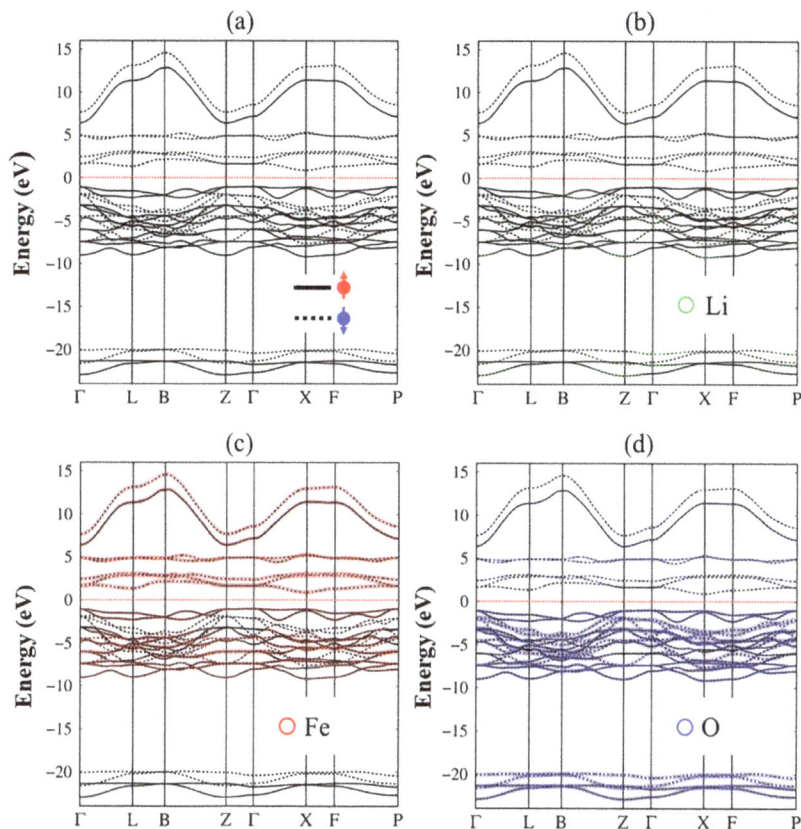

FIGURE 12.4   (a) Band structure along with the high-symmetry points in the wave-vector space, with (b) Li-, (c) Fe- and (d) O-atom dominances (green, red, and blue circles, respectively).

to 5 eV, which is co-dominated by Fe and O atoms; (iii) from −12 eV to −15 eV, which is mainly derived from Fe and O atoms; and (iv) below −20 eV, which is mostly formed by the O atom. The orbital hybridizations related to Li-O, Fe-O chemical bonding could be clarified through the close combination of the current dominated-energy spectrums with spin/charge density distribution and the spin-polarized density of states.

From the experimental aspect, the information about a sizable indirect gap in the 3D ternary LiFeO$_2$ compound could be observed by the optical excitation spectrum as done for other 3D materials, such as Li$_2$GeO$_3$ and Li$_2$SiO$_3$ compounds. Furthermore, the energy dispersion in the valence band can be observed by the ARPESS equipment. Such experiments were very successful in observing the occupied states in graphene and its related systems—for example, the monolayer/bilayer-like energy dispersions at the K/H symmetry points in graphite [72], the parabolic/parabolic and linear in AB stacked bilayer/AB stacked tri-layer graphene [73], and the parabolic dispersion in the energy band structure of graphene nanoribbons [74]. Similar measurements for ternary $LiFeO_2$ have been absent up to now. Apparently, the various energy sub-band accompanied with rather complicated energy dispersions will create a high challenge in the ARPES measurements.

The charge density distributions are capable of providing diversified chemical environments. As clearly shown in Figures 12.5(a)–(b), the strong orbital overlaps are easy to identify in Fe-O bonds, while the opposite is true for the Li-O bonds. The ratio of the former versus the latter is more than five times what could be understood from the different scales. As result, the Fe-O strength is much stronger than that of the Li-O bond and thus explains why the latter is much longer than the former. Such phenomena are very important for the charging or discharging of lithium ions from the cathode material during the battery's operation, since the migration of lithium atoms is supported by the strong Fe-O frames. Very interestingly, the first-step orbital-hybridizations in the chemical bonds can be identified through the delicate analysis of the spatial charge density in the specific chemical bonds. As for the O atom, the blue-white (inner) and the red (outer) regions arise from the (2s), $(2p_x, 2p_y, 2p_z)$ orbitals, respectively. Similarly, the inner and outer parts of the Fe atom correspond to the $(3d_{xy}, 3d_{yz}, 3d_{xz}, 3d_{x^2-y^2}, 3d_{z^2})$ and 4s orbitals. When they combine together to form the Fe-O chemical bond, the serious distortion of these spherical charges (Figure 12.5(a)) clearly evidences the appearance of the important

Charge density distribution   **Spin density distribution**

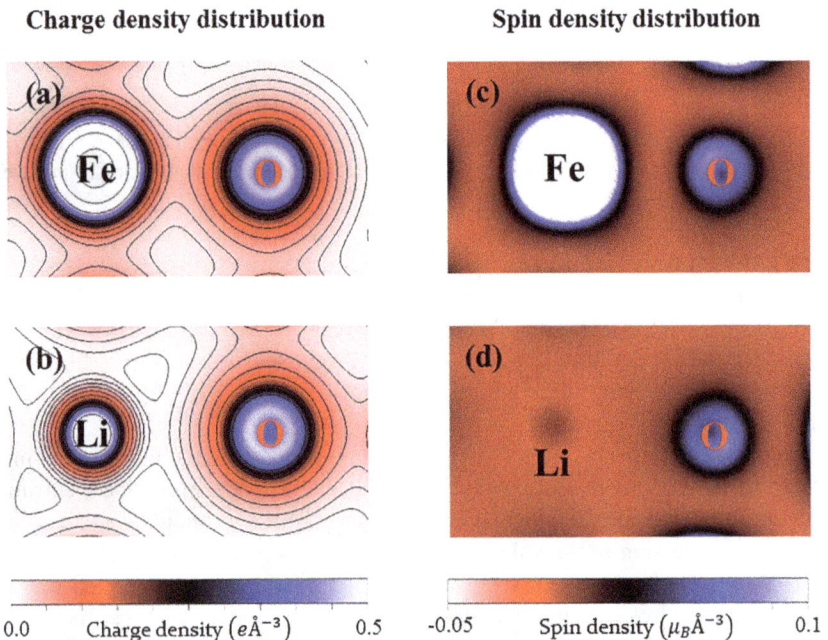

| 0.0 | Charge density $(e\text{Å}^{-3})$ | 0.5 | -0.05 | Spin density $(\mu_B\text{Å}^{-3})$ | 0.1 |

FIGURE 12.5 (a)/(b) Charge density distributions and (c)/(d) spin density distributions related to the significant orbital hybridizations in the Fe-O bonds and Li-O ones, respectively.

multi hybridization of Fe-(4s, $3d_{xy}$, $3d_{yz}$, $3d_{xz}$, $3d_{x^2-y^2}$, $3d_{z^2}$) and O-(2s, $2P_x$, $2P_y$, $2P_z$) orbitals. For the Li-O chemical bond, the charge density distribution around the Li atom can also be divided into two parts, the inner part corresponding to the 1s orbitals while the outer ones belong to the 2s one. As shown in Figure 12.5(b), the slight deformation of Li and O spatial charge densities between Li-O chemical bonds presents the single Li-2s and O-2s and multi Li-2s and O-($2p_x$, $2p_y$, $2p_z$) orbital combinations. The combination of this information with further analyses on the projected density of states could provide the fully physical/chemical pictures contained in the LiFeO$_2$ compound.

The spin density distribution and the magnetic moment could provide more information about the magnetic properties of the anode/cathode/electrolyte compounds (Table 12.1), in which the competition between the spin-up and components will determine the net magnetic moment in a unit cell. The LiFeO$_2$ compound presents the ferromagnetic configuration, with the total magnetic moment being 4.624 μ_B as expected, since the Fe atom introduces four spin-up electrons. As clearly seen in Figure 12.5(c), the spin-up density is the most dominant part and relies on the Fe atom, with its typical magnetic moment being equal to 4.06 μ_B, which is directly reflected by the fully occupied state in strongly dispersive energy bands below the Fermi level (the third region). Furthermore, O and Li atoms, which are non-magnetic before combination, also present the partial minor magnetic contributions (Figure 12.5(d)), with these values being lower than the Fe atom under one and two orders, respectively. As a result, the magnetic properties of LiFeO$_2$ are expected and will be sensitive to change during Li$^+$ migration.

The density of states (DOS) is determined as the number of electronic states within the rather small energy range of dE and directly reflects the main properties of valence and conduction bands simultaneously. Moreover, the orbital hybridizations and spin-polarizations in Li-O and Fe-O bonds could be clarified by the atom and orbital DOS. As revealed in Figures 12.6(a)–(c), the density of states of a 3D ternary LiFeO$_2$ compound mostly present the shoulder and asymmetric van Hove singularities since they mainly originated from the oscillatory, local minimum/maximum, or almost dispersionless relation energy sub-bands. The vanishing of the electronic states around the Fermi level creates a sizable band gap of 1.9 eV. The merge well of various atoms and orbitals is evident in the complicated

TABLE 12.1

The Magnetic Moment of the Specific Orbital and Atom of LiFeO$_2$

Ion	Magnetic moment (μ_B)			
	s	p	d	Tot
Li	0.009	0	0	0.009
Fe	0.023	0.026	4.011	4.060
O	0.018	0.260	0	0.277
O	0.018	0.260	0	0.277
Total	0.061	0.551	4.011	4.624

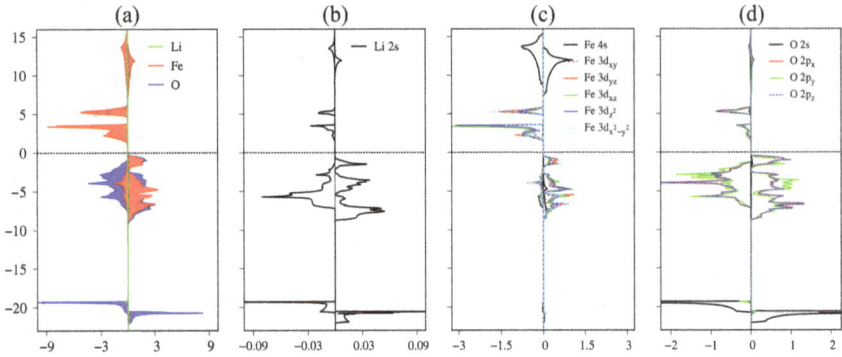

FIGURE 12.6 (a) The atom- and orbital-projected density of states: those coming from (a) Li, Fe, and O atoms, (b) Li-2s orbitals, (c) Fe-(4s, $3d_{xy}$, $3d_{yz}$, $3d_{xz}$, $3d_{x^2-y^2}$, $3d_{z^2}$) orbitals, and (d) O-(2s, $2p_x$, $2p_y$, $2p_z$) orbitals.

orbital hybridizations in Li-O and Fe-O bonds. Very interestingly, the domination and energy of the spin-up and spin-down are extremely different; this indicates that LiFeO$_2$ possesses very strong ferromagnetic behavior, being good in agreement with spin-splitting band structure and spin-density distribution. According to the distribution of specific atoms and orbitals, the orbital hybridization of four energy ranges we mentioned before can be identified: (i) Li-2s and O-($2p_x$, $2p_y$, $2p_z$) and Fe-4s and O-($2p_x$, $2p_y$, $2p_z$) orbital hybridizations; (ii) Li-2s and O-($2p_x$, $2p_y$, $2p_z$) and Fe-(4s, $3d_{xy}$, $3d_{xz}$, $3d_{yz}$, $3d_{x^2-y^2}$, $3d_{z^2}$) orbitals mixing; (iii) Li-2s and O-($2p_x$, $2p_y$, $2p_z$) and Fe-(4s, $3d_{xy}$, $3d_{xz}$, $3d_{yz}$, $3d_{x^2-y^2}^2$, $3d_{z^2}$) and O-($2p_x$, $2p_y$, $2p_z$) orbital hybridizations; and (iv) combined of Li-2s and O-2s and Fe- (4s, $3d_{xy}$, $3d_{xz}$, $3d_{yz}$, $3d_{x^2-y^2}$, $3d_{z^2}$) and O-($2p_x$, $2p_y$, $2p_z$) hybridized.

The number, form, energy, intensity, and polarizations of van Hove singularities in the vicinity of the Fermi level can be measured by the STS and spin-polarized STS measurements. These techniques have successfully discovered the diverse electronic and magnetic properties in carbon-nanotubes [75], few-layer graphene [76], and carbon nanoribbons [40]—for example, various divergence structures under the square root relations in graphene nanoribbons and carbon nanotubes [77], the V-shape vanishing at the Fermi level in monolayer graphene [78], the logarithmically symmetric peaks near E$_F$ in twisted bi-layers graphene [79]. The main features in electronic and magnetic properties in the ternary LiFeO$_2$ compound, including the spacing of occupied and unoccupied states, the asymmetric/symmetric structures, the asymmetry of electron and hole states as well as their widths, the splitting of spin-up and spin-down states, could be further investigated by STS and spin-polarized STS measurements. The information about the van Hove singularities from the numerical calculations combined with STS/spin-polarized STS measurements is very worthy in understanding the complicated orbital hybridizations in Li-O and Fe-O bonds and thus the electronic and magnetic properties of the LiFeO$_2$ compound.

Up till now, the LIBs have been quickly developed owing to their vital importance. The 3D ternary LiFeO$_2$ can be adopted as a candidate for cathode materials. The batteries with this material can reduce the cost and environmental effects while improving ion transport efficiency. Based on the spin-splitting electronic band structures, the spin and charge density distributions, the spin-polarized orbital density of states, the critical orbital hybridizations in Li-O and Fe-O bonds are obtained. Apparently, the ternary LiFeO$_2$ compound has various meta-stable configurations; during the battery's operation, the transformation from the current to the other phases can be established. Therefore, to find out the most efficient evolution paths, it is necessary to systematically investigate the other meta-stables.

The last remark concerns the numerical studies-phenomenological models' combination. The electronic structure achieved by the first-principles method might be too difficult to simulate by the phenomenological approaches due to the presence of an octahedral configuration in geometric structure, complicated energy dispersions with significantly spin-split/degeneracy, the rather complex spatial charge density in each chemical bond and a lot of strong van Hove singularities in the density of states. Particularly, the complex chemical/physical environments, which contain the polyhedral structures of the 3D ternary LiFeO$_2$ compound govern the electronic and magnetic features. Such important factors cover various single and multi-orbital hybridizations in the Li-O and Fe-O chemical bonds. Therefore, the significant Hamiltonian should include various hoping integral and side energies simultaneously. It is very difficult to achieve a concise physical picture for a full understanding of the features of electronic energy spectra.

12.5 CONCLUDING REMARKS

In the current chapter, density functional theory (DFT) was used to investigate the geometric, electronic, and magnetic properties for LiFeO$_2$. Based on the delicate analysis of the geometric, spin-splitting band structure, the spin and charge density distributions, the spin-polarization density of states, the complicated orbital hybridization in Li-O and Fe-O can be successfully achieved.

The 3D LiFeO$_2$ compound presents unique properties: the Moiré superlattice with high atomic ordering and anisotropy geometric, the presence of 18 Li-O/18 Fe-O chemical bonds with identical bond-length, the complicated spin-polarization band structure with various atom dominations, the sizable indirect gap of 1.9 eV, the spatial spin and charge distribution, and a lot of van Hove singularities due to the spin-polarizations and extreme point dispersions. As a consequence, the important multi-orbital (4s, 3p$_x$, 3p$_y$, 3p$_z$, 3d$_{xy}$, 3d$_{yz}$, 3d$_{xz}$, $d_{x^2-y^2}$, 3d_{z^2})-(2s, 2p$_x$, 2p$_y$, 2p$_z$) hybridizations and the multi-orbital 2s-(2p$_x$, 2p$_y$, 2p$_z$) hybridizations and single-orbital 2s-2s orbital hybridizations in Fe-O and Li-O bonds, respectively, could be well achieved. The theoretical predictions on the geometric structure are predicted by PXRD, TEM, SEM, or STM, the electronic band structure can be detected by ARPES/spin-polarized ARPES, while the information of the spin-polarization van Hove singularities can be examined by the STS/spin-polarized STS measurements.

The calculated results clearly indicate that the 3D ternary LiFeO$_2$ could serve as a candidate cathode component in lithium-based batteries. Our prediction provides useful information about the critical physical/chemical pictures in LIBs. Such state-of-the-art analysis is appropriate for fully comprehending the diversified properties in anode/cathode/electrolyte and other emerging materials.

ACKNOWLEDGMENTS

This work is supported by the Hi-GEM Research Center and the Taiwan Ministry of Science and Technology under grant number MOST 108–2212-M-006–022-MY3, MOST 109–2811-M-006–505 and MOST 108–3017-F-006–003.

REFERENCES

[1] Devabhaktuni V, Alam M, Depuru S S S R, Green II R C, Nims D and Near C 2013 Solar energy: Trends and enabling technologies *Renewable and Sustainable Energy Reviews* **19** 555–64

[2] Deng D 2015 Li-ion batteries: Basics, progress, and challenges *Energy Science & Engineering* **3** 385–418

[3] Whittingham M S 2012 History, evolution, and future status of energy storage *Proceedings of the IEEE* **100** 1518–34

[4] Yilmaz M and Krein P T 2012 Review of battery charger topologies, charging power levels, and infrastructure for plug-in electric and hybrid vehicles *IEEE Transactions on Power Electronics* **28** 2151–69

[5] Boyer M J and Hwang G S 2016 Recent progress in first-principles simulations of anode materials and interfaces for lithium ion batteries *Current Opinion in Chemical Engineering* **13** 75–81

[6] Sygletou M, Petridis C, Kymakis E and Stratakis E 2017 Advanced photonic processes for photovoltaic and energy storage systems *Advanced Materials* **29** 1700335

[7] Baxter J, Bian Z, Chen G, Danielson D, Dresselhaus M S, Fedorov A G et al 2009 Nanoscale design to enable the revolution in renewable energy *Energy & Environmental Science* **2** 559–88

[8] Singh A and Kalra V 2019 Electrospun nanostructures for conversion type cathode (S, Se) based lithium and sodium batteries *Journal of Materials Chemistry A* **7** 11613–50

[9] Lu X, Yu M, Zhai T, Wang G, Xie S, Liu T et al 2013 High energy density asymmetric quasi-solid-state supercapacitor based on porous vanadium nitride nanowire anode *Nano Letters* **13** 2628–33

[10] Hall P J, Mirzaeian M, Fletcher S I, Sillars F B, Rennie A J, Shitta-Bey G O et al 2010 Energy storage in electrochemical capacitors: Designing functional materials to improve performance *Energy & Environmental Science* **3** 1238–51

[11] Stambouli A B and Traversa E 2002 Solid oxide fuel cells (SOFCs): A review of an environmentally clean and efficient source of energy *Renewable and Sustainable Energy Reviews* **6** 433–55

[12] Von der Kammer F, Ferguson P L, Holden P A, Masion A, Rogers K R, Klaine S J et al 2012 Analysis of engineered nanomaterials in complex matrices (environment and biota): General considerations and conceptual case studies *Environmental Toxicology and Chemistry* **31** 32–49

[13] Gao Z, Sun H, Fu L, Ye F, Zhang Y, Luo W et al 2018 Promises, challenges, and recent progress of inorganic solid-state electrolytes for all-solid-state lithium batteries *Advanced Materials* **30** 1705702

[14] Zhang Z, Shao Y, Lotsch B, Hu Y-S, Li H, Janek J et al 2018 New horizons for inorganic solid state ion conductors *Energy & Environmental Science* **11** 1945–76

[15] Hochgatterer N S, Schweiger M R, Koller S, Raimann P R, Wöhrle T, Wurm C et al 2008 Silicon/graphite composite electrodes for high-capacity anodes: Influence of binder chemistry on cycling stability *Electrochemical and Solid State Letters* **11** A76

[16] Yi T-F, Xie Y, Zhu Y-R, Zhu R-S and Shen H 2013 Structural and thermodynamic stability of Li4Ti5O12 anode material for lithium-ion battery *Journal of Power Sources* **222** 448–54

[17] Lyu Y, Wu X, Wang K, Feng Z, Cheng T, Liu Y et al 2020 An overview on the advances of LiCoO2 cathodes for lithium-ion batteries *Advanced Energy Materials* 2000982

[18] Mandal S, Amarilla J M, Ibanez J and Rojo J M 2001 The role of carbon black in LiMn2O4-based composites as cathodes for rechargeable lithium batteries *Journal of the Electrochemical Society* **148** A24

[19] Lopez J A, González F, Bonilla F A, Zambrano G and Gómez M E 2010 Synthesis and characterization of Fe3O4 magnetic nanofluid *Revista Latinoamericana de Metalurgia y Materiales* **30** 60–6

[20] Maslen E, Streltsov V, Streltsova N and Ishizawa N 1994 Synchrotron X-ray study of the electron density in α-Fe2O3 *Acta Crystallographica Section B: Structural Science* **50** 435–41

[21] Billaud J, Clément R J, Robert Armstrong A, Canales-Vázquez J, Rozier P, Grey C P and Bruce P G 2014 β-NaMnO2: A high-performance cathode for sodium-ion batteries *Journal of the American Chemical Society* **136**(49) 17243–8

[22] Chaudhari N K and Yu J-S 2008 Size control synthesis of uniform β-FeOOH to high coercive field porous magnetic α-Fe2O3 nanorods *The Journal of Physical Chemistry C* **112** 19957–62

[23] Singh S, Tovstolytkin A and Lotey G S 2018 Magnetic properties of superparamagnetic β-NaFeO2 nanoparticles *Journal of Magnetism and Magnetic Materials* **458** 62–5

[24] Kikkawa S, Ohkura H and Koizumi M 1987 Ion exchange of layered α-NaFeO2 *Materials Chemistry and Physics* **18** 375–80

[25] Nitta N, Wu F, Lee J T and Yushin G 2015 Li-ion battery materials: Present and future *Materials Today* **18**(5) 252–64.

[26] Hu Y, Zhao H and Liu X 2018 A simple, quick and eco-friendly strategy of synthesis nanosized α-LiFeO2 cathode with excellent electrochemical performance for lithium-ion batteries *Materials* **11** 1176

[27] Guan J, Jia C, Li Y, Liu Z, Wang J and Yang Z et al 2018 Direct single-molecule dynamic detection of chemical reactions *Science Advances* **4** eaar2177

[28] Petersen M, Hafner J and Marsman M 2006 Structural, electronic and magnetic properties of Gd investigated by DFT+ U methods: Bulk, clean and H-covered (0001) surfaces *Journal of Physics: Condensed Matter* **18** 7021

[29] Boukhvalov D W, Dreyer D R, Bielawski C W and Son Y W 2012 A computational investigation of the catalytic properties of graphene oxide: Exploring mechanisms by using DFT methods *ChemCatChem* **4** 1844–9

[30] Medeiros P V, Mascarenhas A J, de Brito Mota F and de Castilho C M 2010 A DFT study of halogen atoms adsorbed on graphene layers *Nanotechnology* **21** 485701

[31] Gao N, Zheng W T and Jiang Q 2012 Density functional theory calculations for two-dimensional silicene with halogen functionalization *Physical Chemistry Chemical Physics* **14** 257–61

[32] Maroulis G, Begué D and Pouchan C 2003 Accurate dipole polarizabilities of small silicon clusters from ab initio and density functional theory calculations *The Journal of Chemical Physics* **119** 794–7

[33] da Silveira Lacerda L H and de Lazaro S R 2018 Multiferroism and magnetic ordering in new NiBO3 (B= Ti, Ge, Zr, Sn, Hf and Pb) materials: A DFT study *Journal of Magnetism and Magnetic Materials* **465** 412–20

[34] Giorgi G, Fujisawa J-I, Segawa H and Yamashita K 2014 Cation role in structural and electronic properties of 3D organic—inorganic halide perovskites: A DFT analysis *The Journal of Physical Chemistry C* **118** 12176–83

[35] Mirtamizdoust B, Ghaedi M, Hanifehpour Y, Mague J T and Joo S W 2016 Synthesis, structural characterization, thermal analysis, and DFT calculation of a novel zinc (II)-trifluoro-β-diketonate 3D supramolecular nano organic-inorganic compound with 1, 3, 5-triazine derivative *Materials Chemistry and Physics* **182** 101–9

[36] Nguyen T D H, Pham H D, Lin S-Y and Lin M-F 2020 Featured properties of Li+-based battery anode: Li 4 Ti 5 O 12 *RSC Advances* **10** 14071–9

[37] Han N T, Dien V K, Tran N T T, Nguyen D K, Su W-P and Lin M-F 2020 First-principles studies of electronic properties in Lithium metasilicate (Li2SiO3) *arXiv pre-print arXiv:2001.07128*

[38] Dien V K, Han N T, Nguyen T D H, Huynh T M D, Pham H D and Lin M F 2020 Geometric and electronic properties of Li2GeO3 *Frontiers in Materials* **7** 288

[39] Lin S-Y, Chang S-L, Shyu F-L, Lu J-M and Lin M-F 2015 Feature-rich electronic properties in graphene ripples *Carbon* **86** 207–16

[40] Nguyen D K, Tran N T T, Nguyen T T and Lin M-F 2018 Diverse electronic and magnetic properties of chlorination-related graphene nanoribbons *Scientific Reports* **8** 1–12

[41] Banerjee S, Periyasamy G and Pati S K 2014 Possible application of 2D-boron sheets as anode material in lithium ion battery: A DFT and AIMD study *Journal of Materials Chemistry A* **2** 3856–64

[42] Jónsson E and Johansson P 2015 Electrochemical oxidation stability of anions for modern battery electrolytes: A CBS and DFT study *Physical Chemistry Chemical Physics* **17** 3697–703

[43] Zhang H, Gong Y, Li J, Du K, Cao Y and Li J 2019 Selecting substituent elements for LiMnPO4 cathode materials combined with density functional theory (DFT) calculations and experiments *Journal of Alloys and Compounds* **793** 360–8

[44] Sakurai Y, Arai H and Yamaki J-I 1998 Preparation of electrochemically active α-LiFeO2 at low temperature *Solid State Ionics* **113** 29–34

[45] Wang X, Gao L, Zhou F, Zhang Z, Ji M, Tang C et al 2004 Large-scale synthesis of α-LiFeO2 nanorods by low-temperature molten salt synthesis (MSS) method *Journal of Crystal Growth* **265** 220–3

[46] Büyükyazi M and Mathur S 2015 3D nanoarchitectures of α-LiFeO2 and α-LiFeO2/C nanofibers for high power lithium-ion batteries *Nano Energy* **13** 28–35

[47] Damascelli A 2004 Probing the electronic structure of complex systems by ARPES *Physica Scripta* **2004** 61

[48] Feenstra R M 1994 Scanning tunneling spectroscopy *Surface Science* **299** 965–79

[49] Ortmann F and Roche S 2013 Splitting of the zero-energy Landau level and universal dissipative conductivity at critical points in disordered graphene *Physical Review Letters* **110** 086602

[50] Do T-N, Shih P-H, Chang C-P, Lin C-Y and Lin M-F 2016 Rich magneto-absorption spectra of AAB-stacked trilayer graphene *Physical Chemistry Chemical Physics* **18** 17597–605

[51] Liu L, Zhou H, Cheng R, Yu W J, Liu Y, Chen Y et al 2012 High-yield chemical vapor deposition growth of high-quality large-area AB-stacked bilayer graphene *Acs Nano* **6** 8241–9

[52] Sun G, Kürti J, Rajczy P, Kertesz M, Hafner J and Kresse G 2003 Performance of the Vienna ab initio simulation package (VASP) in chemical applications *Journal of Molecular Structure: THEOCHEM* **624** 37–45

[53] Baerends E J and Gritsenko O V 1997 A quantum chemical view of density functional theory *The Journal of Physical Chemistry A* **101** 5383–403

[54] Kresse G and Joubert D 1999 From ultrasoft pseudopotentials to the projector augmented-wave method *Physical Review B* **59** 1758

[55] Sikam P, Moontragoon P, Jumpatam J, Pinitsoontorn S, Thongbai P and Kamwanna T 2016 Structural, optical, electronic and magnetic properties of Fe-doped ZnO nanoparticles synthesized by combustion method and first-principle calculation *Journal of Superconductivity and Novel Magnetism* **29** 3155–66

[56] Wisesa P, McGill K A and Mueller T 2016 Efficient generation of generalized Monkhorst-Pack grids through the use of informatics *Physical Review B* **93** 155109

[57] Aspnes D E 1982 Optical properties of thin films *Thin Solid Films* **89** 249–62

[58] Charlier J-C, Blase X and Roche S 2007 Electronic and transport properties of nanotubes *Reviews of Modern Physics* **79** 677

[59] Lin C-Y, Wu J-Y, Ou Y-J, Chiu Y-H and Lin M-F 2015 Magneto-electronic properties of multilayer graphenes *Physical Chemistry Chemical Physics* **17** 26008–35

[60] Dresselhaus G 1956 Effective mass approximation for excitons *Journal of Physics and Chemistry of Solids* **1** 14–22

[61] Lin Y-P, Wang J, Lu J-M, Lin C-Y and Lin M-F 2014 Energy spectra of ABC-stacked trilayer graphene in magnetic and electric fields *RSC Advances* **4** 56552–60

[62] Shih P-H, Lin C-Y, Do T-N, Wu J-Y, Lin S-Y, Ho C-H et al 2019 6 AA-bottom-top bilayer silicene systems *Diverse Quantization Phenomena in Layered Materials* 131

[63] Shih P H, Chiu Y-H, Wu J-Y, Shyu F-L and Lin M-F 2017 Coulomb excitations of monolayer germanene *Scientific Reports* **7** 40600

[64] Chen S-C, Wu C-L, Wu J-Y and Lin M-F 2016 Magnetic quantization of s p 3 bonding in monolayer gray tin *Physical Review B* **94** 045410

[65] Chen S-C, Wu J-Y and Lin M-F 2018 Feature-rich magneto-electronic properties of bismuthene *New Journal of Physics* **20** 062001

[66] Pantha N, Chauhan B, Sharma P and Adhikari N 2020 Tuning structural and electronic properties of phosphorene with vacancies *Journal of Nepal Physical Society* **6** 7–15

[67] Boufelfel A 2013 Electronic structure and magnetism in the layered LiFeO2: DFT+ U calculations *Journal of Magnetism and Magnetic Materials* **343** 92–8

[68] Sakurai Y, Arai H, Okada S and Yamaki J-I 1997 Low temperature synthesis and electrochemical characteristics of LiFeO2 cathodes *Journal of Power Sources* **68** 711–15

[69] Kim Y, Hayashi T, Osawa K, Dresselhaus M and Endo M 2003 Annealing effect on disordered multi-wall carbon nanotubes *Chemical Physics Letters* **380** 319–24

[70] Shen Y and Wu H 2012 Interlayer shear effect on multilayer graphene subjected to bending *Applied Physics Letters* **100** 101909

[71] Sun Y, Zheng Z, Cheng J, Liu J, Liu J and Li S 2013 The un-symmetric hybridization of graphene surface plasmons incorporating graphene sheets and nano-ribbons *Applied Physics Letters* **103** 241116

[72] Yan J-A, Ruan W and Chou M 2008 Phonon dispersions and vibrational properties of monolayer, bilayer, and trilayer graphene: Density-functional perturbation theory *Physical Review B* **77** 125401

[73] Lin C-Y, Do T-N, Wu J-Y, Shih P-H, Lin S-Y, Ho C-H et al 2019 5 stacking-configuration-modulated bilayer graphene *Diverse Quantization Phenomena in Layered Materials* 101

[74] Dutta S and Pati S K 2010 Novel properties of graphene nanoribbons: A review *Journal of Materials Chemistry* **20** 8207–23

[75] Ateia M, Koch C, Jelavić S, Hirt A, Quinson J, Yoshimura C et al 2017 Green and facile approach for enhancing the inherent magnetic properties of carbon nanotubes for water treatment applications *PLoS One* **12** e0180636

[76] Do T-N, Chang C-P, Shih P-H, Wu J-Y and Lin M-F 2017 Stacking-enriched magneto-transport properties of few-layer graphenes *Physical Chemistry Chemical Physics* **19** 29525–33

[77] Plaut R, Borum A and Dillard D 2012 Analysis of carbon nanotubes and graphene nanoribbons with folded racket shapes *Journal of Engineering Materials and Technology* **134**

[78] Adhikari S, Perello D J, Biswas C, Ghosh A, Van Luan N, Park J et al 2016 Determining the Fermi level by absorption quenching of monolayer graphene by charge transfer doping *Nanoscale* **8** 18710–17

[79] Campos-Delgado J, Algara-Siller G, Santos C, Kaiser U and Raskin J P 2013 Twisted bi-layer graphene: Microscopic rainbows *Small* **9** 3247–51

13 Stacking-Enriched Quasiparticle Properties of Bilayer HfX$_2$ (X = S, Se, or Te)

Thi My Duyen Huynh and Ming-Fa Lin

CONTENTS

13.1 FROM MONOLAYER TO BILAYER HFX$_2$

After exploration of graphene from graphite in 2004, two-dimensional (2D) materials, particularly 2D-transition metal dichalcogenides (2D-TMDs), are ideal candidates for engineering a potential class of materials. Thus, many examples of them [1–7] have been constructed in both theoretical and experimental studies. They have opened up new opportunities for applications in electronic devices because of their diverse structures [7], which can be constructed as bulk, monolayers, bilayers, or multilayers. Layered structures form a hexagonal packing in TMDs, in which each layer is sandwiched by transition metal and chalcogen atoms [7]. The crystallographic orientations range from the hexagonal sulfides to the orthorhombic tellurides [7]. Various polymorphs can be found depending on the relative position of the chalcogen atoms in each sandwiched structure. The general properties of TMDs have been discussed in our previous study as well as other studies that are not emphasized in this chapter. In this chapter, the probability of constructing a bilayer with varying stacking is considered and discussed.

Bilayer TMDs are technological structures that can be created from their layered bulk form (monolayer) [3, 8–11]. A monolayer is directly constructed from its bulk phase by adding a vacuum to avoid interaction between layers as shown in Figure 13.1. Hence, there only exist the T and H phases in the monolayer structures corresponding to trigonal prismatic and hexagonal arrangements instead of various

DOI: 10.1201/9781003322573-13

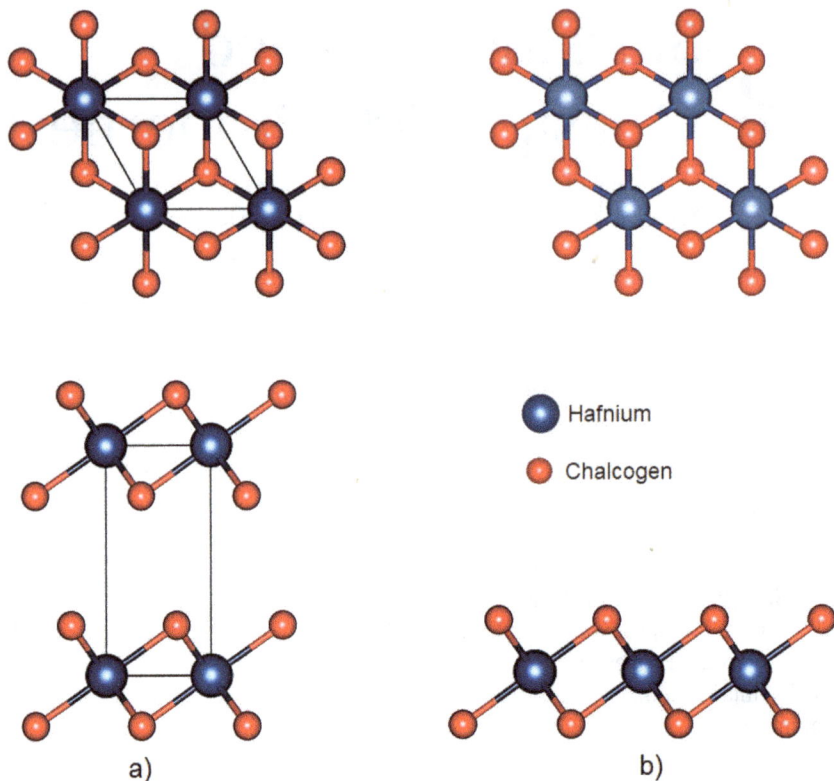

FIGURE 13.1 Top and side views of (a) bulk and (b) monolayer HfX_2 in the T phase.

phases in bulk. Thus, a bilayer TMD is constructed from the monolayer by adding a second layer, considering the stacking to obtain possible and stable structures and maintaining the vacuum as in the monolayer. The different stacking orders and their effects will be discussed in the next section.

Provided that 2D-TMDs show distinct behavior that respects their bulk phase [12, 13], thicknesses down to two (bilayer) or one (monolayer) atomic layers are of particular concern based on their unique electronic and optical properties [4, 12–17]. In particular, the diverse properties from monolayer to bilayer are typically more essential than change achieved using adding more layers (multilayers) [18–21]. Bilayer TMDs have been successfully synthesized [22–30] and provide a range of new applications. Their ultrathin body and anisotropic properties allow a wide range of modifications for the electronic properties [24]. One of these properties is the tuning of band gap that is investigated in MX_2 (M = Mo, W; X = S, Se, or Te) [24, 27, 28], indicating a reduction of the band gap by applying a sufficiently strong electric field across the two layers. Besides, the band structure changes have been considered in WS_2 with increased contribution from indirect transitions [23]. Furthermore, bilayer TMDs have advantages over the monolayer structures that have been demonstrated in previous studies that considered external factors. These advantages have been illustrated in photocatalytic water splitting of MoSSe [25]. Additionally, the noise measurements on bilayer MoS$_2$ transistors show a noise peak in the gate-voltage dependence [30]. Therefore, bilayer TMDs show a high potential for expanding the chemical and physical properties, which could lead to new applications in electronic devices.

As mentioned, TMDs are emergent materials that are now considered in research due to their divergent and sensitive properties. Among them, MoS$_2$ [2, 3, 12, 24, 30–36] is one of the most interesting candidates and shows a wide range of rich essential features. To enlarge the range of potential materials, other groups of TMDs have attracted attention and shown promise of new physical and chemical properties in the application of devices, for example, HfX$_2$ (X = S, Se, or Te). Concerning structure, the bulk [37, 38] and monolayer [37, 39–41] of HfX$_2$ have been reported to show not only the common characteristics of TMDs but also distinct properties. Considering their stability, the T phase structures are found in both bulk and monolayer HfX$_2$ and are predicted to be the most favorable phase [37, 40–45]. Previous studies show that HfTe$_2$ is a semi-metal [46], while both HfS$_2$ [37, 41, 47] and HfSe$_2$ [37, 39, 43, 48, 49] are indirect-gap semiconductors. On the subject of electronic properties, the thickness and strain can tune the band gap of these materials [17, 38, 39, 41, 44]. Furthermore, the chalcogen atoms play an important role in the valence band, while hafnium shows a dominant contribution in the conduction band [37]. Simultaneously, experimental studies reveal HfS$_2$ [50–52] as a candidate for use in transistors, while bilayer HfSe$_2$ [6, 53–56] has a low lattice thermal conductivity, and HfTe$_2$ has been successfully grown on a substrate [57], suggesting a topological material.

Due to these emergent and satisfactory findings, HfX$_2$ is a candidate for investigation and applications. In order to further reduce the size of electronic devices, monolayer and bilayer HfX$_2$ have been explored and compared to bulk HfX$_2$, which are already used in some devices [54–56]. Bilayer HfX$_2$ should therefore be analyzed in order to enhance this group. Although bilayer HfSe$_2$ [49] has been investigated promisingly for thermal conductivity, the perspective of all these materials in the bilayer is still limited. Moreover, theoretical framework using VASP calculations is accurate in solving the quasiparticle problems related to geometry, electron or phonon, including the lattice constant, chemical bonding, and interaction between atoms and layers define the symmetric geometry. Thus, the symmetry of the structure, by changing the stacking, determines the electronic properties of these materials. Through the theoretical framework, the strong relation between quasiparticles and theoretical calculations is illustrated. In this chapter, we therefore focus on the electronic properties of bilayer HfX$_2$ (X = S, Se, or Te) using VASP calculations to provide further information about their features.

13.2   STACKING EFFECT IN BILAYER TMDS

The structural symmetry of 2D materials plays an essential role in their electronic and optical properties. Therefore, engineering geometry has become a promising banner to achieve novel properties. The stacks of 2D materials determine the symmetry of structures and their physical properties, for example, graphene and TMDs. The different stacking modes might lead to interlayer hopping and hybridization of the band energy spectra characteristic of the many emergent properties of bilayer graphene [58–61]. The AB configuration [62] of this bilayer displays a quadratic band dispersion with massive chiral quasiparticles, while a massless Dirac spectrum and gapless band structure were displayed in the AA stacking [63]. Furthermore, the stacking of two monolayer TMDs changes the band gap direction from direct to indirect. The stacking of the identical monolayer TMDs generates large-scale bilayers, which provide an excellent platform to investigate and design bilayer TMDs. Further information about this will be accordingly discussed.

As previously mentioned, the stacking order in structural engineering can be used to manipulate the electronic properties of TMDs. Varying the stacking order provides an alternative to tuning the electronic properties. By changing the stacking, most physical properties, such as the geometry and electronic and optical properties can be manifested [5, 64]. As a potential candidate, MoS_2 represents the emergent TMDs with almost electronic, elastic, optical, transport, and other chemical-physical properties already investigated. The Berry curvature dependent on the stacking shows that MoS_2 is highly tunable [65]. Varying the stacking sequence can control the inversion symmetry in bilayer MoS_2 [31] and WS_2 [66], which provides a new avenue for controlling valleytronics. By using 4D scanning transmission electron microscopy combined with a multi-slice diffraction simulation, the stacking order effects on the local band structure were once again demonstrated for MoS_2 [67]. In terms of identical characteristics, other TMDs have been compared with MoS_2 and were simultaneously considered under the same conditions. The interlayer coupling effects investigated for the different stacking configurations of MoS_2 and WSe_2 [68] show a crucial impact on the valley polarization. The elastic properties are slightly affected by the stacking order of bilayer MX_2 (M = Mo, W; X = O, S, Se, or Te) [69] when alternating five modes. The distinct stacking of bilayer in two phases, T and H, leads to a difference in the interplanar interaction, which is governed by the nature of the X-X bonding [70]. Therefore, versatile stacked structures might exist based on employing the phase approximation. Considering the 2H and 3R phases, MX_2 (M = Mo, W; X = S, Se) exhibits the influence of the stacking on the band energy and values of the optical excitations [64]. The stacking effects have been considered in other TMDs based on MoS_2 to enlarge the range of applications. The interlayer coupling in bilayer SnS_2, with alternative stacking orders that have similar structural parameters, is weaker than that of typical TMDs [71]. The piezoelectricity in WSe_2 with changing stacking orders has been illustrated via CVD and shows good mechanical stability [72]. The experimental and simulated electron diffraction were used to study the impact of different stacking in TaS_2, revealing the information about the stacking orders when the crystal structure is unresolved in real space [73].

In addition, external factors combined with varying stacking cause significant effects in a bilayer. The barrier height can increase when the electric field increases the coupling between the layers of SnS_2 [71]. Applying pressure also changes the stacking order of this material. ZrX_2 (X = S, Se, or Te) has also shown similar features and provides an effective strategy to modulate the electronic properties by implementing electric field and pressure [70].

On the other hand, bilayer TMDs can be constructed by two identical layers or distinct layers, which provides great flexibility for band structure engineering and the design of photoelectronic devices. Due to the effective role of the interlayer charge transfer with different stackings, the band gap of SiS$_2$/WSe$_2$ [74] can be modulated by applying an electric field. The WS$_2$/WSe$_2$ hetero-structural stacking order reveals an alternative for controlling the edges [75]. Moreover, the direct gap can be achieved in the bilayer MoSe$_2$/WSe$_2$ heterostructure by alternating the stacking modes [76].

Typically, there are two main types of TMD bilayers that are either composed of two identical monolayers, called homogeneous bilayers, and the heterogeneous bilayers formed by stacking two different monolayers. Bilayer TMDs show interesting and unique properties compared to monolayer TMDs, such as a higher density of states and carrier mobility [2, 33, 34, 77]. This phenomenon results in superior performance in thin-film transistors and sensors. Importantly, the stacking orders in bilayers provide a great way to design materials and control the physical properties. As mentioned in the previous section, HfX$_2$ (X = S, Se, or Te) are potential candidates for constructing bilayers. By alternating the stacking effect in bilayers, it is possible to modulate the electronic properties of 2D materials. Combined with the emergent features of HfX$_2$, different stacking modes in bilayer HfX$_2$ will be investigated and discussed in the next section.

13.3 STRUCTURAL OPTIMIZATION OF BILAYER HFX$_2$

This section presents the geometry of different stacking in bilayer HfX$_2$, which is divided into three parts: structural configurations, stability, and structural parameters including chemical bonding, layered and interlayer distance (separated distance of two layers), and buckling. The structural configurations are shown by changing the stacking in the T phase with six modes. To construct a bilayer, a monolayer named layer 1 is used as a substrate, and then a bilayer is created by adding another monolayer named layer 2 to layer 1. By arranging the stacking between the two layers in a supercell, different stackings of the bilayer are created, which are predicted to have various physical or chemical features.

Consequently, varying the stacking in bilayers might tune the electronic properties for promising thermoelectric materials [49, 78, 79]. In this work, six stacked structures in the bilayer are constructed labeled AA, AA1, AA2, AB, AB1, and AB2 with the T phase, as shown in Figure 13.2. AA is the bilayer structure that has a similar structure to the bulk 1T phase, while all other structures have a translated or rotated layer 2 (upper layer). As shown in this figure, the AA mode was directly created from the bulk by adding a vacuum. The AA1 and AA2 modes have a translated layer 2 in the positive x direction. Rotating layer 2 of the AA mode by an angle of 180° results in the AB structure. Similar to AA1 and AA2, AB1 and AB2 are constructed by translating layer 2 from the AB mode along the negative x direction. It should be noted that there are at least six types of stackings that can theoretically exist in a bilayer of the T phase structure.

To discuss the energetics of bilayers with the six configurations, the formation energy Ef is defined by:

$$E_f = E_{tot} - 2E_{mono}$$

where Ef and E_{tot} are, respectively, the formation energy and ground state energy in bilayer; E_{mono} is the ground state energy of monolayer. The relative value of the formation energy is also calculated by setting the formation energy of the AA case as a reference value in order to show the results in Figure 13.3. The formation energies in different stacking configurations are all positive, indicating that the bilayer HfX_2 is less stable compared to the monolayer. This result is similar to the calculated values of nanoribbons in bilayer MoSSe and MoS_2 [25, 32]. As in the case of the diamond, which has positive formation energy, external factors, such as pressure and temperature, may be required for two layers to combine into a bilayer. Besides, the formation energy in the six cases can be arranged based on their values as $E_f\_AA1 > E_f\_AB > E_f\_AA2 > E_f\_AB2 > E_f\_AB1 > E_f\_AA$; $E_f\_AA1 > E_f\_AB > E_f\_AB1 > E_f\_AA > E_f\_AA2 > E_f\_AB2$ and $E_f\_AB > E_f\_AA1 > E_f\_AB1 > E_f\_AA > E_f\_AA2 > E_f\_AB2$ in HfS_2, $HfSe_2$, and $HfTe_2$, respectively. The trend of formation energies changes with the different stackings in the three compounds, corresponding to the relative values as shown in Figure 13.3b. It shows that AA and AB2 are the most stable ones in the six types of HfS_2 and $HfSe_2$, $HfTe_2$, respectively. Generally speaking, the formation energies in the six bilayer HfX_2 cases indicate that the sequence of the stability might be some kind of intrinsic property for the bilayer.

Table 13.1 lists the optimized parameters of bilayer HfX_2 for the six types of stackings including chemical bonding, buckling, layered distance, interlayer distance, and band gap. In these stackings, the same initial lattice constant is used to construct all structures.

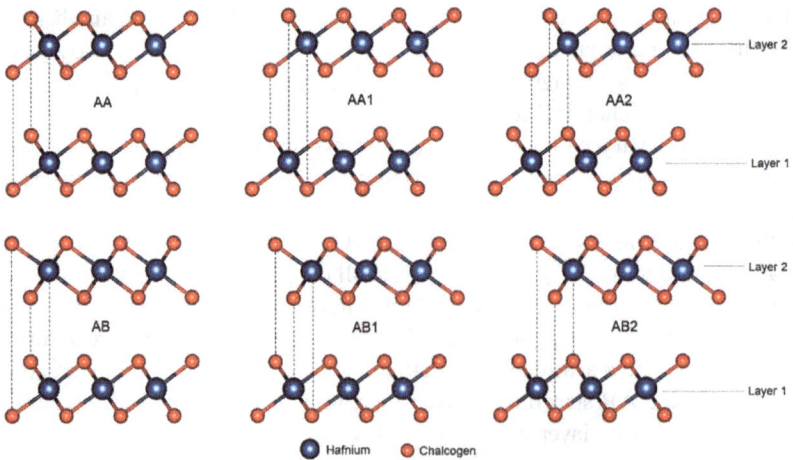

FIGURE 13.2 The structural optimization of bilayer HfX_2 with six modes of stacking.

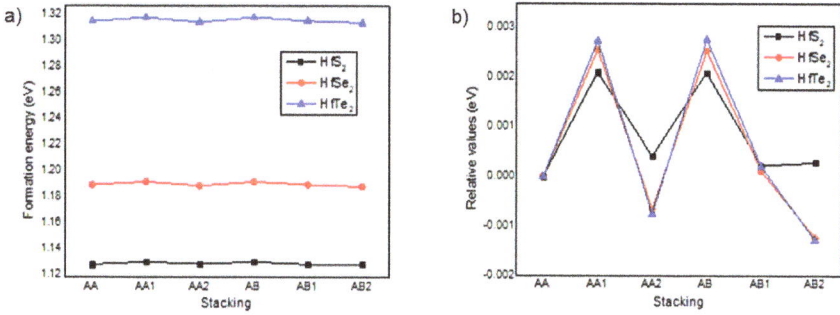

FIGURE 13.3   (a) The binding energy of bilayer HfX$_2$ with six stacking types; (b) the relative value of formation energy with six types of stacking.

TABLE 13.1
The Chemical Bonding (d$_{M-X}$ (Å)), Buckling (Δ (Å)), Layer Distance (d$_{12}$ (Å)), Interlayer Distance δ_{12} (Å), and Band Gap (E$_g$ (eV)) of Bilayer HfX$_2$ for the Six Stacking Modes

HfX$_2$	Stacking	d$_{M-X}$ (Å)	Δ (Å)	d$_{12}$ (Å)	δ_{12} (Å)	E$_g$ (eV)
HfS$_2$	AA	2.55	2.91	6.57	3.66	1.07
	AA1	2.554	2.904	6.583	3.676	1.04
	AA2	2.553	2.907	6.565	3.665	1.048
	AB	2.555	2.904	6.583	3.675	1.014
	AB1	2.552	2.907	6.571	3.661	1.08
	AB2	2.551	2.906	6.569	3.659	1.096
HfSe$_2$	AA	2.686	3.139	6.923	3.784	0.625
	AA1	2.683	3.134	6.93	3.795	0.598
	AA2	2.686	3.147	6.823	3.672	0.612
	AB	2.683	3.133	6.93	3.795	0.58
	AB1	2.685	3.142	6.918	3.772	0.625
	AB2	2.684	3.144	6.793	3.651	0.651
HfTe$_2$	AA	2.913	3.555	7.647	4.099	−0.149
	AA1	2.905	3.525	7.639	4.11	−0.173
	AA2	2.905	3.537	7.536	4.001	−0.141
	AB	2.892	3.511	7.638	4.133	−0.18
	AB1	2.899	3.516	7.635	4.105	−0.131
	AB2	2.914	3.539	7.52	3.967	−0.129

To clarify the chemical bonds in these materials, it should be pointed out that the electron configuration of hafnium and the chalcogen atoms characterizes the contribution of the bonds to the electronic properties. As listed in Table 13.1, the bond lengths increase when replacing the chalcogen atom starting from S to Se and Te. The d-orbital of the hafnium atoms and the lone-pair electrons of the chalcogen atoms determine the rich chemistry of these materials. That means the atomic electronic configuration changes depending on the transition metal present in the compound when the chalcogen atoms take part in chemical bonding. In this group, the structures show an octahedral form with six-fold coordination in which all metal-chalcogen bonds are equal. Furthermore, the chalcogen atoms are sp^3 hybridized with the LP electrons and the atomic orbitals of hafnium characterizing the geometry and properties of this group. Concerning the different stackings in these bilayers, the sequence of chemical bond of HfS_2, $HfSe_2$, and $HfTe_2$ are respectively AB > AA1 > AA2 > AB1 > AB2 > AA, AA = AA2 > AB1 > AB2 > AA1 = AB and AB2 > AA > AA1 = AA2 > AB1 > AB. Although the six stacking modes are simultaneously constructed in the three compounds, the resulting chemical bond is totally different. This is caused by the different sizes of the chalcogen atoms and the interaction forces between hafnium and the chalcogen atoms, as well as two layers. However, each compound shows values approaching six types, which point to a similar possibility of the existence of these stackings.

Unlike graphene, TMDs form a sandwich structure in the individual layers in which the buckling is determined as a fluctuation of the transition metal and chalcogen atoms. The buckling of bilayer HfX_2 is defined in the same way as the monolayer, which is the distance of two sub-planes in a triple layer. Because the bilayer is constructed from two monolayers by substituting each other, each layer has the same buckling. As listed in Table 13.1, the buckling increases by changing chalcogen atoms from S to Se and Te. This feature is caused by the different sizes of the chalcogen atoms, which leads to a change of chemical bonds between hafnium and the chalcogen atoms. In each material with the six types of stacking, the sequence of the buckling is AA > AA2 = AB1 > AB2 > AA1 = AB, AA2 > AB2 > AB1 > AA > AA1 > AB and AA > AB2 > AA2 > AA1 > AB1 > AB for HfS_2, $HfSe_2$, and $HfTe_2$, respectively. In bilayer, this fluctuation is tuned by the weak interaction between two layers. This feature can be precisely correlated to the Moiré periodicity [35].

In multilayer, the concepts of layered distance and interlayer distance are introduced to analyze the geometry. In this work, these parameters should be considered to give further information about the crystalline structure. d_{12} and δ_{12} represent the layered distance and interlayer distance describing the distance between two layers in the bilayer, respectively, as shown in Figure 13.4. With regarding the different stackings, for HfS_2, the sequence of the layered distance is AA1 = AB > AB1 > AA > AB2 > AA2, while the interlayer distance is AA1 > AB > AA2 > AB1 > AA > AB2. In $HfSe_2$, the results for the two kinds of distance are the same as AA1 = AB > AA > AB1 > AA2 > AB2, while the orders of the layered distance and interlayer distance are, respectively, AA > AA1 > AB > AB1 > AA2 > AB2 and AB > AA1 > AB1 > AA2 > AA > AB2 for $HfTe_2$. This finding indicates that each compound has a different sequence of layered and interlayer distances order for the six stacking modes. It should be noted that different stacking types affect the chemical bonding of hafnium and the chalcogen atoms, resulting in the two distance kinds. The values listed in Table 13.1 show that the layered and interlayer distances increase when progressing from sulfur to selenium and tellurium. Since the size of three atoms is arranged as Te > Se > S, therefore, the two kinds of distances are mainly determined by what kind of atom is at the interface. Besides, the attraction forces, as well as van der Waals interaction between the two layers, might tune these distances. It may be because the

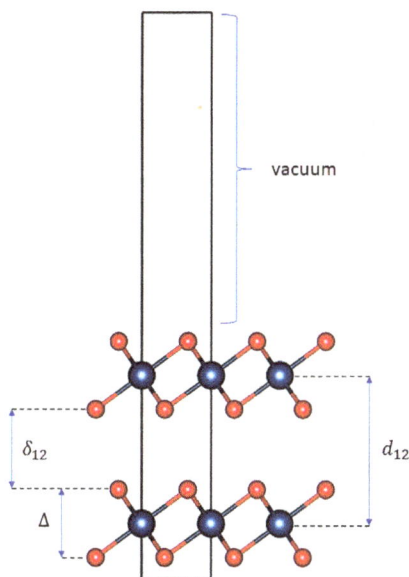

FIGURE 13.4 Bilayer with buckling (Δ) interlayer distance (d_{12}) and separated distance (δ_{12} ($\mathring{A}$)).

interaction is mainly contributed by the accumulated electrons in the middle of the bilayer and depends on which atom provides more electrons.

Obviously, the chemical bonds and the distance between the two layers are partially governed by vdW interaction. In bilayer HfX$_2$, there exist strong covalent bonds in triple layers and weak vdW interaction between the layers. It is well known that the vdW interaction is necessary for describing the geometric properties of layered systems. Therefore, the structural properties of bilayers are investigated using DFT-D3 approximation. Bonding between two layers can be described by the vdW. Due to this bonding, individual layers can be easily constructed, anticipating the possibility to fabricate small devices.

In addition, considering the vdW interaction provides unique advantages and a significant factor to investigate 2D materials. Heterostructures can be built by assembling individual layers into a functional multilayer by employing the vdW forces in the individual layers' TMDs. Using vdW epitaxy, ultrathin layers on substrates [80–85] can be grown. However, the interlayer vdW interaction is not strong enough to form the lattice-matched coherent heterostructure that is found in bilayer MoS$_2$/MoSe$_2$ [1]. In bilayer WSe$_2$/MoS$_2$, the weak intermolecular vdW forces were also involved in which donor and acceptor layers are bound [1, 30, 86, 87]. The weak vdW interaction in the bilayer remains the monolayers stacked together while retaining their intrinsic properties. Based on the atomic interaction between two layers, bilayers exhibit distinctive optical and electronic properties [36, 88–90]. In short, vdW interaction provides a degree of controlling properties that should be considered.

In general, bilayer HfX$_2$ reveals six possible stacking types that might tune the electronic properties by their effects. These structures are formed by adding another monolayer to a substrate monolayer in the T phase, which is the favorable monolayer phase. There are two main types of stacked structures followed by bilayer graphene that are AA and AB modes. From AA and AB, four stacked forms labeled AA1, AA2, and AB1, AB2 are, respectively, constructed by translating the upper layer along the positive and negative x direction. The electron configuration and size of the chalcogen atoms, as well as the d-orbital of the hafnium atom, influence the chemical bonding, the layered and interlayer distance in bilayers, leading to the effect of vdW interaction. These parameters can be controlled by changing stacking order as the results shown.

13.4   ELECTRONIC PROPERTIES

The electronic band structures have been calculated along with the high symmetry point Γ-M-K-Γ belonging to the hexagonal lattice in the Brillouin zone. The system band gaps of bilayer HfX_2 for the six stacking types are shown in Table 13.1. Although they all are bilayer structures in the T phase, the band gap values are considerably different for varying the stackings. Based on the band gap, bilayer HfX_2 is divided into two groups, namely semiconductors (HfS_2 and $HfSe_2$) and semi-metal ($HfTe_2$). The same result is found in the monolayer of these materials [39–41]. As shown in this table, the AB mode has the smallest band gap, while the largest band gap is found in AB2 for both HfS_2 and $HfSe_2$. A similar feature is found in $HfTe_2$; the AB mode shows the largest overlap between the conduction band and valence band, while the AB2 mode has the smallest overlap. Varying stacking orders result in different band gap values. It was also found that the band gap trend in semiconductors and two kinds of distance is the same, namely, S > Se as the size of the Se atom is larger than that of the S atom.

The system band energy spectra of bilayer HfX_2 are shown in Figures 13.5, 13.6, and 13.7, which presents the atom dominances of hafnium and the chalcogens. Both semiconductors share as a common feature that the valence band maximum (VBM) is located at the Γ point (k = 0, 0, 0) and the conduction band minimum (CBM) is at the M point (k = ½, 0, 0) of the high symmetry k-points, while the semi-metal $HfTe_2$ shows an overlap between the conduction and valence bands, which is demonstrated by the negative values of the band gap calculation. The unoccupied conduction bands are asymmetric to the occupied valence bands about the Fermi level. The energy bands have a strong dispersion relation, with the parabolic dispersion being exhibited in both the conduction and the valence bands. However, each material is anticipated to have an unusual energy band structure. There are many distinguishing features between the semiconductor and semi-metal, as well as among the six stacking modes in the electronic band structures.

As previously mentioned, bilayer HfX_2 (X = S, Se, or Te) is constructed with the T phase from the monolayer structures by varying the stacking. The electronic bands of HfS_2 as shown in Figure 13.5 with atom dominance ranging from −15 to 6 eV. In this figure, the blue circles describe hafnium atoms and the brown circles represent sulfur atoms. In general, the VBM and CBM locate at Γ and M points in all stackings, respectively, resulting an indirect band gap semiconductor. The occupied hole band is strongly asymmetric to the unoccupied electron band. The valleys are exhibited

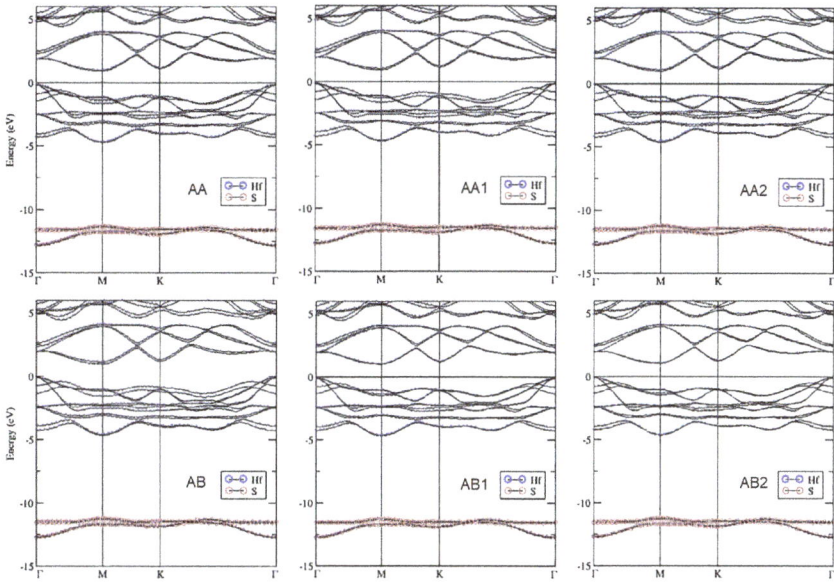

FIGURE 13.5 Band structures of bilayer HfS$_2$.

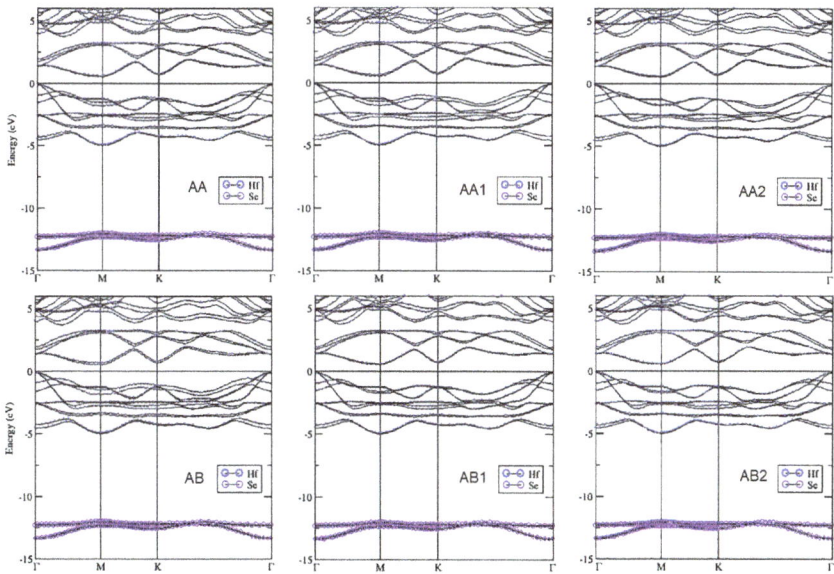

FIGURE 13.6 Band structures of bilayer HfSe$_2$.

at the high-symmetry points M, K, and Γ at the band edge. The low-lying valence bands also change into parabolic bands. The σ and σ^* bands (in the valence and conduction bands) initiate from Γ point at −2.5, 0, 2.5, and 5 eV. Apart from the special valleys located at the band edges, there exist flat bands in the low-lying valence bands around −11.5 eV. With regard to the atom dominance in the bands, the sulfur atom in all structures dominates in the valence band ranging from 0 to −15 eV, especially in the low-lying energy region, while the contribution of hafnium atoms is smaller in this band. In contrast, hafnium atoms contribute to the conduction band more than the valence band. Both hafnium and sulfur atoms contribute to the whole band in the ranges from −15 to 6 eV; sulfur shows a sharper dominance in the valence band, while a sharper dominance of hafnium is found in the conduction band. Furthermore, the crossing bands are found in the low-lying valence band between Γ and M point as well as Γ and K points. Particularly, the crossing point is found at K point. Compared to monolayer, the bilayer band structure shows the double degeneracy in both bands. This feature is clearly revealed in the valence band near the Fermi level and the conduction band ranging from 4.5 to 6 eV. Especially, the AB stacked structure shows the double degeneracy at the band edge states as shown in Figure 13.5.

Most features of the semiconductor HfS_2 are also found in $HfSe_2$, as shown in Figure 13.6. The band structure ranges from −15 to 6 eV, with atom dominance by the hafnium and selenium atoms described by the blue and purple circles, respectively. The occupied hole band is strongly asymmetric to the unoccupied electron band, which shows that the indirect band gap of the semiconductor shifted from Γ point to M point. All valley, parabolic band, and crossing band values are exhibited in the valence band of this material. The σ and σ^* bands initiate from the Γ point at −2.5, 0, 2, and 5 eV. The flat bands at −11.5 eV are found in the low-lying valence bands with the dominance of selenium atoms. Hafnium contributes to the conduction band more than to the valence band. Both hafnium and sulfur contribute to the whole band, which ranges from −15 to 6 eV. Selenium shows a sharper dominance in the valence band, while a sharper dominance of hafnium is found in the conduction band. Furthermore, the crossing bands are found in the low-lying valence band between the Γ and M points, as well as Γ and K points. The double degeneracy ranging from 4.5 to 6 eV is shown in the valence band near the Fermi level and the conduction band. However, there are two structures, namely AA1 and AB, that show the double degeneracy at the band edge states.

In contrast, bilayer $HfTe_2$ demonstrates very distinct properties compared to the two semiconductors in the HfX_2 family. The atom dominance is also considered, with blue circles and red circles representing hafnium atoms and tellurium atoms as shown in Figure 13.7. Firstly, the semi-metallic feature found in the band structure spectrum of the six configurations is the overlap between the valence band and the conduction band, which presents a gapless material. In addition, the CBM of all stackings is below the Fermi level, while those of HfS_2 and $HfSe_2$ are above the Fermi level. The AB stacked structure clearly demonstrates this finding. This feature illustrates the difference between two subgroups of semiconductors and semimetal in HfX_2 as previously mentioned. Except for this characteristic, bilayer $HfTe_2$ with the six modes of stacking shows similar features to the two semiconductors. Like monolayer $HfTe_2$, a semi-metallic property has been shown, which is investigated by the pressure-dependence conductivity and the Hall coefficient. As shown in Figure 13.7, the σ and σ^* bands initiate from the Γ point at −2.5, 0, 1, and 4 eV. The valley, parabolic band, energy crossing, and degeneracy can be found in the energy spectrum of this material. The double degeneracy found in both AA1 and AB

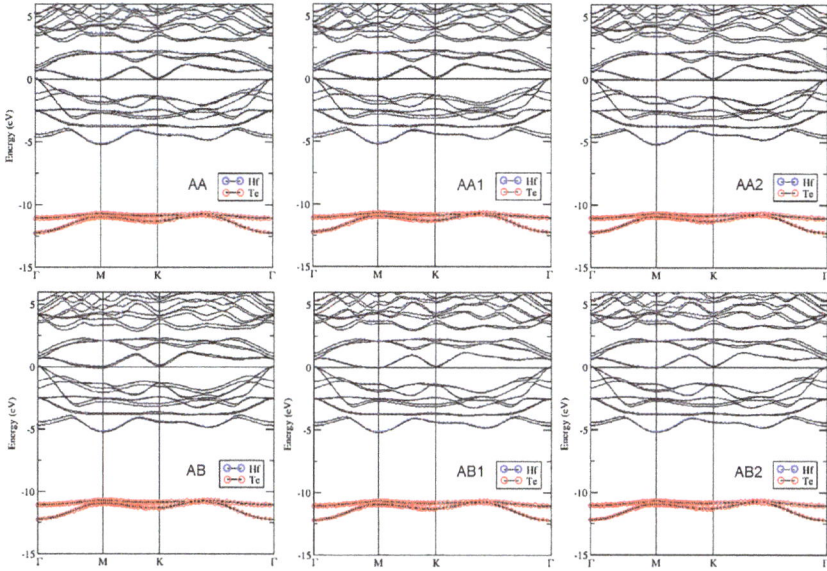

FIGURE 13.7 Band structures of bilayer HfTe$_2$.

structures is a common feature of HfSe$_2$ and HfTe$_2$. With regard to the atom dominance, the contribution of the hafnium atoms is smaller than that of the tellurium in
the valence band, while the opposite is true for the conduction band.

In summary, the band gap opening in the two semiconductors is associated with
the quantum confinement effect, while the overlap between the valence and conduction bands in HfTe$_2$ correlates to the semi-metal feature. With the band energy spectra, a lot of asymmetric peaks are exhibited due to the parabolic form of the energy
bands. A pair of asymmetric peaks is located near the Fermi level that defines the
indirect energy gaps. Furthermore, the parabolic energy dispersions lead to most van
Hove singularities (vHs). The peaks near the Fermi level are found to mainly arise
from the band-edge states, which is a signature feature unique to vHs. With varying
stackings including the six modes, the band energy not only shows common properties similar to their monolayer but also reveals distinct features that can characterize
the stacking effect of bilayer HfX$_2$.

To directly reflect the main features of the band structures, the total density of
states (TDOS) is considered, as shown in Figure 13.8. In this figure, the six stacked
structures of each material are divided into two groups corresponding to the AA
group (left panel) and the AB group (right panel) to compare the effects of these
types. The TDOS ranging from −15 to 6 eV shows a pair of asymmetric peaks centered at the Fermi level. This asymmetry of the two bands is demonstrated by the
large energy difference between the valence and conduction peaks. Furthermore, the
linear bands, the parabolic bands, and the initial band-edge states of the parabolic
bands originate in these structures. Firstly, the two groups of HfS$_2$ are described
in Figure 13.8(a). A logarithmic divergence is exhibited at middle energy. Besides,

shoulder structures appear at the lower and higher energy. Partially flat energy is revealed in the valence bands of all structures and in the conduction bands, while shoulder structures evidently appear in the conduction band. With regard to the stacking order in this material, each line that represents a type of stacking can be realized. This reveals that a slight but remarkable distinct TDOS is formed by varying stackings in the bilayer. Second, the TDOS of $HfSe_2$ with six stackings is shown in Figure 13.8(b) and exhibits similar features to those of HfS_2. The logarithmic divergence, shoulder structures, and flat energy are also found in two bands. Especially, the flat energy in the conduction band appears more obviously compared to that of HfS_2. Another difference between the two materials is that an empty range at low-lying energy of $HfSe_2$ occupies larger than that of HfS_2. Finally, Figure 13.8(c) describes the TDOS of $HfTe_2$, which shows DOS larger than that of both HfS_2 and $HfSe_2$. The logarithmic divergence at the Fermi level is evidence of the overlap between the valence and conduction bands. This is characteristic of a semi-metal. Moreover,

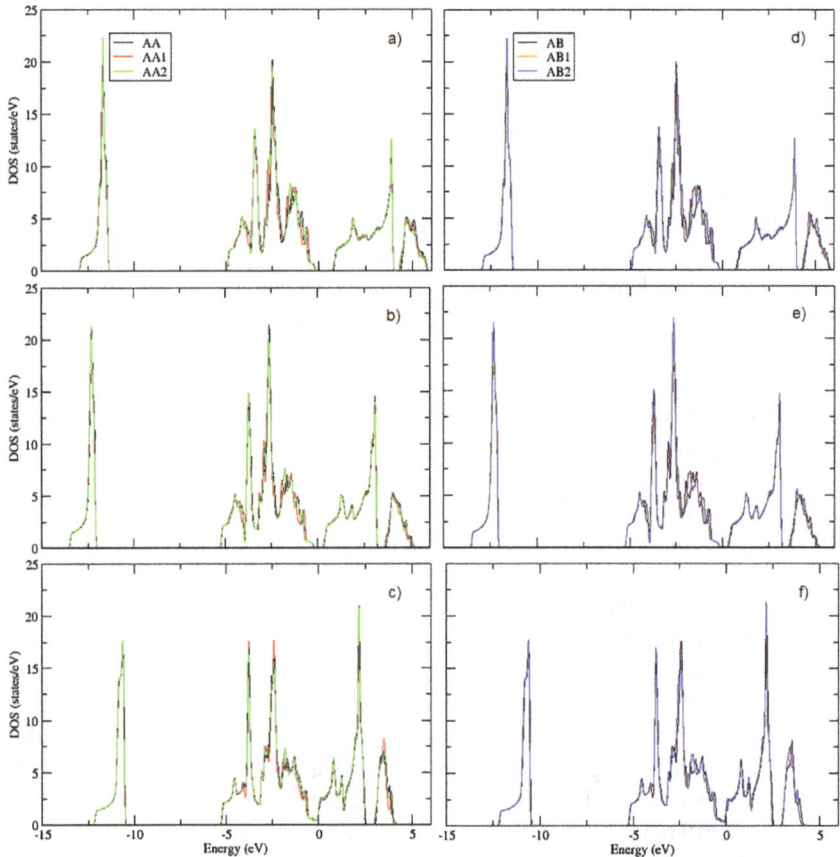

FIGURE 13.8 Total DOS of bilayer (a, b) HfS_2, (c, d) $HfSe_2$, and (e, f) $HfTe_2$ with six stacking modes.

this divergence is found in the low-lying energy at −5 eV and disappears in two semiconductors.

In general, the three compounds share common features as well as distinct ones in the TDOS. The logarithmic divergence and energy difference of the two bands at the Fermi level illustrates the asymmetry that characterizes the indirect band gap. Moreover, the low free carrier density of these compounds is predicted from the low DOS at the Fermi level. At lower and higher energy, the DOS reveals peaks as well as divergent and shoulder structures. It should be noted that the peak and shoulder structures correspond to the $3p_z$, $4p_z$, $5p_z$, and $(3p_x, 3p_y)$, $(4p_x, 4p_y)$, $(5p_x, 5p_y)$ of HfS$_2$, HfSe$_2$, and HfTe$_2$, respectively, dominating the energy bands, which is discussed in the orbital-projected DOS. The flat energy in the DOS near the Fermi level, particularly in HfTe$_2$, indicates the feature of vHs accompanied by the parabolic energy dispersions and saddle points. This characteristic is also found in the monolayer of these materials.

The orbital-projected DOS for the six stacked structures is shown in Figures 13.9, 13.10, and 13.11 and used to verify the configuration effect. The figures describe the density of states with multi-orbital hybridization bilayer HfX$_2$. They show that the main contributions are due to the s and d-orbitals of the hafnium atoms and s, p-orbitals of the chalcogen atoms.

In Figure 13.9, the PDOS of bilayer HfS$_2$ is discussed with multi-hybridization, presenting the dominance of the multi-orbital hafnium and sulfur atoms. Hafnium atoms are dominated by $5d_{xy}$, $5d_{z^2}$, and $5d_{x^2-y^2}$ orbitals in the conduction band, while the DOS depends on the bonding of $3p_x$, $3p_y$, $3p_z$ of sulfur atoms in the valence band, respectively. This is consistent with the atom dominance in the energy band. Hafnium atoms mainly contribute to the conduction band, while sulfur atoms dominate in the valence band. Moreover, these features are the same as those of bulk structures and monolayers. However, the DOS covers in a larger range compared to their monolayer. Although six stacking types share this common feature, they have distinct aspects as shown in the figure. For example, the $5d_{z^2}$ orbital in AA and AB has the largest dominated range in the conduction band of all stackings; for valence band, $3p_x$ orbital in AA shows a strong dominance, which almost cannot be found in the AB stacked type.

The PDOS of bilayer HfSe$_2$ is accordingly presented with multi-hybridization shown in Figure 13.10, showing the dominance of the multi-orbital hafnium and sulfur atoms. Similar to HfS$_2$, hafnium atoms are dominated by $5d_{yz}$, $5d_{z^2}$, and $5d_{x^2-y^2}$ orbitals in the conduction band, whereas the DOS depends on the contribution of $4p_x$, $4p_y$, $4p_z$ of selenium atoms, respectively, in the valence band. However, the DOS covers the larger range compared to HfS$_2$. In the conduction band, the $5d_{yz}$ orbital in AA1 and AB1 cover the larger range of DOS, and $4p_y$ of AB mode occupies the larger DOS. Meanwhile, the $5d_{yz}$ and $4p_y$ orbitals of AA and AB2 make up a big range of DOS in the valence band. These features demonstrate the distinct aspects of the different bilayer HfSe$_2$ stackings.

As discussed in the band energy spectrum, bilayer HfTe$_2$ has different characteristics compared to two semiconductors HfS$_2$ and HfSe$_2$. Figure 13.11 shows the PDOS with multi-hybridization of this material. The overlap between the conduction and valence band is clearly illustrated in DOS about the Fermi level. It should be

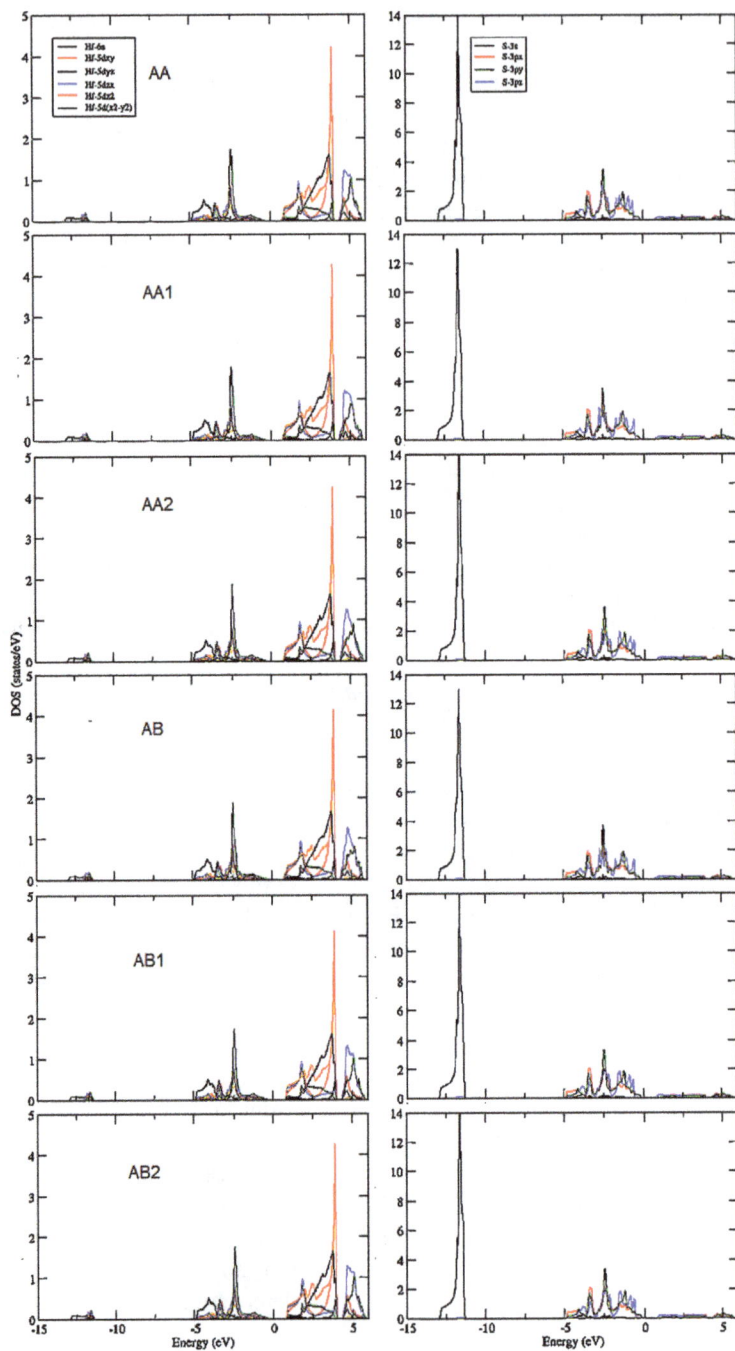

FIGURE 13.9   Orbital projected DOS of bilayer HfS$_2$.

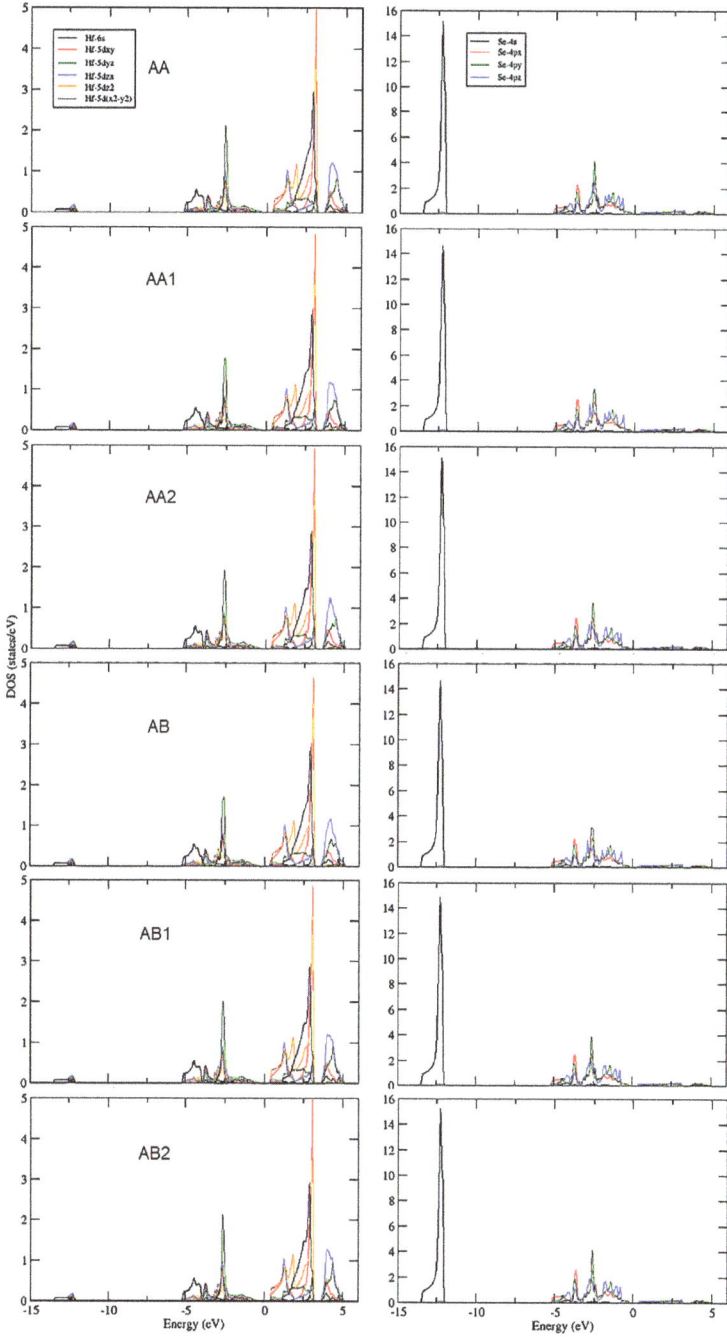

FIGURE 13.10   Orbital projected DOS of bilayer HfSe$_2$.

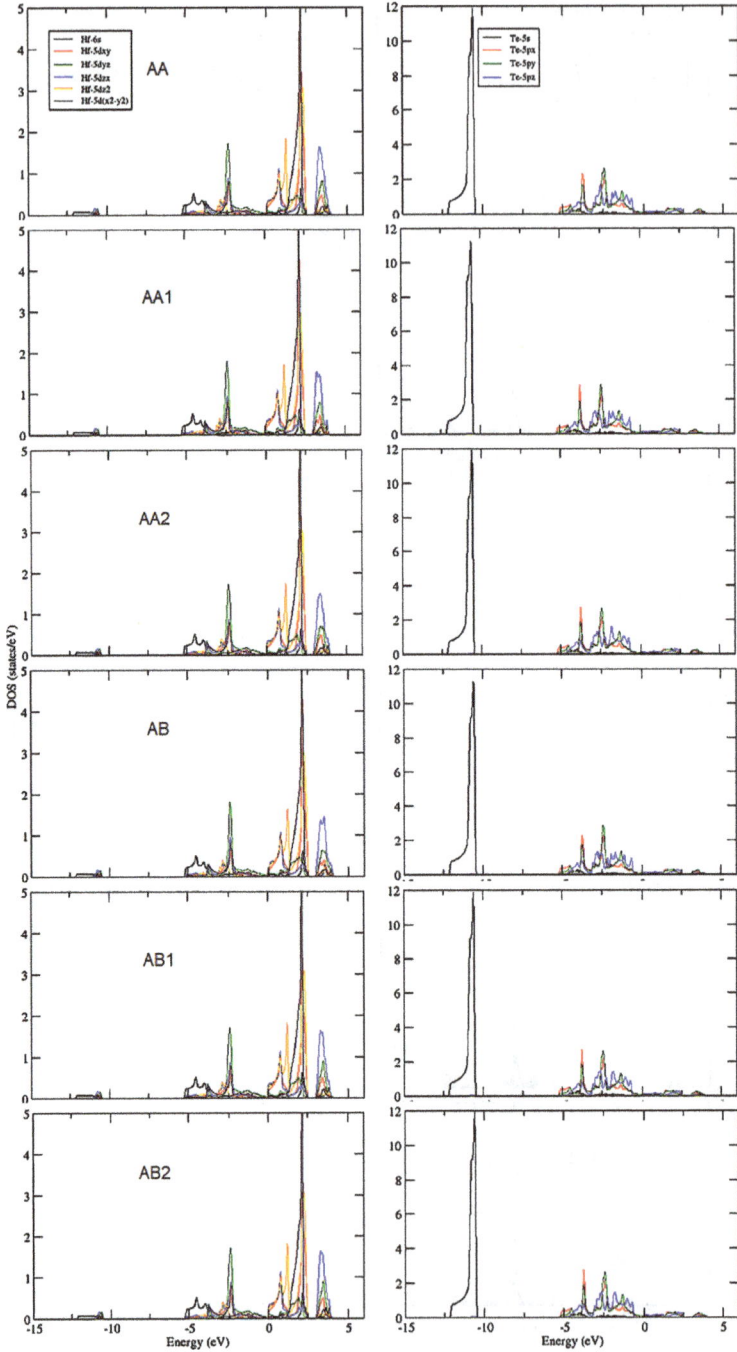

FIGURE 13.11   Orbital projected DOS of bilayer HfTe$_2$.

noted that both band and PDOS support that is a semi-metal property. With regarding orbital hybridization, the contribution of the $5d_{yz}$, $5d_{xz}$, $5d_{z^2}$, and $5d_{x^2-y^2}$ orbitals of hafnium and the $5p_x$, $5p_y$, $5p_z$ orbitals of tellurium atoms, respectively, show the orbital dominance in conduction and valence bands. Additionally, the 6s orbital of the hafnium atoms remarkably occupies the range of (5, 6) eV that cannot be found in the two mentioned semiconductors. This feature can be clearly seen in AB1 and AB2 modes. Considering the stacking in this material, the $5d_{z^2}$ and $5d_{x^2-y^2}$ orbitals of hafnium exhibit a similar contribution in the range of (0.5, 3.5) eV in the conduction band for the six modes. Although the $5p_x$, $5p_y$, $5p_z$ orbitals of the tellurium atoms show a smaller contribution in this band, the $5p_z$ shows much dominance in AB, AB1, and AB2 structures, as shown in Figure 13.11. In general, the orbital hybridization might be changed when the stacking is varied. This sequence of stacking effect in the orbital contribution is the same as in the above materials.

In addition, the charge density can further provide useful information about chemical bonding, thus apprehend the essential features of energy bands well. Taking bilayer HfS₂ as a representative, the charge density along [10] plane is described in Figure 13.12 with six types of stacking. Obviously, changing symmetry structure by varying stacking affects the chemical bonding due to the contribution of hafnium and sulfur atoms in the charge plot. Hafnium atoms contribute four valence electrons taking part in chemical bonding, leading to reduced charge density near the atomic site. Based on the hybridized p_z orbitals, more charge densities are normal to the surface. As shown in the figure, the charge distribution strongly depends on the atomic position of stackings, indicating the modification of geometric structure in electronic properties, further supported by PDOS.

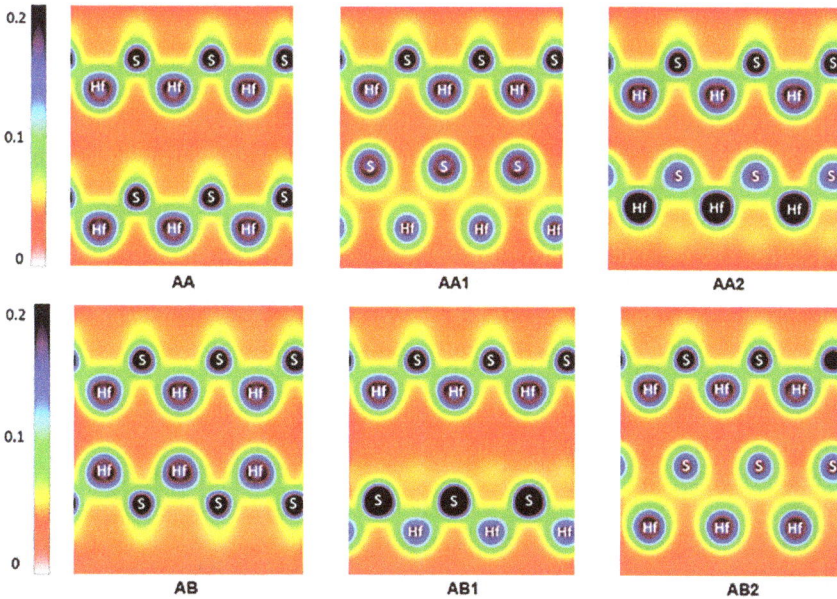

FIGURE 13.12   Charge density along [10] plane of bilayer HfS₂ with six stackings.

The contributions of their orbitals in PDOS are comparable to each other, being feature-rich energy bands that are mainly determined by the d and p orbitals. The strong hybridizations of $(5d_{xy}, 5d_{xy}, 5d_{z^2}, 5d_{x^2-y^2})$ Hf-orbitals and $(3p_x, 3p_y, 3p_z)$, $(4p_x, 4p_y, 4p_z)$, $(5p_x, 5p_y, 5p_z)$ chalcogen-orbitals exist in the conduction and valence bands, respectively. These orbital contributions are evident for the multi-orbital hybridization in these materials, which is the same as the respective monolayers. Furthermore, the p-orbitals of the chalcogens cause significant contributions to the bonding. In consistence of DOS, PDOS, and the band structure, a pair of asymmetric peaks centered at the Fermi level characterizes the energy gap. In addition, the large energy difference between the valence and conduction peaks obviously demonstrates the asymmetry of the two bands. Apart from the general feature in PDOS of the bilayer, varying stacking order can also affect the orbital hybridization. Each stacking mode shows the distinct aspect so that characteristics of the stacking can be realized via the contribution of the orbitals, as mentioned.

13.5   REMARKS

In summary, bilayer HfX_2 reveals that the d-orbitals and p-orbitals of hafnium atoms and X atoms (chalcogen atoms), respectively, take part in the chemical bonding of M-X. All structural parameters depend on the electron configurations and the size of atoms and are affected by varying stacking modes. The sequences of chemical bonding and the layered and interlayer distances can be attributed to the electron configuration and vdW interaction in each individual layer and between two layers. A sharp buckle characterizes the sandwich structures and chemical bonds and shows height fluctuations at equilibrium. These parameters feature the quasiparticle characteristics in solid states that are described by theoretical framework through VASP calculations. Furthermore, six possible structures of stackings can be constructed, promising a wide range of engineering possibilities and design structures. Varying stacking orders leads to a versatile dispersion in electronic properties. The energy gap trend suggests that the atomic species and stacking orders play a dominant role. A lot of asymmetric peaks and a few symmetric ones due to the parabolic form are shown in the energy bands and density of states (DOS). Especially, the van Hove singularities mainly arising from the band-edge states in the DOS were found in these structures. The logarithmic divergence at middle energy and shoulder structures at lower and higher energy are exhibited in the DOS corresponding to the $3p_z$, $4p_z$, $5p_z$, and $(3p_x, 3p_y)$, $(4p_x, 4p_y)$, $(5p_x, 5p_y)$ of chalcogen atoms in the compounds. In addition, these orbitals demonstrate multi-orbital hybridizations, as shown in the orbital-projected DOS and charge density.

Through characteristics defined in geometry and electronic properties using the theoretical framework, bilayer with various stacking HfX_2 is clearly synthesized based on the quasiparticle point of view. These characteristics were slightly but remarkably changed when applying different stacking, as shown in the band energy spectra and orbital projected DOS. They present many potential structures for the applications. So far, most studies on bilayer TMDs have focused on group-VI TMDs. Investigation bilayer group-IV compounds open up opportunities to construct and design new platforms for applications of TMD materials. However, there exist some challenges in naturally constructing different stacking configurations. The stability of all stacking modes should be thoroughly examined to accurately define the structures and investigate their properties.

REFERENCES

[1] Kang J, Li J B, Li S S, Xia J B and Wang L W 2013 Electronic structural Moire pattern effects on MoS2/MoSe2 2D heterostructures *Nano Letters* **13** 5485–90

[2] Das S, Chen H Y, Penumatcha A V and Appenzeller J 2013 High performance multi-layer MoS2 transistors with scandium contacts *Nano Letters* **13** 100–5

[3] Radisavljevic B, Radenovic A, Brivio J, Giacometti V and Kis A 2011 Single-layer MoS2 transistors *Nature Nanotechnology* **6** 147–50

[4] Ataca C, Sahin H and Ciraci S 2012 Stable, single-layer MX2 transition-metal oxides and dichalcogenides in a honeycomb-like structure *Journal of Physical Chemistry C* **116** 8983–99

[5] Xia J, Yan J X and Shen Z X 2017 Transition metal dichalcogenides: Structural, optical and electronic property tuning via thickness and stacking *Flatchem* **4** 1–19

[6] Aretouli K E, Tsipas P, Tsoutsou D, Marquez-Velasco J, Xenogiannopoulou E, Giamini S A, Vassalou E, Kelaidis N and Dimoulas A 2015 Two-dimensional semiconductor HfSe2 and MoSe2/HfSe2 van der Waals heterostructures by molecular beam epitaxy *Applied Physics Letters* **106**

[7] Kolobov A V and Tominaga J 2016 *Two-Dimensional Transition Metal Dichalcogenides* (Switzerland)

[8] Geim A K and Novoselov K S 2007 The rise of graphene *Nature Materials* **6** 183–91

[9] Li H, Xu C, Srivastava N and Banerjee K 2009 Carbon nanomaterials for next-generation interconnects and passives: Physics, status, and prospects *IEEE Transactions on Electron Devices* **56** 1799–821

[10] Fang H, Chuang S, Chang T C, Takei K, Takahashi T and Javey A 2012 High-performance single layered WSe2 p-FETs with chemically doped contacts *Nano Letters* **12** 3788–92

[11] Liu W, Kang J H, Sarkar D, Khatami Y, Jena D and Banerjee K 2013 Role of metal contacts in designing high-performance monolayer n-type WSe2 field effect transistors *Nano Letters* **13** 1983–90

[12] Kumar A and Ahluwalia P K 2012 A first principle comparative study of electronic and optical properties of 1H—MoS2 and 2H—MoS2 *Materials Chemistry and Physics* **135** 755–61

[13] Kumar A and Ahluwalia P K 2012 Tunable dielectric response of transition metals dichalcogenides MX2 (M=Mo, W; X=S, Se, Te): Effect of quantum confinement *Physica B-Condensed Matter* **407** 4627–34

[14] Kumar A and Ahluwalia P K 2012 Electronic structure of transition metal dichalcogenides monolayers 1H-MX2 (M = Mo, W; X = S, Se, Te) from ab-initio theory: New direct band gap semiconductors *European Physical Journal B* **85**

[15] Lebegue S and Eriksson O 2009 Electronic structure of two-dimensional crystals from ab initio theory *Physical Review B* **79**

[16] Wu M S, Xu B, Liu G and Ouyang C Y 2013 First-principles study on the electronic structures of Cr- and W-doped single-layer MoS2 *Acta Physica Sinica* **62**

[17] Yun W S, Han S W, Hong S C, Kim I G and Lee J D 2012 Thickness and strain effects on electronic structures of transition metal dichalcogenides: 2H-M X-2 semiconductors (M = Mo, W; X = S, Se, Te) *Physical Review B* **85**

[18] Puretzky A A, Liang L B, Li X F, Xiao K, Sumpter B G, Meunier V and Geohegan D B 2016 Twisted MoSe2 bilayers with variable local stacking and interlayer coupling revealed by low-frequency Raman spectroscopy *Acs Nano* **10** 2736–44

[19] Samad L, Bladow S M, Ding Q, Zhuo J Q, Jacobberger R M, Arnold M S and Jin S 2016 Layer-controlled chemical vapor deposition growth of MoS2 vertical heterostructures via van der Waals epitaxy *Acs Nano* **10** 7039–46

[20] Ye H, Zhou J D, Er D Q, Price C C, Yu Z Y, Liu Y M, Lowengrub J, Lou J, Liu Z and Shenoy V B 2017 Toward a mechanistic understanding of vertical growth of van der Waals stacked 2D materials: A multiscale model and experiments *Acs Nano* **11** 12780–8

[21] He Y M, Sobhani A, Lei S D, Zhang Z H, Gong Y J, Jin Z H, Zhou W, Yang Y C, Zhang Y, Wang X F, Yakobson B, Vajtai R, Halas N J, Li B, Xie E Q and Ajayan P 2016 Layer engineering of 2D semiconductor junctions *Advanced Materials* **28** 5126–32

[22] Coleman J N et al 2011 Two-dimensional nanosheets produced by liquid Exfoliation of layered materials *Science* **331** 568–71

[23] He Z, Xu W, Zhou Y, Wang X, Sheng Y, Rong Y, Guo S, Zhang J, Smith J M and Warner J H 2016 Biexciton formation in bilayer tungsten disulfide *ACS Nano* **10** 2176–83

[24] Chu T, Ilatikhameneh H, Klimeck G, Rahman R and Chen Z 2015 Electrically tunable bandgaps in bilayer MoS(2) *Nano Lett* **15** 8000–7

[25] Wei S, Li J, Liao X, Jin H and Wei Y 2019 Investigation of stacking effects of bilayer MoSSe on photocatalytic water splitting *The Journal of Physical Chemistry C* **123** 22570–7

[26] Liu H J, Jiao L, Xie L, Yang F, Chen J L, Ho W K, Gao C L, Jia J F, Cui X D and Xie M H 2015 Molecular-beam epitaxy of monolayer and bilayer WSe$_2$: A scanning tunneling microscopy/spectroscopy study and deduction of exciton binding energy *2D Materials* **2**

[27] Kumar A and Ahluwalia P K 2013 Semiconductor to metal transition in bilayer transition metals dichalcogenides MX2 (M= Mo, W; X= S, Se, Te) *Modelling and Simulation in Materials Science and Engineering* **21**

[28] Ramasubramaniam A, Naveh D and Towe E 2011 Tunable band gaps in bilayer transition-metal dichalcogenides *Physical Review B* **84**

[29] Liu C X 2017 Unconventional superconductivity in bilayer transition metal dichalcogenides *Phys Rev Lett* **118** 087001

[30] Xie X, Sarkar D, Liu W, Kang J H, Marinov O, Deen M J and Banerjee K 2014 Low-frequency noise in bilayer MoS2 **8** 5633–40

[31] Yan A, Ong C S, Qiu D Y, Ophus C, Ciston J, Merino C, Louie S G and Zettl A 2017 Dynamics of symmetry-breaking stacking boundaries in bilayer MoS2 *The Journal of Physical Chemistry C* **121** 22559–66

[32] Xiao S L, Yu W Z and Gao S P 2016 Edge preference and band gap characters of MoS2 and WS2 nanoribbons *Surface Science* **653** 107–12

[33] Li H, Yin Z Y, He Q Y, Li H, Huang X, Lu G, Fam D W H, Tok A I Y, Zhang Q and Zhang H 2012 Fabrication of single- and multilayer MoS2 film-based field-effect transistors for sensing NO at room temperature *Small* **8** 63–7

[34] Kim S, Konar A, Hwang W S, Lee J H, Lee J, Yang J, Jung C, Kim H, Yoo J B, Choi J Y, Jin Y W, Lee S Y, Jena D, Choi W and Kim K 2012 High-mobility and low-power thin-film transistors based on multilayer MoS2 crystals *Nature Communications* **3**

[35] Zhu S and Johnson H T 2018 Moire-templated strain patterning in transition-metal dichalcogenides and application in twisted bilayer MoS2 *Nanoscale* **10** 20689–701

[36] Xia M, Yin K B, Capellini G, Niu G, Gong Y J, Zhou W, Ajayan P M and Xie Y H 2015 Spectroscopic signatures of AA and AB stacking of chemical vapor deposited bilayer MoS2 *Acs Nano* **9** 12246–54

[37] Zhao Q Y, Guo Y H, Si K Y, Ren Z Y, Bai J T and Xu X L 2017 Elastic, electronic, and dielectric properties of bulk and monolayer ZrS2, ZrSe2, HfS2, HfSe2 from van der Waals density-functional theory *Physica Status Solidi B-Basic Solid State Physics* **254**

[38] Kreis C, Werth S, Adelung R, Kipp L, Skibowski M, Krasovskii E E and Schattke W 2003 Valence and conduction band states of HfS2: From bulk to a single layer *Physical Review B* **68**

[39] Setiyawati I, Chiang K R, Ho H M and Tang Y H 2019 Distinct electronic and transport properties between 1T-HfSe2 and 1T-PtSe2 *Chinese Journal of Physics* **62** 151–60

[40] Chen Q Y, Liu M Y, Cao C and He Y 2019 Engineering the electronic structure and optical properties of monolayer 1T-HfX2 using strain and electric field: A first principles study *Physica E-Low-Dimensional Systems & Nanostructures* **112** 49–58

[41] Wu N H, Zhao X, Ma X, Xin Q Q, Liu X M, Wang T X and Wei S Y 2017 Strain effect on the electronic properties of 1T-HfS2 monolayer *Physica E-Low-Dimensional Systems & Nanostructures* **93** 1–5

[42] Toh R J, Sofer Z and Pumera M 2016 Catalytic properties of group 4 transition metal dichalcogenides (MX2; M = Ti, Zr, Hf; X = S, Se, Te) *Journal of Materials Chemistry A* **4** 18322–34

[43] Abdulsalam M and Joubert D P 2016 Optical spectrum and excitons in bulk and monolayer MX2 (M=Zr, Hf; X=S, Se) *Physica Status Solidi (B)* **253** 705–11

[44] Salavati M 2019 Electronic and mechanical responses of two-dimensional HfS2, HfSe2, ZrS2, and ZrSe2 from first-principles *Frontiers of Structural and Civil Engineering* **13** 486–94

[45] Reshak A H and Auluck S 2005 Ab initio calculations of the electronic and optical properties of 1T-HfX2 compounds *Physica B-Condensed Matter* **363** 25–31

[46] Sun X and Wang Z 2017 Ab initio study of adsorption and diffusion of lithium on transition metal dichalcogenide monolayers *Beilstein J Nanotechnol* **8** 2711–18

[47] Abdulsalam M and Joubert D P 2016 Optical spectrum and excitons in bulk and monolayer MX2 (M=Zr, Hf; X=S, Se) *Physica Status Solidi B-Basic Solid State Physics* **253** 705–11

[48] Ding G Q, Gao G Y, Huang Z S, Zhang W X and Yao K L 2016 Thermoelectric properties of monolayer MSe2 (M = Zr, Hf): Low lattice thermal conductivity and a promising figure of merit *Nanotechnology* **27**

[49] Yan P, Gao G Y, Ding G Q and Qin D 2019 Bilayer MSe2 (M= Zr, Hf) as promising two-dimensional thermoelectric materials: A first-principles study *Rsc Advances* **9** 12394–403

[50] Glebko N, Aleksandrova I, Tewari G C, Tripathi T S, Karppinen M and Karttunen A J 2018 Electronic and vibrational properties of TiS2, ZrS2, and HfS2: Periodic trends studied by dispersion-corrected hybrid density functional methods *Journal of Physical Chemistry C* **122** 26835–44

[51] Kanazawa T, Amemiya T, Ishikawa A, Upadhyaya V, Tsuruta K, Tanaka T and Miyamoto Y 2016 Few-layer HfS2 transistors *Scientific Reports* **6**

[52] Xu K, Wang Z X, Wang F, Huang Y, Wang F M, Yin L, Jiang C and He J 2015 Ultrasensitive phototransistors based on few-layered HfS2 *Advanced Materials* **27** 7881–7

[53] Yue R Y, Barton A T, Zhu H, Azcatl A, Pena L F, Wang J, Peng X, Lu N, Cheng L X, Addou R, McDonnell S, Colombo L, Hsu J W P, Kim J, Kim M J, Wallace R M and Hinkle C L 2015 HfSe2 thin films: 2D transition metal dichalcogenides grown by molecular beam epitaxy *Acs Nano* **9** 474–80

[54] Yin L, Xu K, Wen Y, Wang Z X, Huang Y, Wang F, Shifa T A, Cheng R, Ma H and He J 2016 Ultrafast and ultrasensitive phototransistors based on few-layered HfSe2 *Applied Physics Letters* **109**

[55] Wang Y Y, Huang S M, Yu K, Jiang J, Liang Y, Zhong B, Zhang H, Kan G F, Quan S F and Yu J 2020 Atomically flat HfO2 layer fabricated by mild oxidation HfS2 with controlled number of layers *Journal of Applied Physics* **127**

[56] Mleczko M J, Zhang C F, Lee H R, Kuo H H, Magyari-Kope B, Moore R G, Shen Z X, Fisher I R, Nishi Y and Pop E 2017 HfSe2 and ZrSe2: Two-dimensional semiconductors with native high-kappa oxides *Science Advances* **3**

[57] Aminalragia-Giamini S, Marquez-Velasco J, Tsipas P, Tsoutsou D, Renaud G and Dimoulas A 2017 Molecular beam epitaxy of thin HfTe2 semimetal films *2d Materials* **4**

[58] Fang S and Kaxiras E 2016 Electronic structure theory of weakly interacting bilayers *Physical Review B* **93**

[59] Morell E S, Correa J D, Vargas P, Pacheco M and Barticevic Z 2010 Flat bands in slightly twisted bilayer graphene: Tight-binding calculations *Physical Review B* **82**

[60] Bistritzer R and MacDonald A H 2011 Moire bands in twisted double-layer graphene *Proceedings of the National Academy of Sciences of the United States of America* **108** 12233–7

[61] McCann E and Koshino M 2013 The electronic properties of bilayer graphene *Reports on Progress in Physics* **76**

[62] McCann E and Fal'ko V I 2006 Landau-level degeneracy and quantum hall effect in a graphite bilayer *Physical Review Letters* **96**

[63] Tabert C J and Nicol E J 2012 Dynamical conductivity of AA-stacked bilayer graphene *Physical Review B* **86**

[64] He J, Hummer K and Franchini C 2014 Stacking effects on the electronic and optical properties of bilayer transition metal dichalcogenides MoS2, MoSe2, WS2, and WSe2 *Physical Review B* **89**

[65] Kormányos A, Zólyomi V, Fal'ko V I and Burkard G 2018 Tunable Berry curvature and valley and spin Hall effect in bilayer MoS2 *Physical Review B* **98**

[66] Wang Y, Cong C, Shang J, Eginligil M, Jin Y, Li G, Chen Y, Peimyoo N and Yu T 2019 Unveiling exceptionally robust valley contrast in AA- and AB-stacked bilayer WS2 *Nanoscale Horiz* **4** 396–403

[67] Nalin Mehta A, Gauquelin N, Nord M, Orekhov A, Bender H, Cerbu D, Verbeeck J and Vandervorst W 2020 Unravelling stacking order in epitaxial bilayer MX2 using 4D-STEM with unsupervised learning *Nanotechnology* **31** 445702

[68] Ciccarino C J, Chakraborty C, Englund D R and Narang P 2019 Carrier dynamics and spin-valley-layer effects in bilayer transition metal dichalcogenides *Faraday Discuss* **214** 175–88

[69] Zeng F, Zhang W-B and Tang B-Y 2015 Electronic structures and elastic properties of monolayer and bilayer transition metal dichalcogenides MX2 (M = Mo, W; X = O, S, Se, Te): A comparative first-principles study *Chinese Physics B* **24**

[70] Kumar A, He H, Pandey R, Ahluwalia P K and Tankeshwar K 2015 Pressure and electric field-induced metallization in the phase-engineered ZrX2 (X = S, Se, Te) bilayers *Phys Chem Chem Phys* **17** 19215–21

[71] Bacaksiz C, Cahangirov S, Rubio A, Senger R T, Peeters F M and Sahin H 2016 BilayerSnS2: Tunable stacking sequence by charging and loading pressure *Physical Review B* **93**

[72] Lee J H, Park J Y, Cho E B, Kim T Y, Han S A, Kim T H, Liu Y, Kim S K, Roh C J, Yoon H J, Ryu H, Seung W, Lee J S, Lee J and Kim S W 2017 Reliable piezoelectricity in bilayer WSe2 for piezoelectric nanogenerators *Adv Mater* **29**

[73] Hovden R, Liu P, Schnitzer N, Tsen A W, Liu Y, Lu W, Sun Y and Kourkoutis L F 2018 Thickness and stacking sequence determination of exfoliated dichalcogenides (1T-TaS2, 2H-MoS2) using scanning transmission electron microscopy *Microsc Microanal* **24** 387–95

[74] Guan Y, Li X, Niu R, Zhang N, Hu T and Zhang L 2020 Tunable electronic properties of type-II SiS2/WSe2 hetero-bilayers *Nanomaterials (Basel)* **10**

[75] Ghatak K, Kang K N, Yang E H and Datta D 2020 Controlled edge dependent stacking of WS2-WS2 homo- and WS2-WSe2 hetero-structures: A computational study *Sci Rep* **10** 1648

[76] Hu X, Kou L and Sun L 2016 Stacking orders induced direct band gap in bilayer MoSe2-WSe2 lateral heterostructures *Sci Rep* **6** 31122

[77] Liu X C, Qu D S, Ryu J J, Ahmed F, Yang Z, Lee D Y and Yoo W J 2016 P-type polar transition of chemically doped multilayer MoS2 transistor *Advanced Materials* **28** 2345–51

[78] Fang L M, Liang W Z, Feng Q G and Luo S N 2019 Structural engineering of bilayer PtSe2 thin films: A first-principles study *Journal of Physics-Condensed Matter* **31**

[79] Terrones H and Terrones M 2014 Bilayers of transition metal dichalcogenides: Different stackings and heterostructures *Journal of Materials Research* **29** 373–82

[80] Koma A, Sunouchi K and Miyajima T 1985 Fabrication of ultrathin heterostructures with Vander Waals epitaxy *Journal of Vacuum Science & Technology B* **3** 724

[81] Koma A and Yoshimura K 1986 Ultrasharp interfaces grown with Vander Waals epitaxy *Surface Science* **174** 556–60

[82] Koma A 1992 Van der Waals epitaxy—a new epitaxial-growth method for a highly lattice-mismatched system *Thin Solid Films* **216** 72–6

[83] Koma A 1999 Van der Waals epitaxy for highly lattice-mismatched systems *Journal of Crystal Growth* **201** 236–41

[84] Ohuchi F S, Shimada T, Parkinson B A, Ueno K and Koma A 1991 Growth of mose2 thin-films with Vander Waals epitaxy *Journal of Crystal Growth* **111** 1033–7

[85] Ueno K, Saiki K, Shimada T and Koma A 1990 Epitaxial-growth of transition-metal dichalcogenides on cleaved faces of mica *Journal of Vacuum Science & Technology a-Vacuum Surfaces and Films* **8** 68–72

[86] Chiu M H, Li M Y, Zhang W J, Hsu W T, Chang W H, Terrones M, Terrones H and Li L J 2014 Spectroscopic signatures for interlayer coupling in MoS2-WSe2 van der Waals stacking *Acs Nano* **8** 9649–56

[87] Lui C H, Ye Z P, Ji C, Chiu K C, Chou C T, Andersen T I, Means-Shively C, Anderson H, Wu J M, Kidd T, Lee Y H and He R 2015 Observation of interlayer phonon modes in van der Waals heterostructures *Physical Review B* **91**

[88] Wang Q H, Kalantar-Zadeh K, Kis A, Coleman J N and Strano M S 2012 Electronics and optoelectronics of two-dimensional transition metal dichalcogenides *Nature Nanotechnology* **7** 699–712

[89] Huang S X, Ling X, Liang L B, Kong J, Terrones H, Meunier V and Dresselhaus M S 2014 Probing the interlayer coupling of twisted bilayer MoS2 using photoluminescence spectroscopy *Nano Letters* **14** 5500–8

[90] Jin C H, Regan E C, Yan A M, Utama M I B, Wang D Q, Zhao S H, Qin Y, Yang S J, Zheng Z R, Shi S Y, Watanabe K, Taniguchi T, Tongay S, Zettl A and Wang F 2019 Observation of Moire excitons in WSe2/WS2 heterostructure superlattices (vol 567, pg 76, 2019) *Nature* **569** 76–80

14 Geometric and Electronic Properties of Ternary Compound $Li_4Ti_5O_{12}$

Thi Dieu Hien Nguyen and Ming-Fa Lin

CONTENTS

14.1 INTRODUCTION

Due to industrial requirements, lithium-ion batteries (LIBs) have become an important research topic, with many applications such as electronic devices [1, 2] (mobile, laptops, digital cameras, and so on), electric and hybrid vehicles (EV) [2, 3], and as emergency power backup [4] for communication and medical technology. The first commercial Li-ion based battery was delivered by a Sony and Asahi Kasei research group under the guidance of Yoshio Nishi in 1991 [5]. LIBs materials are remarkable for their great contributions to overcoming environmental and energy problems, leading to the Nobel Prize in Chemistry in 2019 for Yoshino, Goodenough, and Whittingham for the development of lithium-ion batteries. Consequently, LIBs show lots of merits in storage of energy, e.g., high power, energy densities, and high reliability, as well as affordable price, long life cycle, and friendly to environment. In general, LIBs consist of three main components, which are a cathode (positive electrode), an anode (negative electrode), and an electrolyte. Especially, the ion transports of positive lithium ions between anodes and cathodes are based on the electrolyte part [1]. There are many developments in the anode materials, which can be commonly constructed by graphite and lithium titanium oxide. The commercial graphite-based anode material possesses a high capacity and cheap price. On the contrary, in this material, the volume expansion during many rapid intercalations and deintercalation processes might induce a huge drawback. To surmount the difficulty of graphitic materials in the drastic volume change, lithium titanium oxide material has been developed to serve as the negative electrode due to a zero-strain material. Furthermore, lithium titanate presents many advantages such as long life, rapid charging, high input/output power performance, wide temperature resistance,

DOI: 10.1201/9781003322573-14

and so on. Besides, lithium titanium oxide provides a high potential plateau at 1.55 V versus Li/Li$^+$ [6]. This will prevent the occurrence of lithium dendrites, which create fire in rechargeable batteries. As a result, this secondary lithium-based battery could enhance the safety problem.

Theoretical analysis using the first-principles calculations under the quasiparticle framework has proposed significant methods to thoroughly understand the geometric and electronic properties. The contributions of multi- or single-orbital hybridizations in various chemical bonds are obtained from the optimal lattice symmetry, the atom-dominated band structures, the spatial charge densities and their changing under chemical modifications, and the atom- and orbital-decomposed density of states. Many examples use this developed quasiparticle framework such as silicene- and graphene-related systems. Also, it could be generalized to other emergent materials and other complicated cathode [7–10], anode [10–13], and electrolyte materials [14–16] of lithium-ion batteries. Hopping, integral parameters are key factors to connect theoretical and phenomenological approach. Furthermore, all of these calculations need to be thoroughly tested in further experimental investigations.

In previous time, there were several research using first-principles calculations, which focus on the unusual superconducting properties of LiTiO-related materials. However, significant quasiparticle properties that are related to the physical and chemical points of view are still absent in the Li-Ti-O compound, e.g., they lack significant orbital hybridizations in various chemical bonds. This work focuses on the geometric and electronic quasiparticle properties of Li$_4$Ti$_5$O$_{12}$, which can serve as an anode of Li$^+$-ion batteries. The completed picture of multi-hybridization in Li-O and Ti-O bonds could be identified in the optimal Moiré superlattice, the atom-dominated electronic energy spectrum, the spatial charge densities, and the atom- and orbital-decomposed van Hove singularities. There exist combined hybridizations of Li-2s, Ti-[4s,3$d_{x^2-y^2}$, 3d$_{xy}$, 3d$_{yz}$, 3d$_{zx}$, 3d_{z^2}] and O-[2s, 2p$_x$, 2p$_y$, 2p$_z$] orbitals. This quasiparticle theory is developed and generalized in comprehending cathode/anode/electrolyte materials in rechargeable batteries.

14.2   NUMERICAL CALCULATION

All calculations in this chapter are given based on the Vienna Ab Initio Simulation package (VASP). The geometric and electronic quasiparticle properties of lithium titanate are investigated under density functional theory (DFT) [17, 18]. The Perdew-Burke-Ernzerhof (PBE) functional [19] under the general gradient approximation (GGA) [20] is used for the calculation of the many-particle exchange and correlation energies, which describe the electron-electron Coulomb interactions [21]. The projector-augmented wave (PAW) [22] pseudopotentials are used to identify the electron-ion interactions. The plane waves are chosen as a complete set with a maximum energy cutoff of 520 eV. The first Brillouin zone is sampled 12 × 12 × 12 k-point meshes within a Gamma scheme along the three-dimensional periodic direction for electronic quasiparticle structures. The convergence criterion was determined by conjugate gradient minimization of the total energy of 10^{-6} eV between two consecutive simulation steps. The maximum Hellmann-Feynman force acting on each atom is less than 0.01 eV Å^{-1} during the ionic relaxations.

14.3   AN OPTIMAL STRUCTURE

Three-dimensional ternary compound Li$_4$Ti$_5$O$_{12}$ possesses a specific geometric symmetry. The stable structure is tested by numerical calculation, which achieves the lowest energy at -459.908 eV with a volume of $619.372 \ \mathring{A}^3$ (clearly illustrated in Figure 14.1). This structure belongs to the cubic crystal system (shown in Figure 14.2, where (Li, Ti, O) atoms are green, blue, and orange, respectively), whose primitive cell consists of 56 atoms (8-Li, 16-Ti, 32-O). The space group symmetries are Fd $\overline{3}$m Hermann Mauguin, F 4d 2 3 $\overline{1}$ d, and m $\overline{3}$ m point group. By using the delicate Vienna Ab Initio Simulation Package (VASP) calculation, the lattice constants

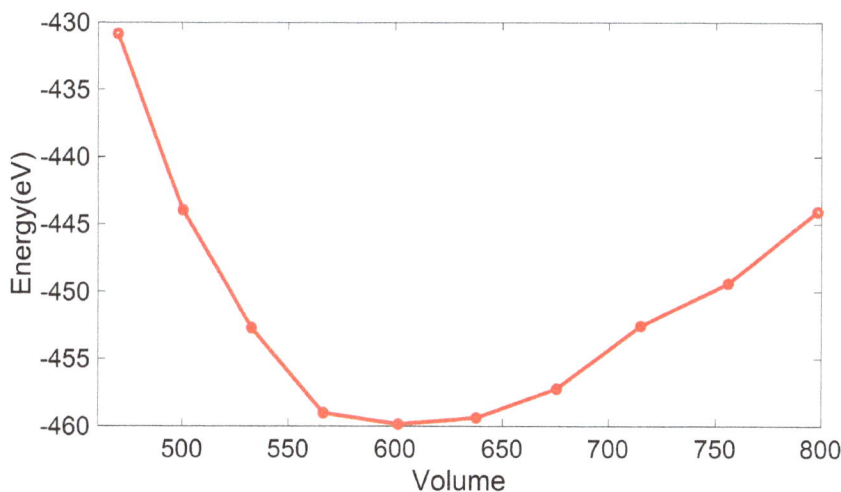

FIGURE 14.1   The expression of lowest energy versus volume of the ternary compound.

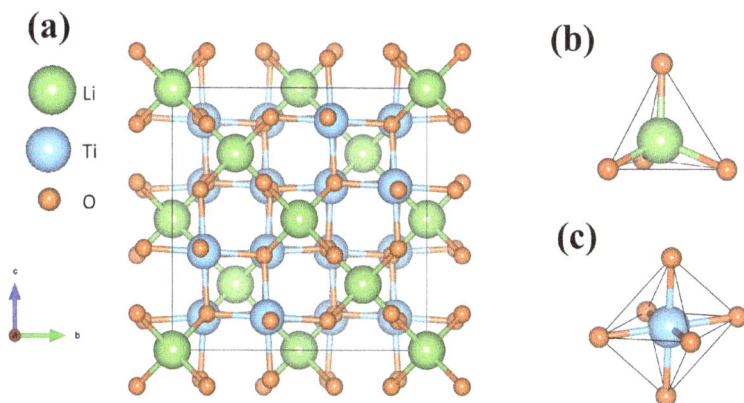

FIGURE 14.2   (a) The optimal geometry of Li$_4$Ti$_5$O$_{12}$, with triclinic symmetry, where a primitive unit cell has 56 atoms. Lattice constants: a = b = c = 8.357 Å, tilted angles: $\alpha = \beta = \gamma = 90^O$. (b) LiO$_4$, (c) TiO$_6$ structures.

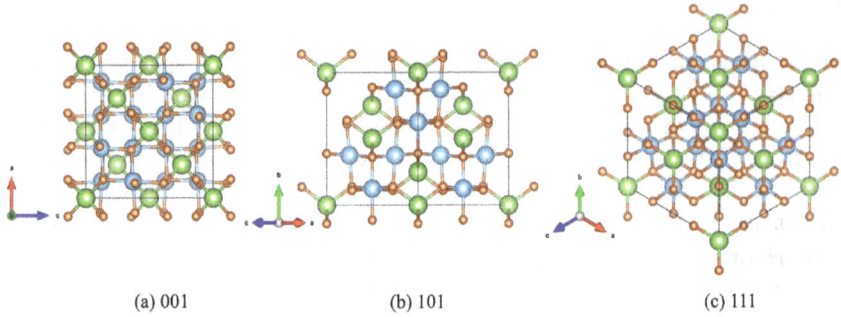

(a) 001   (b) 101   (c) 111

FIGURE 14.3 The geometric structure of $Li_4Ti_5O_{12}$ along different projections: (a) (001), (b) (101), and (c) (111), in which Li, Ti, and O atoms are represented by green, blue, and orange balls, respectively.

TABLE 14.1
The Chemical Bond Lengths of Li-O and Ti-O

Chemical bonds	Numbers of bonds	Bond length
Li-O	32	2.0418–2.0428 Å
Ti-O	96	2.0234–2.0239 Å

are given by a = b = c = 8.357 Å and the tilted angles of $\alpha = \beta = \gamma = 90°$. Previous research results [6, 23–25] provide the range of 8.352 Å to 8.370 Å for a, b, c. The whole system exists in an anisotropic and slightly nonuniform chemical environment, which forms LiO_4 and TiO_6 structures (indicated in Figure 14.2(b–c)). The diverse projected arrangements of this compound are (a) (001), (b) (101), and (c) (111) in Figure 14.3(a–c). It is noted that there exists some symmetric structures in some directions; therefore, the similar form can be ignored.

The essential quasiparticle properties of lithium titanate $Li_4Ti_5O_{12}$ could be determined by the rich and unique chemical bonds. In the primitive cell, there mainly exist two kinds of bonds, including Li-O and Ti-O, which are listed in Table 14.1. In general, Li-O bond lengths lie in the range of 2.0418–2.0428 Å, while Ti-O ones are 2.0234–2.0239 Å. The slight modulation in $Li_4Ti_5O_{12}$ strongly confirms the zero-strain characteristic, which stays the same in lattice dimension during the change from the initial state to the final state ($Li_4Ti_5O_{12}$ to $Li_7Ti_5O_{12}$). The bond lengths of this system could be understood in detail by referring to the charge density (in Figure 14.7). All the bond lengths can be explained under the multi-orbital hybridizations. This can be discussed further in the partial density of state seen in Figure 14.6. The complicated multi-hybridization might be explored further in the tight-binding theory. However, it is difficult to achieve reliable parameters for the diagonal matrix in the Hamiltonian. Several experimental measurements, including high-resolution in situ structural measurements and ex-situ XRD are used to comprehend the lattice parameter. Besides X-ray diffraction [26], low-energy electron diffraction (LEED)

[27] is an efficient technique to determine the 3D lattice symmetries. The diffracted electrons in LEED, as spots on a fluorescent screen, could be observed to determine the surface morphology. Scanning electron microscopy (SEM [26, 28]) can provide information about surface properties, as revealed from defects. Remarkably, STM [29] and tunneling electron microscopy (TEM [28]) are not applied in this case, e.g., use for the nanoscaled top- and side-view structures, respectively. As a result, the diversified lengths in Li-O and Ti-O bonds can be clearly investigated. Therefore, this will be very useful in indirectly identifying the complicated multi-orbital hybridizations in all chemical bonds.

14.4 ORBITAL-HYBRIDIZATION-ENRICHED ELECTRONIC PROPERTIES

The electronic quasiparticle properties of Li$_4$Ti$_5$O$_{12}$ could be investigated through the band structures, atom dominances, charge densities, and orbital-decomposed density of states. Li$_4$Ti$_5$O$_{12}$ presents a rich and unique electronic characteristic. Figure 14.5(a) provides the band structure by using the numerical first-principles calculations. The wave vector in the reciprocal space is utilized to reveal the unoccupied states and occupied states along with the high symmetric points Γ-X-M-Γ-R-X|M-R (Figure 14.4). This 3D ternary Li$_4$Ti$_5$O$_{12}$ compound possesses highly anisotropic properties. From Figure 14.5(a), we see that the occupied electron energy spectrum (conduction band) is asymmetric to the unoccupied hole one (valence band) about the Fermi level (E$_F$ = 0), which is originated from the various multi-orbital

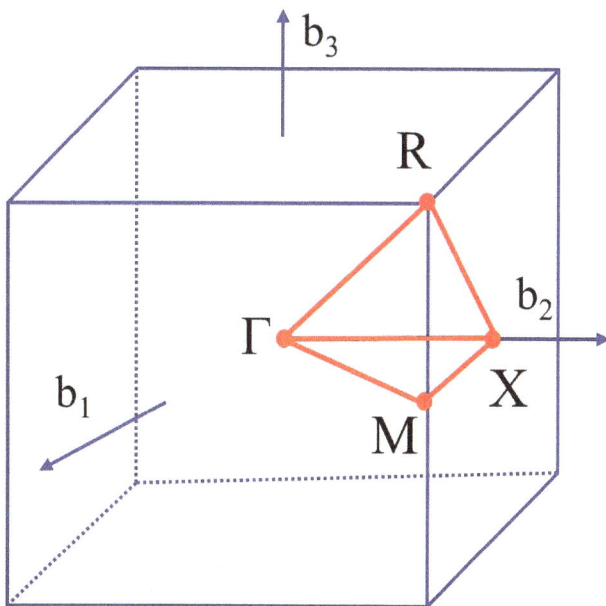

FIGURE 14.4 The first Brillouin zone with the paths of high-symmetry points.

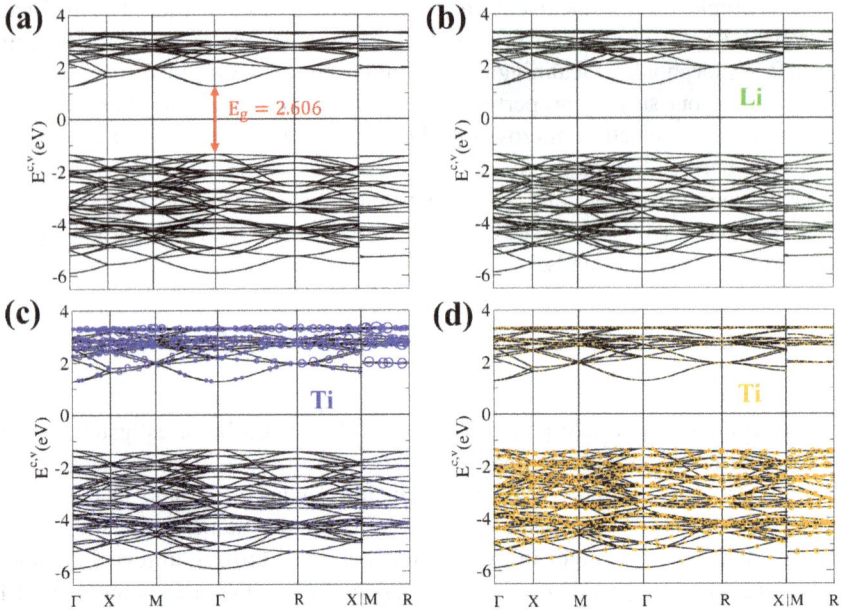

FIGURE 14.5   (a) Electronic energy spectrum for $Li_4Ti_5O_{12}$ along the high-symmetry points in the first Brillouin zone within the range of -6.0 eV $\leq E^{c,v} \leq 4.0$ eV; for the specific (b) lithium, (c) titanium and (d) oxygen dominances (green, blue, and orange balls, respectively).

hybridizations in the nonuniform chemical bonds (Figure 14.2). The current lithium titanate structure possesses the direct band gap of $E_g^i = 2.606$ Å semiconductor. This gap is directly measured from the highest occupied state to the lowest unoccupied state (shown by the red arrow in Figure 14.5). Theoretically, this gap is equivalent to threshold optical frequency. Therefore, the UV-visible absorption spectra can be used in experimental measurement to fully test the result [30]. The absorption edge is observed at ~400–420 nm with the experimental band gap of 2.95 eV by using spectroscope ellipsometry (SE). The current calculation of band gap is underestimated compared with the experimental value. Moreover, the band energy dispersion along the synthetic dimension needs to be identified from the ARPES [29, 31], which is a powerful experimental technique directly measuring the single-particle spectral function $A(k, \omega)$. In ARPES experiments, electrons are probed inside solids in the energy and momentum space and are able to provide some physical information, such as Fermi surface, energy gap, and many-body interactions (e.g., electron–phonon, electron–electron interactions). As an advanced technique, high-resolution ARPES microscopy using a well-focused light source has recently attracted a lot of interest based on its potential to achieve local electronic information at micro- or nanoscale.

Due to the large band gap, the spin-split electronic states around the Fermi level, which is originated from the spin-degenerate valence and conduction bands, might be difficult to achieve. Consequently, the net magnetic moment vanishes, and the spin-dependent interactions could be uncounted in the numerical calculation. The band structure presents various subbands because of a big number of atoms in the primitive cell and the outer s, d, p-orbitals, corresponding to Li, Ti, O atoms. In general, the electronic structures are complex in terms of oscillatory dispersion relations, the highly anisotropic behaviors, and the frequent crossings or anti-crossings. As a result, the width of each energy sub-band is difficult to identify, e.g., the various valley structures for the different subbands are impossible to distinguish. Many band-edge states come into existence along the various wave vectors, including the high-symmetric points or the value between them. These critical points can be seen in the energy-wave-vector space, associated with the unique van Hove singularities (discussed later in the section on density of states).

The atom dominances establish the beginning understanding about the chemical bonding, according to the contributions of lithium, titanium, and oxygen atoms by the green, blue, and orange balls, respectively. Figure 14.5(b) shows the Li-atoms with a small green ball, spread in the whole range of the energy band spectrum of about 10 eV along with the high-symmetric points. Even though the contribution of Li is weak but it is significant to the chemical bonds of Li-O. This unique result is caused by the single-orbital contribution of 2s of the outer cell. Titanium atoms (blue circles) are presented in the valence- and conduction-band ranges of -5.87 eV $\leq E^v \leq$ -1.30 eV and 1.30 eV $\leq E^c \leq 3.32$ eV. Most importantly, they dominate in the unoccupied electronic states, which take part in the Li-O bonds. It is noted that this might originate from the outer orbitals of titanium atoms. On the contrary, oxygen atoms have a higher contribution to the occupied bands. In this case, oxygen dominances might be existed because of many simultaneous O-related chemical bonds in deeper ($2s$, $2p_x$, $2p_y$, $2p_z$) orbital energies. In general, partial information about the critical multi-orbital hybridizations could be achieved from the atom-dominated electronic structures.

In the experimental point of view, the electronic band structures along the high-symmetric points below Fermi level are commonly investigated by high-resolution angular resolved photoelectron spectroscopy (ARPES). This technique is successful in confirming the rich band structures of graphene-related systems in the different dimensions. The measurements cover the linear Dirac-cone structure in the monolayer system [32, 33], the parabolic bands in bilayer AB stacking [33–35], the linear and parabolic dispersion in tri-layer ABA stacking, and the partial flat, linear, and sombrero-shaped in tri-layer ABC stacking [36–39]. They are useful in understanding the interlayer $2p_z$-$2p_z$ orbital hybridizations in the honeycomb lattices. The theoretical prediction of the 3D ternary $Li_4Ti_5O_{12}$ can be tested through similar experimental measurements. For example, the sensitive wave-dependent vector, the large band gap, the highly asymmetric occupied and unoccupied states, the various energy dispersions such as parabolic, linear, partial flat band, and the frequent non-crossing/crossing/anti-crossing behaviors can be investigated further. In general, this could provide a fundamental technique for solving the critical orbital hybridizations of chemical bonds. One of the significant parts of the electronic quasiparticle properties of the 3D ternary compound is the spatial change density, which is exhibited in Figure 14.6(a–e), including the isolated atom and Li-O, Ti-O chemical bonds. This directly relates to the atom- and orbital-decomposed DOS with van Hove singularities (further investigated in Figure 14.7) and the atom-dominated band structures (Figure 14.5). The non-uniform chemical environment in the cubic structures (Figure 14.2) is present through the tetrahedral of Li-O and octahedral of Ti-O. The chemical bond lengths survive in the $Li_4Ti_5O_{12}$, which is determined by the orbital hybridization. They depend on the distance between the nearest atoms. Figure 14.6(d) shows the Li-O bonds (2.0418 Å), where the effective range in charge density of Li atoms due to 2s orbitals is approximately 0.41 Å from the deep-red region of the Li^+-ion core. In the Li-O bonds, the 1s orbitals are deep and stable; thus, it is independent of the critical orbital hybridizations. Apparently, the distance between lithium and oxygen atoms is quite large, so the overlap range might be difficult to observe. In the O orbitals, there exists a significant contribution of ($2p_x$, $2p_y$, $2p_z$), which are represented by the light-green and yellow colors. It is noted that 2s orbitals are almost negligible in the contribution of the chemical bonds between lithium and oxygen atoms. Moreover, the multi-orbital hybridizations of $2s$-($2p_x$, $2p_y$, $2p_z$) in Li-O bonds show the various hopping integrals in the phenological model [40]. However, this problem might face difficulty in providing suitable parameters in the matrix Hamiltonian.

Ti-O bonds are stronger than Li-O ones because their distance is shorter (Figure 14.6(e)), which clear overlap is identified from 0.53–1.33 from the O-orbital core. Apparently, the Ti atoms possess a large atomic number with the ($3d_{x^2-y^2}$, $3d_{xy}$, $3d_{yz}$, $3d_{zx}$, $3d_{z^2}$) outer orbitals and the O mainly by ($3p_x$, $3p_y$, $3p_z$) ones. Therefore, the effective contribution of transition metal titanium atoms is a little higher than that of oxygen atoms. Figure 14.6(e) shows three main regions of Ti atoms, including heavy-red, light-red, and yellow-green regions, which corresponds to ($3s$, $3p_x$, $3p_y$, $3p_z$), ($4s$), and ($3d_{x^2-y^2}$, $3d_{xy}$, $3d_{yz}$, $3d_{zx}$, $3d_{z^2}$). The most important parts that determine

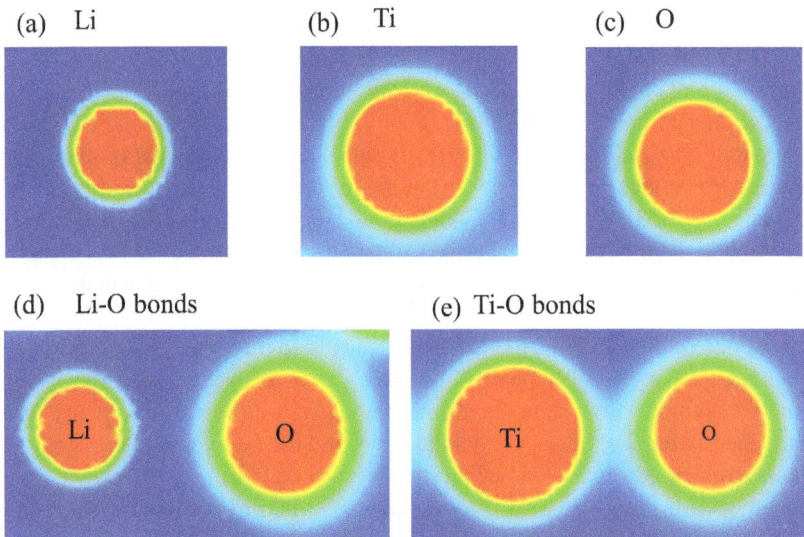

FIGURE 14.6 Diversified spatial charge density distributions of the isolated atoms (a) Li, (b) Ti, and (c) O under the longest/middle/shortest Li-O for (b)/(c)/(d) and Ti-O for (e)/(f)/(g), respectively.

the multi-orbital hybridizations in the Ti-O bonds are the two latter, while the former ones are obviously independent of these factors. In this case, the charge will transfer from titanium to oxygen atoms in the deformed carrier distributions. The diversified bond lengths of Li-O and Ti-O chemical bonds will create difficulty in finding the reliable hopping integral parameters in the tight-binding model.

The atom- and orbital-projected density of states (in Figure 14.7(a–d)) is essentially calculated by the number of electronic states in a small range dE, which can completely provide a full understanding of the multi-orbital hybridizations on the Li-O and Ti-O chemical bonds. In the experimental measurement, STS is one of the powerful methods to investigate the significant characteristic of DOS. In the values from −1.30 eV to 1.30 eV, DOS is uncounted because this range corresponds to the large band gap of the ternary compounds, e.g., E^i_g =2.606 eV (Figure 14.4(a-d)). DOS presents remarkably asymmetric just above and below the Fermi level. The conduction and valence spectra exist in many specific forms such as shoulders and asymmetric/symmetric peaks in about 10 eV. These come from a lot of oscillatory energy subbands, with frequent crossing or anti-crossings in the compound. Also, the existence of the dense sub-bands might form the merged band-edge states which can create a special structure. The unusual structures, van Hove singularities, originate from the zero-group velocity band-edge states of energy sub-bands. These critical points can be classified into extrema (including local maximum and minimum), saddle points, and dispersionless points in the energy wave-vector subbands. They are located in the high symmetric points or somewhere along with the wave vector energies. Very quirkily, the various van Hove singularities might depend on the different dimensions. For example, in 0D discrete energy levels, the delta-function-like peaks are prominent, the square-root asymmetric peaks or the plateau structures show in the 1D parabolic/linear energy sub-bands, and the shoulder structures/V-shapes/logarithmic peaks/square-root asymmetric peaks present in the 2D parabolic/Dirac-cone/saddle-point/constant-energy-loop band structures. In Figure 14.7(a), the contribution of lithium (green curve) is lowest but very important in the whole occupied and unoccupied spectrum range. Apparently, 1s core orbitals do not take part in DOS contribution. Consequently, the titanium and oxygen atoms' density of states play a dominating role in the energy range of −5.87 eV ≤ E^v ≤ −1.30 eV and 1.30 eV≤ E^c ≤ 3.32 eV (blue and orange curves), respectively. As we discussed before in atom dominances, DOSs in Figure 14.7(a) totally agree with each other.

The orbital-projected van Hove singularities can be revealed through Figures 14.6(b–d), which determine the important multi-orbital hybridizations in the Li-O and Ti-O. Apparently, the main orbital contributions create various van Hove singularities due to their number, energies, and intensities. There are three groups, which correspond to lithium, titanium, and oxygen atoms, including (i) Li-2s orbital (green curve in Figure 14.7(b)), (ii) Ti-($4s$, $3d_{x^2-y^2}$, $3d_{xy}$, $3d_{yz}$, $3d_{zx}$, $3d_{z^2}$) orbitals (pink, light blue, blue, red, light green, and yellow curves in Figure 14.7(c)), and (iii) O-($2s$, $2p_x$, $2p_y$, $2p_z$) orbitals (purple, pink, green, and orange curves in Figure 14.7(d)). Only the single Li-2s and three O-($2p_x$, $2p_y$, $2p_z$) contribute to the Li-O bonds, which is displayed by the DOS of these atoms. Among the current orbitals, O-2s exhibits low DOS intensity, thus can be negligible in the energy range (−6.0 eV ≤ E ≤ 4.0 eV).

The essential reason comes from the completely filled electrons in the occupied and unoccupied states associated with spin-up and spin-down configurations. In general, the rich chemical bonds Li-O are attributed to the complicated multi orbitals $2s$-($2p_x$, $2p_y$, $2p_z$), which imply multi-orbital hopping integrals of on-site Coulomb potentials.

Very interestingly, the oxygen atoms ($2p_x$, $2p_y$, $2p_z$) also make a significant hybridization with the transition-metal titanium atoms ($4s$, $3d_{x^2-y^2}$, $3d_{xy}$, $3d_{yz}$, $3d_{zx}$, $3d_{z^2}$) (Figure 14.6(c)) because their van Hove singularities come into existence with each other. Oxygen atoms contribute higher intensity in the valence energy, while the opposite is true for titanium atoms. The orbital-dependent contributions are almost comparable in the whole energy spectrum except for the small $4s$ orbital density of state with the conduction-band range (pink curve). In general, the initial conduction- or valence-state energy spectrum is mainly determined by the titanium and oxygen atoms. In summary, the 3D ternary compound consists of the Li-O and Ti-O

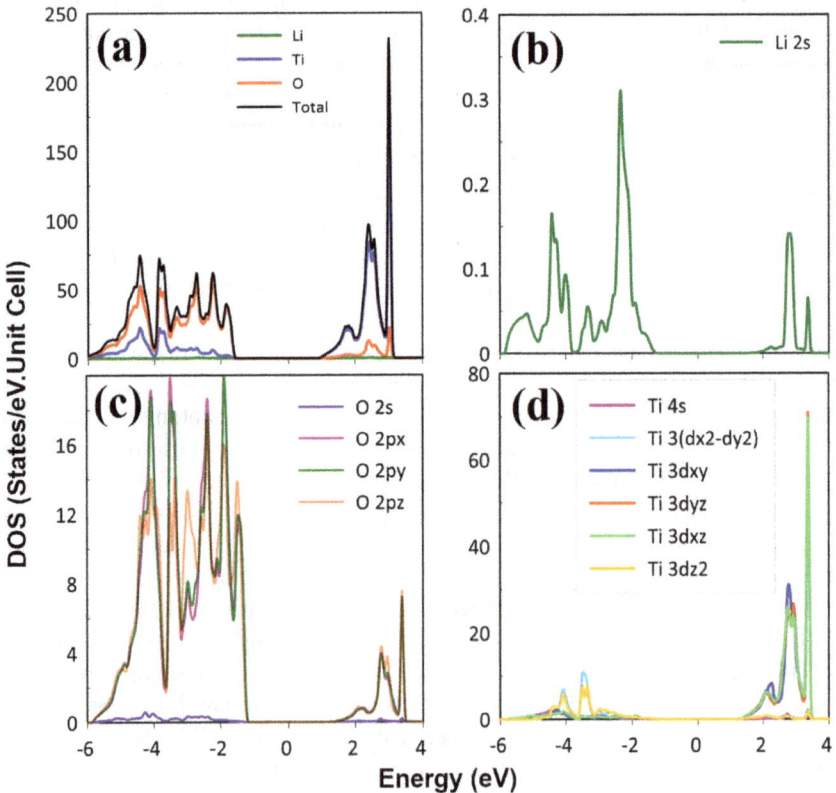

FIGURE 14.7 The atom- and orbital-decomposed density of states due to (a) Li, Ti, and O atoms (green, blue, and orange curves); (b) Li-2s orbital (green curve); (c) Ti-($4s$, $3d_x^2-y^2$, $3d_{xy}$, $3d_{yz}$, $3d_{xz}$, $3d_z^2$) orbitals (purple, light blue, dark blue, red, light green, yellow curves); and (d) O-($2s$, $2p_x$, $2p_y$, $2p_z$) orbitals (blue, purple, green, and pink-orange curves).

chemical bonds, which are deduced to have multi-orbital hybridizations of $2s$-$(2p_x,$ $2p_y, 2p_z)$, $(4s, 3d_{x^2-y^2}, 3d_{xy}, 3d_{yz}, 3d_{zx}, 3d_{z^2})$-$(2p_x, 2p_y, 2p_z)$, respectively.

As mentioned, the high-resolution STS technique provides an important measurement in identifying van Hove singularities of DOS due to the band-edge states from the tunneling current-voltage characteristic. Principally, the experimental measurement records I-V curves under a weak tunneling quantum current. The differential conductance (dI/dV)/(I/V) is useful to achieve the spatial distribution of DOS under current imaging tunneling spectroscopy (CITS). This technique has been successful in comprehending the electronic quasiparticle properties of graphene-related systems, including a nearly symmetric [41] V-shape structure vanishing at the Fermi level for a monolayer system (a zero-gap semiconductor) [42], a gate-voltage-induced band gap in bilayer AB and tri-layer ABC stacking [43], a prominent peak centered around the Fermi level due to partially flat bands (surface states) in tri-layer and penta-layer ABC stacking [44], and a sharp dip structure near E_F combined with a pair of asymmetric peaks in tri-layer AAB stacking (a narrow-gap semiconductor with low-lying constant-energy loops) [45, 46]. The experimental examinations, which directly apply the 3D ternary Li₄Ti₅O₁₂, could confirm the theoretical calculations with van Hove singularities. This can lead to a full understanding of complicated multi-orbital hybridizations of Li-O and Ti-O.

Another approach to understand essential properties of materials is phenomenological model with the reliable hopping integrals. However, this method might face difficulty due to the complicated structures of the 3D Li₄Ti₅O₁₂ compound. For example, the primitive cell consists of a large Moiré superlattice (56 atoms) and a big number of chemical bonds (32 Li-O and 92 Ti-O bonds in Figure 14.1). Besides, the band structures possess complicated subbands in various forms, e.g., weak dispersions, parabolic, and associated with the band-edge states. The large band gap $E_g \sim 2.606$ eV (Figure 14.5(a)), nonhomogeneous spatial charge densities in diverse chemical bonds (Figure 14.6(a–e)), and complicated multi-orbital hybridizations in van Hove singularities (Figure 14.6(a–d)) are thoroughly discussed. As a result, due to the non-uniform hopping integral with the diverse orbital hybridizations, the parameterized tight-binding model might be difficult to involve in the first-principles calculations. However, in the other Li⁺-based anode [10, 11, 13, 47, 48], cathode [7, 9, 10, 49], and electrolyte [14, 50, 51] materials, this combination can become a further issue to understand the significant quasiparticle characteristic of the materials.

REFERENCES

[1] Pistoia G 2013 *Lithium-Ion Batteries: Advances and Applications* (Amsterdam: Elsevier)

[2] Deng D 2015 Li-ion batteries: Basics, progress, and challenges *Energy Science & Engineering* **3** 385–418

[3] Li M, Lu J, Chen Z and Amine K 2018 30 years of lithium-ion batteries *Advanced Materials* **30** 1800561

[4] Suzuki I, Shizuki T and Nishiyama K 2003 High power and long life lithium-ion battery for backup power sources in *The 25th International Telecommunications Energy Conference, 2003. INTELEC'03* (Yokohama: IEEE) pp 317–22

[5] Nazri G-A and Pistoia G 2008 *Lithium Batteries: Science and Technology* (Berlin: Springer Science & Business Media)

[6] Zhao B, Ran R, Liu M and Shao Z 2015 A comprehensive review of Li4Ti5O12-based electrodes for lithium-ion batteries: The latest advancements and future perspectives *Materials Science and Engineering: R: Reports* **98** 1–71

[7] Takahashi M, Tobishima S, Takei K and Sakurai Y 2001 Characterization of LiFePO4 as the cathode material for rechargeable lithium batteries *Journal of Power Sources* **97** 508–11

[8] Qiu X-Y, Zhuang Q-C, Zhang Q-Q, Cao R, Ying P-Z, Qiang Y-H and Sun S-G 2012 Electrochemical and electronic properties of LiCoO 2 cathode investigated by galvanostatic cycling and EIS *Physical Chemistry Chemical Physics* **14** 2617–30

[9] He P, Yu H and Zhou H 2012 Layered lithium transition metal oxide cathodes towards high energy lithium-ion batteries *Journal of Materials Chemistry* **22** 3680–95

[10] Mekonnen Y, Sundararajan A and Sarwat A I 2016 A review of cathode and anode materials for lithium-ion batteries in *SoutheastCon 2016* (Virginia: IEEE) pp 1–6

[11] Nguyen T D H, Pham H D, Lin S Y and Lin M F 2020 Featured properties of Li+-based battery anode: Li4Ti5O12 *Rsc Advances* **10** 14071–9

[12] Wu Z-S, Ren W, Xu L, Li F and Cheng H-M 2011 Doped graphene sheets as anode materials with superhigh rate and large capacity for lithium ion batteries *ACS nano* **5** 5463–71

[13] Sandhya C P, John B and Gouri C 2014 Lithium titanate as anode material for lithium-ion cells: A review *Ionics* **20** 601–20

[14] Khuong Dien V, Thi Han N, Nguyen T D H, Huynh T M D, Pham H D and Lin M-F 2020 Geometric and electronic properties of Li2GeO3 *Frontiers in Materials* **7**

[15] Li Q, Chen J, Fan L, Kong X and Lu Y 2016 Progress in electrolytes for rechargeable Li-based batteries and beyond *Green Energy & Environment* **1** 18–42

[16] Wang Y, Song S, Xu C, Hu N, Molenda J and Lu L 2019 Development of solid-state electrolytes for sodium-ion battery—A short review *Nano Materials Science* **1** 91–100

[17] Argaman N and Makov G 2000 Density functional theory: An introduction *Am J Phys* **68** 69–79

[18] Kohn W, Becke A D and Parr R G 1996 Density functional theory of electronic structure *The Journal of Physical Chemistry* **100** 12974–80

[19] Paier J, Hirschl R, Marsman M and Kresse G 2005 The Perdew—Burke—Ernzerhof exchange-correlation functional applied to the G2–1 test set using a plane-wave basis set *The Journal of Chemical Physics* **122** 234102

[20] Perdew J P, Burke K and Ernzerhof M 1996 Generalized gradient approximation made simple *Phys Rev Lett* **77** 3865

[21] Fuchs M 1999 *Exchange-Correlation Energy: From LDA to GGA and Beyond* (Berlin: Faradayweg)

[22] Larsen A H, Vanin M, Mortensen J J, Thygesen K S and Jacobsen K W 2009 Localized atomic basis set in the projector augmented wave method *Physical Review B* **80** 195112

[23] Wolfenstine J and Allen J 2008 Electrical conductivity and charge compensation in Ta doped Li4Ti5O12 *Journal of Power Sources* **180** 582–5

[24] Mosa J, Vélez J F, Lorite I, Arconada N and Aparicio M 2012 Film-shaped sol—gel Li4Ti5O12 electrode for lithium-ion microbatteries *Journal of Power Sources* **205** 491–4

[25] Leonidov I, Leonidova O, Perelyaeva L, Samigullina R, Kovyazina S and Patrakeev M 2003 Structure, ionic conduction, and phase transformations in lithium titanate Li 4 Ti 5 O 12 *Physics of the Solid State* **45** 2183–8

[26] Chauque S, Oliva F Y, Visintin A, Barraco D, Leiva E P M and Camara O R 2017 Lithium titanate as anode material for lithium ion batteries: Synthesis, post-treatment and its electrochemical response *Journal of Electroanalytical Chemistry* **799** 142–55

[27] VanHove M A, Weinberg W H and Chan C-M 2012 *Low-Energy Electron Diffraction: Experiment, Theory and Surface Structure Determination* vol 6 (Berlin: Springer Science & Business Media)

[28] Egerton R, Li P and Malac M 2004 Radiation damage in the TEM and SEM *Micron* **35** 399–409

[29] Bussolotti F, Chi D, Goh K J, Huang Y L and Wee A T 2020 *2D Semiconductor Materials and Devices* (Armsterdam: Elsevier) pp 199–220

[30] Zhao M, Lian J, Jia Y, Jin K, Xu L, Hu Z, Yang X and Kang S 2016 Investigation of the optical properties of LiTi 2 O 4 and Li 4 Ti 5 O 12 spinel films by spectroscopic ellipsometry *Optical Materials Express* **6** 3366–74

[31] Damascelli A 2004 Probing the electronic structure of complex systems by ARPES *Physica Scripta* **2004** 61

[32] Yue S, Zhou H, Geng D, Sun Z, Arita M, Shimada K, Cheng P, Chen L, Meng S and Wu K 2020 Experimental observation of Dirac cones in artificial graphene lattices *Physical Review B* **102** 201401

[33] Tran N T T, Lin S-Y, Lin C-Y and Lin M-F 2017 *Geometric and Electronic Properties of Graphene-Related Systems: Chemical Bonding Schemes* (Boca Raton: CRC Press)

[34] Nguyen V L, Perello D J, Lee S, Nai C T, Shin B G, Kim J G, Park H Y, Jeong H Y, Zhao J and Vu Q A 2016 Wafer-scale single-crystalline AB-stacked bilayer graphene *Advanced Materials* **28** 8177–83

[35] Lu C, Chang C-P, Huang Y-C, Chen R-B and Lin M 2006 Influence of an electric field on the optical properties of few-layer graphene with AB stacking *Physical Review B* **73** 144427

[36] Coletti C, Forti S, Principi A, Emtsev K V, Zakharov A A, Daniels K M, Daas B K, Chandrashekhar M, Ouisse T and Chaussende D 2013 Revealing the electronic band structure of trilayer graphene on SiC: An angle-resolved photoemission study *Physical Review B* **88** 155439

[37] Lu C L, Chang C P, Huang Y C, Ho J H, Hwang C C and Lin M F 2007 Electronic properties of AA-and ABC-stacked few-layer graphites *J Phys Soc Jpn* **76** 024701

[38] Yelgel C 2016 Electronic structure of ABC-stacked multilayer graphene and trigonal warping: a first principles calculation *Journal of Physics: Conference Series* 012022

[39] Hattendorf S, Georgi A, Liebmann M and Morgenstern M 2013 Networks of ABA and ABC stacked graphene on mica observed by scanning tunneling microscopy *Surf Sci* **610** 53–8

[40] Foulkes W M C and Haydock R 1989 Tight-binding models and density-functional theory *Physical Review B* **39** 12520

[41] Kahn L and Ying S 1975 Alkali-metal chemisorption *Solid State Commun* **16** 799–801

[42] Deshpande A, Bao W, Miao F, Lau C N and LeRoy B J 2009 Spatially resolved spectroscopy of monolayer graphene on SiO 2 *Physical Review B* **79** 205411

[43] Gao L 2014 Probing electronic properties of graphene on the atomic scale by scanning tunneling microscopy and spectroscopy *Graphene and 2D Materials* **1**

[44] Pierucci D, Sediri H, Hajlaoui M, Girard J-C, Brumme T, Calandra M, Velez-Fort E, Patriarche G, Silly M G and Ferro G 2015 Evidence for flat bands near the Fermi level in epitaxial rhombohedral multilayer graphene *ACS Nano* **9** 5432–9

[45] Lin C-Y, Huang B-L, Ho C-H, Gumbs G and Lin M-F 2018 Geometry-diversified Coulomb excitations in trilayer AAB stacking graphene *Physical Review B* **98** 195442

[46] Do T-N, Lin C-Y, Lin Y-P, Shih P-H and Lin M-F 2015 Configuration-enriched magneto-electronic spectra of AAB-stacked trilayer graphene *Carbon* **94** 619–32

[47] Rahman M M, Sultana I, Yang T, Chen Z, Sharma N, Glushenkov A M and Chen Y 2016 Lithium germanate (Li2GeO3): A high-performance anode material for lithium-ion batteries *Angewandte Chemie* **128** 16293–7

[48] Lu J, Chen Z, Pan F, Cui Y and Amine K 2018 High-performance anode materials for rechargeable lithium-ion batteries *Electrochemical Energy Reviews* **1** 35–53

[49] Wang L, He X, Sun W, Wang J, Li Y and Fan S 2012 Crystal orientation tuning of LiFePO4 nanoplates for high rate lithium battery cathode materials *Nano Letters* **12** 5632–6

[50] Li X, Liu J, Banis M N, Lushington A, Li R, Cai M and Sun X 2014 Atomic layer deposition of solid-state electrolyte coated cathode materials with superior high-voltage cycling behavior for lithium ion battery application *Energy & Environmental Science* **7** 768–78

[51] Li S, Luo Z, Li L, Hu J, Zou G, Hou H and Ji X 2020 Recent progress on electrolyte additives for stable lithium metal anode *Energy Storage Materials* **32** 306–19

15 Zero-Point Vibration of the Adsorbed Hydrogen on the Pt(110) Surface

Tran Thi Thu Hanh and Nguyen Van Hoa

CONTENTS

15.1 INTRODUCTION

The adsorption on the platinum (Pt) surfaces has been paid special attention either under the ultra-high vacuum (UHV) [1, 2] or in contact with the solution [3–11] because of its applications. In the last decades, the adsorptions of hydrogen on different Pt surfaces are in focus by both experimental and theoretical studies [8–11].

Hydrogen adsorption occurs from acidic or aqueous solutions. It also can be accomplished from non-aqueous solutions that are able to dissolve hydrogen containing acids. As we know, the proton H^+ cannot exist by itself in aqueous acidic solution. However, it can easily combine with a nonbonding electron pair of a H_2O molecule to form H_3O^+ [12–16]. The H_3O^+ is further hydrated to forma $H^+.4H_2O$ ion. Then this ion encounters the region close to the electrode surface, where the H^+ discharge takes the place with formation of the adsorbed hydrogen atom [12, 16–19] (Volmer step):

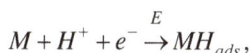

$$M + H^+ + e^- \xrightarrow{E} MH_{ads},$$

DOI: 10.1201/9781003322573-15

where M is a surface atom of the metal and E is electrode potential. Then, two adsorbed hydrogen atoms recombine to yield an H_2 molecule as the following (Tafel step):

$$2MH_{ads} \rightarrow 2M + H_2,$$

or (Heyrovsky step)

$$H^+ + e^- + MH_{ads} \xrightarrow{E} M + H_2.$$

In many interaction studies of hydrogen with transition metal surfaces, the important role of studying dynamical processes has been shown [20, 21, 22]. Dino et al. showed that the links between theory and experiment are more convincing when quantum effect calculations are taken into account [20]. The molecular dynamics simulation results show that the quantum effect is very sensitive to pore dimensions in the research of Kurma et al. [22].

Besides, in our recent study, we showed that the most stable absorption site of the H on the Pt(111) surface was the top site when the phonon effect is not taken into account. But when including the influence of zero-point energy, the most stable site is the fcc site [25]. This is in contrast to a few previous theoretical studies [26, 27] but is consistent with the experimental results studied [28]. This result indicates that the quantum effect is significant on the adsorption order of H on the Pt surface.

Kunimatsu et al. used surface-enhanced infrared spectroscopy to study the hydrogen adsorption on an acid-immersed Pt electrode at the potential at which H_2 is evolved [12, 23]. They found that the decreasing electrode potential shifts the Pt-H stretching mode at around 2090 cm^{-1} to a lower wavenumber at a rate of 90–180 cm^{-1}/V [12, 13]. Several possible reasons for the frequency shift were indicated: the vibrational Stark effect [12], the change of Pt-H bond strength with changing potential, and the H-coverage effects [13]. It could provide insight into the microscopic processes occurring at the electrode if we clearly understand the mechanism of this shift. Tomonari et al. carried out DFT calculations for the H/Pt(111) surface model under external fields to explain the experimentally reported large potential dependence of the Pt-H vibrational frequency at Pt electrodes. With increasing field strength, the Pt-H stretching frequency increases at a rate of 210 cm^{-1}/(V/Å) [24].

However, the frequencies of the H atom of all possible adsorption sites were not investigated in detail in other types of Pt surfaces (Pt(110) and Pt(100)), which are also very intriguing to explain the surface bonding mechanism for hydrogen adsorption. Furthermore, the H/Pt(110) system is known to be an excellent model system for Pt particles in the hydrogen fuel cells [29, 30].

A precise knowledge of the H bonding geometry, the vibrational frequency of all possible adsorption sites, and the energetics are therefore mandatory for a detailed understanding of both the properties of H/Pt(110) and the technological application of Pt in the fuel cells and catalysis. As a step toward understanding, the density functional theory (DFT) calculations that take into account the quantum effect were done for the hydrogen electro-adsorption on the Pt(110) surface model. Furthermore, the quasiparticle frequency calculation in previous studies was based on calculations for anharmonic oscillation of a hydrogen atom on the surface [37, 38], but in this study, we use the calculations for harmonic oscillations of hydrogen atoms on each axis. And the calculation results show agreement between theory and experiment for several calculated adsorption sites.

15.2   CALCULATION METHODS

15.2.1   DENSITY FUNCTIONAL THEORY (DFT) CALCULATION

According to our previous investigations [25, 31] we also used the SIESTA package [32, 33] with the linear combination of the atomic orbitals (LCAO) and pseudo-potential scheme to study. Figure 15.1 shows the H/Pt(110) model with four hydrogen adsorption sites. The basic set used in the DFT calculation had similarly been the generalized gradient approximation (GGA) to the exchange-correlation functional due to Perdew-Burke-Ernzerhof (PBE) [34]. We used the Brillouin zone surface of the k-point mesh, which was generated by the Monkhorst-Pack (MP) scheme [35]. The double-zeta polarized (DZP) basic set, the mesh-cutoff of 200Ry, were also used. We used the electronic temperature of 300 K and the energy shift of 200 meV for the Pt surface. These choosing parameters provided reasonable accuracy in the previous DFT calculation of the hydrogen adsorption on the Pt(111) surface [25].

Also, for simulating the Pt surface, the repeated slab was built. For reducing the interaction energy between the slabs to 1 meV, a vacuum equivalent to a 20-layer slab was used to separate the Pt slabs, where the interlayer spacing was taken as 1.387 Å. A bottom Pt-layer was fixed for relaxing all the H and Pt atoms to calculate the

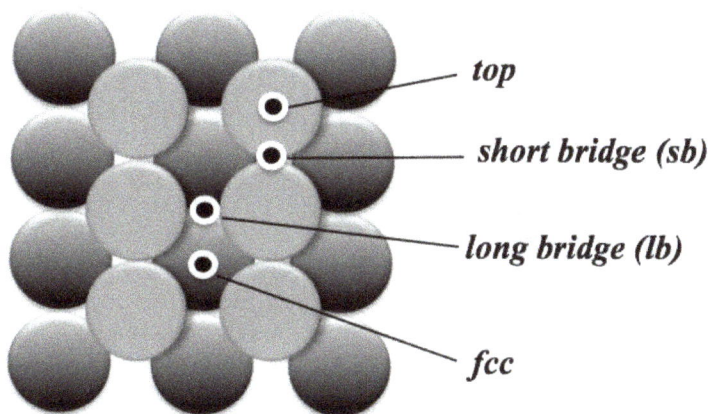

FIGURE 15.1   The Pt(110) simulation model, on which H atoms were adsorbed on the top, fcc, short bridge, and long bridge sites while the H coverage ranges from 0 to 1.

total energy. And we also took the optimized lattice constant of the unreconstruction Pt(110) of 3.9247 Å, which is close to the experimental bulk lattice constant (3.9242 Å) [36]. For the adsorbed H atoms, similar data were used: the energy shift of 60 meV and the split norm of 0.53 for the second zeta. This ensures obtaining correct bond length and energy of H_2 molecule.

For calculating the H adsorption on the Pt(110) surface, we firstly allowed one H atom adsorbing on the Pt(110) surfaces with the (1×1) and (2×2) lateral unit cells with the corresponding H-coverage of $\Theta_H \sim$ 1ML and ~ 1/4ML. Second, the surface of the lateral unit cell (1×1) was used to investigate the convergence property with respect to the number of the Pt layers and the k-point mesh. A 61-special-k-points ((11×11×1) MP grid, which is proved as the converged k-point in section 3.2) was used for studying the system with the (1×1) and (2×2) lateral unit cells. Then we calculated the zero-point energy (ZPE) for the H/Pt(110) system. our calculating ZPE is the quantum effect energy, which has not been theoretically calculated in detail before.

15.2.2 ZERO-POINT ENERGY (ZPE) CALCULATION

Although the oscillation of the H atom on the Pt metal surface is known as the anharmonic oscillation, in this chapter, we use the harmonic approximation of hydrogen for each axis to calculate ZPE. Because the Pt surface structure is evenly distributed, Pt atoms can be considered as symmetrically placed around the equilibrium position of the H atom. In addition, the distance between Pt and H atoms is quite large compared to the oscillation of H around the equilibrium position. Therefore, the oscillation of H around the equilibrium position on each axis may be considered as the harmonic approximation oscillation, and we performed the ZPE calculation.

First, the equilibrium position is selected as the adsorption position after optimizing the H/Pt model. Next, we displace the H atom around the equilibrium position along the x, y, and z-axes and calculate its energies. Approximation obtained energy equals the energy of a harmonic oscillation given by the formula:

$$E = \frac{1}{2}kx^2,$$

where x (Å) is the displacement from the equilibrium position of the hydrogen atom and the coefficient k ($eV/Å^2$) is the force constant. Using the method of minimum square calculation, we find the value of force constant k. Then we use k to continue calculating the vibrational frequencies in the interaction of hydrogen with the metal surface Pt(110). The expression of the frequency has the following form:

$$\tilde{v} = \frac{1}{2\pi c}\sqrt{\frac{k\left(m_1 + m_2\right)}{m_1 m_2}},$$

where c is the speed of light, m_1 is the mass of the hydrogen atom and m_2 is the mass of the platinum atom. Finally, ZPE is calculated through:

$$\varepsilon = \frac{h\nu}{2} = \frac{hc}{2\lambda} = \frac{hc\tilde{\nu}}{2},$$

where h is Planck's constant.

15.3 RESULTS AND DISCUSSION

15.3.1 COMPARISON OF HYDROGEN MOLECULE WITH PREVIOUS CALCULATIONS

To better understand the relevance of the selected DFT calculation parameters, the hydrogen model was computed and compared again with the previous experiment. We performed a series of total energies of the hydrogen molecule. The equilibrium bond length (l_{eq}) and the binding energy (E_b) of an isolated H_2 molecule were obtained by using a cube unit cell of length ~7.5 Å and by allowing the spin polarization [35]. The bond length of the hydrogen molecule is 0.754 Å, which is in good agreement with experimental data ~0.74 Å [39] (see Figure 15.2). We calculate the ZPE of the

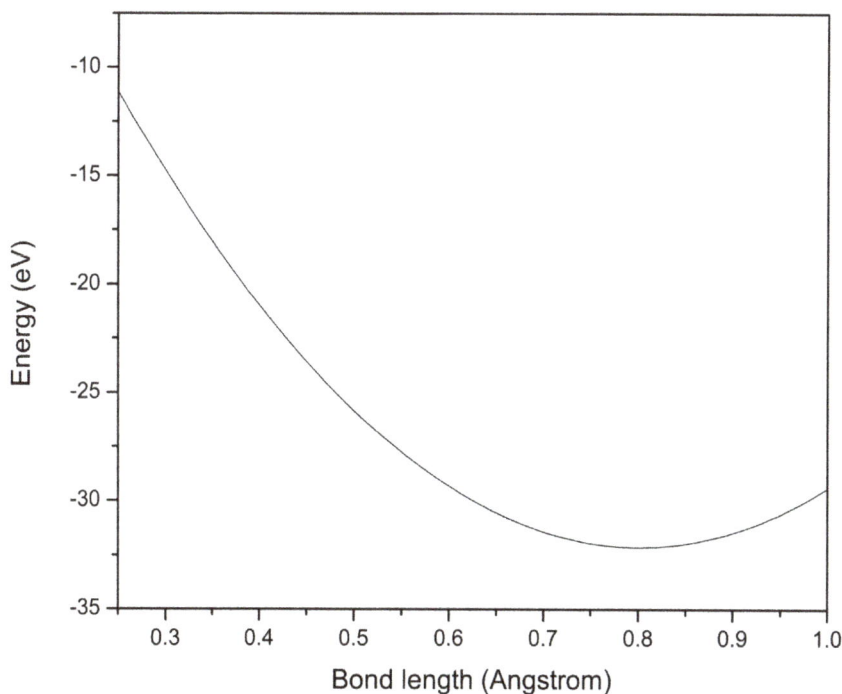

FIGURE 15.2 The relative of the energy (eV) and the bond length (Å) of the hydrogen molecule.

TABLE 15.1
The Energies of Hydrogen (eV)

	Simulation data	Experimental data [40]
Rotational free energy of H_2	−0.035	
Translational free energy of H_2	−0.300	
Total energy of H_2	−31.621	
Total energy of H	−13.548	−13.600
Binding energy	4.525	4.530
Zero-point energy of H_2	0.269	0.270

hydrogen molecule by displacing one H atom around its equilibrium position along the H-H bonding axis. The binding and the zero-point energies also agree well with experimental data (Table 15.1) [40]. These results once again show the accuracy of our basic set data used in SIESTA calculations for studying the adsorption of the H on the Pt(110) surface.

15.3.2 CONVERGENCE PROPERTY

In this session, firstly we tested the stable sites of hydrogen atoms, which were adsorbed on the Pt(110) surface by using the similar calculation adsorption energy from the ref. [25]:

$$E_{ads} = E_{tot}\left(N_H\right) - E_{tot}\left(0\right) - \frac{N_H}{2}E_{H_2},$$

where $E_{tot}\left(N_H\right)$ is the total energy of the Pt surface adsorbed with (N_H) H atoms and E_{H2} is the total energy of the isolated H_2 molecule. E_{ads} shows that the short bridge site (sb) is the strongest adsorption site, then the top site, the long bridge site (lb), and finally the fcc site when the H-coverage $\Theta_H \leq 1ML$ (see Table 15.2). The result of the most stable sb site is consistent with the previous experimental and theoretical research results [41, 42]. Table 15.2 also shows a large difference in stable energies between the sb and the top sites: 344 meV for $\Theta_H \sim 1ML$ and 64 meV for $\Theta_H \sim 1/4ML$ when using 10 Pt layers.

Then we checked the convergence energy with respect to the computational parameters: the Pt layers and the k-points. The calculations were done using (1×1) lateral unit cell with different Pt layers and k-points, on which one H atom was let adsorb on the most stable sites: on the top, and the short bridge (sb) sites. Table 15.3 shows the adsorption energies of the hydrogen on the Pt(110) surface with the MP grids changed from (3×3×1) to (15×15×1). Figure 15.3 plots the dependence of the ΔE_{ads} (the adsorption energy of the sb relative to that on the top) on the k-points. It shows that from the k-point of (11×11×1) MP grid, the value of ΔE_{ads} becomes convergence. And the dependence of the energy ΔE_{ads} on the number of the Pt layers was shown in Figure 15.4. We found that the results for various numbers of the k-points

TABLE 15.2

The Adsorption Energy (eV) of the H/Pt(110) System When Using (3×3×1) MP Grid

Cell	Pt layers	top	lb	sb	fcc
1ML					
(1×1)	4	−0.619	−0.429	−0.911	−0.369
	5	−0.723	−0.390	−1.030	−0.237
	6	−0.773	−0.490	−1.057	−0.389
	7	−0.529	−0.262	−0.874	−0.187
	10	−0.675	−0.268	−1.019	−0.186
1/4ML					
(2x2)	4	−0.657	−0.421	−0.786	−0.197
	5	−0.725	−0.525	−0.797	−0.115
	6	−0.803	−0.559	−0.855	−0.272
	7	−0.715	−0.490	−0.777	−0.198
	10	−0.712	−0.444	−0.776	−0.143

TABLE 15.3

The Adsorption Energy (eV) of the H/Pt(110) Model with Different Pt Layers and k-Points

Pt layers	(3×3×1) MP		(5×5×1) MP		(7×7×1) MP		(11×11×1) MP		(13×13×1) MP		(14×14×1) MP		(15×15×1) MP	
	top	sb	top	sb	top	sb	top	sb	top	sb	top	sb	top	sb
3	−0.99	−1.27	−0.96	−0.97	−0.92	−0.87	−0.94	−0.92	−0.93	−0.91	−0.94	−0.92	−0.94	−0.91
4	−0.62	−0.91	−0.68	−0.78	−0.66	−0.76	−0.69	−0.82	−0.68	−0.81	−0.68	−0.81	−0.68	−0.80
5	−0.72	−1.03	−0.71	−0.81	−0.69	−0.77	−0.69	−0.80	−0.68	−0.78	−0.69	−0.79	−0.69	−0.78
6	−0.77	−1.06	−0.74	−0.85	−0.73	−0.79	−0.74	−0.83	−0.74	−0.83	−0.75	−0.83	−0.75	−0.83
7	−0.53	−0.87	−0.61	−0.79	−0.63	−0.77	−0.65	−0.80	−0.64	−0.78	−0.64	−0.78	−0.64	−0.78
8	−0.70	−1.02	−0.72	−0.84	−0.71	−0.78	−0.72	−0.82	−0.72	−0.82	−0.72	−0.81	−0.72	−0.81
9	−0.58	−0.89	−0.64	−0.82	−0.67	−0.79	−0.67	−0.81	−0.66	−0.79	−0.67	−0.80	−0.67	−0.80
10	−0.68	−1.02	−0.67	−0.81	−0.67	−0.79	−0.68	−0.80	−0.68	−0.80	−0.67	−0.80	−0.67	−0.79
11	−0.66	−0.93	−0.67	−0.82	−0.71	−0.83	−0.70	−0.83	−0.70	−0.82	−0.71	−0.83	−0.69	−0.81
12	−0.67	−1.00	−0.65	−0.80	−0.66	−0.78	−0.67	−0.81	−0.66	−0.79	−0.66	−0.80	−0.66	−0.78

(from (11×11×1) MP grid to (15×15×1) MP grid) become very close to each other when using 10 Pt layers. From these results, we concluded that the converged value of ΔE_{ads} is reached when using (11×11×1) MP grid and 10 Pt layers.

15.3.3 ZERO-POINT VIBRATION

After carefully checking the converged data, we used 61 k-point numbers ((11×11×1) MP grid) for all our continuous calculating models. For investigating the vibrational

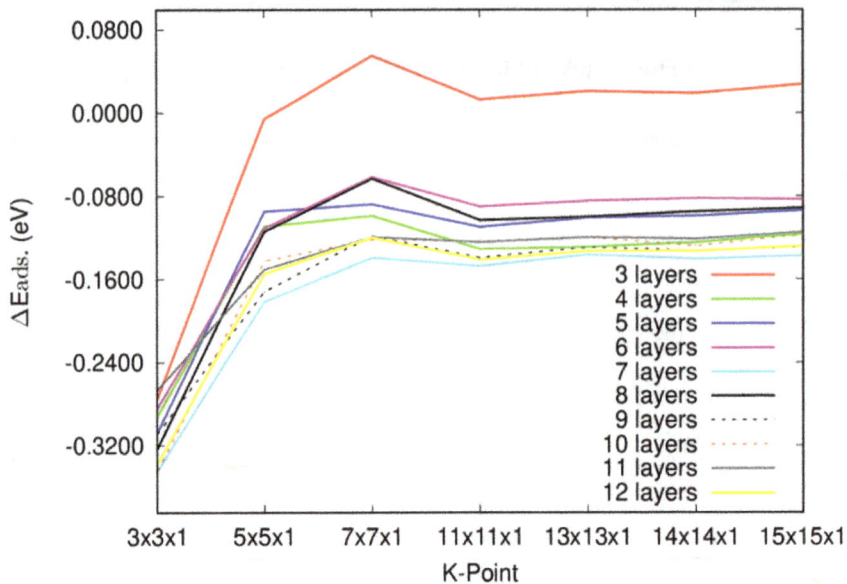

FIGURE 15.3 k-point dependence of the adsorption energy on the short bridge relative to that on the top (ΔE_{ads}).

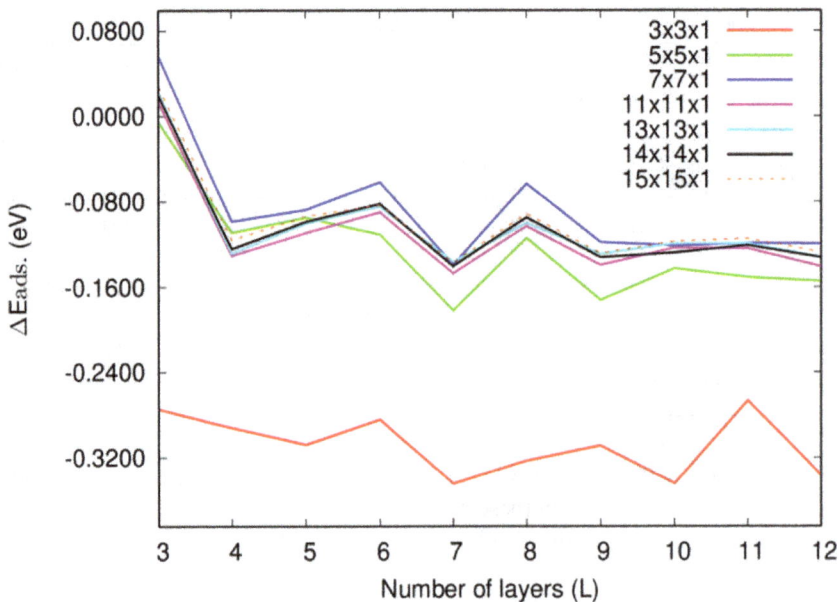

FIGURE 15.4 The dependence of the energy value ΔE_{ads} on the number of the Pt layers.

frequency and zero-point energy, the (1×1) and (2×2) lateral unit cells were used, on which one H atom was let adsorb on the top, lb, sb, and fcc sites. The optimized Pt-H bond lengths for the H adsorption on the Pt(110) surface are calculated in Table 15.4. It's shown that the results were affected by less than 1% when we change the Pt layers from 4 layers to 10 layers. From the calculation, we also found that the H atoms are kept almost at the ideal high symmetry position as in previous studies for the H/Pt systems [25, 31, 43].

The vibrational frequencies of the H on the Pt(110) surface are listed in Table 15.5. The frequency value is highest when the H atom vibrates on the top site, then on the fcc, the sb, and finally on the lb site. The frequency of a hydrogen atom on the

TABLE 15.4
The Optimized Pt-H Bond Length (Å) of the H/Pt(110) Model

Cell	Pt layers	fcc	lb	sb	top
1ML (1×1)	4	1.67	1.98	1.77	1.57
	5	1.67	1.98	1.77	1.58
	6	1.67	1.98	1.77	1.58
	7	1.66	1.98	1.77	1.58
	10	1.67	1.98	1.77	1.58
1/4ML (2×2)	4	1.65	1.88	1.76	1.58
	5	1.64	1.87	1.77	1.59
	6	1.65	1.86	1.76	1.58
	7	1.64	1.86	1.76	1.58
	10	1.65	1.87	1.76	1.58

TABLE 15.5
The Pt-H Vibrational Frequency (cm⁻¹)

Cell	Pt layers	fcc	lb	sb	top
1ML (1×1)	4	1701	676	1421	2272
	5	1661	651	1423	2237
	6	1635	673	1426	2263
	7	1707	675	1421	2250
	10	1666	667	1413	2247
1/4ML (2×2)	4	1751	885	1381	2251
	5	1783	894	1366	2214
	6	1755	918	1373	2242
	7	1811	886	1371	2240
	10	1792	886	1373	2236

top site for the Pt(110) surface is around 2200 cm⁻¹, which shows good agreement with the values of the experimental calculation using SEIRAS for the H/Pt system (2080–2095 cm⁻¹) [11] and the DFT calculation for the Pt(111) and the missing row Pt(110)-(1×2) surfaces (~ 2100 cm⁻¹) [25, 31, 43]. And the obtained frequency of the hydrogen atom on the bridge site is nearly 1380 cm⁻¹, which agrees well with the previous theoretical calculation (~1320 cm⁻¹) [42]. The ZPE calculation of the H on the Pt(110) surface when $\Theta_H \leq 1ML$ is shown in Table 15.6. It shows that the quantum effect energy is highest (~ 140 meV) when the H atom is adsorbed on the top site. And for the most stable short bridge site, the ZPE is ~ 88 meV. These values are in good agreement with the experimental study using HREELS (149 meV for top, 83 meV for bridge) [44].

Using the ZPE of the H/Pt system, we calculated more accurately about the adsorption energy:

$$E_{ads} = E_{tot}\left(N_H\right) - E_{tot}\left(0\right) - \frac{N_H}{2}E_{H_2} + \sum_{\alpha \in \{top, lb, sb, fcc\}} N_H^\alpha E_{ZPE}\left(Pt - H^\alpha\right).$$

Table 15.7 shows the adsorption energies when the zero-point energies were added. Although the ZPE of the H atom on the sb site is smaller than that on the top site, the result of the total adsorption energy is still reaffirmed once again as the most stable site of the short bridge site for the adsorption of the hydrogen on the Pt(110) surface. However, the difference in the energy between the sb site and the top site is now about 71 meV for $\Theta_H \sim 1ML$ and of 29 meV for $\Theta_H \sim 1/4ML$. This proves that the influence of quantum effect on the adsorption energy of H on Pt(110) is significant. And this also proves that the population of the H atom on the sb and on the top sites might not be significantly different when the H coverage is small ($\Theta_H \sim 1/4ML$), but the H atoms on the sb site will be the majority when the hydrogen coverage is large ($\Theta_H \sim 1ML$).

TABLE 15.6
The Zero-Point Energy (eV) of the H/Pt(110) System

Cell	Pt layers	fcc	lb	sb	top
1ML (1×1)	4	0.105	0.042	0.088	0.141
	5	0.103	0.040	0.088	0.139
	6	0.101	0.042	0.088	0.140
	7	0.106	0.042	0.088	0.140
	10	0.103	0.041	0.088	0.139
1/4ML (2×2)	4	0.109	0.055	0.086	0.140
	5	0.111	0.055	0.085	0.137
	6	0.109	0.057	0.085	0.139
	7	0.112	0.055	0.085	0.139
	10	0.111	0.055	0.085	0.139

TABLE 15.7
The Adsorption Energy (eV) of the H on the Electrode Pt(110) Surface at the Coverage $\Theta_H \leq 1ML$

Cell	Pt layers	fcc	lb	sb	top
1ML (1×1)	4	−0.361	−0.354	−0.905	−0.827
	5	−0.237	−0.298	−0.886	−0.827
	6	−0.295	−0.358	−0.921	−0.882
	7	−0.283	−0.334	−0.887	−0.791
	10	−0.262	−0.329	−0.889	−0.818
1/4ML (2×2)	4	−0.363	−0.524	−0.901	−0.865
	5	−0.263	−0.539	−0.897	−0.895
	6	−0.338	−0.617	−0.954	−0.948
	7	−0.317	−0.554	−0.880	−0.856
	10	−0.306	−0.561	−0.906	−0.877

15.4 CONCLUSION

Density functional theory calculations were carried out on the hydrogen adsorption on the Pt(110) electrode in ultra-high vacuum by taking into account the vibrational frequencies, the quantum zero-point effects, and the adsorption energies. We found that the most stable adsorption site is the short bridge site, then the top, the long bridge, and the fcc sites. The highest stretching frequency ~ 2200 cm^{-1} and the highest zero-point energy ~ 140 meV were shown when the adsorption hydrogen atom is on the top site. The results presented in this work show that the quantum effect has a significant effect on the adsorption energy, indicating the simultaneous existence of adsorbed hydrogen at the top and short bridge sites.

OPEN ISSUES

The nature of hydrogen interaction on the surface can be calculated in the follow-up study by using the adsorption energy including the ZPE for this H/Pt system.

ACKNOWLEDGMENTS

This research is funded by the Vietnam National Foundation for Science and Technology Development (NAFOSTED) under grant number 103.01–2017.50.
    This research is published in the journal *Adsorption*, 2020, 25, 1–7.

REFERENCES

[1] Christmann K, Ertl G and Pignet T 1976 Adsorption of hydrogen on a Pt (111) surface *J Surf Sci* **54** 365
[2] Nordlander P, Holloway S and Nørskov J K 1984 Hydrogen adsorption on metal surfaces *J Surf Sci* **136** 59

[3]   Clavilier J, Rodes A, Achi K E and Zamakhchari M A 1991 Electrochemistry at plati-
      num single crystal surfaces in acidic media: Hydrogen and oxygen adsorption *J Chim
      Phys* **88** 1291
[4]   Marković N M, Grgur B N and Ross P N 1997 Temperature-dependent hydrogen elec-
      trochemistry on platinum low-index single-crystal surfaces in acid solutions *J Phys
      Chem B* **101** 5405
[5]   Jerkiewicz G 1998 Hydrogen sorption ATIN electrodes *Prog Surf Sci* **57** 137
[6]   Zolfaghari A and Jerkiewicz G 1999 Temperature-dependent research on Pt (111) and
      Pt (100) electrodes in aqueous H2SO4 *J Electroanal Chem* **467** 177
[7]   Marković N M, Schmidt T J, Grgur B N, Gasteiger H A, Behm R J and Ross P N 1999
      Effect of temperature on surface processes at the Pt (111)– liquid interface: Hydrogen
      adsorption, oxide formation, and CO oxidation *J Phys Chem B* **103** 8568
[8]   Conway B E and Jerkiewicz G 2000 Relation of energies and coverages of underpo-
      tential and overpotential deposited H at Pt and other metals to the "volcano curve" for
      cathodic H2 evolution kinetics *J Electrochim Acta* **45** 4075
[9]   Marković N M and Ross P N 2002 Surface science studies of model fuel cell electro-
      catalysts *J Surf Sci Rep* **45** 117
[10] Kita H 2003 Horiuti's generalized rate expression and hydrogen electrode reaction *J
      Mol Catal A Chem* **199** 161
[11] Kunimatsu K, Senzaki T, Samjeske G, Tsushima M and Osawa M 2007 Hydrogen
      adsorption and hydrogen evolution reaction on a polycrystalline Pt electrode studied by
      surface-enhanced infrared absorption spectroscopy *J Electrochim Acta* **52** 5715
[12] Kunimatsu K, Senzaki T, Tsushima M and Osawa M 2005 A combined surface-
      enhanced infrared and electrochemical kinetics study of hydrogen adsorption and evo-
      lution on a Pt electrode *J Chem Phys Lett* **401** 451
[13] Conway B E 1981 *Ionic Hydration in Chemistry and Biophysics* (New York: Elsevier)
[14] Conway B E and Jerkiewicz G 1993 Thermodynamic and electrode kinetic factors in
      cathodic hydrogen sorption into metals and its relationship to hydrogen adsorption and
      poisoning *J Electroanal Chem* **357** 47
[15] Conway B E and Jerkiewicz G 1994 Factors in the electrolytic sorption of H into metals
      and its relation to cathodic hydrogen evolution kinetics *Zeit Phys Chem Bd* **183** 281
[16] Jerkiewicz G and Zolfaghari A 1996 Comparison of hydrogen electroadsorption from
      the electrolyte with hydrogen adsorption from the gas phase *J Electrochem Soc* **143**
      1240
[17] Breiter M W and Kennel B 1960 Über den Einfluß der Anionen und der Zeit nach
      der Aktivierung auf die Adsorptionswarme von Wasserstoff an Platinelektroden *Z
      Elektrochem* **64** 1180
[18] Will F G and Knorr C A 1960 Untersuchung von adsorptionserscheinungen an rhodium,
      iridium, palladium und gold mit der potentiostatischen dreieckmethode *Z Electrochem*
      64 258
[19] Frumkin N 1963 *Advances of Electrochemistry and Electrochemical Engineering* P
      Delahey (Ed) Vol 3 (New York: Interscience Publishers)
[20] Dino W A, Kasaia H and Okiji A 2000 Orientational effects in dissociative adsorp-
      tion/associative desorption dynamics of H2 (D2) on Cu and Pd *J Progress in Surface
      Science* **63** 63–134
[21] Kallen G and Wahnstrom G 2001 Quantum treatment of H adsorbed on a Pt (111) sur-
      face *J Phys Rev B* **65** 033406
[22] Kumar A V, Jobic H and Bhatia S K 2006 Quantum effects on adsorption and diffusion
      of hydrogen and deuterium in microporous materials *J Phys Chem B* **110** 16666

[23] Kunimatsu K, Uchida H, Osawa M and Watanabe M 2006 In situ infrared spectroscopic and electrochemical study of hydrogen electro-oxidation on Pt electrode in sulfuric acid *J Electroanal Chem* **587** 299

[24] Tomonari M and Sugino O 2007 DFT calculation of vibrational frequency of hydrogen atoms on Pt electrodes: Analysis of the electric field dependence of the Pt—H stretching frequency *J Chem Phys Lett* **437** 170

[25] Hanh T T T, Takimoto Y and Sugino O 2014 First-principles thermodynamic description of hydrogen electroadsorption on the Pt (111) surface *J Surf Sci* **625** 104

[26] Olsen R A, Kores G J and Baerends E J 1999 Atomic and molecular hydrogen interacting with Pt (111) *J Chem Phys* **111** 11155

[27] Ford D C, Xu Y and Mavrikakis M 2005 Atomic and molecular adsorption on Pt (111) *J Surf Sci* **587** 159

[28] Lasia A 2004 Modeling of hydrogen upd isotherms *J Electroanal Chem* **562** 23

[29] Lu C and Masel R I 2001 The effect of ruthenium on the binding of CO, H2, and H2O on Pt(110) *J Phys Chem B* **105** 9793

[30] Chrzanowski W and Wieckowski A 1998 Surface structure effects in platinum/ruthenium methanol oxidation electrocatalysis *Langmuir* **14** 1967

[31] Hanh T T T and Hang N T T 2017 A DFT study of hydrogen electroadsorption on the missing row Pt (1 1 0)-(1× 2) surface *J Comp Mat Sci* **13** 295

[32] Ordejón P, Artacho E and Soler J M 1996 Self-consistent order-N density-functional calculations for very large systems *J Phys Rev B* **53** R10441

[33] Soler J M, Artacho E, Gale J D, García A, Junquera J, Ordejón P and Sánchez-Portal D 2002 The SIESTA method for ab initio order-N materials simulation *J Phys Condens Matter* **14** 2745

[34] Perdew J P, Burke K and Ernzerhof M 1996 Generalized gradient approximation made simple *J Phys Rev Lett* **77** 3865 (ibid. 78: 1396, 1997)

[35] Monkhorst H J and Pack D 1976 Special points for Brillouin-zone integrations *Phys Rev B* **13** 5188

[36] Waseda Y, Hirata K and Ohtani M 1975 High-temperature thermal expansion of platinum, tantalum, molybdenum, and tungsten measured by X-ray diffraction *J High Temp High Pressures* **7** 221

[37] Ihm J, Zunger A and Cohen M L 1979 Momentum-space formalism for the total energy of solids *J Phys C* **12** 4409

[38] Frank W 1995 Ab initio force-constant method for phonon dispersions in alkali metals *J Phys Rev Lett* **74** 1791

[39] Roger L D and Harry G 1989 *Chemical Structure and Bonding* (California Institute of Technology, University Science Books) p 199

[40] Landau L D and Lifshitz E M 1958 *Quantum Mechanics* (Oxford: Elsevier Butterworth-Heinemann) p 319

[41] Furuya N and Koide S 1989 Hydrogen adsorption on iridium single-crystal surfaces *Surf Sci* **220** 18

[42] Yu Y, Yang J, Hao C and Zhao X 2009 The adsorption, vibration and diffusion of hydrogen atoms on platinum low-index surfaces *J Comput Theor Nanosci* **6** 439

[43] Hamada I and Morikawa Y 2008 Density-functional analysis of hydrogen on Pt (111): Electric field, solvent, and coverage effects *J Phys Chem C* **112** 10889

[44] Stenzel W, Jahnke S A, Song Y and Conrad H 1990 HREELS investigations of adsorbed hydrogen on transition metals *Prog Surf Sci* **35** 159

16 Magnetotransport Properties of Bismuth Chalcogenide Topological Insulators

Le Thi Cam Tuyen, Phuoc Huu Le, and Ming-Fa Lin

CONTENTS

16.1 INTRODUCTION TO BISMUTH CHALCOGENIDE TOPOLOGICAL INSULATORS AND THEIR FASCINATING PROPERTIES

In recent years, a novel electronic state called the topological surface state (TSS) has been predicted and observed in topological insulators (TIs) [1–8]. Unlike the trivial insulator, TIs have a spin-degenerate and fully gapped bulk state but exhibit a spin-polarized and gapless electronic state on the surface [8]. This metallic surface state has a linear energy-momentum dispersion relation in the low-energy region, which is known as a Dirac cone. Unlike the Dirac cone of graphene, the Dirac cone of a TI is protected by the time-reversal symmetry. The robust TSSs that are manifest as chiral Dirac fermion quasiparticles can survive under time-reversal-invariant perturbations, such as surface pollution, crystalline defects, and distortions of the surface [6]. Additionally, because of the fully spin-polarized characteristics of the surface state, TIs have a high potential for the development of spintronic devices and quantum computation [6, 9].

Bismuth chalcogenide compounds (Bi_2Ch_3, Ch = Se, Te) have been extensively investigated in material science and condensed-matter physics because of their

DOI: 10.1201/9781003322573-16

317

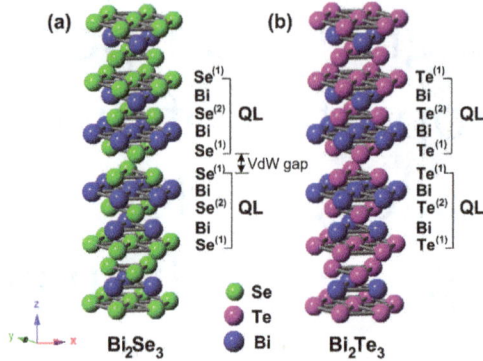

FIGURE 16.1   The crystal structures of Bi_2Se_3 and Bi_2Te_3.

intriguing properties regarding thermoelectricity and three-dimensional TIs [10]. Bi_2Ch_3 is a narrow-gap semiconductor with a rhombohedral crystal structure belonging to the $D_{3d}^5 (R\overline{3}m)$ space group. The Bi_2Ch_3 crystal structure is constructed from repeated quintuple layers (QL) arranged along the c axis. The unit lattice cell of a Bi_2Ch_3 crystal is composed of three QLs. Each QL is stacked in a sequence of atomic layers Ch(1)-Bi-Ch(2)-Bi-Ch(1) and is weakly bonded to the next QL via Van der Waals (VdW) interaction. Here, we take Bi_2Se_3 as an example. The crystal structure of Bi_2Se_3 is shown in Figure 16.1(a). For convenience, the crystal structure is also described by a hexagonal lattice, where the a-axis (the xy plane) lattice constant is 4.138 Å and the c-axis (along the z axis) lattice constant is 28.64 Å [11]. For Bi_2Te_3, the a axis (xy plane) lattice constant is 4.384 Å and the c axis (along the z axis) lattice constant is 30.487 Å Figure 16.1(b) [11].

Binary Bi_2Se_3 crystals was confirmed as strong TIs, and the electronic structure of Bi_2Se_3 crystal with spin-orbit coupling was also calculated [10]. By tuning the spin-orbit coupling in the system, band inversion occurred around the Γ point. As these two levels, which are closest to the Fermi energy, have opposite parity, the inversion between them drives the system into a TI phase [10]. In transport experiments, the surface signal is usually washed out by the bulk contribution. Thus, the isolation of the surface states from the bulk electronic states is important for a functional electronic structure of a TI. Ternary bismuth chalcogenide compounds (e.g., Bi_2Te_2Se) have been found to possess high bulk resistivity; thus, they are of great interests for studying TSSs by transport method [12]. The surface transport results give band structure properties and parameters such as Fermi momentum, Fermi velocity, etc. Neupane et al. studied angle-resolved photoemission spectroscopy (ARPES) electronic structures on three batches of as-grown Bi_2Te_2Se ("native 1," "native 2," and "n type") with slightly different growth parameters (Figure 16.2) [13]. A single Dirac cone on the cleaved (111) surface was revealed via ARPES. The three Bi_2Te_2Se batches had different chemical potentials, as shown in Figure 16.2(a). The Bi_2Te_2Se "native 1" had the Fermi level at 0.3 eV above the Dirac node (E_D) and an average Fermi momentum (k_F) of 0.05 Å^{-1} [13]. Meanwhile, the "native 2" batch had Fermi level at 0.3 eV above E_D and a larger $k_F = 0.1$ Å^{-1}. In addition, the Fermi level of

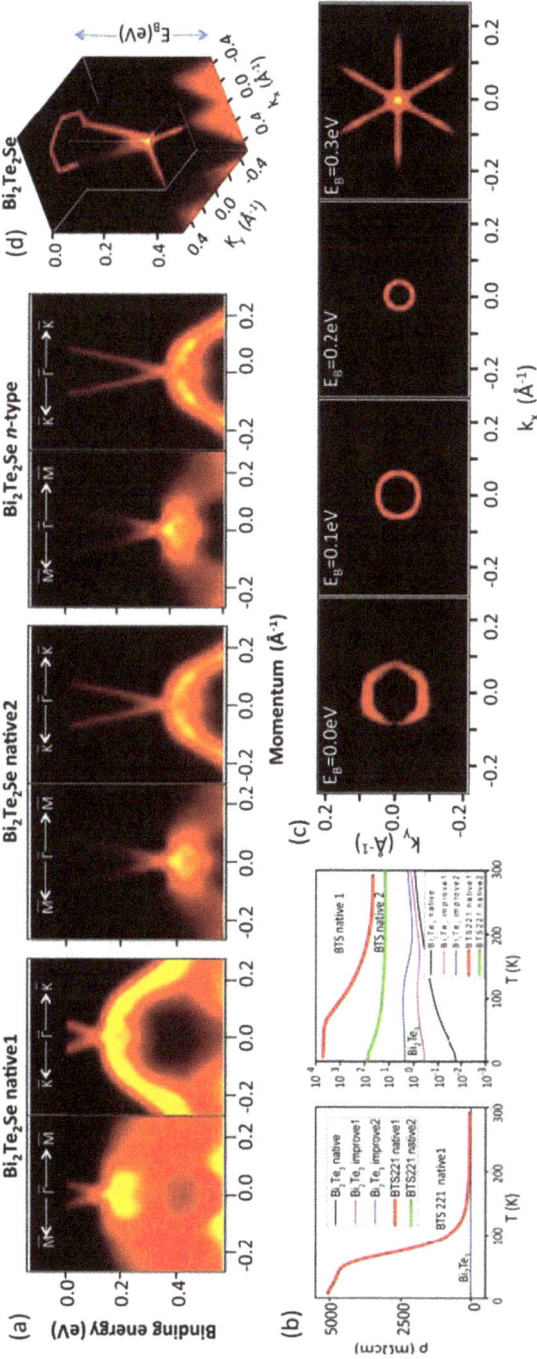

FIGURE 16.2 (a) ARPES k-E maps along the high symmetry directions $\bar{\Gamma} - \bar{M}$ and $\bar{\Gamma} - \bar{K}$ of Bi$_2$Te$_2$Se for three different samples: Bi$_2$Te$_2$Se "native 1," Bi$_2$Te$_2$Se "native 2," and Bi$_2$Te$_2$Se "n type," respectively. A single Dirac cone on the cleaved (111) surface is observed. (b) Resistivity profiles of native Bi$_2$Te$_3$, compared to native and improved Bi$_2$Te$_3$, are presented in linear (left) and logarithmic scale (right). (c) ARPES maps of constant energy contours presented at several binding energies for the "native 2" sample. (d) Three-dimensional illustration of the electronic structure in (c) [13].

Bi_2Se_2Te "n type" batch was near bulk conduction band minimum, and thus its bulk band gap was approximately 0.3 eV [13]. Figure 16.2(b) shows temperature- dependent resistivities [$\rho_{xx}(T)$] of Bi_2Se_2Te "native 1, 2" and Bi_2Se_3 samples. Clearly, ρ_{xx} increase substantially upon cooling from room temperature, which is an indication of an insulating behavior. This behavior corresponds to the Bi_2Se_2Te batch with Fermi level crossing the surface state and within the bulk gap.

Figure 16.2(c) presents two-dimensional constant energy contour plots of the ARPES intensity at various binding energies (E_B) [13]. Obviously, the "native 2" batch exhibited a hexagonal shape within the bulk band gap (the first panel), and the contours gradually revert from hexagon to a circle as approaching the Dirac node. The valence band feature with a sixfold petal-like intensity pattern was observed at E_B ~0.3 eV in the vicinity of Dirac point (Figure 16.2(c), right). Moreover, it is clear from three-dimensional representation of the electronic structure of the "native 2" batch (Figure 16.2(d)) that the Dirac point of Bi_2Te_2Se is not exposed, since it is buried in the bulk valence bands. High Fermi velocity (v_F) of Bi_2Te_2Se was estimated to be 6×10^5 m/s the $\overline{\Gamma} - \overline{M}$ direction, and 8×10^5 m/s the $\overline{\Gamma} - \overline{K}$ direction, making it favorable for a long mean-free path ($L = v_F\tau$) on the surface.

He et al. report the results from ARPES on Bi_2Se_3 films of various thicknesses grown by molecular beam epitaxy [14]. The energy gap opening is observed when the thickness is below six quintuple layers (QLs). This can preliminarily be attributed to the effect that the surface-state wavefunctions of the top and bottom surfaces of the film become overlapping when the film thickness is small enough (several nanometers). In this case, the spin-polarized surface states at one surface will be mixed up with the components of opposite spin from the other surface, leading to a hybridization gap at the Dirac point to avoid crossing of bands with the same quantum numbers [14]. It has been found that the measured gap size is a function of thickness, in which the gap size enlarges when the film thickness is decreased from 5 to 2 QL. Meanwhile, one Dirac cone of the TSS ($\Delta = 0$ eV) is observed for the Bi_2Se_3 films with thicknesses d ≥ 6 QL.

Since the TSS is protected by the time-reversal symmetry [1–8], the TSS can coexist with the nonmagnetic impurities doped into the TI matrix. Thus, the coexistence of the TSS and other broken-symmetry states becomes possible. This possibility offers the chance to study Majorana quasiparticles, which is of great interest in fundamental physics and quantum computation [15]. The first superconducting TI was discovered by the research groups of Hasan [15] and Cava [16]. In the Cu intercalated TI Bi_2Se_3, the superconducting transition temperature T_c is approximately 3.8 K for $Cu_{0.12}Bi_2Se_3$ crystal (Figure 16.3(a)) [15, 16].

The Cu atoms in $Cu_xBi_2Se_3$ act as an ambipolar dopant. That is, a Cu atom may substitute for a Bi atom or intercalate in the Van der Waals gap. However, the superconducting transition has only been observed in Cu-intercalated $Cu_xBi_2Se_3$. Figure 16.3(b) shows the crystal structure of Cu-intercalated $Cu_xBi_2Se_3$. The doped Cu forming the Cu layer is shown in Figure 16.3(b). Another superconducting TI is Bi_2Te_3 under high pressure. The superconducting phase transition occurs above 3 K when the pressure approaches 3.2 GPa (Figure 16.4(a)) [17]. The phase transition temperature T_c reaches nearly 9.5 K at 13.6 GPa [18]. The dR_H/dP in the lower panel of Figure 16.4(b) exhibits a dramatic change of up to 1.5 GPa, reflecting the pronounced

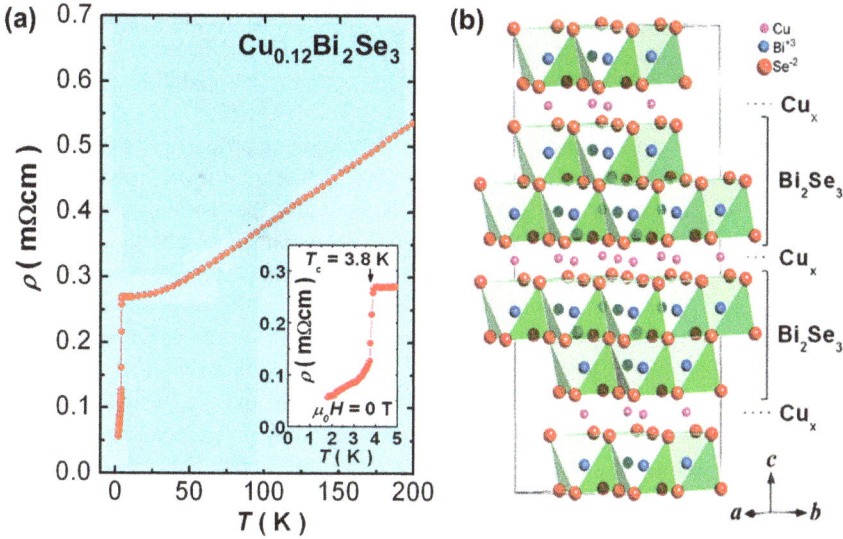

FIGURE 16.3   (a) Temperature-dependent resistivity of a $Cu_{0.12}Bi_2Se_3$ crystal with applied current in the ab. The inset shows that the superconducting transition occurs at ~3.8 K [16]. (b) The crystal structure of $Cu_xBi_2Se_3$ [16].

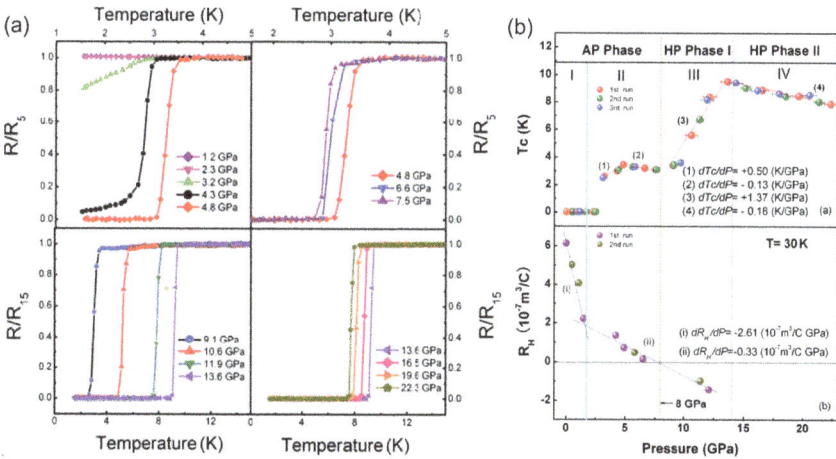

FIGURE 16.4   (a) The temperature-dependent resistance of Bi_2Te_3 under different pressures. (b) The superconducting transition temperature T_c and the Hall coefficient R_H under different pressures in Bi_2Te_3 [18].

change in band structure [18]. In pressure region II, the superconducting transition also begins to occur. A calculation of the electronic structure of Bi_2Te_3 under 4.0 GPa also displays a distinct change [17]. Analogous to this idea of Cu-intercalated $Cu_xBi_2Se_3$, the modulation of the electronic structure is also observed [15]. Moreover, the superconducting transition temperature T_c decreased smoothly with the increase in pressure for $Cu_xBi_2Se_3$ [19]. This behavior is similar to the superconductivity of Bi_2Te_3 under pressures ranging from 13.6 GPa to 22.7 GPa [18]. From these experimental results, we believe that the structural strain may be related to the origin of superconductivity in TIs.

Because of the salient electronic, optical, and magnetic properties of the TSS, TIs offer innovative opportunities to illuminate many potentially revolutionary applications. However, the native Se vacancies in Bi_2Se_3 cause the rising of a Fermi level. The increasing bulk carrier density will suppress the contribution of the TSS in carrier transportation. To solve this problem, Chen et al. [20], Checkelsky et al. [21], and Steinberg et al. [22] used the gate voltage to manipulate the position of the Fermi level in Bi_2Se_3. By controlling the gate voltage, the Fermi level near the surface can be renormalized. The conduction behavior of the device is dominated by the TSS below 100 K [21]. However, the temperature limitations and the complicated fabrication process limit the applications of a TI controlled by the gate voltage method.

To overcome this limitation, some researchers have applied chemical doping to compensate for the bulk carrier density of a bismuth chalcogenide TI. Based on defect chemistry, the carrier density of the TI can be reduced by doping hole carriers in the TI matrix. For example, Ca atoms can be used to substitute for the Bi atoms in Bi_2Se_3 [23, 24]. However, this approach can generate the enormous charge impurities and cause a reduction of the carrier mobility. Thus, the reduction of intrinsic defects such as Se vacancies or anti-site defects becomes an important approach for compensating for the high bulk carrier density. In 2010, Analytis et al. successfully reduced the defect density of Se vacancies in Bi_2Se_3 by substituting Bi atoms with Sb atoms [25]. At the same time, it was found that one of the ternary tetradymite compounds, Bi_2Te_2Se, exhibited a suppression of the bulk conductivity because the well-confined Se atoms in the central layer are expected to suppress the Se vacancy as well as the anti-site defects between Bi and Te atoms [12, 26]. Indeed, the basic quintuple-layer unit of Bi_2Te_2Se is Te-Bi-Se-Bi-Te, as shown in Figure 16.5(a) [12]. Moreover, Figure 16.5(b) presents the XRD patterns of Bi_2Te_2Se and Bi_2Te_3 crystals in powder forms, which confirms the ordering of the chalcogen layers of Bi_2Te_2Se single crystals in the study [12]. The high-quality Bi_2Te_2Se single crystals exhibited a high resistivity exceeding 1 Ω cm. Furthermore, the temperature dependence of the resistivity ρ_{xx} and the Hall coefficient R_H for three Bi_2Te_2Se samples are shown in Figure 16.5(c). Clearly, the ρ_{xx} increases roughly two orders of magnitude upon cooling from room temperature, which is an indication of an insulating behavior [12]. At temperature below 20 K, the resistivity exhibits saturation that implies a finite metallic conductivity at T = 0 K and suggests for the origin of the topological surface state (see later by SdH oscillations) [12].

Moreover, Ando et al. proposed the quaternary TI $Bi_{2-x}Sb_xTe_{3-y}Se_y$, which may have the lowest bulk carrier density [12, 27]. This crystal has the same lattice structure as Bi_2Ch_3, but the outer Se layer in each QL is substituted by a Te atom, and a portion of Bi will be substituted by Sb. Because the Se atoms are trapped between two Bi atomic layers, the Se vacancies can be suppressed. At the same time, because of the stronger chemical bonding between Bi and Se, the anti-site defect between Bi and Te is also suppressed. This type of material has the lowest bulk carrier density of all TIs [12, 27].

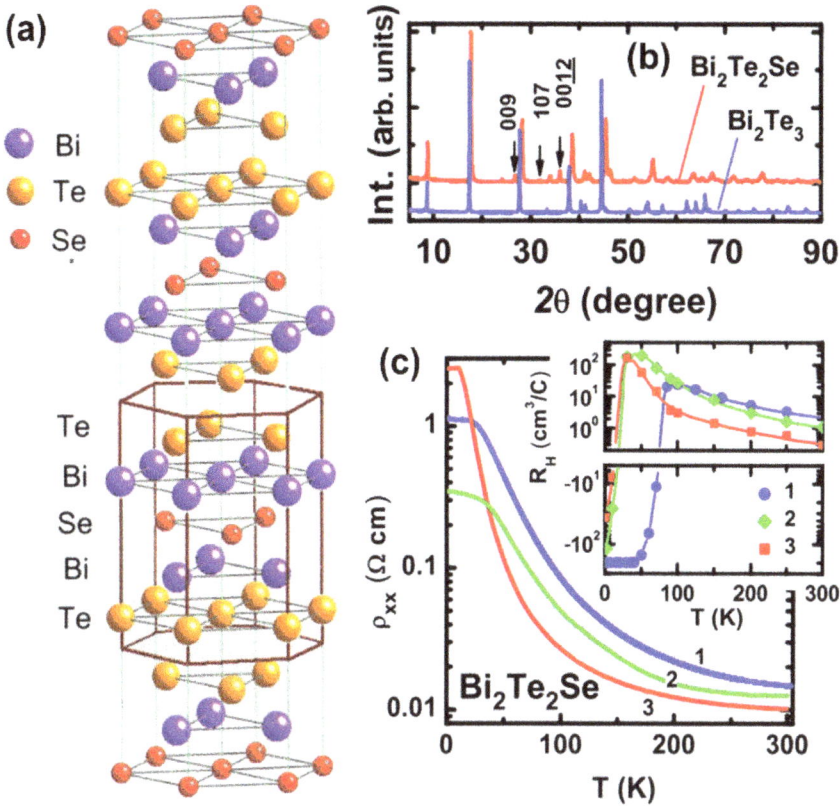

FIGURE 16.5   (a) Layered crystal structure of Bi_2Te_2Se, showing the ordering of Te and Se atoms. (b) Comparison of the x-ray powder-diffraction patterns of Bi_2Te_2Se and Bi_2Te_3. Arrows indicate the peaks characteristic of Bi_2Te_2Se. (c) Temperature dependence of resistivity ρ_{xx} for Bi_2Te_2Se samples 1–3; inset shows $R_H(T)$ for the same samples [12].

16.2   MAGNETOTRANSPORT PROPERTIES OF BISMUTH CHALCOGENIDE TOPOLOGICAL INSULATORS

TIs are exotic materials that are an insulating in their interior but can support the flow of electrons on their surface. TIs can be used in different applications, such as spintronics and quantum computing [3–6]. The topological surface states (TSSs) possess Dirac linear energy dispersion inside the bulk gap, spin-polarization by spin-momentum locking nature, and weak anti-localization (WAL) in low magnetic field due to the strong spin-orbit coupling [3–6] and possible quantum (or Shubnikov-de Haas, SdH) oscillations in a high magnetic field regime [12, 28–33].

16.2.1   WEAK ANTI-LOCALIZATION IN TIs

The weak anti-localization (WAL) is caused by the wave nature of electrons, and it is a negative quantum correction to classical magnetoresistance (MR). In TIs, WAL is induced by both the helicity of the surface state and the spin-orbit coupling of bulk [29, 34–36]. In a low magnetic field (B), two-dimensional (2D) WAL MR of a system at the strong spin-orbit interaction is usually described very well by the Hikami-Larkin-Nagaoka equation [29, 34, 36, 37]:

$$\frac{\Delta R_\square(B)}{[R_\square(0)]^2} = -\alpha \frac{e^2}{2\pi^2\hbar}\left[\Psi(\frac{1}{2}+\frac{B_\phi}{B})-\ln(\frac{B_\phi}{B})\right] \qquad (16.1)$$

Where $R_\square$ is the sheet resistance, $\Delta R_\square = R_\square(B) - R_\square(0)$, $\Psi(x)$ is the digamma function, $B_\varphi = \hbar/(4eL_\varphi^2)$ is a magnetic field varying with the coherence length L_φ, α is a parameter and reflects the number of conduction channels. In a 3D TI, $\alpha = -1/2$ for a single coherent transport channel in the 2D surface states, and $\alpha = -1$ for two independent coherent transport channels with similar L_ϕ in the 2D surface states [34, 37].

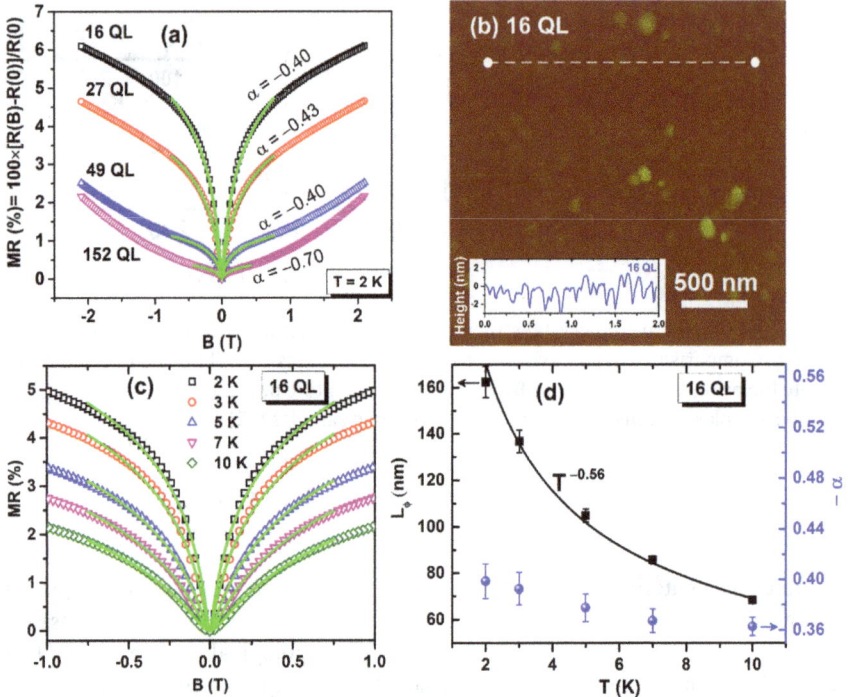

FIGURE 16.6   (a) Magnetoresistance [MR (%), $B = \pm 2.1$ T at 2 K] of Bi_2Te_3 films with various thicknesses of 16, 27, 49, 152 QLs (1 QL ≈ 1 nm). (b) AFM image of the 16 QL-thick Bi_2Te_3 thin film. (c) The low field MR curves ($B = \pm 1$ T) of the 16 QL-thick Bi_2Te_3 film taken at different temperatures from 2 to 15 K reveals the weak anti-localization (WAL) effect. The solid green lines in panels (a, c) are the theoretical predictions of 2D WAL using Eq. (16.1). (d) Variation of the extracted electron dephasing length $L\phi$ and parameter -α as a function of temperature for the 16 QL-thick film. The solid (black) curve shows the power law dependence of temperature for the $L\phi$ of the 16-QL film [38].

Figure 16.6(a) shows the MR results of the various-thick Bi_2Te_3 thin films from 16 to 152 QL (1 QL ≈ 1 nm) grown on c-plane sapphire substrates using pulsed laser deposition [38]. Obviously, in a low B field regime (± 0.75 T), the MR curves can be fitted well with Eq. (16.1). Moreover, the fitted α values at 2 K are −0.4, −0.43, −0.4, and −0.7 for 16, 27, 49, 152 QL, respectively [38], indicating that the 16−49 QL-thick Bi_2Te_3 films exhibit a single phase-coherent channel of the 2D quantum-interference effect [36]; meanwhile, the 152 QL-thick film presents two decoupled conduction channels. Here, we report the detailed MR result of the 16 QL-thick Bi_2Te_3 film. As shown in Figure 16.6(b), the film has a granular flat surface morphology with the average surface roughness (R_a) of 1.2 nm [38].

Figures 16.6(c, d) show thickness-dependent MR curves of the 16 QL-thick Bi_2Te_3 film at various temperatures from 2 to 10 K and the extracted $\alpha(T)$ and $L_\varphi(T)$ results using Eq. (16.1). At 2 K, L_ϕ is 162.4 nm, which is comparable to L_ϕ~331 nm for a 50-nm-thick Bi_2Te_3 film grown by molecular beam epitaxy (MBE) [39], and L_ϕ~280nm for Bi_2Te_3 microflakes [37]. In addition, as shown in Figure 16.6(d), L_ϕ decreases monotonically with increasing T with the power law of L_ϕ~$T^{-0.56}$. Theoretically, the power law dependence of the coherence length is L_ϕ~$T^{-0.5}$ for the predominant e-e scattering in 2D weakly disordered systems, and it is L_ϕ~$T^{-0.75}$ for the 3D systems with the dominant dephasing sources of e-e and electron-phonon (e-ph) scatterings [29, 36, 38]. Hence, the present $L_\phi(T)$ result is originated from the predominant e-e scattering in 2D weakly disordered systems, suggesting the 2D surface states for the observed WAL [38].

Furthermore, the −α decreases slightly from 0.40 to 0.36 that implies a stronger coupling between charge transport channels when temperature increases from 2 to 10 K (Figure 16.6(d)). This is an unclear phenomenon that has also been observed in the Bi_2Te_3 microflakes [37]. Nevertheless, the −α value varies within a narrow range of 0.36−0.4 (close to −α = 0.5), indicating the existence of a single coherent transport channel (i.e., likely a single surface state) [36, 38]. The observation of the 2D WAL effect suggests the occurrence of entangled phase-coherent channels, creating from a 2D TSS and a 3D bulk state [38]. However, it is important to note that WAL is not conclusive evidence to directly probe the TSS since it simultaneously reflects the 3D contribution of spin-orbit coupling in bulk and the Dirac nature of the 2D TSSs [37, 38].

To gain further insight into the origins of WAL, the MR in tilted magnetic field should be studied. J-N. Wang et al. studied the WAL in Bi_2Te_3 thin films grown on GaAs(111) substrate with an un-doped ZnSe buffer layer using MBE [39]. In low perpendicular B fields, a WAL effect in Bi_2Te_3 films is clearly observed as a sharp dip MR curves. In tilted magnetic fields, the WAL weakens gradually for changing B field direction from 90° (perpendicular field) toward 0° (in-plane field), as shown in Figure 16.7(a, c). In addition, WAL is still observed for 50-nm films at θ = 0°, which indicates the bulk contribution to the observed WAL owing to both strong spin-orbit scattering in TI and TSSs. The bulk contribution to WAL is reasonably assumed to be weakly dependent on the B field direction, and thus the bulk contribution to the observed WAL at each θ can be subtracted, that is, ΔG (θ, B) = 1/$R(\theta$, B) − 1/$R(0$, B). Figure 16.7(b, d) shows the obtained magnetoconductance ΔG (θ, B) curves as a function of the normal B component, and they all collapse nicely onto a single curve in low fields [39]. This observation reveals the TSSs induced- the 2D WAL after subtraction of bulk contribution to WAL. Moreover, this 2D WAL effect is fitted well

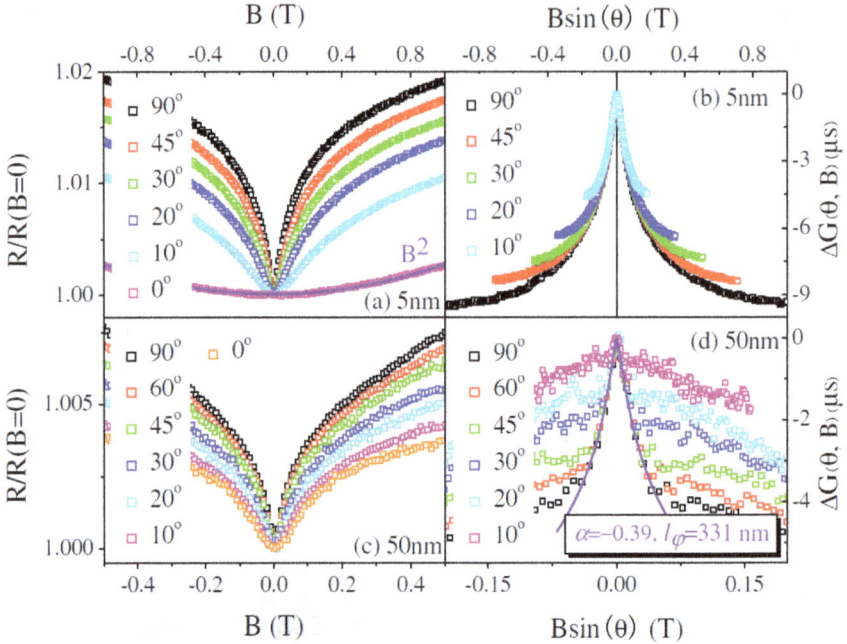

FIGURE 16.7 (a) Normalized MR of a 5 nm Bi_2Te_3 film (a) and a 50 nm Bi_2Te_3 film (c), measured in tilted B fields at $T = 2$ K. Solid curve in (a) is a parabolic fit to the MR data measured at $\theta = 0°$. Magnetoconductance as a function of the normal B component with the $\theta = 0°$ magnetoconductance subtracted for 5 nm sample (b) and 50 nm sample (d). Solid curve in (d) is a fit to ΔG ($\theta = 90$, B) in the low B-field region [39].

with equation (16.1) to yield $\alpha = -0.39$ (a single phase-coherent channel). This 2D WAL is ascribed to the top surface of Bi_2Te_3 film, and this result is similar to the MBE-grown Bi_2Se_3 thin films [40].

16.2.2 QUANTUM OSCILLATIONS IN BISMUTH CHALCOGENIDE TIs

Magnetic quantum oscillations occurring in conductivity are called Shubnikov-de Haas (SdH) oscillations. The origin of SdH oscillations is the Landau quantization of the electronic energy spectrum. SdH has been used to verify experimentally the Dirac spectrum in graphene [41]. For 3D TIs, SdH oscillations (if observable) plays an important role in identifying the conduction contribution of the TSSs, which is usually hidden by the bulk conduction. SdH oscillations have been observed and analyzed to probe the TSSs transport in bismuth chalcogenide bulk crystals [12, 28–30], thin films [31, 32, 42], and nanostructures [33]. SdH oscillations in a semi-classical magneto-oscillation description are given by [43, 44].

$$\Delta G_{xx} = G(B,T)\cos\left[2\pi\left(\frac{F}{B} - \frac{1}{2} + \gamma\right)\right] \tag{16.2}$$

where $G(B, T)$ is the temperature- and magnetic field-dependent SdH oscillation amplitude, F is the SdH oscillation frequency in $1/B$, and γ is the associated Berry phase ($0 \leq \gamma \leq 1$). Detailed analysis of SdH oscillations allows separating bulk and surface conduction, obtaining parameters such as surface mobility, cyclotron mass, and Fermi velocity, and finally verifying the Dirac nature of TSSs.

According to the Onsager relation [45]:

$$F = \hbar/2\pi e A(E_F) \qquad (16.3)$$

The frequency of the SdH oscillation (F) is proportional to the cross-sectional area A of the Fermi surface (FS) normal to the applied magnetic field. For the 2D Fermi surface case, $F \propto 1/\cos\theta$, where θ is the angle between the direction of the magnetic field and the normal to the 2D plane. It means that the F only depends on the perpendicular component of the applied magnetic fields. The SdH oscillations originate from a 2D system if the $F \propto 1/\cos\theta$ is found. For example, the SdH oscillation data for a single crystal Bi_2Se_3 sample with a high carrier concentration of $\sim10^{19}$ cm^{-3} exhibits clearly the θ-dependent F of a 2D Fermi surface (Figure 16.8(c)) [45]. In TIs, however, besides the TSSs, the 2D character can also be attributed to the other possible trivial 2D electron gas. Indeed, the 3D bulk energy states will be modified to form 2D quantum-well states as a result of the quantum confinement effect if the thickness of TI thin films is comparable to the de Broglie wavelength of the carriers. Furthermore, a 2D electron gas on the surface of 3D TIs can be created by the surface bending. The 2D states in both cases can produce the observed 2D SdH oscillations instead of the TI surface states. Consequently, the angular dependence of the SdH oscillations alone cannot distinguish between a genuine TI surface state and a trivial 2D state [44].

In Eq. (16.2), Berry phase $\gamma = 1/2$ indicates the existence of Dirac particles, while $\gamma = 0$ (or, equivalently, $\gamma = 1$) corresponds to the trivial case. The Dirac dispersion relation of the TSSs is associated with the π Berry phase. The best verification for whether the observed SdH oscillations arise from TSSs or bulk states is the accurate determination of the 1/2-shift associated with the Berry phase. The standard phase analysis is done by the Landau-level (LL) fan diagram, which is a plot of $1/B_n$ as a function of the oscillation index n [12, 28, 46]. The intercept of the linear fit to the data with the n-index axis yields the Berry phase. By convention, the "index field" B_n is defined as the field at which the Fermi energy E_F lies between two LLs, that is, at the minima in ΔG_{xx}. The correct extraction of index field B_n is the key to pinning down the 1/2-shift in the fan diagram [44]. The states are quantized into LLs with quantum numbers N = 0, 1, . . . in finite B normal to the 2D plane. In the Schrödinger case, there are n filled LLs below E_F as $B = B_n$ (the highest filled LL has $N_{max} = n-1$). Meanwhile, for Dirac electrons, we have $n + 1/2$ filled LLs as $B = B_n$ (now $N_{max} = n$). The additional 1/2 derives from the N = 0 LL or, equivalently, from the π-Berry phase intrinsic to each Dirac cone [28]. Hence, as $1/B \rightarrow 0$, the plot of $1/B_n$ versus n intercepts the n-axis at the value $\gamma = 1/2$ for the Dirac case, whereas the intercept $\gamma = 0$ in the Schrödinger case. The 1/2-shift is experimentally verified for the Dirac spectrum in graphene and expressed equivalently as a Berry-phase π shift [28, 41]. In both cases, G_{xx} is a local minimum at B_n [44]. Figure 16.9(a) shows the LL fan diagram

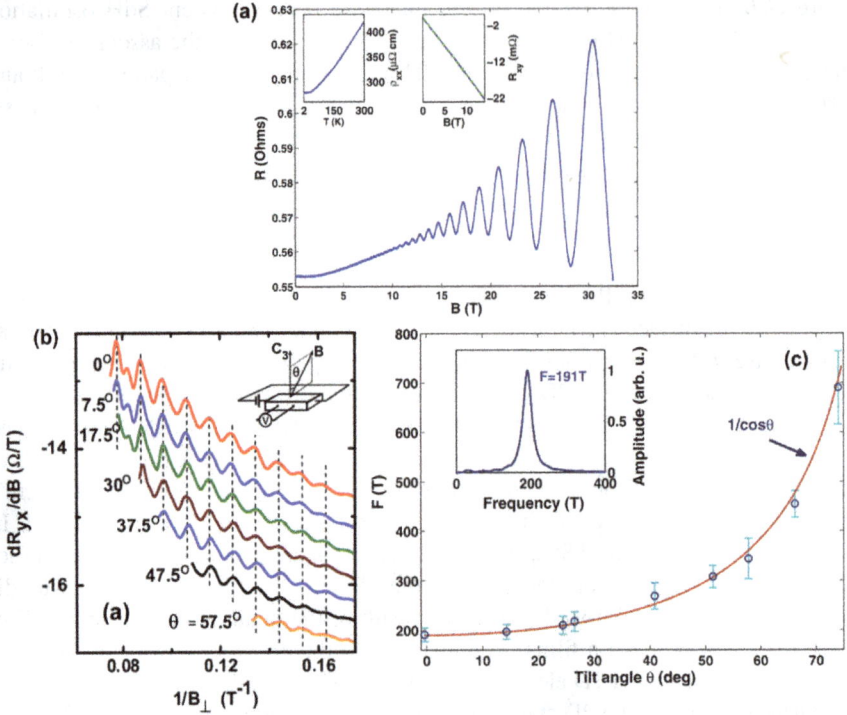

FIGURE 16.8 (a) Resistance as a function of field for a Bi_2Se_3 single crystal at T = 4.2K in a perpendicular magnetic field up to 32.5 T [45] (b) Surface SdH oscillations in the 10-QL Bi_2Se_3 film: dR_{xy}/dB in tilted magnetic fields, plotted as a function of $1/B_\perp(=1/Bcos\theta)$ [42]; curves are shifted vertically for clarity. Dashed lines mark the positions of maxima. Inset shows the geometry of the experiment. Frequency of the oscillations measured in a Bi_2Se_3 single-crystal as a function of as a function of θ. The bars represent the full-width at half-maximum of the fast Fourier transforms, which are an upper bound for the error. The solid line is the expected 2D $\dfrac{1}{cos\,\theta}$ behavior [45].

derived from the SdH oscillations conductance G_{xx} observed in a 10-QL MBE-grown Bi_2Se_3 film [42]. The upper inset in Figure 16.9(a) presents ΔG_{xx} versus $1/B$ after subtracting a smooth background, and the lower inset shows its Fourier transform giving F = 106.8 T. In the fan diagram, the positions of the minima in ΔG_{xx} are plotted as a function of n (solid squares), while the positions of the maxima in ΔG_{xx} are plotted as a function of half integers n + 1/2 (open circles). The slope of the linear fitting is fixed at F = 106.8 T to minimize the error occurring from extrapolation. As a result, the intercept is determined as n = 0.40 ± 0.04 (solid line in Figure 16.9(a)), which is very close to the ideal value of 0.5 for Dirac electrons bearing the π Berry phase. This result confirms the origin of the SdH oscillations from the TSSs. For comparison, in Figure 16.9(a), a red dashed line with the same slope yields zero Berry phase, which is obviously inconsistent with the experimental data [42].

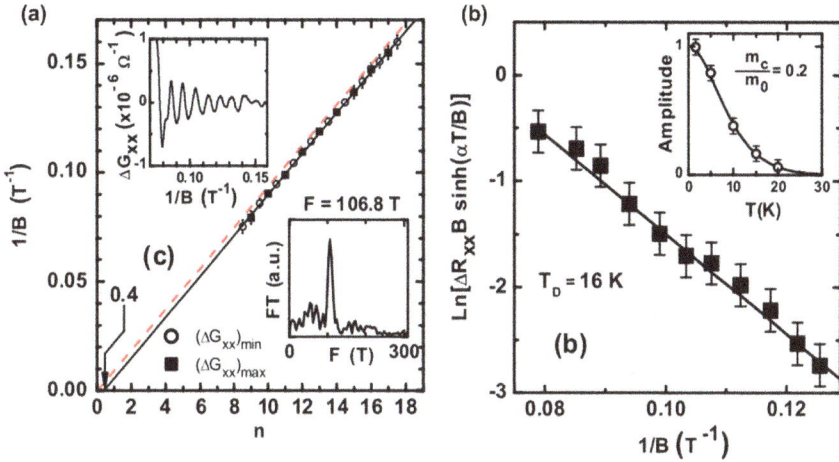

FIGURE 16.9 (a) Landau-level fan diagram for oscillations in G_{xx} measured at T = 1.6 K and $\theta = 0°$. Integers n (solid squares) and half-integers $n + 1/2$ (open circles) are assigned to the minima and maxima in ΔG_{xx}, respectively. The solid line is a linear fitting to the data with the slope fixed at $F = 106.8$ T; the dashed line has the same slope and extrapolates to zero. Upper inset shows ΔG_{xx} versus $1/B$ after subtracting a smooth background; lower inset shows its Fourier transform, giving $F = 106.8$ T. (b) Dingle plot of the oscillations in ΔR_{xx} at 1.6 K, obtained after subtracting a smooth background from $R_{xx}(B)$, giving the Dingle temperature of 16 K, which in turn gives the surface electron mobility $\mu_s = 1330$ cm^{2}V^{-1}s^{-1}. Inset: T-dependence of the SdH amplitude for $\theta = 0°$ giving the cyclotron mass $m_c = 0.2m_e$ (m_e is the free electron mass), which in turn gives the Fermi velocity $v_F = 3.3 \times 10^7$ cm.s^{-1} [42].

On the other hand, it is possible to distinguish the TSSs from bulk states or the 2D QW states by systematic investigation of SdH oscillations over a wide range of sample thicknesses. The Fermi wave vector (k_F) can be deduced from the measured frequency of SdH oscillation, the Fermi surface cross section for 2D carriers is given as $A = \pi k_F^2$. Moreover, no spin degeneracy can be reasonably assumed for the case of TSSs; thus, the density of 2D carriers is given by $n_s = k_F^2 / 4\pi$. A way to distinguish TSSs from the trivial 2D electron gas is to make a comparison between the measured Fermi wavenumber with other transport data. k_F may require the bulk Fermi level at a value that is inconsistent with other transport data, and therefore, it leaves surface Dirac fermions as the only origin of the observed oscillations [47].

A careful analysis of SdH oscillations is needed to separate the topological surface in a low magnetic field from the trivial bulk states in a metallic topological insulator. It is also of great interests to determine the microscopic parameters defining the quantum oscillations. These purposes can be achieved through the Lifshitz-Kosevich analysis of the SdH oscillations [44], in which the oscillation amplitude in ΔR_{xx}, $R(B, T)$ is given as:

$$R(B,T) = R_T R_D; \text{ where } R_T = \frac{2\pi^2 k_B T}{\hbar\omega_c \sinh\frac{2\pi^2 k_B T}{\hbar\omega_c}}; R_D = e^{-\frac{2\pi^2 k_B T_D}{\hbar\omega_c}} \quad (16.4)$$

Here, R_T is a thermal damping factor and R_D is the Dingle damping factor. k_B is the Boltzmann constant, ω_c (= eB/m_c, m_c is the cyclotron mass) is cyclotron frequency, and T_D (= $\hbar/2\pi k_B \tau$, τ is the quantum scattering time) is the Dingle temperature. As R_D does not change with T, a fitting of the temperature dependence of the SdH amplitude at a fixed B through R_T allows one to determine cyclotron mass m_c (Figure 16.9(b) inset). For 2D Dirac fermions, $m_c = \hbar k_F/v_F$, where v_F is Fermi velocity. By combining k_F deduced from the measured frequency of SdH oscillations, one can calculate $v_F = \hbar k_F/m_c$. This result can be compared with the slope of the Dirac dispersion known from the ARPES experiment to serve as an evidence of Dirac fermion origin of the observed SdH oscillations [44]. The Dingle temperature can be determined from the B-dependence of the SdH amplitude at fixed T. The so-called Dingle plot appears as $\ln(\Delta R_{xx} B \sinh 2\pi^2 k_B T / \hbar \omega_c)$ is plotted against $1/B$, as shown in Figure 16.9(b) [42]. This plot is expected to generate a straight line. Its slope contains the Dingle temperature and is given by $-2\pi^2 k_B m_c T_D / \hbar e$. From T_D, one obtains the quantum scattering time τ, which in turn allows one to estimate the surface carrier mobility $\mu_s^{SdH} = e\tau / m_c = e\ell_s^{SdH} / \hbar k_F$ and the surface state conductance $G_s = e\mu_s^{SdH} n_s = (e^2 / h)\ell_s^{SdH} k_F$, where $\ell_s^{SdH} = v_F \tau$ is the mean free path of surface carrier [44]. It is worth mentioning that the real transport mobility of surface carriers is generally higher than the value of μ_s^{SdH} estimated from SdH oscillations because large-angle scattering events play a more important role in transport, while oscillation scattering events in all directions are accounted equally in SdH.

16.3   ISSUES AND PERSPECTIVES

Bismuth chalcogenide topological insulators usually suffer from high-doping carrier concentrations due to the sufficient high defects of Se, Te vacancies and/or anti-sites, and thus the Fermi level is often located in the bulk conduction or valence band. Consequently, it is relatively challenging to probe the TSSs via transport experiments. Nevertheless, by detailed and careful, deep analysis of the well-collected transport data, the transport properties of TSSs are still extracted successfully. More importantly, the investigation of the topological transport properties of TIs might finally lead to the real application of TIs in the near future.

Since the discovery of TIs, especially 3D TI materials including bismuth chalcogenides (i.e., Bi_2Se_3, Bi_2Te_3, Bi_2Te_2Se, . . .), many new opportunities have emerged in physics and device structures [48]. First, the topologically robust surface states in TIs protected by time-reversal-symmetry enable the coherent transport for the massless Dirac fermion with an ultrahigh mobility on the surfaces. In addition, the quantum spin Hall effect occurs in TI systems without the need of applied magnetic field that induces spatial separation between the traffic lanes of electrons. Importantly, it prohibits any elastic backscattering from nonmagnetic impurities due to the topological phase as a result of a large intrinsic spin-orbit coupling. Thus, TIs have potential for applications in low-power-dissipation electronic devices. Second, the spin-momentum-lock mechanism promises potential applications using an effective spin-polarized current, which can be generated by simply applying a lateral electric-field across a TI film. Hence, it is believed that TIs are a suitable platform for spintronic applications.

ACKNOWLEDGMENTS

Financial support from the Vietnam National Foundation for Science and Technology Development (NAFOSTED) under grant number 103.02–2019.374 is gratefully acknowledged.

REFERENCES

[1] Fu L, Kane C and Mele E J 2007 Topological insulators in three dimensions *Phys Rev Lett* **98** 106803
[2] Moore J E and Balents L 2007 Topological invariants of time-reversal-invariant band structures *Phys Rev B* **75** 121306
[3] Hsieh D, Qian D, Wray L, Xia Y, Hor Y S, Cava R J and Hasan M Z 2008 A topological Dirac insulator in a quantum spin Hall phase *Nature* **452** 970–4
[4] Hsieh D, Xia Y, Qian D, Wray L, Meier F, Dil J H, Osterwalder J, Patthey L, Fedorov A V, Lin H, Bansil A, Grauer D, Hor Y S, Cava R J and Hasan M Z 2009 Observation of time-reversal-protected single-Dirac-cone topological-insulator states in Bi_2Te_3 and Sb_2Te_3 *Phys Rev Lett* **103** 146401
[5] Moore J E 2010 The birth of topological insulators *Nature* **464** 194–8
[6] Hasan M Z and Kane C L 2010 Colloquium: Topological insulators *Rev Mod Phys* **82** 3045–67
[7] Hasan M Z and Moore J E 2011 Three-dimensional topological insulators *Annu Rev Condens Matter Phys* **2** 55–78
[8] Qi X-L and Zhang S-C 2011 Topological insulators and superconductors *Rev Mod Phys* **83** 1057–110
[9] Xue Q K 2011 Nanoelectronics: A topological twist for transistors *Nat Nanotechnol* **6** 197–8
[10] Zhang H, Liu C-X, Qi X-L, Dai X, Fang Z and Zhang S-C 2009 Topological insulators in Bi_2Se_3, Bi_2Te_3 and Sb_2Te_3 with a single Dirac cone on the surface *Nat Phys* **5** 438–42
[11] Miller G R, Li C-Y and Spencer C W 1963 Properties of Bi_2Te_3-Bi_2Se_3 alloys *J Appl Phys* **34** 1398
[12] Ren Z, Taskin A A, Sasaki S, Segawa K and Ando Y 2010 Large bulk resistivity and surface quantum oscillations in the topological insulator Bi_2Te_2Se *Phys Rev B* **82** 241306
[13] Neupane M, Xu S-Y, Wray L A, Petersen A, Shankar R, Alidoust N, Liu C, Fedorov A, Ji H, Allred J M, Hor Y S, Chang T-R, Jeng H-T, Lin H, Bansil A, Cava R J and Hasan M Z 2012 Topological surface states and Dirac point tuning in ternary topological insulators *Phys Rev B* **85** 235406
[14] Zhang Y, He K, Chang C-Z, Song C-L, Wang L-L, Chen X, Jia J-F, Fang Z, Dai X, Shan W-Y, Shen S-Q, Niu Q, Qi X-L, Zhang S-C, Ma X-C and Xue Q-K 2010 Crossover of the three-dimensional topological insulator Bi_2Se_3 to the two-dimensional limit *Nat Phys* **6** 584–8
[15] Wray L A, Xu S-Y, Xia Y, Hor Y S, Qian D, Fedorov A V, Lin H, Bansil A, Cava R J and Hasan MZ 2010 Observation of topological order in a superconducting doped topological insulator *Nat Phys* **6** 855–9
[16] Hor Y S, Williams A J, Checkelsky J G, Roushan P, Seo J, Xu Q, Zandbergen HW, Yazdani A, Ong N P and Cava R J 2010 Superconductivity in $Cu_xBi_2Se_3$ and its implications for pairing in the undoped topological insulator *Phys Rev Lett* **104** 057001
[17] Zhang J L, Zhang S J, Weng H M, Zhang W, Yang L X, Liu Q Q, Feng S M, Wang X C, Yu R C, Cao L Z, Wang L, Yang W G, Liu H Z, Zhao W Y, Zhang SC, Dai X, Fang Z

and Jin C Q 2011 Pressure-induced superconductivity in topological parent compound Bi$_2$Te$_3$ *PNAS* **108** 24–8

[18] Zhang C, Sun L, Chen Z, Zhou X, Wu Q, Yi W, Guo J, Dong X and Zhao Z 2011 Phase diagram of a pressure-induced superconducting state and its relation to the Hall coefficient of Bi$_2$Te$_3$ single crystals *Phys Rev B* **83** 140504

[19] Bay T V, Naka T, Huang Y K, Luigjes H, Golden M S and de Visser A 2012 Superconductivity in the doped topological insulator Cu$_x$Bi$_2$Se$_3$ under high pressure *Phys Rev Lett* **108** 057001

[20] Chen J, Qin H J, Yang F, Liu J, Guan T, Qu F M, Zhang G H, Shi J R, Xie X C, Yang C L, Wu K H, Li Y Q and Lu L 2010 Gate-voltage control of chemical potential and weak antilocalization in Bi$_2$Se$_3$ *Phys Rev Lett* **105** 176602

[21] Checkelsky J G, Hor Y S, Cava R J and Ong N P 2011 Bulk band gap and surface state conduction observed in voltage-tuned crystals of the topological insulator Bi$_2$Se$_3$ *Phys Rev Lett* **106** 196801

[22] Steinberg H, Laloë J-B, Fatemi V, Moodera J S, and Jarillo-Herrero P 2011 Electrically tunable surface-to-bulk coherent coupling in topological insulator thin films *Phys Rev B* **84** 233101

[23] Chen Y L, Analytis J G, Chu J-H, Liu Z K, Mo S-K, Qi X L, Zhang H J, Lu DH, Dai X, Fang Z, Zhang S C, Fisher I R, Hussain Z and Shen Z-X 2009 Experimental realization of a three-dimensional topological insulator, Bi$_2$Te$_3$ *Science* **325** 178–81

[24] Hor Y S, Richardella A, Roushan P, Xia Y, Checkelsky J G, Yazdani A, Hasan M Z, Ong N P and Cava R J 2009 p-type Bi$_2$Se$_3$ for topological insulator and low-temperature thermoelectric applications *Phys Rev B* **79** 195208

[25] Analytis J G, McDonald R D, Riggs S C, Chu J-H, Boebinger G S and Fisher IR 2010 Two-dimensional surface state in the quantum limit of a topological insulator *Nat Phys* **6** 960–4

[26] Miyamoto K, Kimura A, Okuda T, Miyahara H, Kuroda K, Namatame H, Taniguchi M, Eremeev S V, Menshchikova T V, Chulkov E V, Kokh K A and Tereshchenko O E 2012 Topological surface states with persistent high spin polarization across the Dirac point in Bi$_2$Te$_2$Se and Bi$_2$Se$_2$Te *Phys Rev Lett* **109** 166802

[27] Ren Z, Taskin A A, Sasaki S, Segawa K and Ando Y 2011 Optimizing Bi$_{2-x}$Sb$_x$Te$_{3-y}$Se$_y$ solid solutions to approach the intrinsic topological insulator regime *Phys Rev B* **84** 165311

[28] Xiong J, Luo Y, Khoo Y, Jia S, Cava R J and Ong N P 2012 High-field Shubnikov-de Haas oscillations in the topological insulator Bi$_2$Te$_2$Se *Phys Rev B* **86** 045314

[29] Bao L, He L, Meyer N, Kou X, Zhang P, Chen Z, Fedorov A V, Zou J, Riedemann T M, Lograsso T A, Wang K L, Tuttle G and Xiu F 2012 Weak anti-localization and quantum oscillations of surface states in topological insulator Bi$_2$Se$_2$Te *Sci Rep* **2** 726

[30] Analytis J G, Chu J-H, Chen Y, Corredor F, McDonald R D, Shen Z X and Fisher I R 2010 Bulk Fermi surface coexistence with Dirac surface state in Bi$_2$Se$_3$: A comparison of photoemission and Shubnikov—de Haas measurements *Phys Rev B* **81** 205407

[31] Qu D-X, Hor Y S, Xiong J, Cava R J and Ong N P 2010 Quantum oscillations and hall anomaly of surface states in the topological insulator Bi$_2$Te$_3$ *Science* **329** 821–4

[32] Wang K, Liu Y, Wang W, Meyer N, Bao L H, He L, Lang M R, Chen Z G, Che X Y, Post K, Zou J, Basov D N, Wang K L and Xiu F 2013 High-quality Bi$_2$Te$_3$ thin films grown on mica substrates for potential optoelectronic applications *Appl Phys Lett* **103** 031605

[33] Tang H, Liang D, Qiu R L J and Gao X P A 2011 Two-dimensional transport-induced linear magneto-resistance in topological insulator Bi$_2$Se$_3$ nanoribbons *ACS Nano* **5** 7510–16

[34] Hikami S, Larkin A I and Nagaoka Y 1980 Spin-orbit interaction and magnetoresistance in the two dimensional random system *Prog Theor Phys* **63** 707–10

[35] Suzuura H and Ando T 2002 Crossover from symplectic to orthogonal class in a two-dimensional honeycomb lattice *Phys Rev Lett* **89** 266603

[36] Cha J J, Kong D, Hong S-S, Analytis J G, Lai K and Cui Y 2012 Weak antilocalization in $Bi_2(Se_xTe_{1-x})_3$ nanoribbons and nanoplates *Nano Lett* **12** 1107–11

[37] Chiu S P and Lin J-J 2013 Weak antilocalization in topological insulator Bi_2Te_3 microflakes *Phys Rev B* **87** 035122

[38] Le P H, Liu P-T, Luo C W, Lin J-Y and Wu K H 2017 Thickness-dependent magneto-transport properties and terahertz response of topological insulator Bi_2Te_3 thin films *J Alloys Compd* **692** 972–9

[39] He H-T, Wang G, Zhang T, Sou I-K, Wong G K L, Wang J-N, Lu H-Z, Shen S-Q and Zhang F-C 2011 Impurity effect on weak antilocalization in the topological insulator Bi_2Te_3 *Phys Rev Lett* **106** 166805

[40] Chen J, Qin H J, Yang F, Liu J, Guan T, Qu F M, Zhang G H, Shi J R, Xie X C, Yang C L, Wu K H, Li Y Q and Lu L 2010 Gate-voltage control of chemical potential and weak anti-localization in bismuth selenide *Phys Rev Lett* **105** 176602

[41] Zhang Y, Tan Y-W, Stormer H L and Kim P 2005 Experimental observation of the quantum Hall effect and Berry's phase in graphene *Nature* **438** 201–4

[42] Taskin A A, Sasaki S, Segawa K and Ando Y 2012 Manifestation of topological protection in transport properties of epitaxial Bi_2Se_3 thin films *Phys Rev Lett* **109** 066803

[43] Shoenberg D 1984 *Magnetic Oscillations in Metals* (Cambridge: Cambridge University Press)

[44] He H and Wang J 2015 *Topological Insulators: Fundamentals and Perspectives* (Germany: Wiley-VCH) pp 331–55

[45] Petrushevsky M, Lahoud E, Ron A, Maniv E, Diamant I, Neder I, Wiedmann S, Guduru V K, Chiappini F, Zeitler U, Maan J C, Chashka K, Kanigel A and Dagan Y 2012 Probing the surface states in Bi_2Se_3 using the Shubnikov—de Haas effect *Phys Rev B* **86** 045131

[46] Xiong J, Khoo Y, Jia S, Cava R J and Ong N P 2013 Tuning the quantum oscillations of surface Dirac electrons in the topological insulator Bi_2Te_2Se by liquid gating *Phys Rev B* **88** 035128

[47] Matsuo S, Koyama T, Shimamura K, Arakawa T, Nishihara Y, Chiba D, Kobayashi K, Ono T, Chang C Z, He K, Ma X C and Xue Q K 2012 Weak antilocalization and conductance fluctuation in a submicrometer-sized wire of epitaxial Bi_2Se_3 *Phys Rev B* **85** 075440

[48] He L, Kou X and Wang K L 2013 Review of 3D topological insulator thin-film growth by molecular beam epitaxy and potential applications *Phys Status Solidi—RRL* **7** 50–63

17 Applications

*Thi Dieu Hien Nguyen, Vo Khuong Dien,
Thi My Duyen Huynh, Wei-Bang Li, Nguyen
Thanh Tien, Phuoc Huu Le, and Ming-Fa Lin*

The quasiparticle properties of various emerging materials have been investigated in theoretical and experimental studies. They concerned applications with significant roles in many technologies in industries, e.g., energy storages, electronic devices, and so on. The full examinations from the 1D-3D materials cover penta-graphene nanoribbons, halogen-adsorbed silicene nanoribbons, metals/transition metals-adsorbed graphene nanoribbons, zigzag carbon and silicon nanotubes, boron/carbon and nitrogen-substituted silicene systems, germanene and its structural defects, plumbene adsorption hydrogen, stage-2/3/4 in $AlCl_4/Al_2Cl_7$ graphite intercalation compounds, $LiFeO_2$, bilayer HfX_2, $Li_4Ti_5O_{12}$, adsorbed hydrogen, bismuth chalcogenide topological insulators and will be provided in detail.

Penta-graphene nanoribbons (PGNRs) have many potential applications in the field of electronics. A PGNR-based field-effect transistor (FET) consists of four terminals: gate, source, drain, and substrate together with insulating dielectric over a conducting channel, as shown in Figure 17.1.

The semiconducting nature of the doped sawtooth penta-graphene nanoribbons (SSPGNRs) with band gap of few eV (2.2–2.6 eV) means it can be fully switched on-off by tuning the Fermi level. Current intensity of N-, P-doped and B-doped SSPGNRs increases about 10^8 times compared with the one of pure SSPGNR due to raising the number of free electrons and the number of transport channels. Depending on the type of carbon hybridization (sp^3 or sp^2) at the doping sites, the currents under bias are classified into three types: oscillatory, parabolic, and Ohm. This is convenient for various device applications.

Moreover, with the ability of p-type or n-type doping, SSPGNRs can develop applications for heterojunction devices (Figure 17.2).

It possibly used PGNRs for fabrication of electrochemical devices, such as biosensors, because of their physicochemical properties such as large surface area, warping, and electroconductivity. Biosensors are devices that use biological molecules for the detection of certain biochemicals such as genes, peptides, proteins, etc. Further, PGNRs emerged as a suitable material for fabrication of biosensors for electroanalytical sensing.

DOI: 10.1201/9781003322573-17

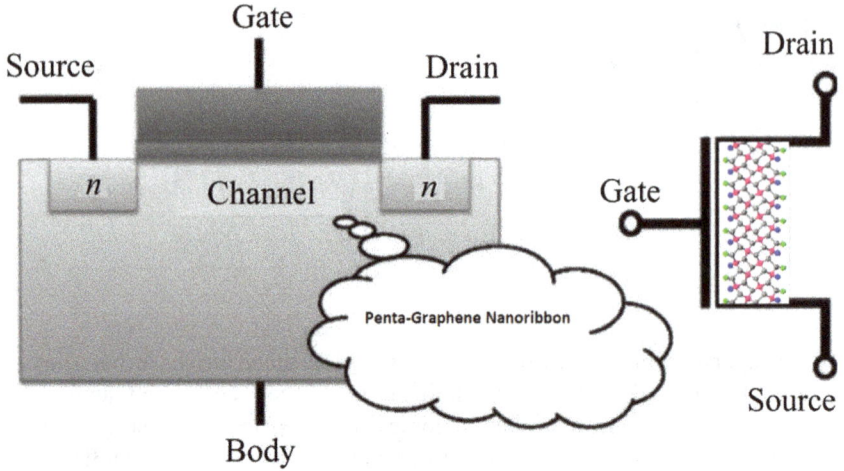

FIGURE 17.1   PGNR-based field-effect transistors.

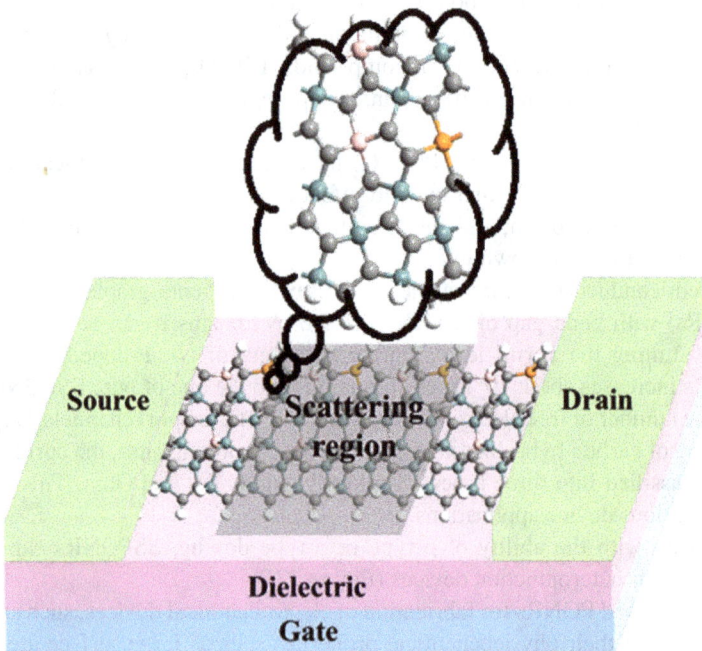

FIGURE 17.2   Geometry schematic illustration of the model device based on SSPGNR heterojunction.

2D materials with an ultra-thin nature, large surface area, high mobility ratio, bipolar, anisotropy, adjustable band gap, and flexibility character hold great potential to enhance performance for many practical applications at nano-scale [1, 2]. The emergence of 2D materials has indeed brought about a new era of science and technology. To date, the 2D materials have been utilized in various applications, including chemical sensors, biosensors, solar cells, batteries, water filtration, flexible displays, conductive inks, airplane components, high-speed transistors, and photodetectors [3, 4]. Besides, the fast growth of the semiconductor industry requires higher integration in electronic components. However, the traditional bulk semiconductors have reached the limit of Moore's law in the microchip [5]. Meanwhile, the advantageous features at the nano regime of the 2D materials show superior integration that opens a new path to revolutionize the semiconductor technology, in which the 2D materials are considered the promising candidate to replace traditional semiconductor materials in the past-Moore microchip era. Among various 2D materials, silicene with the main component of silicon elements is preferred for electronic applications due to its ideal compatibility with the current silicon-based technology [6]. However, to be suitable in the various electronic applications requires that the electronic properties of silicene are further tuned. For field-effect transistors (FETs), it requires the band gap of silicene to have a good switching behavior that can be achieved through various approaches, including external electric fields, chemical dopings, and quantum confinements [7]. Specifically, various silicene systems used as channel materials in FETs have been reported in some recent studies. Ni, Zeyuan, et al., have reported a silicene-based FET with a dual-gate structure, in which a vertical electric field is used to open the band gap of silicene [8]. The opened band gap of silicene has resulted in a good switching effect observed by an applied gate voltage. The opening band gap of silicene through alkali adsorption to apply for a bottom-gate FET has been investigated by Quhe, Ruge, et al., where sodium-adsorbed silicene exhibits a maximum transport gap of 0.5 eV that result in a high on/off current ratio [9]. The silicene nanoribbon-based FET has been studied by Salimian et al., in which the opened band gap of silicene nanoribbons creates a significant transport gap, resulting in the effective switching behavior [10]. Furthermore, with rich magnetic states induced by various edge structures, silicene nanoribbons with proper modulations can be applied for spintronics, spin FET, and magneto-resistance devices [11]. The diverse electronic and magnetic properties induced by various halogen adsorptions revealed in this chapter will be very useful for many aforementioned high-performance applications.

Recently, nanospintronic devices have gained much attention due to their advanced applications in information processing and data storage [12–15]. Finding a way to effectively induce magnetic moments is necessary for nanospintronic devices. In our work (Chapter 6), the spin arrangement and magnetic moments have strong relations with the carbons' edge and adsorbates of graphene nanoribbons (GNRs). Most of Ti-/Fe-/Co-adsorbed asymmetric GNRs exhibit the ferromagnetic configurations under the induced spin states themselves and the greatly reduced edge-carbon magnetic moments, being in sharp contrast with the pristine GNR magnetic configuration. Specifically, the Ti- and Fe-adsorbed systems with strong magnetic moments might bring potential applications in spintronic devices.

1D zigzag nanotubes, with a unique cylindrical symmetry and electronic properties, are very useful in various fields. Mechanical properties of carbon nanotubes exhibit the strongest and stiffest materials analyzed by methods of tensile strength and elastic modulus. The special mechanical properties are because of the sp2 bonds formed by each individual carbon atom. Yu et al. show the strength results of carbon nanotubes have a tensile strength of 9,100,000 psi [16]. The density of carbon nanotubes is relatively low about 1.3–1.4 g/cm^3, so its specific strength can be up to 48000 kNmkg^{-1} compared with high-carbon steel 154 kN mkg^{-1} [17]. Electronic properties of carbon nanotubes can be either metallic or semiconducting depending on chiral vectors. The metallic carbon nanotubes are able to carry an electric current density of 4*10^9 A/cm^2, which is more than 1,000 times greater than current materials like copper [18]. Optical properties of carbon nanotubes are potentially useful. They show beneficial results of absorption and photoluminescence properties, which use light-emitting diodes (LEDs) [19, 20] and photo-detectors [21]. More, the thermal conductivity of carbon nanotubes is extremely high, around 3,500 Wm^{-1}K^{-1} compared with coper 385 Wm^{-1}K^{-1} [22]. The applications of silicon nanotubes are no less than carbon nanotubes. Due to the significant results of ballistic conductivity, silicon nanotubes can be used in electronics such as thermoelectric generators [23]. Besides, silicon nanotubes can be good candidates to be used in lithium-ion batteries. Conventional Li-ion batteries' anode materials are graphite, and the capacity of graphitic carbon is 372 mA h/g. Silicon nanotubes show very high reversible charge capacity of 3247 mA h/g with Coulombic efficiency of 89% [24].

In recent years, scientists, as well as commercial electronics companies, have paid great attention to two-dimensional (2D) materials. In the general tendency of competition, the goal is how to create new 2D materials from which to produce high-performance electronic devices in extremely small sizes (in the nanoscale regime). Among these emergent 2D materials, silicene (2D of silicon) possesses a lot of advantages because of its ability to integrate with silicon-based technology. Besides, another advantage is that monolayer silicene has been successfully synthesized both on substrates and in the freestanding form by various experimental methods [25–28]. Obviously, minimizing and eliminating the influence of the substrate is one of the important steps in bringing silicene to practical application.

Most of the previous studies discussed the applications of silicene based on the electrical properties, e.g., increasing the free carrier density or widening its band gap [6, 11]. Besides, some optical properties of silicene are also studied and developed [29]. Very interestingly, previous studies have shown that silicene systems have great potential to be applied to the production of chemical sensors [30, 31]. The active environment on the surface of silicene makes it highly sensitive to certain gas molecules and chemical species so silicene-based chemical sensors can be investigated. They found that silicene had a high sensitivity to the gas molecules NO and NH$_3$ because these two molecules could be chemisorbed on silicene with moderate adsorption energies, a certain amount of charge transfer, and opening of a small band gap. Compared with the intrinsic charge carrier concentrations of pristine graphene at room temperature, the calculated charge carrier concentrations of NO$_2$-adsorbed silicene are more than 3 times larger. These results show that the effective enhancement of hole conductivity in silicene is beneficial for nanoelectronic devices. In general, the strong binding of gas molecules to silicene and the flexible electronic properties of the adsorbed-silicene could lead to potential applications in gas detection and catalysis. In the initial step of studying the fundamental properties of silicene substitutes, Chapter 8 of this book has systematically investigated optical properties including absorption, the real and imaginary part of the dielectric function, the electron loss function, the reflectivity, and the refractive index of pristine silicene and substitution silicene compounds in-plane ($\perp$) and out of the plane ($\parallel$) polarizations. The knowledge of the optical properties is critically important for both the spectroscopic study of silicene and optoelectronic applications. Moreover, it can also help understand the growing or preparation process of such 2D systems on certain substrates or in a freestanding form during rapid degradation synthesis.

Hydrogen storage is an important tool enabling the advancement of hydrogen technology and natural batteries in applications including electricity, mobile power, and transportation. Hydrogen has the highest energy per mass of any fuel data. However, its low ambient temperature density results in a low energy per unit volume, thus responding to the development of advanced storage methods that have the potential for higher density qualities. Hydrogen can be physically stored as a gas or a liquid. Storing hydrogen as a gas usually requires normal high pressure (pressure of 350–700 bar [5,000–10,000 psi]). Hydrogen storage as a liquid requires cryogenic temperatures because the boiling point of hydrogen at 1 atmospheric pressure is −252.8°C. Hydrogen can also be stored on the surface of solids (by adsorption) or inside solids (by absorption).

High-density hydrogen storage is a challenge for stationary and mobile applications and remains a significant challenge for transportation applications. Existing storage options often require large-volume systems to store hydrogen in the gaseous state. This is less of an issue for stationary applications where the footprint of compressed air tanks may be less important.

However, fuel cell vehicles need enough hydrogen to provide a driving range of more than 300 miles with the ability to refuel the vehicle quickly and easily. While a number of light-duty hydrogen fuel cell electric vehicles capable of reaching this range are already on the market, these vehicles will rely on onboard compressed air storage using composite tanks with high pressure and large volume. The large storage volume required may be less of an issue for larger vehicles, but providing sufficient hydrogen reserves on all light platforms remains a challenge. The importance of the 300-mile range goal can be gauged, which shows that most vehicles sold today are capable of exceeding this minimum. Therefore, hydrogen electrosorption reaction is one of the most fundamental processes in electrocatalysis.

It is an admitted fact that although 2D materials such as graphene, silicene, germanene, and stanene (the so-called xenes) are very prominent materials, all of them are virtually unusable in pristine forms. Apparently, the interesting properties of these 2D materials originate from defects or doping impurities that can create local resistivity (opening its band gap), enhancement of its free-carrier density, or appearance of anomalous magnetic properties. Chapter 10 studies the substitution of boron, carbon, and nitrogen on germanene. These three guest atoms have three, four, and five outermost electrons, so it is very suitable to have a complete understanding of the doping phenomenon of one of the xenes. Compared with other xenes, an important advantage of germanene is that it can be more easily functionalized so that synthesis can be more easily accomplished by chemical methods. Also for that reason, the physical and chemical properties of germanane can be easily adjusted, and it is expected to be used in the field of energy conversion and storage.

One of the most notable applications of germanene-based materials is spintronics. All modern electronic devices are based on discrete charge flows of moving electrons. But all electrons carry spin, which is a combination of the two fundamental quantum states spin-up and spin-down. Spintronics also uses the motion of electrons but is now monitoring the spin properties of the charge cloud. According to the results of Chapter 10, the nitrogen-substituted germanene exhibits the rich feature of the magnetic state. In particular, the band gap expansion of ~0.25 eV along with the anti-ferromagnetic (AFM) state could lead to applications in spintronic devices. On the other hand, carbon-substituted germanene exhibits an extended band gap of 2.12 eV. This result can be further investigated in applications that rely on optical transition. With direct semiconductors, the band transition turns out to be most useful for light-emitting diodes (LEDs) or lasers. Besides the notable applications mentioned already, functionalized germanane can also be applied in many fields such as an artificial substrate for group-III/V solar cells, field-effect transistors, and so on.

Plumbenes appear to be quite different from graphene and other group-IV systems in terms of their electronic properties, which is due to the buckled honeycomb lattices, multi-orbital hybridization, and strong spin-orbital coupling. The chemical modifications, as revealed in the experimental and theoretical investigations [32, 33], are one of the most efficient approaches in dramatically changing the geometric, electronic band structure and density of states through the orbital hybridization modification and thus strongly affect the main profile of optical excitations. In particular, plumbanes, hydrogenations of plumbene, present the short Pb-H bond and sub-lattice height difference. The dramatic change from sp^2-sp^3 mixing to sp^3s clearly indicated in the absence of π band, the slightly modified σ bands. Furthermore, the extremely strong Pb-H chemical bond contributes to the flat bands at deep energy and created similar van Hove singularities. The hydrogen adsorptions and significantly spin-orbital interactions induce a sizable energy band gap. The optical properties cover the expansion of the optical spectrum into the visible region, the stable exciton states, the strong plasmon modes, and the sensitive change of the absorbance and reflection spectrum. In this work, we have predicted a new class of 2D quantum spin hall insulators in Pb-H monolayers with a giant-gap of ~ 1.5 eV, allowing for viable applications at room temperature. Furthermore, the rich optical excitation events of the monolayer plumbene/plumbane can be utilized for electro-optic devices. Under current investigations, the state-of-the-art theoretical framework could be generated for other emergent 2D material systems.

The large-scale energy storage systems that are inexpensive, robust, and highly efficient are necessary for the development and integration of green energy like solar and wind into the electrical grid [34]. There are many candidate materials for electrodes of batteries, e.g., alkali metal, aluminum, iron, phosphide, Sn, and so on. The rechargeable lithium-ion battery is the current solution of choice for energy sources in portable electronic devices and electric vehicles, but the high cost and the limited distribution of lithium resources have motivated scientists to look for alternative solutions.

The aluminum-chloride-GICs battery is based on inexpensive and abundant materials and is eco-friendly. However, there has not been a significant development fully covering the essential properties of the aluminum-chloride-based electrode on batteries in spite of its advantages. The synthesis and electrochemical properties of charge/discharge, capacity, and cycling stability between graphite and aluminum-based GICs have arisen in past years [35]. The aluminum-based battery has superior features among many candidate materials. The aluminum-based battery possesses faster charge/discharge, longer lifetime, wider temperature tolerance, and greater safety than the well-known lithium-ion battery and lead-acid battery. Some previous studies including theory and experiment display the fundamental features, such as voltage or cycling stability [36–38]. The theoretical calculations are necessary and capable of providing the vital information to investigate the essential mechanics within batteries. Our work in this chapter is mainly focused on constructing the framework of the essential physical and chemical properties of stage-1/2/3/4 aluminum-chloride-GICs. The calculations include critical and significant features, such as the optimized geometric structures of lattice, spatial charge distributions and variations, the kz-depedent energy bands, and the density of states. These vital results can provide important information for us to investigate the mechanics of stage-1/2/3/4 aluminum-chloride-GICs in detail, e.g., orbital hybridizations. The calculations also help us to compare the different properties between intercalation and deintercalation (so-called charge and discharge). The unique structure and stage index play an important role in aluminum-chloride-GICs.

Our calculations provide the essential physical and chemical results for stage-1/2/3/4 aluminum-chloride-GICs. However, there are still some unusual problems that must be solved. Take iron-chloride-GICs as an example. The experiment displays that the $FeCl_3$-few-layer graphene (less than four layers) performs with better specific capacity compared to other cases; the stage-2 $FeCl_3$-GIC exhibits higher reversible capacity than a stage-1 one [39]. These phenomena are crucial and needed to be completed. Also, the related issues covering all unique geometry structures are complicated and need much more time to accomplish, e.g., the higher stage index and different configurations. It is worthy to calculate and research all cases in aluminum-chloride-GICs to completely construct the theoretical framework that clearly illustrates the phenomenon in applications.

The layered rock-salt-type structure materials, which consist of $LiTO_2$ (T = transition metal), have received a great deal of attention, both experimentally [40–42] and theoretically [43, 44], owing mainly to their high potential for application as a cathode material in lithium-ion-based batteries [42, 45]. For example, ternary $LiFeO_2$ compounds are utilized frequently as commercial solid-state cathodes [42]. During the charging and discharging processes, the lithium ions escape the original cathode, migrate along with the electrolyte, transfer into the graphite-based anode, and then return through the opposite path. The significant and flexible transformations in the geometric structure can be observed; that is, the deformed lattice structure plays a critical role in battery performance [46]. Most theoretical predictions present geometric and electronic properties through the first-principles calculations [43]. However, the delicate results and analyses are thoroughly absent up to now. That is to say, the calculated results are insufficient, and there are no critical mechanisms in comprehending the diversified phenomena. The identification of the real orbital hybridization in various chemical bonds can help us to comprehend the lithium migration mechanism. The $LiFeO_2$ compound has been chosen as a typical material in current VASP calculations. This system has the smallest unit cell, with 12 atoms in a triangular lattice, and it consists of three iron oxide layers intercalated by Li atoms along the z-axis.

The LiFeO$_2$ compound is a semiconductor with an energy band gap of Eg ~ 1.9 eV. The rich and unique phenomena include the atom- and spin-dominated valence and conduction bands, the complex charge density distributions in various chemical bonds, the spin magnetic configurations, and the atom-, orbital-decomposed spin-polarized densities of states. The highly accurate calculations and delicate analyses can provide the first step toward a full understanding of this functional material and promote its application. To deeply understand the lithium migrations, the delicate calculations on other LiFeO meta-stable configurations are required. The current research could be expanded to explore the rich essential properties of not only cathode but also anode and electrolyte battery materials.

As mentioned before, two-dimensional (2D) semiconductors with a layer structure as transition metal dichalcogenides (TMDs) are now among the most investigated in both experimental and theoretical approaches. As an emergent TMD material group, the HfX$_2$ (X = S, Se, or Te) family is one of a new generation of candidates for nanodevices, such as transistors [47, 48], photodetectors [49–51], and sensors [52, 53]. Being the most attractive among these materials, HfS$_2$ [47–49, 51, 54–58] exhibits such potential and wide-range applications in devices due to its experimental measurement and theoretical synthesis. For transistors, back-gate and top-gate HfS$_2$ field-effect transistors (FETs) were fabricated with the Ti/Au contact [47] forming a smoother film that indicates higher carrier injection and transport efficiency. Due to the high carrier, the multilayer HfS$_2$ back-gate FETs also display competitive mobility [48], leading to ultra-high responsivity for phototransistors. It is anticipated to be higher than the responsivity of the multilayer MoS$_2$ in near-infrared photodetection [48]. Furthermore, ultrathin HfS$_2$ flakes demonstrate the high-quality crystal structure at high deposition temperature with a fast photo-response time (55 ms) [58], which is excellent optoelectronic performance for photodetectors. Apparently, by growing on substrates [52], layered HfS$_2$ exhibits an atomically sharp interface and the epitaxial relationship between this material and substrates. Photodetectors based on HfS$_2$ on h-BN substrate [51] reveal excellent visible-light sensing performance. HfS$_2$ on sapphire [49] similarly shows a high-performance photodetector. Because of similar characteristics, HfSe$_2$ is expected to have corresponding applications like HfS$_2$. HfSe$_2$ on SiO$_2$ [59] promises applications for FETs due to its high carrier mobility [60]. This material might be used as a high-κ dielectric that could replace silicon in electronic devices [61]. HfSe$_2$ is highly sensitive under pressure [62] and could propose a flexible material for nano electronic devices. Moreover, HfSe$_2$ performs as a saturable adsorber (SA) device [63], indicating its excellent nonlinear optical adsorption properties for the ultra-fast photonics field. With the contrast feature that is semi-metal, HfTe$_2$ now attracts investigations to explore new potential material for various fields. HfTe$_2$ nanosheet reveals the sensing capabilities for the environmental hazardous (CO, CO$_2$, NO, NO$_2$, NH$_3$) and common environmental gases (O$_2$, N$_2$, H$_2$) [53] that could be an efficient sensor at low voltage. Thus, HfTe$_2$ has been predicted to be less powerful and can be sensitive to the low concentration of environment, being a suitable candidate for the medical field. Generally, HfX$_2$

(X = S, Se, or Te) layered materials offer enormous candidates for various applications in electronic and photoelectronic devices. Understanding their fundamentals provides a perspective for development of these materials and opens opportunities for applications.

Being a core material in Li-ion batteries, lithium titanate ($Li_4Ti_5O_{12}$) possesses many advantages, especially the high-power properties. We know that rechargeable batteries play an important role in energy resource technologies, including electronic devices, i.e., laptops, cell phones, iPods, modulated planes, electric vehicles (EVs), and hybrid electric vehicles (HEVs). The critical mechanism of batteries is operated by the electric supply and chemical energy, which is characterized through the unique charging and discharging process (shown in Figure 17.3). These processes mainly arise from the rapid ion transports in the internal circuit, based on their special mechanism. The up-to-date batteries come into existence with common advantages, e.g., high capacity, large output voltage, long-term stability, and friendliness with the chemical environments. A typical battery generally consists of the critical anode, cathode, and electrolyte materials. The diverse combinations of three significant components would form many advantages and disadvantages based on the high-performance criteria.

In this work, the LiTiO anode compound is thoroughly investigated for the geometric, electronic, and optical properties. Maybe this system can provide an excellent

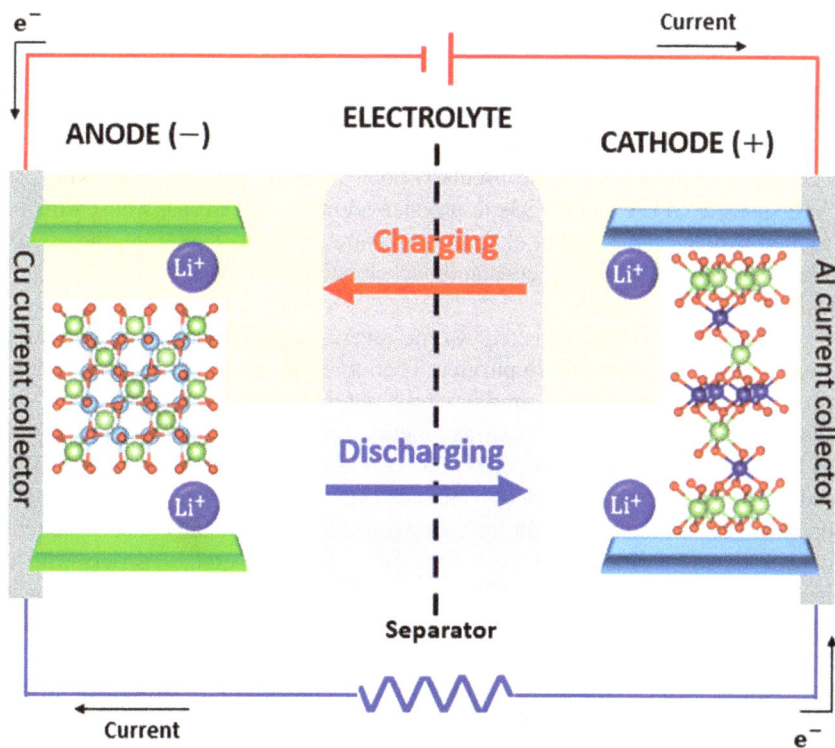

FIGURE 17.3   Charging and discharging processes of lithium-ion-related batteries.

heterojunction using the electrolyte of a lithium oxide compound. The significant differences with graphite are worthy of detailed discussions. Lithium titanate also provides outstanding characteristics, with the voltage plateau approximately 1.55 V versus Li/Li$^+$. Lithium atoms insert or extract during the charging and discharging process can form two phases $Li_4Ti_5O_{12}$ and $Li_7Ti_5O_{12}$. Based on the significant properties, lithium-ion batteries' (LIBs) materials are also considered as important emergent materials to solve environmental and energy issues. The developed research of LIBs earned the Chemistry Nobel Prize in 2019 for M. S. Whittingham, J. B. Goodenough, and A. Yoshinowon. Therefore, further improvements in the electrode and electrolyte materials of LIBs are attracting much attention to satisfy the rapidly increasing demand.

Bismuth chalcogenide topological insulators (BiCh-TIs) have demonstrated the dominant applications of semiconductor devices such as photodetectors, magnetic devices, field effect transistors (FETs), and lasers [64]. Herein, we present the recent advances in photodetector development using BiCh-TIs.

A photodetector converts an optical signal into an electrical signal. It works based on the absorption of incident radiation and heat or directly absorbs the photons via a BiCh-TI. Owing to a broad band of absorption spectrum, a graphene-based photodetector with the responsivity from mA·W^{-1} to A·W^{-1} has been developed [65, 66]. A drawback of monolayer graphene is its lack of band gap, while BiCh-TIs exhibit a certain narrow band. BiCh-TIs have wide absorption spectrum and good transparency to become good photoelectric materials [67]. The TI-based photodetector has high responsivity, fast response time, and broadband detection ranging from ultraviolet to optical telecommunication wavelengths [67]. Indeed, in 2014, the Bi_2Se_3 nanoribbons were synthesized by CVD and applied to develop photodetectors [68]. A photocurrent was induced by the spin-polarized topological surface state (TSS) of the Bi_2Se_3 nanoribbons under the circularly polarized light (CPL). The working principle is that the CPL excited TSSs to produce additional electrons, whose direction of motion was the same as that of the temperature gradient. Therefore, the oriented motions of electron spin were constantly accelerated to produce high voltage up to 400 µV [68].

The TSS of BiCh-TIs is protected by time-reversal symmetry to prevent from scattering by other non-magnetic impurities. Therefore, the materials possess excellent absorption properties (even infrared or ultraviolet light) [69]. In 2015, Yao et al. used Bi_2Te_3-Si heterostructure thin film to produce a photodetector with high responsivity working in ultra-wide band from ultraviolet to terahertz at room temperature [70]. In addition, the Bi_2Te_3-Si heterojunction photodetector exhibited excellent stability after exposure to the air or strong light for a long time [70]. Additionally, a heterojunction was believed to be efficient in the separation of photo-generated carriers with the help of a built-in electric field, and these carriers could be collected quickly owing to the high surface mobility of the Bi_2Te_3 film [70]. As a result, the response time of the photodetector was greatly reduced (below 100 ms) [70]. Furthermore, the Bi_2Te_3-based photodetector could work without bias owing to its low dark current, high sensitivity and detection, and low energy dissipation [70]. Zhang et al. fabricated a Bi_2Se_3-Si heterojunction photodetector with high detectivity 4.39×10^{12} Jones (Jones = cm·Hz$^{1/2}$·W^{-1}) and a fast response speed of approximately a few microseconds [67].

These device parameters represent the highest values for topological insulator-based photodetectors.

Overall, BiCh-TIs have been applied in photodetectors, but some problems still remain. An ideal BiCh-TIs has high surface electron mobility, but the real mobility of TSS carriers does not currently reach the theoretical value because BiCh-TIs usually have substantial defects including intrinsic defects and doping defects. Nevertheless, the problem is likely to be solved in the near future.

REFERENCES

[1] Zhao J L, Ma D T, Wang C, Guo Z N, Zhang B, Li J Q, Nie G H, Xie N and Zhang H 2021 Recent advances in anisotropic two-dimensional materials and device applications *Nano Research* **14** 897–919

[2] Glavin N R, Rao R, Varshney V, Bianco E, Apte A, Roy A, Ringe E and Ajayan P M 2020 Emerging applications of elemental 2D materials *Advanced Materials* **32**

[3] Khan K, Tareen A K, Aslam M, Wang R H, Zhang Y P, Mahmood A, Ouyang Z B, Zhang H and Guo Z Y 2020 Recent developments in emerging two-dimensional materials and their applications *Journal of Materials Chemistry C* **8** 387–440

[4] Cai F, Deng G S, Li X X and Lin F J 2020 Contact resistance parallel model for edge-contacted 2D material back-gate FET *Electronics* **9**

[5] Keyes R W 2006 The impact of Moore's Law *IEEE Solid-State Circuits Society Newsletter* **11**(3) 25–7

[6] Molle A, Grazianetti C, Tao L, Taneja D, Alam M H and Akinwande D 2018 Silicene, silicene derivatives, and their device applications *Chemical Society Reviews* **47** 6370–87

[7] Nguyen D K, Tran N T T, Chiu Y H and Lin M F 2019 Concentration-diversified magnetic and electronic properties of halogen-adsorbed silicene *Scientific Reports* **9**

[8] Ni Z Y, Liu Q H, Tang K C, Zheng J X, Zhou J, Qin R, Gao Z X, Yu D P and Lu J 2012 Tunable bandgap in silicene and germanene *Nano Letters* **12** 113–18

[9] Ye M, Quhe R, Zheng J X, Ni Z Y, Wang Y Y, Yuan Y K, Tse G, Shi J J, Gao Z X and Lu J 2014 Tunable band gap in germanene by surface adsorption *Physica E-Low-Dimensional Systems & Nanostructures* **59** 60–5

[10] Salimian F and Dideban D 2019 Comparative study of nanoribbon field effect transistors based on silicene and graphene *Materials Science in Semiconductor Processing* **93** 92–8

[11] Kharadi M A, Malik G F A, Khanday F A, Shah K R A, Mittal S and Kaushik B K 2020 Review-silicene: From material to device applications *Ecs Journal of Solid State Science and Technology* **9**

[12] Rezapour M R, Lee G and Kim K S 2020 A high performance N-doped graphene nanoribbon based spintronic device applicable with a wide range of adatoms *Nanoscale Advances* **2** 5905–11

[13] Husain S, Gupta R, Kumar A, Kumar P, Behera N, Brucas R, Chaudhary S and Svedlindh P 2020 Emergence of spin-orbit torques in 2D transition metal dichalcogenides: A status update *Applied Physics Reviews* **7**

[14] Ahn E C 2020 2D materials for spintronic devices *Npj 2d Materials and Applications* **4**

[15] Liu H M, Kondo H and Ohno T 2016 Spintronic transport in armchair graphene nanoribbon with ferromagnetic electrodes: Half-metallic properties *Nanoscale Research Letters* **11**

[16] Yu M F, Lourie O, Dyer M J, Moloni K, Kelly T F and Ruoff R S 2000 Strength and breaking mechanism of multiwalled carbon nanotubes under tensile load *Science* **287** 637–40

[17] Collins P G and Avouris P 2000 Nanotubes for electronics *Scientific American* **283** 62–9

[18] Hong S and Myung S 2007 Nanotube electronics—A flexible approach to mobility *Nature Nanotechnology* **2** 207–8

[19] Misewich J A, Martel R, Avouris P, Tsang J C, Heinze S and Tersoff J 2003 Electrically induced optical emission from a carbon nanotube FET *Science* **300** 783–6

[20] Chen J, Perebeinos V, Freitag M, Tsang J, Fu Q, Liu J and Avouris P 2005 Bright infrared emission from electrically induced excitons in carbon nanotubes *Science* **310** 1171–4

[21] Freitag M, Martin Y, Misewich J A, Martel R and Avouris P H 2003 Photoconductivity of single carbon nanotubes *Nano Letters* **3** 1067–71

[22] Pop E, Mann D, Wang Q, Goodson K E and Dai H J 2006 Thermal conductance of an individual single-wall carbon nanotube above room temperature *Nano Letters* **6** 96–100

[23] Morata A, Pacios M, Gadea G, Flox C, Cadavid D, Cabot A and Tarancon A 2018 Large-area and adaptable electrospun silicon-based thermoelectric nanomaterials with high energy conversion efficiencies *Nature Communications* **9**

[24] Xiao Q Z, Zhang Q, Fan Y, Wang X H and Susantyoko R A 2014 Soft silicon anodes for lithium ion batteries *Energy & Environmental Science* **7** 2261–8

[25] Vogt P, De Padova P, Quaresima C, Avila J, Frantzeskakis E, Asensio M C, Resta A, Ealet B and Le Lay G 2012 Silicene: Compelling experimental evidence for graphene-like two-dimensional silicon *Physical Review Letters* **108**

[26] Meng L, Wang Y L, Zhang L Z, Du S X, Wu R T, Li L F, Zhang Y, Li G, Zhou H T, Hofer W A and Gao H J 2013 Buckled silicene formation on Ir(111) *Nano Letters* **13** 685–90

[27] Gao N, Li J C and Jiang Q 2014 Tunable band gaps in silicene-MoS2 heterobilayers *Physical Chemistry Chemical Physics* **16** 11673–8

[28] Ge M, Zong M, Xu D, Chen Z, Yang J, Yao H, Wei C, Chen Y, Lin H and Shi J 2021 Freestanding germanene nanosheets for rapid degradation and photothermal conversion *Materials Today Nano* **15**

[29] Matthes L, Pulci O and Bechstedt F 2014 Optical properties of two-dimensional honeycomb crystals graphene, silicene, germanene, and tinene from first principles *New Journal of Physics* **16**

[30] Feng J W, Liu Y J, Wang H X, Zhao J X, Cai Q H and Wang X Z 2014 Gas adsorption on silicene: A theoretical study *Computational Materials Science* **87** 218–26

[31] Hu W, Xi N, Wu X J, Li Z Y and Yang J L 2014 Silicene as a highly sensitive molecule sensor for NH3, NO and NO2 *Physical Chemistry Chemical Physics* **16** 6957–62

[32] Son J, Lee S, Kim S J, Park B C, Lee H-K, Kim S, Kim J H, Hong B H and Hong J 2016 Hydrogenated monolayer graphene with reversible and tunable wide band gap and its field-effect transistor *Nature Communications* **7** 13261

[33] Whitener Jr K E 2018 Review article: Hydrogenated graphene: A user's guide *Journal of Vacuum Science & Technology A* **36** 05G401

[34] Kempton W and Tomic J 2005 Vehicle-to-grid power implementation: From stabilizing the grid to supporting large-scale renewable energy *Journal of Power Sources* **144** 280–94

[35] Rani J V, Kanakaiah V, Dadmal T, Rao M S and Bhavanarushi S 2013 Fluorinated natural graphite cathode for rechargeable ionic liquid based aluminum-ion battery *J Electrochem Soc* **160** A1781

[36] Wu M, Xu B, Chen L and Ouyang C 2016 Geometry and fast diffusion of AlCl4 cluster intercalated in graphite *Electrochimica Acta* **195** 158–65

[37] Novko D, Zhang Q and Kaghazchi P 2019 Nonadiabatic effects in Raman spectra of Al Cl 4–-graphite based batteries *Physical Review Applied* **12** 024016

[38] Bhauriyal P, Mahata A and Pathak B 2017 The staging mechanism of AlCl 4 intercalation in a graphite electrode for an aluminium-ion battery *Physical Chemistry Chemical Physics* **19** 7980–9

[39] Wang L, Zhu Y, Guo C, Zhu X, Liang J and Qian Y 2014 Ferric chloride-graphite intercalation compounds as anode materials for Li-ion batteries *ChemSusChem* **7** 87–91

[40] Abdel-Ghany A E, Mauger A, Groult H, Zaghib K and Julien C M 2012 Structural properties and electrochemistry of α-LiFeO2 *Journal of Power Sources* **197** 285–91

[41] Layek S, Greenberg E, Xu W, Rozenberg G K, Pasternak M P, Itié J-P and Merkel D G 2016 Pressure-induced spin crossover in disordered α-LiFeO$_2$ *Physical Review B* **94** 125129

[42] Armstrong A R, Tee D W, La Mantia F, Novák P and Bruce P G 2008 Synthesis of tetrahedral LiFeO2 and its behavior as a cathode in rechargeable lithium batteries *Journal of the American Chemical Society* **130** 3554–9

[43] Boufelfel A 2013 Electronic structure and magnetism in the layered LiFeO2: DFT+U calculations *Journal of Magnetism and Magnetic Materials* **343** 92–8

[44] Dien V K, Han N T, Su W-P and Lin M-F 2021 Spin-dependent optical excitations in LiFeO2 *ACS Omega* **39** 25664–71

[45] Lyu Y, Wu X, Wang K, Feng Z, Cheng T, Liu Y, Wang M, Chen R, Xu L, Zhou J, Lu Y and Guo B 2021 An overview on the advances of LiCoO2 cathodes for lithium-ion batteries *Advanced Energy Materials* **11** 2000982

[46] Khuong Dien V, Thi Han N, Nguyen T D H, Huynh T M D, Pham H D and Lin M-F 2020 Geometric and electronic properties of Li2GeO3 *Frontiers in Materials* **7**

[47] Nie X R, Sun B Q, Zhu H, Zhang M, Zhao D H, Chen L, Sun Q Q and Zhang D W 2017 Impact of metal contacts on the performance of multilayer HfS2 field-effect transistors *Acs Applied Materials & Interfaces* **9** 26996–7003

[48] Fu L, Wang F, Wu B, Wu N, Huang W, Wang H L, Jin C H, Zhuang L, He J, Fu L and Liu Y Q 2017 Van der Waals epitaxial growth of atomic layered HfS2 crystals for ultrasensitive near-infrared phototransistors *Advanced Materials* **29**

[49] Wang D G, Zhang X W, Liu H, Meng J H, Xia J, Yin Z G, Wang Y, You J B and Meng X M 2017 Epitaxial growth of HfS2 on sapphire by chemical vapor deposition and application for photodetectors *2d Materials* **4**

[50] Ulaganathan R K, Sankar R, Lin C Y, Murugesan R C, Tang K C and Chou F C 2020 High-performance flexible broadband photodetectors based on 2D hafnium selenosulfide nanosheets *Advanced Electronic Materials* **6**

[51] Wang D G, Meng J H, Zhang X W, Guo G C, Yin Z G, Liu H, Cheng L K, Gao M L, You J B and Wang R Z 2018 Selective direct growth of atomic layered HfS2 on hexagonal boron nitride for high performance photodetectors *Chemistry of Materials* **30** 3819–26

[52] Lu Y and Warner J H 2020 Synthesis and applications of wide bandgap 2D layered semiconductors reaching the green and blue wavelengths *Acs Applied Electronic Materials* **2** 1777–814

[53] Chakraborty D and Johari P 2020 First-principles investigation of the 1T-HfTe2 nanosheet for selective gas sensing *Acs Applied Nano Materials* **3** 5160–71

[54] Mattinen M, Popov G, Vehkamaki M, King P J, Mizohata K, Jalkanen P, Raisanen J, Leskela M and Ritala M 2019 Atomic layer deposition of emerging 2D semiconductors, HfS2 and ZrS2, for optoelectronics *Chemistry of Materials* **31** 5713–24

[55] Cao Y Y, Zhu S and Bachmann J 2021 HfS2 thin films deposited at room temperature by an emerging technique, solution atomic layer deposition *Dalton Transactions* **50** 13066–72

[56] Ye J F, Liao K, Ge X, Wang Z, Wang Y, Peng M, He T, Wu P S, Wang H L, Chen Y F, Cui Z Z, Gu Y, Xu H Y, Xu T F, Li Q, Zhou X H, Luo M, Li N, Zubair M, Wu F, Wang P, Shan C X, Wang G, Miao J S and Hu W D 2021 Narrowing bandgap of HfS2

by Te substitution for short-wavelength infrared photodetection *Advanced Optical Materials* **9**

[57]   Wang D G, Zhang X W and Wang Z G 2018 Recent advances in properties, synthesis and applications of two-dimensional HfS2 *Journal of Nanoscience and Nanotechnology* **18** 7319–34

[58]   Yan C Y, Gan L, Zhou X, Guo J, Huang W J, Huang J W, Jin B, Xiong J, Zhai T Y and Li Y R 2017 Space-confined chemical vapor deposition synthesis of ultrathin HfS2 flakes for optoelectronic application *Advanced Functional Materials* **27**

[59]   Kang M, Rathi S, Lee I, Lim D, Wang J, Li L, Khan M A and Kim G H 2015 Electrical characterization of multilayer HfSe2 field-effect transistors on SiO2 substrate *Applied Physics Letters* **106**

[60]   Yin L, Xu K, Wen Y, Wang Z X, Huang Y, Wang F, Shifa T A, Cheng R, Ma H and He J 2016 Ultrafast and ultrasensitive phototransistors based on few-layered HfSe2 *Applied Physics Letters* **109**

[61]   Mleczko M J, Zhang C F, Lee H R, Kuo H H, Magyari-Kope B, Moore R G, Shen Z X, Fisher I R, Nishi Y and Pop E 2017 HfSe2 and ZrSe2: Two-dimensional semiconductors with native high-kappa oxides *Science Advances* **3**

[62]   Andrada-Chacon A, Morales-Garcia A, Salvado M A, Pertierra P, Franco R, Garbarino G, Taravillo M, Barreda-Argueso J A, Gonzalez J, Baonza V G, Recio J M and Sanchez-Benitez J 2021 Pressure-driven metallization in hafnium diselenide *Inorganic Chemistry* **60** 1746–54

[63]   Li L, Wang Y, Liu W J, Wang H Z, Wang J, Ren W and Wang Y G 2019 Hafnium diselenide as a Q-switcher for fiber laser application *Optical Materials Express* **9** 4597–604

[64]   Tian W C, Yu W B, Shi J and Wang Y K 2017 The property, preparation and application of topological insulators: A review *Materials* **10**

[65]   Furchi M, Urich A, Pospischil A, Lilley G, Unterrainer K, Detz H, Klang P, Andrews A M, Schrenk W, Strasser G and Mueller T 2012 Microcavity-integrated graphene photo-detector *Nano Letters* **12** 2773–7

[66]   Zhang Y Z, Liu T, Meng B, Li X H, Liang G Z, Hu X N and Wang Q J 2013 Broadband high photoresponse from pure monolayer graphene photodetector *Nature Communications* **4**

[67]   Zhang H B, Zhang X J, Liu C, Lee S T and Jie J S 2016 High-responsivity, high-detectivity, ultrafast topological insulator Bi2Se3/Silicon heterostructure broadband photodetectors *Acs Nano* **10** 5113–22

[68]   Yan Y, Liao Z M, Ke X X, Van Tendeloo G, Wang Q S, Sun D, Yao W, Zhou S Y, Zhang L, Wu H C and Yu D P 2014 Topological surface state enhanced photothermoelectric effect in Bi2Se3 nanoribbons *Nano Letters* **14** 4389–94

[69]   Zhang X A, Wang J and Zhang S C 2010 Topological insulators for high-performance terahertz to infrared applications *Physical Review B* **82**

[70]   Yao J D, Shao J M, Wang Y X, Zhao Z R and Yang G W 2015 Ultra-broadband and high response of the Bi2Te3-Si heterojunction and its application as a photodetector at room temperature in harsh working environments *Nanoscale* **7** 12535–41

18 Concluding Remarks

Vo Khuong Dien, Tran Thi Thu Hanh,
Wei-Bang Li, Ngoc Thanh Thuy Tran,
Duy Khanh Nguyen, and Ming-Fa Lin

This book is mainly focused on the theoretical development of the quasiparticle framework through the delicate calculations and analyses of the VASP results. Certain emergent materials, covering 1D-3D main-stream systems [1–4], are chosen for a model study. The diverse phenomena are clearly revealed in crystal symmetries, electronic energy spectra/wave functions, spatial charge density distributions, van Hove singularities, magnetic moments, spin arrangements, optical absorption structures with/without excitonic effects, quantum transports, and atomic coherent vibrations [2, 5]. The quasiparticle behaviors are identified to be very sensitive to dimensionalities (quantum confinements [6, 7], Moiré superlattices [8, 9]), and chemical adsorptions/substitutions [10, 11]/[12, 13]. Moreover, the significant multi- and single-orbital hybridizations of chemical bonds and the atom- and orbital-induced spin momentum are thoroughly examined from the various physical quantities. Only some theoretical predictions agree with the experimental observations, while the others require further precise tests. How to combine the theoretical and experimental studies would play critical roles in solving the open issues (details in Chapter 18), e.g., the absence of ARPES/STS detections on very complicated band structures/ density of states in multi-component lithium oxides [2, 4]. Very obviously, when the first-principles calculations (Chapter 2.1 [14]) are thoroughly assisted by the phenomenological models (Chapter 2.2 [15, 16]), this will greatly promote the understanding of various quasiparticle properties [17, 18].

The quasiparticles viewpoints, which had been proposed in the initial studies of basic physics, are further developed in understanding the rich and unique phenomena [17, 18]. Their main features are thoroughly examined through the VASP simulations for certain emergent materials, being conducted on the critical mechanisms. The concise physical/chemical/mechanical pictures are clearly identified from the 1D-3D semiconducting, semi-metallic, and metallic systems (details in Chapters 4–17). That is to say, the critical single- and multi-orbital hybridizations and magnetic configurations are fully determined from the geometric, electronic, magnetic, and optical properties, with the different projections of various orbitals and spin arrangements. Some interesting research topics are required in the current investigations, such as the predictions of X-ray diffraction patterns [19], the spin-dependent optical excitation spectra (using ferromagnetic materials [20, 21]), the hydrogen storages on layered surfaces (group-IV and group-V systems in Refs [22–24]), and the momentum-dependent Coulomb excitations [25, 26]. Apparently, there exist a lot of intermediate

DOI: 10.1201/9781003322573-18

states during the charging and discharging process of chemical reactions in lithium/ aluminum/iron-based batteries [2, 4, 27], in which the crystal structures might possess very large Moiré superlattices [2, 4]. This will lead to rather high barriers for the numerical VASP calculations, e.g., the evaluations of optical electric-dipole transitions associated with the initial and final state scatterings [2, 5]. Only Chapter 16 is related to phonon quasiparticles. The first-principles method is available in the further development of the phonon-related properties, such as the collective atom oscillations in 1D cylindrical carbon nanotubes/planar graphene nanoribbons [28]/ [29], the coexistence of intralayer and interlayer coherent vibrations in layered materials [30], the dramatic changes of phonon spectra after the chemical adsorptions/ substitutions [31, 32]/[33, 34], the unusual phonon modes under the various heterojunctions [35], the diverse phonons in composite materials (group-IV-group-IV ones in [36]), the complicated phonon spectra in anode/electrolyte/cathode sub-systems of ion-based batteries [37]/[38, 39]/[40], and the significant electron-phonon interactions with sufficiently strong strengths (the composite quasiparticles, polarons; their energy spectra and lifetimes in [41, 42]). In addition, the previous predictions clearly show that the oscillator couplings are the only model in dealing with the collective atom vibrations at a finite temperature. These are worthy of systematic studies by choosing suitable condensed-matter systems.

Up to now, the high-resolution experimental measurements provide only a few results and conclusions for emergent materials in this book. X-ray/STM/TEM [43]/ [44]/[45], STS/ARPES [46, 47], optical reflectance/transmittance/photoluminescence spectroscopies [48]/[49, 50], and SP-STS/SP-ARPES [51]/[52] are, respectively, available in detecting the geometric, electronic, optical, and magnetic properties. Their examinations are very useful in testing the developed framework of quasiparticle viewpoints (details in Chapter 2 [2, 4]). Very interestingly, successful observations have been done for the stable crystal structures of the 3D cathode/electrolyte/anode materials of ion-related batteries [53, 54]. However, it is rather difficult to determine plenty of intermediate configurations during the charging/discharging processes [2, 4] since they possess low lattice symmetries. How to predict and identify them is an obvious challenge. The high-precision ARPES measurements on such Moiré superlattices have been absent up to now, mainly owing to the non-conservation of the perpendicular transferred momenta and a lot of valence subbands. The undefined wave vectors are also revealed in 1D nanotube systems [55]. Some reflectance, transmittance, and photoluminescence examinations on these core systems display ambiguous results [56, 57], in which the physical/chemical/material pictures cannot be achieved from a lot of absorption structures, e.g., those in explaining the prominent absorption peaks, the stable/quasi-stable excitonic bound state, and an optical threshold frequency [2, 58]. This issue could be solved by the strong support of theoretical predictions in the presence/absence of excitonic effects [59, 60]. Certain research themes on the low-dimensional materials should have thorough investigations, such as the simultaneous observations of STM/TEM and STS measurements on silicene/germanene/tinene/plumbene nanoribbons before/after chemical modifications [61]/[62]/[63]/[64], as well as those on the related nanotubes [65, 66]. These will be very helpful in deterring the critical orbital hybridizations and spin configurations. In addition, reflection/transmission electron energy loss spectroscopy

[67]/[68, 69] are capable of testing the dynamic screening abilities of valence and conduction electrons, e.g., plasmon modes and electron-hole excitations in batteries [2] and graphite intercalation compounds [70, 71]. Furthermore, the Raman spectroscopy could be utilized to explore the coherent atom/molecule vibrations, e.g., the diversified phonon modes due to the significant adsorptions on a metal surface [72, 73].

Simulations of novel structures for nanodevices demonstrate that SSPGNRs might be suitable for digital circuit design in electronic applications. The quantity and quality of research works on penta-graphene have motivated new researchers to contribute to the field. We hope that SSPGNR-based devices will be realized. The scientist will have access to better materials for developing novel devices with enhanced performance. Researchers have been exploring alternative device structures that take advantage of tunneling and negative resistance effects that could be used to overcome fundamental issues presented by conventional approaches. Devices with n-p doping profiles enable deeper testing of the electronic transport properties of SSPGNRs and are very promising for replacing silicon. In addition to the scientific importance of 1D pentagonal materials, we appreciate the educational merits that result from the research on 1D pentagonal materials. We believe that the discovery of 1D pentagonal materials is an emergent direction of 1D materials, where computational simulations, especially DFT tools, are useful for predicting new physical and chemical properties on the foundation of electronic confinement and quasiparticle behavior.

The diverse structural, electronic, and magnetic properties of halogen-adsorbed SiNRs are fully revealed in the spin-polarized DFT calculations. The essential physical quantities have been developed through the first-principles calculations, including the binding energies, optimal lattice parameters, atom-dominated band structures, atom- and orbital-projected DOS, and spin density distributions. These developed physical quantities are sufficient to clarify the rich chemical and physical phenomena induced by the halogenation effects. Specifically, the diversified geometric structures are evaluated through the binding energies and optimal lattice parameters. The feature-rich electronic properties are rigidly analyzed in atom-dominated band structures that can be thoroughly verified through atom- and orbital-projected DOSs. The diverse magnetic configurations are thoroughly comprehended in the spin density distributions. As for halogen-adsorbed ASiNR systems, under the single halogen adsorptions, the direct middle-gap semiconducting feature of the pristine system becomes the p-type metallic behavior for all the halogen adatoms. This is because the halogen adatoms attract electrons from ASiNRs to generate free holes in the systems. This is to say that creating a red shift of EF leads to a p-type metallic behavior. When the adatom concentrations reach the critical value of 50% adsorption, the p-type metallic-semiconducting transition appears, and the band gaps are further opened under higher adatom concentrations. The largest band gap is found at the highest concentration of 100% regardless of any halogen adatoms. On the other side, the unique anti-ferromagnetic configuration of the pristine ZSiNR system is greatly diversified under the various halogen concentrations and distributions. For the single halogen-adsorbed ZSiNR systems, it shows the Si-related ferromagnetic configurations for all halogen adatoms, in which the net magnetic moments are different in edge and one-edge adsorption positions. Under further increasing of concentrations of double adatoms, the Si-related ferromagnetic configuration only remains at specific adatom distribution of edge and non-edge adsorption positions. It is worthy to notice that the diverse electronic and magnetic properties of halogen-adsorbed SiNRs can be suitable for a wide range of applications in electronic and spintronic devices. Furthermore, the developed first-principles theoretical framework in this chapter can be fully generalized to many other emergent layered materials.

The metal/transition-metals-adsorbed graphene nanoribbons (GNRs) clearly display the important differences in the essential properties. The Al/Ti/Fe/Co/Ni adatoms prefer to be adsorbed at the hollow-site optimal positions, while Al adatoms might have the y-direction shifts, especially for the non-symmetric distributions. The adatom chemisorptions might be easily experimentally observed for the Ti and Fe chemisorptions due to the large binding energies. In general, the metal adatoms and transition metals can induce metallic behaviors. As to the magnetic properties, the Al adatoms do not create the spin distributions, but their interactions with the zigzag carbon atoms can destroy the latter's ones and thus create the ferromagnetic (FM) or non-magnetic (NM) configurations. Armchair and zigzag Al-adsorbed GNRs, respectively, belong to the NM and AFM/FM/NM metals. On the other side, most Ti/Fe/Co-adsorbed asymmetric systems exhibit the FM configurations under the induced spin states themselves and the greatly reduced edge-carbon magnetic moments. The anti-ferromagnetic (AFM) spin arrangements are strongly associated with the ribbon-edges and adsorbates. Our complete and reliable results should be very helpful in the design and development of potential device applications-based GNRs.

The 1D zigzag carbon and silicon nanotubes, with unique cylindrical symmetry, are very suitable for studying the basic and applied sciences. The geometric structure and electronic properties of zigzag nanotubes are investigated by the method of first principles. Both nanotubes exhibit diverse geometries and electronic properties as a result of the significant chemical bonding of outer four orbitals and spin-orbital couplings. The band structure dependent of chiralities and diameters, the spatial charge density ρ, and the orbital-projected density of states (PDOSs) are useful in exploring the orbital-dependent essential properties. The latter possess buckled and planar honeycomb lattices of 3.76–3.814 Å, the ground state energy E_0, and buckling distance Δh decrease with an increase of radius, and the E_0 of carbon nanotubes decreases with an increase of radius as well. The bond length b_1 and b_2 of carbon and silicon nanotubes increases when the diameter of the tube turns small. The results of band structure from carbon and silicon nanotubes are quite different. Compared to carbon nanotubes, silicon nanotubes exhibit direct and indirect band gap due to their unsymmetrical structure. The zigzag silicon nanotubes from (4,0) to (9,0) possess metallic behaviors, but the zCNTs behave with gapless metallic features when the chiral vectors are multiples of 3. The energy dispersion of zSiNTs and zCNTs is reflected in PDOSs as many anti-symmetric and symmetric peaks. There are five kinds of low-lying peak structures characteristic of the main features of essential properties. The predicted geometric structure, energy bands, and PDOSs are shown to be reliable by experimental tools of STM, ARPES, and STS, respectively. Besides, ARPES and STS are suitable for the spin resolution. The 1D zigzag carbon and silicon nanotubes, with unique cylindrical symmetry, are very suitable for studying the basic and applied sciences. The chirality- and radius-dependent semiconductors and metals are available in functionalized nanodevices, such as electronic and optoelectronic devices. The intrinsic 1D band structures, with the decoupled angular momenta, dominate the quantized electrical and optical properties, e.g., the geometry-dependent resistance, capacitance, inductance, and optical conductance [74–76]. At present, the field-effect transistors (FETs) based on the semiconducting carbon nanotubes are widely developed, mainly owing to the advantage of high carrier mobility under the low scatterings. The first FETs composed of single- or multi-walled carbon nanotubes were reported in 1998 [77, 78]. They present superior electrical conductance at room temperature using the gate-voltage modulation. A very high carrier mobility (> $105 \, cm^2/Vs$) is identified to be related to the diffusive transport of holes under the electron-phonon scatterings [75, 79]. The nano-scaled radii allow the gate's ability to control the potential of the conducting channel in the ultimate thin FETs while suppressing short-channel effects [80, 81]. Moreover, the semiconducting carbon nanotubes have highly potential applications in optoelectronic devices, including electroluminescent light emitters [82, 83], supercapacitors [84], and photodetectors [85–87]. The band-edge states in 1D nanotubes, with the high PDOSs, can create very prominent optical excitations [74]. The excited electrons and holes on a cylindrical surface are further driven toward each other by using an appropriate bias between the source and drain of the FET. The electron-hole recombination will emit strong electroluminescence. The highly efficient photon-emission process has been extensively utilized in light-emitting diodes (LEDs) [88–90]. The application of the photodetector is based on the electric current generated by resonant excitations. On the other hand, the metallic carbon nanotubes could be selected as candidates in interconnected materials of integrated electronic devices [90, 91]. In addition to the Coulomb scatterings, the electron-phonon scatterings at finite temperatures might play important roles in the electron-electron effective interactions and thus have a strong effect on electronic excitation spectra and quasiparticle lifetimes. So the inelastic scatterings could be taken into account simultaneously under the random-phase approximation of the same order [92]. The armchair nanotubes, which exhibit the metallic characters [93], obviously display exponential decay at short distance and the well-known Friedel oscillations at long distance. The characteristic decay length and the rapid charge oscillations are highly sensitive to the concentration of free carriers and the spatial dimensions.

Chemical substitution plays an important role to enhance the potential of silicene systems. The intrinsic novel properties including geometric structure, electronic band structure, charge density distribution, and spin configuration of substitution systems are investigated, these properties exhibit completely different from the pristine one. While boron-/carbon-substituted silicene systems present the well-behavior of pi and sigma bands in energy spectra, the complicated hybridization of sp^3 chemical bondings is observed in the nitrogen-silicene compound. Roughly speaking, based on the first-principle electronic band structure results, the tight-binding model might easily simulate the DFT results in which further properties can be studied, such as the quantization phenomena, quantum spin Hall effect, and so on. Moreover, the very first results of the guest-atom substitution silicenes are also discussed in Chapter 8. In both parallel and perpendicular polarized directions, the optical properties are calculated. Optical constants such as the dielectric function, the absorption coefficient, and the energy loss function show dependence on the type of guest-atom and the direction of polarization. As a result, B/N substituted silicene does not appear to be as effective as C-silicene when we want to enhance the optical properties of these compounds.

The hydrogen adsorption on the defect-free germanene and the germanene structural defects have been studied using the DFT calculation. The most stable adsorption sites of the hydrogen atoms are the top sites: HT1 and HT2. The localized germanene surface curvature and the hydrogen zero point energy are shown. The vibration of the hydrogen atom along the z-axis is greatest for both T1 and T2 positions, and along the x- and y-axes, the vibration is negligible. We find that the interaction between two adjacent hydrogen atoms on the same type sites of germanene is repulsive, while the interaction between two adjacent hydrogen atoms on the different type sites is attractive. Besides, the germanene defects are demonstrated via four defects: the Stone-Wales (55–77), the divacancies (77–555 and 555–7), and the pentagon-heptagon linear defect (5–7). The lowest formation energy of the pentagon-heptagon linear defect is shown for the first time in this study.

The geometric, electronic, and magnetic properties of guest-atom substituted germanene are investigated in Chapter 10. The intrinsic buckling of pristine germanene becomes less/more serious depending on the chemical bondings of host-guest atoms. The substitution systems exhibit the unusual band structure that is reflected in charge distribution and density of states. Importantly, only nitrogen-substituted silicene possesses the magnetic configuration, while such a feature is absent in two other kinds of guest atoms. Several van Hove singularities can be observed in the atom- and orbital-decomposed density of states, which bear close relation to the band-edge electronic states. Importantly, similar chemical modifications can be achieved from hydrogenation, alkalization, oxidation, and halogenation, as demonstrated in graphene-like compounds. Furthermore, to fully understand the diverse characteristics of quasiparticles through important chemisorptions of adatoms is extremely necessary and can bring high applicability in the future.

Monolayer plumbene exhibits diverse geometric, electronic, and optical properties. The geometric structure is an energy favorite in the buckle structure as the consequence of sp^2-sp^3 mixing hybridizations. The strong σ band/rather weak but active π band is directly exhibited in the electronic band structure and orbital-projected density of states. The presence of spin-orbit coupling induced the opening of a sizable band gap. The optical properties, which include the single and collective optical excitation, reflected the main features of the electronic properties. Very interestingly, the strong bonding between the hydrogen and lead atoms can largely broaden the fundamental physical and chemical properties of plumbene. The geometric, electronic, and optical properties are closely affected by the complicated hybridizations of 1s and ($6p_x$, $6p_y$, $6p_z$, 6s) orbitals. In general, plumbane/plumbone presents the shortest Pb-H bond, the smallest/largest height difference. The dramatic change from sp^2-sp^3 mixing to sp^3s hybridization is clearly indicated in the absence of π band, the slightly/significantly modified the σ bands in the case of fully/partly hydrogenations. Furthermore, the extremely strong Pb-H chemical bond contributes to the flat bands at the deep energy for all systems and creates similar van Hove singularities.

Under the spin and orbital interactions, the energy states are significantly splitting, and thus, the electronic band gap is opened in fully-hydrogenation. The optical transition covers the presence of threshold frequency, the rich and unique single- and collective excitations. The optical excitation events of all systems mainly occur in the visible range, and thus, the monolayer plumbene/plumbane/plumbone can be used for electro-optic devices. Under current investigations, the state-of-the-art theoretical framework could be generated for other emergence 2D material systems.

Interestingly, the stage-n $AlCl_4$-molecule graphite intercalation compounds show diverse quasiparticle properties, being sensitive to the intercalant stackings and distributions. The intercalant concentration could also be modulated/reduced by the planar arrangements, as done in the previous study [94]. The neutral- and ionic-cases intercalations are rather different from each other, such as the semimetal-metal and semimetal-semiconductor transitions during the chemical intercalation/de-intercalation processes. Moreover, the n- or p-type charge transfers occur in intercalants with distinct affinities [95]. For example, $AlCl_4$ molecules and alkali atoms are, respectively, capable of generating free (valence) hole quasiparticles and (conduction) electron quasiparticles (the red and blue shifts of the Fermi levels [94]). According to the current VASP simulations and theoretical predictions on the featured geometric and quasiparticle properties, all the active chemical bonds are clearly identified to include the intralayer/interlayer C-C orbital hybridizations, the interlayer C-intercalant interactions [96], and intra-/inter-intercalant molecular interactions. The concise electron-quasiparticle behaviors [97] are thoroughly tested from the stage-n crystal symmetries with highly non-uniform environments, C-, Al-, and Cl-dominated energy spectra under the distinct energy ranges, spatial charge distributions before and after intercalations/de-intercalations, and atom- and orbital-projected van Hove singularities (especially for their merged special structures due to the various orbitals [98]), as well as their emergences. Most importantly, the diverse quasiparticle phenomena exhibit as follows: the coexistence of honeycomb and intercalant layers, their Moiré superlattices with stage-n structures, the zone-folding-induced many (valence) hole quasiparticles and (conduction) electron quasiparticles energy subbands, the dramatic transformation of the initial high-symmetry valleys, the strengthened asymmetry of hole quasiparticles and electron quasiparticles spectra about $E_F = 0$, the great reduction of band overlap, an obvious redshift of the Femi level [the diversified free carrier densities], the almost isotropic/anisotropic π-electronic states close to/away from $E_F = 0$, the various energy dispersions at the distinct critical points, the frequent band non-crossings/crossings/anti-crossings, and the well-behaved or undefined characteristics of the whole π and σ valence bands, the carbon-, aluminum-, chloride-dominated band structures within the different energy ranges. Furthermore, the specific orbital hybridizations in significant C-C/C-Cl/Al-Cl/Cl-Cl bonds are $(2s, 2p_x, 2p_y)$-$(2s, 2p_x, 2p_y)$ and $2p_z$-$2p_z$/$(2s, 2p_x, 2p_y, 2p_z)$-$(3s, 3p_x, 3p_y, 3p_z)$/$(3s, 3p_x, 3p_y, 3p_z)$-$(3s, 3p_x, 3p_y, 3p_z)$/$(3s, 3p_x, 3p_y, 3p_z)$-$(3s, 3p_x, 3p_y, 3p_z)$. It should be noticed that the clear evidence resulting from the properties of quasiparticles cannot be observed for the survival of C-Al and Al-Al bonds. The numerical VASP results on band structures are available in getting the suitable hopping-integral of the mentioned orbital mixings, and then the combination of the tight-binding model and the modified single-/many-particle theories is capable of fully exploring magnetic quantization phenomena [99], optical absorption spectra with/without excitonic effects (details in Chapter 2.2), Coulomb excitations [100], and quantum Hall conductivities. Finally, their important roles in ion-based batteries, basic researches, and high-performance applications are worthy of a series of accurate investigations.

The 3D ternary $LiFeO_2$ compounds, candidates for cathode materials of lithium-ion batteries, are predicted to exhibit the rich and unique lattice symmetries, energy band structures, charge and spin density distributions, and atom-/orbital-projected density of states. The accurate analyses, which are made from the exact first-principles calculations but not for the phenomenological methods, are available in identifying the significant single-multi-orbital hybridizations and spin configurations in Li-O and Fe-O bonds. Specifically, the 3D $LiFeO_2$ compound presents the Moiré superlattice with high atomic ordering and anisotropy geometric, the presence of 18 Li-O/18 Fe-O chemical bonds with identical lengths, the complicated spin-polarization band structure with various atom dominations, the sizable indirect gap of 1.9 eV, the spatial spin and charge distribution, and a lot of van Hove singularities due to the spin polarizations and extreme point dispersions. As a consequence, the important multi-orbital (4s, $3p_x$, $3p_y$, $3p_z$, $3d_{xy}$, $3d_{yz}$, $3d_{xz}$, $3d_{x2-y2}$, $3d_{z2}$) - (2s, $2p_x$, $2p_y$, $2p_z$) hybridizations and the multi-orbital 2s -($2p_x$, $2p_y$, $2p_z$) hybridizations and single-orbital 2s-2s orbital hybridizations in Fe-O and Li-O bonds, respectively, could be achieved. The calculated results clearly indicate that the 3D ternary $LiFeO_2$ could be served as a candidate cathode component in lithium-based batteries. Our prediction provides certain meaningful information about the critical physical/chemical pictures in LIBs. Such state-of-the-art analysis is very useful for fully comprehending the diversified properties in anode/cathode/electrolytes and other emerging materials. Furthermore, the theoretical predictions on the optimal geometries, the rich occupied bands, and many van Hove singularities and other quasi-particle features could be verified from the high-resolution measurements of the X-ray diffraction, angle-resolved photoemission spectroscopy, and scanning tunneling spectroscopy, respectively.

On the topic of bilayer HfX2, they reveal the chemical bonding of M-X belongs to the d-orbitals and p-orbitals of hafnium atoms and X atoms (chalcogen atoms), respectively. The electron configurations and size of atoms are responsible for all structural parameters such as chemical bonds, buckling, layered, and interlayer distances. The sequences of these parameters also depend on vdW interaction in each individual layer and between two layers. Buckling is characterized by sandwiched structures and chemical bonds featured by a sandwich structure that shows height fluctuations at equilibrium. There are six possible structures of stacking that can be constructed and are promising for a wide range of engineering and design structures. Varying stacking orders lead to versatile dispersion in electronic properties of band gap, energy dispersion, and van Hove singularities. Moreover, d and p orbitals of hafnium and chalcogen atoms demonstrate multi-orbital hybridizations, as shown in the orbital-projected DOS. These characteristics were slightly but remarkably changed when applying different stacking, as shown in band energy spectra and orbital projected DOS, which suggest many potential structures for the investigations and applications. Investigation bilayer group-IV, particularly HfX_2, opens up opportunities to construct and design new flatform materials and find out their significant properties.

Simultaneously, the consistency of the theoretical framework through VASP calculations and quasiparticles is obviously demonstrated.

The essential properties of $Li_4Ti_5O_{12}$ have been investigated based on the first-principles calculations under quasiparticle phenomena. This theoretical framework is mainly built from the critical multi-orbital hybridizations in Li-O and Ti-O chemical bonds, which are accurately identified from the atom-dominated valence and conduction bands, the spatial charge density, and the atom- and orbital-decomposed density of states. In general, their anisotropic and non-uniform properties in a large unit cell show that a reliable tight-binding model, with various hopping integrals and on-site Coulomb potentials, is very difficult to obtain in the VASP band structure. The three-dimensional material has the smallest unit cell of 56 atoms and belongs to a cubic structure. A large direct band gap $E^d_g \sim 2.606$ eV due to the various and strong covalent bondings. This gap equals the optical threshold absorption frequency. The band structure shows asymmetric behaviors between the valence and conduction bands about the Fermi level, in which their energy dispersions come into existence in parabolic, flat, weak forms, and so on. Also, there are many other properties of the sub-bands, e.g. non-crossing, crossing, and anti-crossing in the range of -6.0 eV $\leq E_{c,v} \leq 4.0$ eV. Van Hove singularities in the density of states appear based on the band-edge states, the critical points in the energy-wave-vector spaces. The important multi-orbital hybridizations, such as 2s-($2p_x$, $2p_y$, $2p_z$) and (4s, $3d_{x^2-y^2}$, $3d_{xy}$, $3d_{yz}$, $3d_{xz}$, $3d_z^2$)-($2p_x$, $2p_y$, $2p_z$) in Li-O and Ti-O bonds, are achieved in the atom- and orbital-dependent special structures, the asymmetric or symmetric peaks, and broadening shoulders. The spatial charge density distribution is used to support the significant theoretical mechanisms, while the atom-created energy bands and charge density distributions will provide the information of the diverse covalent bonding. The whole theoretical framework requires high-resolution optical spectroscopy measurements [101–103]. These delicate developments could be applied to Li^+-based battery cathode [104, 105], anode [104–106], and electrolyte materials [104, 105, 107]. In particular, rich oxide compounds deserve systematic investigations of various physical, chemical, and materials science research.

Density functional theory calculations were carried out on the hydrogen adsorption on the Pt(110) electrode in ultra-high vacuum by taking into account the vibrational frequencies, the quantum zero-point effects, and the adsorption energies. We found that the most stable adsorption site is the short bridge site, then the top, the long bridge, and the fcc sites. The highest stretching frequency ~ 2200 cm^{-1} and the highest zero-point energy ~ 140 meV were shown when the adsorption hydrogen atom is on the top site. The results presented in this work show that the quantum effect has a significant effect on the adsorption energy, indicating the simultaneous existence of adsorbed hydrogen at the top and short bridge sites.

This chapter introduces the bismuth chalcogenide TIs and their fascinating topological surface state (TSS) properties via magnetotransport studies. In TIs, since both the spin-orbit coupling of the bulk and the helicity of the surface states can induce WAL, 2D WAL MR is realized as a signature of TSSs. Consequently, further MR study in tilted magnetic fields is necessary to explore the origin of the observed WAL. Indeed, the 2D WAL effect reveals the TSSs after subtraction of the bulk contribution to WAL. On the other hand, Shubnikov-de Haas (SdH) oscillations allow us to probe the surface conductance and the TSSs. Well-analysis of SdH oscillations enables us to extract the carrier concentration, effective mass, Dingle temperature, and Berry phase of TIs. It is demonstrated that transport study is a reliable method to investigate the peculiar properties of TIs toward deepening understanding of the Dirac fermion physics.

REFERENCES

[1] Geim A K and Novoselov K S 2010 *Nanoscience and Technology: A Collection of Reviews from Nature Journals* (Canada: World Scientific) pp 11–19

[2] Han N T, Dien V K and Lin M-F 2020 Excitonic effects in the optical spectra of lithium metasilicate (Li2SiO3) *arXiv preprint arXiv:2010.11621*

[3] Harris P J 2004 Carbon nanotube composites *International Materials Reviews* **49** 31–43

[4] Dien V K, Han N T, Nguyen T D H, Huynh T M D, Pham H D and Lin M-F 2020 Geometric and electronic properties of Li $ _2 $ GeO $ _3$ *arXiv preprint arXiv:2009.02154*

[5] Dien V K, Pham H D, Tran N T T, Han N T, Huynh T M D, Nguyen T D H and Fa-Lin M 2020 Orbital-hybridization-created optical excitations in Li8Ge4O12 *arXiv preprint arXiv:2009.02160*

[6] Bryant G W 1988 Excitons in quantum boxes: Correlation effects and quantum confinement *Physical Review B* **37** 8763

[7] Molle A, Berikaa E R, Pont F M and Bande A 2019 Quantum size effect affecting environment assisted electron capture in quantum confinements *The Journal of Chemical Physics* **150** 224105

[8] McGilly L J, Kerelsky A, Finney N R, Shapovalov K, Shih E-M, Ghiotto A, Zeng Y, Moore S L, Wu W and Bai Y 2020 Visualization of moiré superlattices *Nature Nanotechnology* **15** 580–4

[9] Ni G, Wang H, Wu J, Fei Z, Goldflam M, Keilmann F, Özyilmaz B, Neto A C, Xie X and Fogler M 2015 Plasmons in graphene moiré superlattices *Nature Materials* **14** 1217–22

[10] Kou L, Du A, Chen C and Frauenheim T 2014 Strain engineering of selective chemical adsorption on monolayer MoS 2 *Nanoscale* **6** 5156–61

[11] Pan B and Xing B 2008 Adsorption mechanisms of organic chemicals on carbon nanotubes *Environmental Science & Technology* **42** 9005–13

[12] Pham H D, Gumbs G, Su W-P, Tran N T T and Lin M-F 2020 Unusual features of nitrogen substitutions in silicene *RSC Advances* **10** 32193–201

[13] Pham H D, Nguyen T D H, Vo K D, Huynh T M D and Lin M-F 2020 Rich essential properties of boron, carbon, and nitrogen substituted germanenes *Applied Physics Express* **13** 085502

[14] Yao Y, Ye F, Qi X-L, Zhang S-C and Fang Z 2007 Spin-orbit gap of graphene: First-principles calculations *Physical Review B* **75** 041401

[15] Foulkes W M C and Haydock R 1989 Tight-binding models and density-functional theory *Physical Review B* **39** 12520

[16] Wang C, Pan B, Tang M, Haas H, Sígalas M, Lee G and Ho K 1997 Environment-dependent tight-binding potential model *MRS Online Proceedings Library Archive* **491**

[17] Jalabert R and Sarma S D 1989 Quasiparticle properties of a coupled two-dimensional electron-phonon system *Physical Review B* **40** 9723

[18] Steiner M, Albers R and Sham L 1992 Quasiparticle properties of Fe, Co, and Ni *Physical Review B* **45** 13272

[19] Warren B E 1990 *X-Ray Diffraction* (New York: Courier Corporation)

[20] Steil S, Großmann N, Laux M, Ruffing A, Steil D, Wiesenmayer M, Mathias S, Monti O L, Cinchetti M and Aeschlimann M 2013 Spin-dependent trapping of electrons at spinterfaces *Nature Physics* **9** 242–7

[21] Wilson J S, Dhoot A, Seeley A, Khan M S, Köhler A and Friend R H 2001 Spin-dependent exciton formation in π-conjugated compounds *Nature* **413** 828–31

[22] Huang H-C, Lin S-Y, Wu C-L and Lin M-F 2016 Configuration-and concentration-dependent electronic properties of hydrogenated graphene *Carbon* **103** 84–93

[23] Lin S-Y, Chang S-L, Tran N T T, Yang P-H and Lin M-F 2015 H—Si bonding-induced unusual electronic properties of silicene: A method to identify hydrogen concentration *Physical Chemistry Chemical Physics* **17** 26443–50

[24] Tran N T T, Lin S-Y, Lin C-Y and Lin M-F 2017 *Geometric and Electronic Properties of Graphene-Related Systems: Chemical Bonding Schemes* (Boca Raton: CRC Press)

[25] Lin M-F, Chuang Y-C and Wu J-Y 2012 Electrically tunable plasma excitations in AA-stacked multilayer graphene *Physical Review B* **86** 125434

[26] Shih P-H, Chiu Y-H, Wu J-Y, Shyu F-L and Lin M-F 2017 Coulomb excitations of monolayer germanene *Scientific Reports* **7** 40600

[27] Barpanda P, Ye T, Nishimura S-I, Chung S-C, Yamada Y, Okubo M, Zhou H and Yamada A 2012 Sodium iron pyrophosphate: A novel 3.0 V iron-based cathode for sodium-ion batteries *Electrochemistry Communications* **24** 116–19

[28] Dresselhaus M and Eklund P 2000 Phonons in carbon nanotubes *Advances in Physics* **49** 705–814

[29] Nika D L and Balandin A A 2012 Two-dimensional phonon transport in graphene *Journal of Physics: Condensed Matter* **24** 233203

[30] Yan J-A, Ruan W and Chou M 2008 Phonon dispersions and vibrational properties of monolayer, bilayer, and trilayer graphene: Density-functional perturbation theory *Physical Review B* **77** 125401

[31] Sahin H and Ciraci S 2012 Chlorine adsorption on graphene: Chlorographene *The Journal of Physical Chemistry C* **116** 24075–83

[32] Tiwari J N, Mahesh K, Le N H, Kemp K C, Timilsina R, Tiwari R N and Kim K S 2013 Reduced graphene oxide-based hydrogels for the efficient capture of dye pollutants from aqueous solutions *Carbon* **56** 173–82

[33] Dias A, Khalam L A, Sebastian M T, Paschoal C W A and Moreira R L 2006 Chemical substitution in Ba (RE1/2Nb1/2) O3 (RE= La, Nd, Sm, Gd, Tb, and Y) microwave ceramics and its influence on the crystal structure and phonon modes *Chemistry of Materials* **18** 214–20

[34] Katre A, Carrete J, Dongre B, Madsen G K and Mingo N 2017 Exceptionally strong phonon scattering by B substitution in cubic SiC *Physical Review Letters* **119** 075902

[35] Estreicher S, Gibbons T, Kang B and Bebek M 2014 Phonons and defects in semi-conductors and nanostructures: Phonon trapping, phonon scattering, and heat flow at heterojunctions *Journal of Applied Physics* **115** 012012

[36] Sivek J, Sahin H, Partoens B and Peeters F M 2013 Adsorption and absorption of boron, nitrogen, aluminum, and phosphorus on silicene: Stability and electronic and phonon properties *Physical Review B* **87** 085444

[37] Hu J, Ouyang C, Yang S A and Yang H Y 2019 Germagraphene as a promising anode material for lithium-ion batteries predicted from first-principles calculations *Nanoscale Horizons* **4** 457–63

[38] Duan Y and Sorescu D C 2009 Density functional theory studies of the structural, electronic, and phonon properties of Li 2 O and Li 2 CO 3: Application to CO 2 capture reaction *Physical Review B* **79** 014301

[39] Ramondo F, Bencivenni L and Grandinetti F 1990 The geometries and vibrational patterns of LiClO3 and NaClO3 ion pairs: An ab initio SCF study *Chemical Physics Letters* **173** 562–8

[40] Pan Y, Chen S and Jia Y 2020 First-principles investigation of phonon dynamics and electrochemical performance of TiO2-x oxides lithium-ion batteries *International Journal of Hydrogen Energy* **45** 6207–16

[41] Ivanovska T, Dionigi C, Mosconi E, De Angelis F, Liscio F, Morandi V and Ruani G 2017 Long-lived photoinduced polarons in organohalide perovskites *The Journal of Physical Chemistry Letters* **8** 3081–6

[42] Ribeiro Jr L A, da Cunha W F, Fonseca A L d A, e Silva G M and Stafström S 2015 Transport of polarons in graphene nanoribbons *The Journal of Physical Chemistry Letters* **6** 510–14

[43] Azároff L V, Kaplow R, Kato N, Weiss R J, Wilson A and Young R 1974 *X-Ray Diffraction* vol 3 (New York: McGraw-Hill)

[44] Nguyen H, Cutler P, Feuchtwang T E, Huang Z-H, Kuk Y, Silverman P, Lucas A and Sullivan T E 1989 Mechanisms of current rectification in an STM tunnel junction and the measurement of an operational tunneling time *IEEE Transactions on Electron Devices* **36** 2671–8

[45] Wang Z, Poncharal P and De Heer W 2000 Measuring physical and mechanical properties of individual carbon nanotubes by in situ TEM *Journal of Physics and Chemistry of Solids* **61** 1025–30

[46] Feenstra R M 1994 Scanning tunneling spectroscopy *Surface Science* **299** 965–79

[47] Palczewski A D 2010 Angle-resolved photoemission spectroscopy (ARPES) studies of cuprate superconductors. Ames Laboratory (AMES), Ames, IA (United States)

[48] Kortüm G 2012 *Reflectance Spectroscopy: Principles, Methods, Applications* (Berlin, Heidelberg: Springer Science & Business Media)

[49] Dowell F E, Pearson T C, Maghirang E B, Xie F and Wicklow D T 2002 Reflectance and transmittance spectroscopy applied to detecting fumonisin in single corn kernels infected with Fusarium verticillioides *Cereal Chemistry* **79** 222–6

[50] Gilliland G 1997 Photoluminescence spectroscopy of crystalline semiconductors *Materials Science and Engineering: R: Reports* **18** 99–399

[51] Corbetta M, Ouazi S, Borme J, Nahas Y, Donati F, Oka H, Wedekind S, Sander D and Kirschner J 2012 Magnetic response and spin polarization of bulk Cr tips for in-field spin-polarized scanning tunneling microscopy *Japanese Journal of Applied Physics* **51** 030208

[52] Ichinokura S 2017 *Observation of Superconductivity in Epitaxially Grown Atomic Layers: In Situ Electrical Transport Measurements* (Singapore: Springer)

[53] Okamoto Y, Matsumoto R, Yagihara T, Iwai C, Miyoshi K, Takeuchi J, Horiba K, Kobayashi M, Ono K, Kumigashira H, Saini N L and Mizokawa T 2017 Electronic structure and polar catastrophe at the surface of ${\mathbf{Li}}_{x}{\mathbf{CoO}}_{2}$ studied by angle-resolved photoemission spectroscopy *Physical Review B* **96** 125147

[54] Arango Y C, Huang L, Chen C, Avila J, Asensio M C, Grützmacher D, Lüth H, Lu J G and Schäpers T 2016 Quantum transport and nano angle-resolved photoemission spectroscopy on the topological surface states of single Sb2Te3 nanowires in *Scientific Reports* p 29493

[55] Ohta T, Bostwick A, McChesney J L, Seyller T, Horn K and Rotenberg E 2007 Interlayer interaction and electronic screening in multilayer graphene investigated with angle-resolved photoemission spectroscopy *Physical Review Letters* **98** 206802

[56] Nicodemus F E 1970 Reflectance nomenclature and directional reflectance and emissivity *Applied Optics* **9** 1474–5

[57] Woolley J T 1971 Reflectance and transmittance of light by leaves *Plant Physiology* **47** 656–62

[58] Trukhin A N, Rogulis U and Spingis M 1997 Self-trapped exciton in Li2GeO3 *J. Lumin.* **72** 890–2

[59] Rohlfing M and Louie S G 2000 Electron-hole excitations and optical spectra from first principles *Physical Review B* **62** 4927

[60] Tiago M L, Northrup J E and Louie S G 2003 Ab initio calculation of the electronic and optical properties of solid pentacene *Physical Review B* **67** 115212

[61] Liu J, Yang Y, Lyu P, Nachtigall P and Xu Y 2018 Few-layer silicene nanosheets with superior lithium-storage properties *Advanced Materials* **30** 1800838

[62] Zhao F, Wang Y, Zhang X, Liang X, Zhang F, Wang L, Li Y, Feng Y and Feng W 2020 Few-layer methyl-terminated germanene—graphene nanocomposite with high capacity for stable lithium storage *Carbon* **161** 287–98

[63] Stehle Y, Meyer III H M, Unocic R R, Kidder M, Polizos G, Datskos P G, Jackson R, Smirnov S N and Vlassiouk I V 2015 Synthesis of hexagonal boron nitride monolayer: Control of nucleation and crystal morphology *Chemistry of Materials* **27** 8041–7

[64] Metois J and Le Lay G 1983 Complementary data obtained on the metal-semiconductor interface by LEED, AES and SEM: Pb/Ge (111) *Surface Science* **133** 422–42

[65] Han X, Tong X, Liu X, Chen A, Wen X, Yang N and Guo X-Y 2018 Hydrogen evolution reaction on hybrid catalysts of vertical MoS2 nanosheets and hydrogenated graphene *ACS Catalysis* **8** 1828–36

[66] Poh H L, Šaněk F, Sofer Z and Pumera M 2012 High-pressure hydrogenation of graphene: Towards graphane *Nanoscale* **4** 7006–11

[67] Vicanek M 1999 Electron transport processes in reflection electron energy loss spectroscopy (REELS) and X-ray photoelectron spectroscopy (XPS) *Surface Science* **440** 1–40

[68] Kleebe H J, Turquat C and Sorarù G D 2001 Phase separation in an SiCO glass studied by transmission electron microscopy and electron energy-loss spectroscopy *Journal of the American Ceramic Society* **84** 1073–80

[69] Krivanek O L, Ursin J P, Bacon N J, Corbin G J, Dellby N, Hrncirik P, Murfitt M F, Own C S and Szilagyi Z S 2009 High-energy-resolution monochromator for aberration-corrected scanning transmission electron microscopy/electron energy-loss spectroscopy *Philosophical Transactions of the Royal Society A: Mathematical, Physical and Engineering Sciences* **367** 3683–97

[70] Song S H, Jang M H, Chung J, Jin S H, Kim B H, Hur S H, Yoo S, Cho Y H and Jeon S 2014 Highly efficient light-emitting diode of graphene quantum dots fabricated from graphite intercalation compounds *Advanced Optical Materials* **2** 1016–23

[71] Lin M, Huang C and Chuu D 1997 Plasmons in graphite and stage-1 graphite intercalation compounds *Physical Review B* **55** 13961

[72] Joo T H, Kim M S and Kim K 1987 Surface-enhanced Raman scattering of benzenethiol in silver sol *Journal of Raman Spectroscopy* **18** 57–60

[73] Van Duyne R P 2012 Laser excitation of Raman scattering from adsorbed molecules on electrode surfaces *Chemical and Biochemical Applications of Lasers* **4** 101

[74] Lin M F and Shung K W K 1994 Plasmons and optical properties of carbon nanotubes *Physical Review B* **50** 17744–7

[75] Pennington G and Goldsman N 2003 Semiclassical transport and phonon scattering of electrons in semiconducting carbon nanotubes *Physical Review B* **68** 045426

[76] Dürkop T, Getty S A, Cobas E and Fuhrer M S 2004 Extraordinary mobility in semi-conducting carbon nanotubes *Nano Letters* **4** 35–9

[77] Martel R, Schmidt T, Shea H R, Hertel T and Avouris P 1998 Single- and multi-wall carbon nanotube field-effect transistors *Applied Physics Letters* **73** 2447–9

[78] Shlafman M, Tabachnik T, Shtempluk O, Razin A, Kochetkov V and Yaish Y E 2016 Self aligned hysteresis free carbon nanotube field-effect transistors *Applied Physics Letters* **108** 163104

[79] Tans S J, Verschueren A R M and Dekker C 1998 Room-temperature transistor based on a single carbon nanotube *Nature* **393** 49–52

[80] Myodo M, Inaba M, Ohara K, Kato R, Kobayashi M, Hirano Y, Suzuki K and Kawarada H 2015 Large-current-controllable carbon nanotube field-effect transistor in electrolyte solution *Applied Physics Letters* **106** 213503

[81] Niimi Y, Matsui T, Kambara H, Tagami K, Tsukada M and Fukuyama H 2006 Scanning tunneling microscopy and spectroscopy of the electronic local density of states of graphite surfaces near monoatomic step edges *Physical Review B* **73** 085421

[82] Misewich J A, Martel R, Avouris P, Tsang J C, Heinze S and Tersoff J 2003 Electrically induced optical emission from a carbon nanotube FET *Science* **300** 783–6

[83] Yuksel R, Sarioba Z, Cirpan A, Hiralal P and Unalan H E 2014 Transparent and flexible supercapacitors with single walled carbon nanotube thin film electrodes *ACS Appl Mater Interfaces* **6** 15434–9

[84] Liang S, Ma Z, Wu G, Wei N, Huang L, Huang H, Liu H, Wang S and Peng L M 2016 Microcavity-integrated carbon nanotube photodetectors *ACS Nano* **10** 6963–71

[85] Zhang T-F, Li Z-P, Wang J-Z, Kong W-Y, Wu G-A, Zheng Y-Z, Zhao Y-W, Yao E-X, Zhuang N-X and Luo L-B 2016 Broadband photodetector based on carbon nanotube thin film/single layer graphene Schottky junction *Scientific Reports* **6** 38569

[86] He X, Léonard F and Kono J 2015 Uncooled carbon nanotube photodetectors *Advanced Optical Materials* **3** 989–1011

[87] Mueller T, Kinoshita M, Steiner M, Perebeinos V, Bol A A, Farmer D B and Avouris P 2010 Efficient narrow-band light emission from a single carbon nanotube p-n diode *Nature Nanotechnology* **5** 27–31

[88] Pyatkov F, Fütterling V, Khasminskaya S, Flavel B S, Hennrich F, Kappes M M, Krupke R and Pernice W H P 2016 Cavity-enhanced light emission from electrically driven carbon nanotubes *Nature Photonics* **10** 420–7

[89] Wang S, Zeng Q, Yang L, Zhang Z, Wang Z, Pei T, Ding L, Liang X, Gao M, Li Y and Peng L-M 2011 High-performance carbon nanotube light-emitting diodes with asymmetric contacts *Nano Letters* **11** 23–9

[90] Mittal J and Lin K L 2017 Carbon nanotube-based interconnections *Journal of Materials Science* **52** 643–62

[91] Singh K and Raj B 2015 Temperature-dependent modeling and performance evaluation of multi-walled CNT and single-walled CNT as global interconnects *Journal of Electronic Materials* **44** 4825–35

[92] Mahan G D 2000 *Many-Particle Physics* (New York: Kluwer Academic/Plenum Publisher)

[93] Ho Y H, Ho G W, Chen S C, Ho J H and Lin M F 2007 Low-frequency excitation spectra in double-walled armchair carbon nanotubes *Phys Rev B* **76**

[94] Li W B, Lin S Y, Tran N T T, Lin M F and Lin K I 2020 Essential geometric and electronic properties in stage-ngraphite alkali-metal-intercalation compounds *Rsc Adv* **10** 23573–81

[95] Meng X Q, Tongay S, Kang J, Chen Z H, Wu F M, Li S S, Xia J B, Li J B and Wu J Q 2013 Stable p- and n-type doping of few-layer graphene/graphite *Carbon* **57** 507–14

[96] Duong D L, Yun S J and Lee Y H 2017 van der Waals layered materials: Opportunities and challenges *Acs Nano* **11** 11803–30

[97] Bostwick A, Ohta T, Seyller T, Horn K and Rotenberg E 2007 Quasiparticle dynamics in graphene *Nat Phys* **3** 36–40

[98] Li G H, Luican A, dos Santos J M B L, Castro Neto A H, Reina A, Kong J and Andrei E Y 2010 Observation of Van Hove singularities in twisted graphene layers *Nat Phys* **6** 109–13

[99] Goerbig M O 2011 Electronic properties of graphene in a strong magnetic field *Rev Mod Phys* **83** 1193–243

[100] Ho J H, Lu C L, Hwang C C, Chang C P and Lin M F 2006 Coulomb excitations in AA- and AB-stacked bilayer graphites *Phys Rev B* **74**

[101] Kordyuk A 2014 ARPES experiment in fermiology of quasi-2D metals *Low Temperature Physics* **40** 286–96

[102] Damascelli A 2004 Probing the electronic structure of complex systems by ARPES *Physica Scripta* **2004** 61

[103] Bussolotti F, Chi D, Goh K J, Huang Y L and Wee A T 2020 *2D Semiconductor Materials and Devices* (Singapore: Elsevier) pp 199–220

[104] Nitta N, Wu F, Lee J T and Yushin G 2015 Li-ion battery materials: Present and future *Materials Today* **18** 252–64

[105] Goodenough J B and Park K-S 2013 The Li-ion rechargeable battery: A perspective *J. Am. Chem. Soc.* **135** 1167–76

[106] Goriparti S, Miele E, De Angelis F, Di Fabrizio E, Zaccaria R P and Capiglia C 2014 Review on recent progress of nanostructured anode materials for Li-ion batteries *Journal of Power Sources* **257** 421–43

[107] Augustsson A, Herstedt M, Guo J-H, Edström K, Zhuang G, Ross Jr P, Rubensson J-E and Nordgren J 2004 Solid electrolyte interphase on graphite Li-ion battery anodes studied by soft X-ray spectroscopy *Physical Chemistry Chemical Physics* **6** 4185–9

19 Future Challenges

Ching-Hong Ho, Vo Khuong Dien, Nguyen Thi Han, Tran Thi Thu Hanh, Ngoc Thanh Thuy Tran, and Ming-Fa Lin

Emergent materials, which have been/will be successfully synthesized in experimental laboratories, deserve systematic investigations from viewpoints of both theories and measurements. They cover a lot of unusual condensed-matter systems, such as carbon-created crystals (diamond/graphene/carbon nanotube/graphene nanoribbon/carbon onion/fullerene/carbon disk/carbon chain [1, 2]/[3–5]/[6, 7]/[8, 9]/[10–12]/[13, 14], layered group-IV and group-V materials [15–19]/[20–23], transition metal disulfide-related compounds [24–26], cathode/electrolyte/anode core components of lithium/aluminum/iron-ion-based batteries [27, 28]/[29–31]/[32, 33], quantum topology insulators [34–36], and perovskite solar cells [37–39]). The various physical/chemical/material phenomena have been clearly identified in the previous studies. For example, systematic investigations have been done on the electronic [40–42], optical [43–45], and Coulomb-excitation properties [46] for the graphene-related 2D materials. The theoretical frameworks of quasiparticle viewpoints are examined and reviewed by the diversified phenomena and expressed in several books [46–52]. They could be classified into two branches: the phenomenological models and the first-principles simulations (details in Chapter 2), in which their unifications are very important under certain cases. Apparently, many open issues need to be covered in the current framework development. Certain of them are thoroughly discussed in the following paragraphs, based on the essential quasiparticle properties [46–52].

The optimal crystal structures or meta-stable configurations could be thoroughly examined by the first-principles calculations [53–55]. It is well known that the chemical reactions are available in generating many compounds and thus creating the diversified fundamental properties, e.g., the intercalation and de-intercalation processes in graphene-related systems [56–58] and anode/electrolyte/anode materials in ion-based batteries [59–61]. For example, the systematic investigations on alkali- and halogen-intercalation layered graphene systems (Figures 19.1(a) and 19.1(b) [62, 63]/[64, 65]) have been absent up to now, in which the various arrangements and concentrations of guest adatoms are not thoroughly examined/verified from the theoretical simulations [62, 64–66] and the experimental measurements [57, 67, 68]. The partially previous studies [62, 69, 70] are insufficient in understanding the various graphene intercalation compounds, such as the competition among AA [71], AB [62, 72, 73], ABC [62], and AAB stackings [74] and the existence of stage-n

(a)   (b)

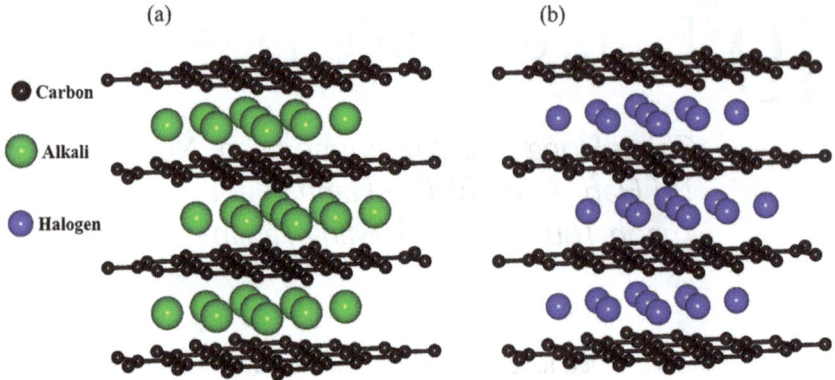

FIGURE 19.1   The optimal geometric structures for (a) alkali- and (b) halogen-intercalation layered graphene systems.

structure [the relative relationship of intercalant and graphitic layers [62, 75]. The close combination of DFT calculations [76] and molecular dynamics [77, 78] would become an efficient way to explore the optimal growth processes. Similar studies could be generalized to graphite/carbon nanotube/fullerene intercalation compounds [62]/[79, 80]/[81, 82], especially for the charging and discharging processes of the first ones [62]. The various crystal symmetries, which will greatly diversify the quasiparticle behaviors, deserve thorough investigation.

It is very interesting in fully understanding the meta-stable crystal structures during the ion transports of lithium- (Figure 19.2 [83–85]), aluminum- [86], or iron-based batteries [87]. Apparently, the intercalations and de-intercalations of host atoms come to exist very quickly, even resulting in the terminations of the continuous x-ray examinations [88, 89]. That is, the high-resolution experimental measurements of x-ray diffraction peaks cannot follow the very rapid variations of crystal structures during the chemical reactions. The first-principles simulations, being based on the chemical viewpoints [83–85], might be suitable and reliable in simulating a lot of intermediate states due to the chemical modifications. If the effective times of charging and discharging processes could be estimated under this approach, this is very useful in detecting the initial and final crystal symmetries of core cathode/electrolyte/anode materials. Most importantly, the numerical simulations could achieve the spatial charge density distributions, in which their time dependences dominate the stationary current density in ion transport. How to efficiently link with the phenomenological models [90] in exploring the x-ray diffraction patterns [91] and the electrical conductivities [92, 93] should be important issues in near-future studies.

There also exist many complicated structures, being required to thoroughly clarify the geometric features by the first-principles simulations, such as the battery boundaries between cathode and electrolyte/electrolyte and anode [83–85], few-layer group-IV/group-V on different substrates [40, 94–96], their various composites (Figure 19.3 [97, 98]), and rare-earth-/transition-metal adatoms chemically adsorbed above them

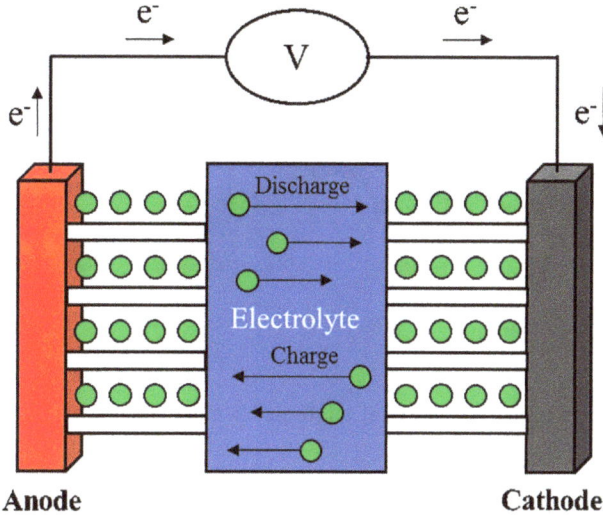

FIGURE 19.2   The charging and discharging processes in lithium-ion-based batteries.

(a) Graphene-silicene

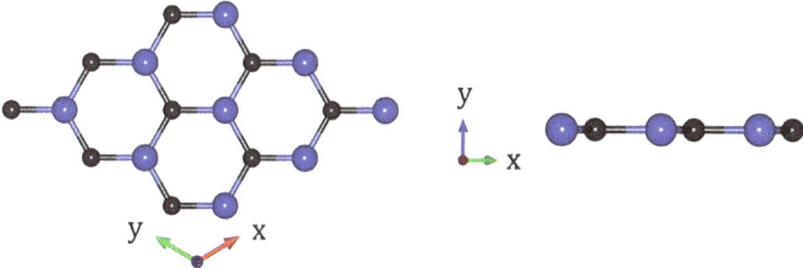

(b) Graphene-silicene

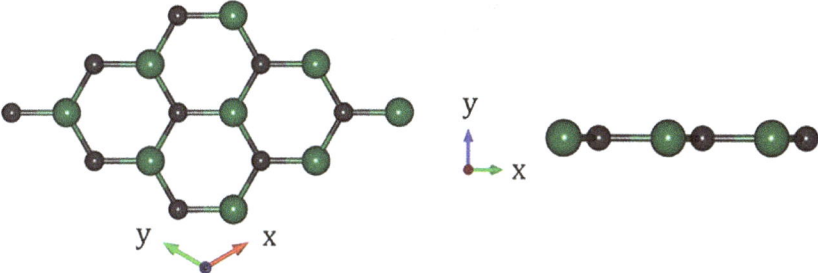

FIGURE 19.3   The composite structures of (a) graphene-silicene and (b) graphene-plumbene.

(a) Transition-metal adatoms on monolayer graphene

(b) Rare-earth-metal adatoms on monolayer graphene

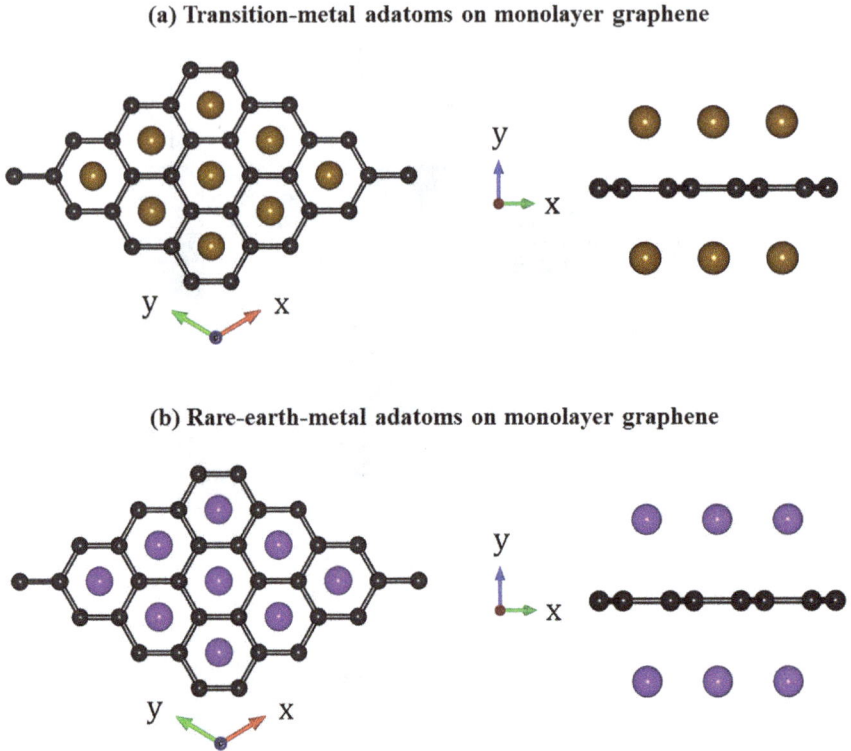

FIGURE 19.4 The chemisorption of (a) transition- and (b) rare-earth-metal adatoms on monolayer graphene under the highest adatom concentration.

(Figure 19.4 [99]). The optimal geometries are very important in comprehending the other fundamental properties, e.g., the reliable ion transport behaviors across two boundaries and the heterojunction effects on the essential properties of 2D systems. It should have the modified chemical bondings [49, 52] and even the spin-dependent on-site Coulomb interactions [49, 52] through the coupling of two different subsystems. The active orbital hybridizations in two subsystems [100] and heterojunction [101, 102], as well as the atom- and orbital-decomposed spin configurations [49, 52], are expected to dominate the optimal geometric structures. How many layers of the substrate is reasonable during the numerical simulations is a very interesting issue in the current investigation [103–105]. The basic rule is to be consistent with the reliable experimental measurements [105]. Very interestingly, the chemisorption/substitution of rare-earth-/transition-metal adatoms on 2D emergent materials is outstanding candidate in thoroughly investigating the competitive and/or cooperative relations of magnetic configurations (ferromagnetic, anti-ferromagnetic, and non-magnetic ones [106–110]) and orbital hybridizations. That is, the critical roles of quasiparticle charges and spins can be clarified from the featured fundamental properties.

While the chemical modifications [111, 112] come into existence at very dilute adatom/guest-atom concentrations, they will clearly show another very interesting topic of charged impurities. It is well known that the 3D diamond crystals of sp^3 bondings, being built from carbon [113], silicon [113], and germanium elements [113], are able to exhibit the impurity-enriched quasiparticle phenomena. Furthermore, group-III B/Al/Ga/In/Tl [113] and group-V N/P/AS/Sb/B [113], respectively, serve as acceptor- and donor-type charged impurities. The attractive Coulomb interaction is formed between the negative/positive ion charge and the extra hole/electron. This could be regarded as a hydrogen-like quasiparticle, in which the effective mass of the kinetic energy and the static dielectric constant of the Coulomb potential energy will determine the stable bound states above/below the top/bottom of the valence/conduction band. Similar topics are associated with the low-dimensional emergent materials. Both 2D germanene and silicene, as well as 1D partners of nanoribbons germanene, are outstanding candidates in diversifying the charged impurity behaviors. How to utilize the 2D and 1D Schrodinger depends on the reliable parameters of m^* and $\epsilon(\acute{E}=0)$. Their values could be estimated from the VASP calculations [114].

Moiré superlattices [115], with many atoms/molecules in a unit cell, frequently come to exist in most multi-component compounds [83, 85] and mono-element materials under an external magnetic field [116]. That is, they are easily generated by the stable/meta-stable crystal structures, the layered/coaxial/spherical couplings, the distinct sample growths on various substrates (heterojunctions), the composite buckled materials of distinct subsystems, the chemical absorptions/substitutions/intercalations, and the external electric/magnetic fields. This will introduce emergent challenges for theoretical predictions and experimental analyses. Most importantly, their chemical environments are highly non-uniform, since the active orbital hybridizations are very sensitive to the atomic positions. This is clearly revealed as the strong fluctuations of the active chemical bonds. For example, the quaternary compound of $LiFePO_4$, which belongs to a highly efficient cathode material of Li^+-based battery [117, 118], presents an observable variation for the various Li-O Fe-O and P-O bonds. This can provide an outstanding chemical environment for ion transports, since the single- and multi-orbital hybridizations are easily modulated by the different atom positions and even concentrations during the charging and discharging processes [62]. The main features of geometric and electronic properties could be fully explored under the VASP simulations but not the tight-binding model. How to further utilize the numerical results of the spatial charge densities in determining the X-ray diffraction patterns deserves near-future systematic investigations. Moreover, the reliable parameters of the tight-binding model and the suitable hopping integrals of the neighboring orbital hybridizations would become very difficult to achieve from the fitting of both methods within the low-lying valence and conduction sub-bands. The main reason is too complicated electronic energy spectra. This would lead to serious suppression for the current development of the theoretical model. Apparently, the previous calculations [83–85], being finished on the core materials of batteries, do not show the theoretical models with the analytic formulas.

While Moiré superlattices are greatly reduced under specific conditions, the relatively simple chemical environments [49, 119], with the concise hopping integrals [46, 47], are responsible for building the reliable phenomenological models. For example, the high concentration ratio of guest and host atoms, being supported by the active orbital hybridizations, can provide the necessary cases, such as the prominent chemisorption of hydrogen/oxygen/alkali adatoms on graphene/group-IV systems (Figures 19.5(a)–19.5(c) [40]/[120]/[121]), $B_xC_yN_z$ binary/ternary planar/tubular compounds (Figures 19.6(a)/19.6(b) [122, 123]/[124, 125]), and Si-/Ge-/Sn-/Pb-C honeycomb crystals [126]. It should be noticed that most of the critical chemical bondings have been thoroughly examined under the theoretical frameworks of quasiparticle particles [49, 52]. For example, the hydrogenated, oxidized, and alkalized graphene, respectively, corresponding to the top-, bridge- and hollow-site chemisorption, present the interlayer s-sp^3 [40], sp^3-sp^3 [120], and s-2p$_z$ [62, 121] hybridizations. Furthermore, the other organic compounds belong to the sp^3-sp^3 bondings. The

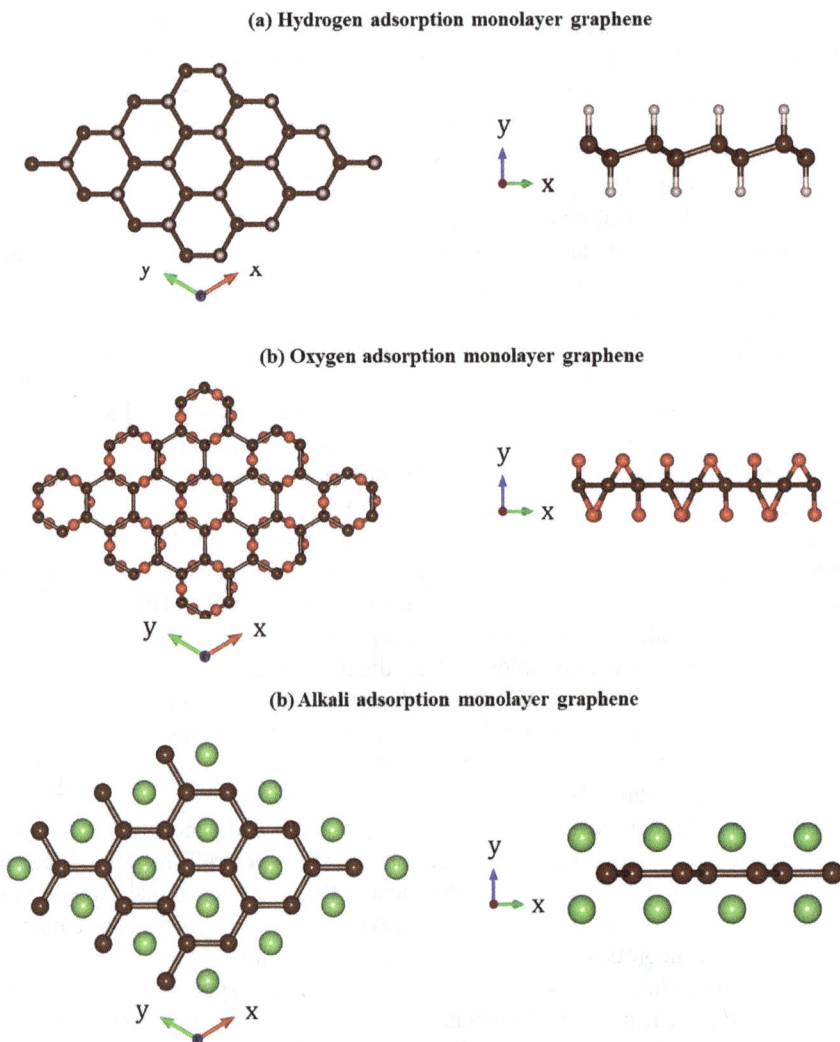

FIGURE 19.5   The optimal geometries for (a) hydrogen, (b) oxygen, and (c) alkali adatoms on graphene, with the largest adsorptions.

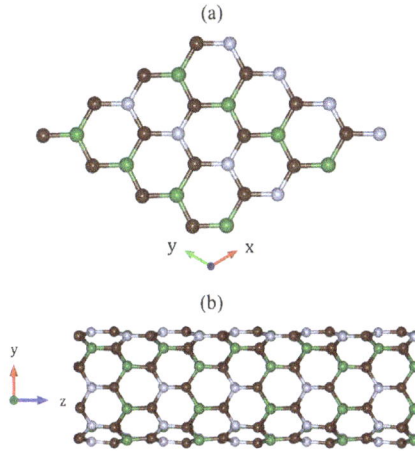

FIGURE 19.6 $B_xC_yN_z$ (a) planar and (b) nanotube compounds.

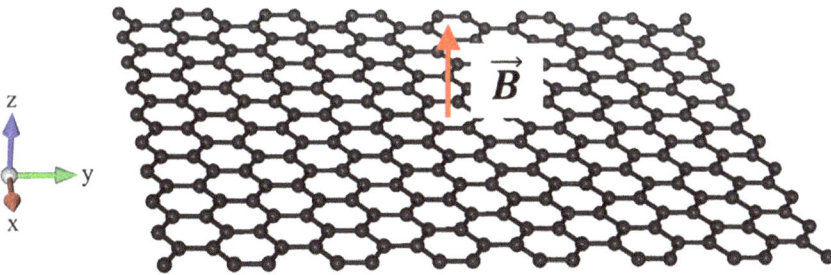

FIGURE 19.7   A greatly enlarged unit cell of monolayer graphene, being created by a uniform perpendicular magnetic field.

strongest hopping integrals, which mainly come from the few nearest neighbors, are expected to be obtained through the suitable fitting of the tight-binding model with the VASP calculations about the low-lying energy spectra. The close combinations of two methods will open other research themes.

The generalized tight-binding model is further developed under the various external fields: (a) a uniform perpendicular magnetic field [127, 128], (b) a spatially modulated magnetic or electric field [127, 129], and a composite field with the homogeneous and non-uniform components [127, 130, 131]. Very interestingly, the diverse magnetic quantization phenomena are easily achieved from the unusual materials and the significant modulations. Obviously, the vector potential can create the well-known Peierls phases in the hopping integrals of active orbital hybridizations and thus another periodical period. In general, the magnetic unit cell is greatly enlarged under a suitable gauge [116], as clearly illustrated in Figure 19.7 for a honeycomb lattice. The magnetic Hamiltonian, being based on a field-dependent Moiré superlattice, is a rather large Hermitian matrix [116]. It would be very difficult to directly finish the efficient diagonalizations. However, under the exact technique, one can overcome the numerical barriers [48]. That is, the optimal sequence of the magnetic tight-binding functions is available in solving the Landau-level spectra and their oscillatory and localized wave functions. The quantum numbers are further characterized from the featured zero-point numbers on a dominating sublattice. A lot of interesting quasi-particle phenomena have been clarified through the systematic studies [all the details in books [48]], such as the rich magnetic-field dependences in AA-, AB- [132], ABC- [129], and AAB-stacked graphene systems [74], their non-crossing/crossing/anti-crossing behaviors [133, 134], the valley splitting due to an extra gate voltage [23, 135], magneto-optical selection rules [95], the various quantum Hall conductances [136], and the single-particle and collective magneto-electronic excitations [137, 138]. Similar calculations and analyses could be utilized to further explore the diversified effects associated with the high-concentration chemisorption [40] and substitution [100]. For example, it is very interesting in understanding the critical roles of the outmost alkali and carbon-$2p_z$ orbitals for the Landau levels in alkalized graphene systems [121]. Many outstanding layered candidates, as mentioned earlier [139], are very suitable in the further explorations of the unusual magnetic quantization phenomena. It should be noticed that as for the core materials of ion-based batteries, the generalized tight-binding model is meaningful only for the layered graphite intercalation compounds but useless for the 3D compounds with large Moiré superlattices. The main reason is that the generalized tight-binding model is capable of examining the featured Landau sub-bands with an observable energy spacing.

Optical excitations belong to one of the mainstream science researches [140] and play critical roles in the frequent applications [141, 142]. Their rich and unique features could be examined/predicted by the numerical simulations [143, 144] and the phenomenological model [119, 145]. The former and the latter, respectively, possess the research merits under Moiré superlattices/excitonic effects [146, 147] and magnetic quantization behaviors [127, 148]. Very interestingly, the systematic investigations of theoretical simulations, being made on the single-particle and many-body optical properties of carbon-sp^2-bonding systems [149, 150], might have been absent up to now. The graphene-related systems, with the hexagonal honeycomb crystals [4], cover the AA- [151], AB- [152], ABC- [153], and AAB-stacked ones [154], single-/multi-walled carbon nanotubes (Figures 19.8(a) and 19.8(b) [155, 156]), their composite nanotube bundles, and graphene nanoribbons [44, 127]. Furthermore, they become the various binary and ternary compounds after the very efficient adsorption, intercalation, and substitution. Apparently, the important mechanisms, which correspond to the geometric symmetries, the chemical modifications, and the excitonic effects, are expected to generate the diverse phenomena of the vertical optical transitions. For example, there are three kinds of single-walled nanotubes because of its radius and chirality [157]; therefore, this will lead to the rather different optical threshold frequencies in the presence/absence of many-particle electron-hole interactions [158, 159]. The attractive Coulomb interactions, with the charge screening from all carriers, are very difficult to deal with using the random-phase approximation. The numerical barriers could be solved by the VASP simulations [76]. Only this method could provide all the featured excitation spectra. Such optical research topics

(a) double-wall armchair nanotube

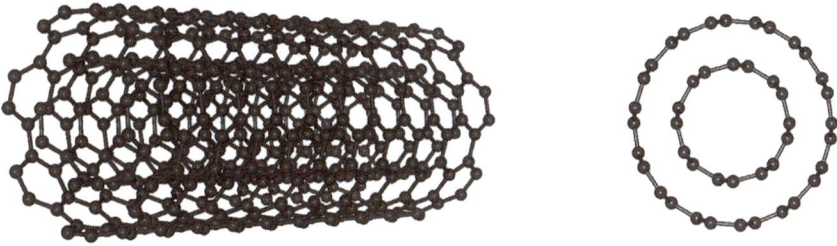

(b) double-wall zigzag nanotube

FIGURE 19.8   The double-wall (a) armchair and (b) zigzag carbon nanotubes.

are under the current explorations, especially for carbon nanotubes [160, 161] and layered graphenes [162] under the various cases.

Quantum-confinement and dimension-crossover effects on optical properties are very interesting research themes [163–165]. The optical scatterings of an electromagnetic wave inside a layered material have been absent up to now, such as the absence of theoretical predictions and experimental examinations about the layer- and frequency-dependent reflectance [166, 167] and transmission spectra [168]. It is well known that the essential properties of graphene systems strongly depend on layer numbers and stacking configurations [169, 170]. While an electromagnetic field is normally incident finite-thickness graphene (Figure 19.9 [171]), simultaneous incidence, reflection, and transmission occur at the top and bottom surfaces. The two independent boundary conditions, the continuities of the tangential electric and magnetic fields [171], are required there during a lot of optical scatterings. Furthermore, the dynamic charge responses are responsible for the significant couplings with the propagation electromagnetic waves. All the orbital-hybridization-related electronic states [172, 173] can effectively absorb photon energies at the distinct frequency ranges, thus creating the absorption coefficient and the decline of the wave packet. First, the generalized tight-binding model is utilized to calculate the electronic energy spectra and wave functions for layered graphene systems with any layers under the

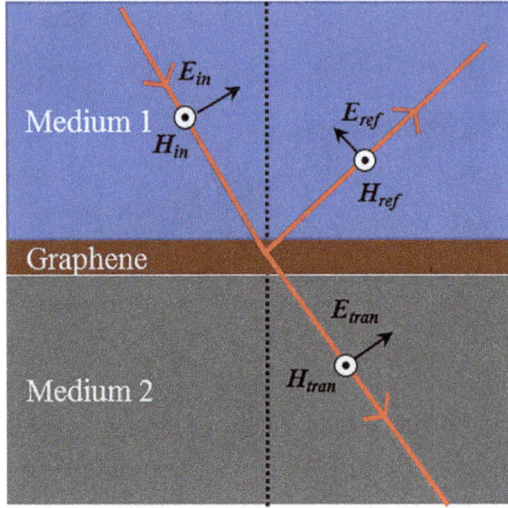

FIGURE 19.9   The incidence, reflectance, and transmission of an electromagnetic wave on a layered graphene.

ABC [153] and AAB stackings [174]. This is also suitable in fully exploring mag-neto-optical properties [95, 175]. Second, the transverse dielectric function, which characterizes the charge screening behaviors [176], is evaluated from the dynamic Kubo formula [177]. Finally, the Helmholtz wave equation, being supported by two independent boundary conditions, is responsible for the optical scatterings within a sample and thus determines reflectance, absorbance, and transmittance [171]. The current investigations are focused on the featured reflectance and transmission spec-tra closely related to dimension crossover behaviors [178] and quantum confinement effects [164, 179].

Obviously, hydrogen storage has become an emergent research theme of green energies [180, 181]. This might be achieved by the significant chemisorptions on 2D layered materials. For example, the hydrogenated group-IV systems are outstand-ing candidates in presenting the rich and unique phenomena. From the first-step VASP calculations [182, 183], the very effective chemical bondings are established between H-adatoms and hosts, being clearly indicated from the created/enhanced non-planar honeycomb crystal with the top-site absorptions. Most importantly, the specific multi-orbital hybridizations need to be thoroughly examined from the atom-dominated band structures, the spatial charge densities, and the orbital- and orbital-decomposed van Hove singularities (details in Chapter 2.1). Furthermore, the spin configurations are solved from the atom- and orbital-projected magnetic moments, the spin-split energy spectra, and spin-related density of states. Since there exist significant differences among C, Si, Ge, Sn, and Pb atoms in the sp^2, sp^3 chemical bondings, the spin-orbit couplings [184, 185], and the hydrogenation effects [186], the hydrogenated group-IV compounds are expected to behave so. The main fea-tures of geometric and electronic properties might lie in energy gaps, destruction of

Dirac-cone structures, serious mixings of π and σ energy sub-bands, the coexistence of spin-split and spin-degenerate electronic states, unusual charge density distributions, and unique van Hove singularities. Such results are directly reflected in the vertical optical transitions [187]. It is also noticed that the excitonic effects, being beyond the single-particle pictures, will strongly modify the electronic excitations near the threshold frequency [188, 189]. How many prominent many-body absorption peaks should be very sensitive to the atomic numbers and the hydrogen absorption cases. Apparently, the optical absorption structures, which come to exist within the observable frequency range, are available in determining the multi-/single-orbital hybridizations of active chemical bonds, the non-magnetic/ferromagnetic/anti-ferro-magnetic configurations, and the excitonic effects. The systematic VASP simulations are under the current investigations, in which they are partially supported by the phenomenological models under small Moiré superlattices. In addition, to clarify the spin-induced optical excitations, similar studies have been generalized to the fer-romagnetic materials, such as the spin-related optical absorption structures in rare-earth and transition-metal-absorbed group-IV compounds [190, 191].

The emergent 2D/1D systems, which have been successfully/will be synthesized by the various synthesis methods, can provide excellent physical/chemical environments in a thorough investigation of the many-particle electron-electron Coulomb interactions, such as graphene- [192, 193], nanotube- [194, 195], and nanoribbon-related ones [196, 197] and their group-IV partners [198, 199]. The complete studies, such as the Friedel oscillations due to charged impurities [200, 201], the dynamic charge responses associated with the perturbation of electron beams [202, 203], and the Coulomb deexcitations for quasi-particle energy spectrum and lifetime, do not achieve up to date. For example, how to generalize/develop the current theoretical frameworks for multi-layer and coaxial crystal structures might be extra challenges. The previous development of the modified random-phase approximation [204] and self-energy method [205, 206] could be utilized to understand the unique electron-electron inelastic scatterings. For example, whether the coupled graphene/carbon nanotube systems could present the beating screening behaviors in the presence/absence of n-/p-type dopings is a very interesting research topic. This will clarify the existence of the composite Friedel oscillations [the multiple Coulomb scatter-ings related to the Fermi surfaces [207, 208]. As to the featured propagation of an electron beam, the significant couplings between charge carriers and time- and position-dependent Coulomb-field perturbations are predicted to display the plasma waves. While one does the delicate calculations and analyses the dynamic behaviors of the spatial screening charge densities and effective potentials, the single-parti-cle and collective excitations will dominate the main features of the spatial wave propagation. However, an open issue is induced by the real case due to the external charge beam; that is, a suitable model for its perturbation will be responsible for creating the diversified phenomena of charge dynamics [209, 210]. It is well known that the electron-hole pair excitations and plasmon modes are the effective deexcita-tion channels [211]. These should be very complicated through the interlayer atomic and Coulomb interactions. The band-structure effects, as well as the intralayer and interlayer electric polarizations [132, 212], must be taken into account for the model calculations simultaneously. The calculated results could be further examined by the high-resolution angle-resolved photoemission spectroscopy [details in Chapter 3], e.g., the quasiparticle energy spectra of few-layer AB- [213] and ABC-stacked graphene systems [214, 215].

Quantum Hall electrical conductivities [216, 217] and magneto-optical selection rules [95, 218] can present the unique theoretical frameworks, namely, the direct combination of the generalized tight-binding model [details in Chapter 2.2 [219, 220]] and the static/dynamic Kubo formulas [221–223]. Any layered materials, with small Moiré superlattices, could be chosen for fully illustrating the various magnetic quantizations. Under this composite method [219, 224], the pristine few-layer graphene systems have been verified to display the layer-number- and stacking-enriched quantum transports. The concise physical scatterings are successfully proposed in explaining the rich and unusual results [95, 225]. In addition, the effective-mass approximation cannot provide the exact Landau levels with the anti-crossing and oscillatory features [226, 227] or the reliable quantum structures under the specific stackings [e.g., ABC and AAB ones [228, 229]]. Similar frameworks would be very useful in exploring the prominent effects due to the chemical modifications. For example, layered $B_xC_yN_z$ honeycomb crystals [123, 230], B/C/N adatom adsorptions on silicene/germanene [111, 231]/[100], AA-, AB-, and ABC-stacked bulk graphites [232, 233], and graphite intercalation compounds [234, 235] are expected to show the doping-, valley-, orbital-hybridization-, spin- and dimension-diversified Hall conductivities, i.e., the significant couplings of the critical mechanisms could induce the diverse quantum Hall effects. Since the initial and final magneto-electronic states dominate the carrier mobility, their spatial distribution symmetries would set the scattering selection rules [116, 236]. At very low temperatures [116, 237], this accounts for the various quantum structures in the gate-voltage- and magnetic-field-dependent conductivities, e.g., the non-uniform heights and widths in the quantized transport properties of few-layer graphene systems with the distinct stackings [94, 238]. On the other side, the dynamic magneto-optical excitations are capable of illustrating another magnetic quantization phenomenon, as done for the AA-, AB-, ABC, and AAB-stacked trilayer graphenes [153, 174, 239]. Obviously, there exist diverse optical absorption structures, covering the various magneto-optical selection rules [at least three kinds in [239]], the single-, pair-, and/or double-like peak structures [95, 153, 174], and the Landau-level subgroup-induced optical excitation categories [151, 240]. How to introduce the other critical factors in observing the diversified phenomena deserves a closer examination within an enlarged framework.

Topological materials currently emerging from research are, from the application viewpoint, much more superior as compared with topologically trivial materials. Specifically speaking, the robustness against disorder implies high mobility of electrons; in particular, because massless fermions are sensitive in response to applied electric and/or magnetic fields, the edge states of Chern insulators can serve as dissipationless electric current-carrying channels [241, 242]. Such advantages could be exciting at the prospect of perfect electric conduction. More strikingly, the Berry phase manifests itself by hardness with electron spin or some kinds of pseudospin so that the application in spin current filtration is promising. This provides the basis of spintronics [243]. However, the state of the art is far from successful in application.

Despite the ongoing discovery of topological materials [34, 244–252], there is yet a lot of room to improve for practice with better materials.

The current development of first-principles techniques has achieved great success in helping researchers explore many topological materials. First principles have the ability to calculate complicated band structures of realistic materials [250]. Besides, the topological robustness against disorder and the single-particle picture weak correlation frequently met make that reasonable. People might conclude that first-principles calculations, together with a helpful analysis for a suitable lattice model by group theory, can be effective in the prediction of almost all topological materials. However, there are several problems that could bring about challenges to such an approach. Firstly, either LDA- (local density approximation) or GGA- (generalized gradient approximation) based first-principle calculation originates from the mean-field theory [250]. Therefore, it cannot be used for strongly correlated systems, such as systems showing fractional quantum Hall effects and high-temperature superconductors, where the single-particle picture fails [252]. Further, the so-called topological superconductors have been identified in various topological classes, with single-particle bands and renormalized high-temperature superconducting pairing terms combined in the Hamiltonians [253, 254]. For this, the present ability of first-principles calculations is yet limited. Next, another shortcoming of first principles is that the calculation results are prone to show band gaps less than their actual values [250]. Hence, the distinction between a gapped phase with a narrow gap and a gapless phase cannot be distinguished clearly. Lastly, one technological problem ought to be pointed out. The Bloch wave functions taken into first-principles calculations are random over all the gauge-dependent eigenstates, though the Berry flux derived from Bloch wave functions and rendering topological invariants is gauge independent. Hence, the calculated topological invariants can be involved and need more complexity in the calculations.

Two-dimensional materials provide many applications in green energy, such as hydrogen storage capacity, catalysts for solar fuels [255–260]. Therefore, the future studies on the hydrogen evolution reaction and the CO_2 reduction are the potential directions with the questions that need further understanding, such as: What are the stable structures of the 2D surfaces with/without defects? What is their stability under the equilibrium thermodynamic conditions? What are the possible intermediate states of the hydrogen on these surfaces, and their thermodynamic stability? What is the activation barrier as the reaction happened? How does the activation energy changes including the salvation effects? Furthermore, most of the recent studies on the interaction of atoms and molecules on 2D materials have neglected the phonon calculation, the quantum effect on the binding capacity. Previous studies for bulk materials showed that the influence of phonon is significant for the bonding of atoms and molecules to the surface of bulk materials [261–263]. Therefore, this is also a research orientation for 2D materials in the next studies.

REFERENCES

[1] Kimball G E 1935 The electronic structure of diamond *The Journal of Chemical Physics* **3** 560–4

[2] Roberts R and Walker W 1967 Optical study of the electronic structure of diamond *Physical Review* **161** 730

[3] Castro Neto A H, Guinea F, Peres N M R, Novoselov K S and Geim A K 2009 The electronic properties of graphene *RvMP* **81** 109–62

[4] Geim A K and Novoselov K S 2010 *Nanoscience and Technology: A Collection of Reviews from Nature Journals* (Singapore: World Scientific) pp 11–19

[5] Rao C, Biswas K, Subrahmanyam K and Govindaraj A 2009 Graphene, the new nano-carbon *J Mater Chem* **19** 2457–69

[6] Jiao L, Zhang L, Wang X, Diankov G and Dai H 2009 Narrow graphene nanoribbons from carbon nanotubes *Nature* **458** 877–80

[7] Son Y-W, Cohen M L and Louie S G 2006 Half-metallic graphene nanoribbons *Nature* **444** 347–9

[8] Choudhury S, Zeiger M, Massuti-Ballester P, Fleischmann S, Formanek P, Borchardt L and Presser V 2017 Carbon onion—sulfur hybrid cathodes for lithium—sulfur batteries *Sustainable Energy & Fuels* **1** 84–94

[9] Pech D, Brunet M, Durou H, Huang P, Mochalin V, Gogotsi Y, Taberna P-L and Simon P 2010 Ultrahigh-power micrometre-sized supercapacitors based on onion-like carbon *Nature Nanotechnology* **5** 651–4

[10] Curl R F and Smalley R E 1991 Fullerenes *Scientific American* **265** 54–63

[11] Hirsch A 1994 *The Chemistry of the Fullerenes* (Wienheim: Wiley Online Library)

[12] Kadish K M and Ruoff R S 2000 *Fullerenes: Chemistry, Physics, and Technology* (Hoboken, New Jersey: John Wiley & Sons)

[13] Zhao G, He Y, Xu Z, Hou J, Zhang M, Min J, Chen H Y, Ye M, Hong Z and Yang Y 2010 Effect of carbon chain length in the substituent of PCBM-like molecules on their photovoltaic properties *Adv Funct Mater* **20** 1480–7

[14] Zhao X, Ando Y, Liu Y, Jinno M and Suzuki T 2003 Carbon nanowire made of a long linear carbon chain inserted inside a multiwalled carbon nanotube *Physical Review Letters* **90** 187401

[15] Johnson N W, Vogt P, Resta A, De Padova P, Perez I, Muir D, Kurmaev E Z, Le Lay G and Moewes A 2014 The metallic nature of epitaxial silicene monolayers on Ag (111) *Adv Funct Mater* **24** 5253–9

[16] Tao L, Cinquanta E, Chiappe D, Grazianetti C, Fanciulli M, Dubey M, Molle A and Akinwande D 2015 Silicene field-effect transistors operating at room temperature *Nature Nanotechnology* **10** 227–31

[17] Gou J, Zhong Q, Sheng S, Li W, Cheng P, Li H, Chen L and Wu K 2016 Strained mono-layer germanene with 1×1 lattice on Sb (111) *2D Materials* **3** 045005

[18] Gupta S K, Singh D, Rajput K and Sonvane Y 2016 Germanene: A new electronic gas sensing material *RSC Advances* **6** 102264–71

[19] Roome N J and Carey J D 2014 Beyond graphene: Stable elemental monolayers of silicene and germanene *ACS Applied Materials & Interfaces* **6** 7743–50

[20] Kecik D, Özçelik V O, Durgun E and Ciraci S 2019 Structure dependent optoelectronic properties of monolayer antimonene, bismuthene and their binary compound *Physical Chemistry Chemical Physics* **21** 7907–17

[21] You B, Wang X, Zheng Z and Mi W 2016 Black phosphorene/monolayer transition-metal dichalcogenides as two dimensional van der Waals heterostructures: A first-principles study *PCCP* **18** 7381–8

[22] Zeng J, Cui P and Zhang Z 2017 Half layer by half layer growth of a blue phosphorene monolayer on a GaN (001) substrate *Physical Review Letters* **118** 046101

[23] Chen R-B, Jang D-J, Lin M-C and Lin M-F 2018 Optical properties of monolayer bis-muthene in electric fields *Optics Letters* **43** 6089–92

[24] Choi W, Choudhary N, Han G H, Park J, Akinwande D and Lee Y H 2017 Recent development of two-dimensional transition metal dichalcogenides and their applications *Mater Today* **20** 116–30

[25] Kong D, Cha J J, Wang H, Lee H R and Cui Y 2013 First-row transition metal dichalcogenide catalysts for hydrogen evolution reaction *Energy & Environmental Science* **6** 3553–8

[26] Tian H, Chin M L, Najmaei S, Guo Q, Xia F, Wang H and Dubey M 2016 Optoelectronic devices based on two-dimensional transition metal dichalcogenides *Nano Research* **9** 1543–60

[27] Marom R, Amalraj S F, Leifer N, Jacob D and Aurbach D 2011 A review of advanced and practical lithium battery materials *J Mater Chem* **21** 9938–54

[28] Van Schalkwijk W and Scrosati B 2002 *Advances in Lithium-Ion Batteries* (New York City: Springer) pp 1–5

[29] Jayaprakash N, Das S and Archer L 2011 The rechargeable aluminum-ion battery *Chem Commun* **47** 12610–2

[30] Yang H, Li H, Li J, Sun Z, He K, Cheng H M and Li F 2019 The rechargeable aluminum battery: Opportunities and challenges *Angew Chem Int Ed* **58** 11978–96

[31] Yang S and Knickle H 2002 Design and analysis of aluminum/air battery system for electric vehicles *J Power Sources* **112** 162–73

[32] Licht S, Tel-Vered R and Halperin L 2002 Direct electrochemical preparation of solid Fe (VI) ferrate, and super-iron battery compounds *Electrochem Commun* **4** 933–7

[33] Licht S, Wang B and Ghosh S 1999 Energetic iron (VI) chemistry: The super-iron battery *Science* **285** 1039–42

[34] Hasan M Z and Kane C L 2010 Colloquium: Topological insulators *Reviews of Modern Physics* **82** 3045

[35] Moore J E 2010 The birth of topological insulators *Nature* **464** 194–8

[36] Qi X-L and Zhang S-C 2010 The quantum spin Hall effect and topological insulators *arXiv preprint arXiv:1001.1602*

[37] Correa-Baena J-P, Saliba M, Buonassisi T, Grätzel M, Abate A, Tress W and Hagfeldt A 2017 Promises and challenges of perovskite solar cells *Science* **358** 739–44

[38] Green M A, Ho-Baillie A and Snaith H J 2014 The emergence of perovskite solar cells *Nature Photonics* **8** 506–14

[39] Jung H S and Park N G 2015 Perovskite solar cells: From materials to devices *Small* **11** 10–25

[40] Huang H-C, Lin S-Y, Wu C-L and Lin M-F 2016 Configuration-and concentration-dependent electronic properties of hydrogenated graphene *Carbon* **103** 84–93

[41] Lin S-Y, Chang S-L, Chen H-H, Su S-H, Huang J-C and Lin M-F 2016 Substrate-induced structures of bismuth adsorption on graphene: A first principles study *PCCP* **18** 18978–84

[42] Thanh Thuy Tran N, Lin S-Y, Lin Y-T and Lin M-F 2015 Chemical bondings induced rich electronic properties of oxygen absorbed few-layer graphenes *arXiv arXiv:1507.07057*

[43] Chen S-C, Chiu C-W, Wu C-L and Lin M-F 2014 Shift-enriched optical properties in bilayer graphene *RSC Advances* **4** 63779–83

[44] Chung H-C, Lin Y-T, Lin S-Y, Ho C-H, Chang C-P and Lin M-F 2016 Magnetoelectronic and optical properties of nonuniform graphene nanoribbons *Carbon* **109** 883–95

[45] Lee M, Chung H, Lu J, Chang C and Lin M-F 2015 Electronic and optical properties in graphane *Philosophical Magazine* **95** 2717–30

[46] Lin C-Y, Wu J-Y, Chiu C-W and Lin M-F 2019 *Coulomb Excitations and Decays in Graphene-Related Systems* (Boca Raton, Florida: CRC Press)

[47] Lin C-Y, Chen R-B, Ho Y-H and Lin M-F 2017 *Electronic and Optical Properties of Graphite-Related Systems* (Boca Raton, Florida: CRC Press)

[48] Lin C-Y, Do T-N, Huang Y-K and Lin M-F 2017 *Optical Properties of Graphene in Magnetic and Electric Fields* (Bristol, England: IOP Publishing)

[49] Tran N T T, Lin S-Y, Lin C-Y and Lin M-F 2017 *Geometric and Electronic Properties of Graphene-Related Systems: Chemical Bonding Schemes* (Boca Raton, Florida: CRC Press)

[50] Chen S, Wu J, Lin C and Lin M 2016 Theory of magnetoelectric properties of 2d systems *New J. Phys* **18** 103024

[51] Lin C-Y, Ho C-H, Wu J-Y, Do T-N, Shih P-H, Lin S-Y and Lin M-F 2019 *Diverse Quantization Phenomena in Layered Materials* (Boca Raton, Florida: CRC Press)

[52] Lin S-Y, Tran N T T, Chang S-L, Su W-P and Lin M-F 2018 *Structure-and Adatom-Enriched Essential Properties of Graphene Nanoribbons* (Boca Raton, Florida: CRC Press)

[53] Freysoldt C, Grabowski B, Hickel T, Neugebauer J, Kresse G, Janotti A and Van de Walle C G 2014 First-principles calculations for point defects in solids *Reviews of Modern Physics* **86** 253

[54] Muscat J, Swamy V and Harrison N M 2002 First-principles calculations of the phase stability of TiO 2 *Physical Review B* **65** 224112

[55] Takeuchi N 2002 First-principles calculations of the ground-state properties and stability of ScN *Physical Review B* **65** 045204

[56] Lee E and Persson K A 2012 Li absorption and intercalation in single layer graphene and few layer graphene by first principles *Nano Lett* **12** 4624–8

[57] Petrović M, Rakić I Š, Runte S, Busse C, Sadowski J, Lazić P, Pletikosić I, Pan Z-H, Milun M and Pervan P 2013 The mechanism of caesium intercalation of graphene *Nature Communications* **4** 1–8

[58] Tian L, Zhuang Q, Li J, Shi Y, Chen J, Lu F and Sun S 2011 Mechanism of intercalation and deintercalation of lithium ions in graphene nanosheets *Chin Sci Bull* **56** 3204

[59] Han N T, Dien V K, Tran N T T, Nguyen D K, Su W-P and Lin M-F 2020 First-principles studies of electronic properties in Lithium metasilicate (Li2SiO3) *arXiv preprint arXiv:2001.07128*

[60] Lin W-B, Tran N T T and Lin S-Y 2019 Diverse fundamental properties in stage-n graphite alkali-intercalation compounds: Anode materials of Li+-based batteries *arXiv preprint arXiv:2001.02042*

[61] Xiang H, Li Z, Xie K, Jiang J, Chen J, Lian P, Wu J, Yu Y and Wang H 2012 Graphene sheets as anode materials for Li-ion batteries: Preparation, structure, electrochemical properties and mechanism for lithium storage *Rsc Advances* **2** 6792–9

[62] Li W-B, Lin S-Y, Tran N T T, Lin M-F and Lin K-I 2020 Essential geometric and electronic properties in stage-n graphite alkali-metal-intercalation compounds *RSC Advances* **10** 23573–81

[63] Li Y, Lu Y, Adelhelm P, Titirici M-M and Hu Y-S 2019 Intercalation chemistry of graphite: Alkali metal ions and beyond *Chem Soc Rev* **48** 4655–87

[64] Ahmad S, Miró P, Audiffred M and Heine T 2018 Tuning the electronic structure of graphene through alkali metal and halogen atom intercalation *Solid State Communications* **272** 22–7

[65] Rudenko A, Keil F, Katsnelson M and Lichtenstein A 2010 Adsorption of diatomic halogen molecules on graphene: A van der Waals density functional study *Physical Review B* **82** 035427

[66] Lin, Shih-Yang et al. 2020 *Silicene-Based Layered Materials* (Bristol, England: IOP Publishing Limited)

[67] Andersen M, Hornekær L and Hammer B 2014 Understanding intercalation structures formed under graphene on Ir (111) *Physical Review B* **90** 155428

[68] Ligato N, Cupolillo A and Caputi L 2013 Study of the intercalation of graphene on Ni (111) with Cs atoms: Towards the quasi-free graphene *Thin Solid Films* **543** 59–62

[69] Nobuhara K, Nakayama H, Nose M, Nakanishi S and Iba H 2013 First-principles study of alkali metal-graphite intercalation compounds *J Power Sources* **243** 585–7

[70] Okamoto Y 2014 Density functional theory calculations of alkali metal (Li, Na, and K) graphite intercalation compounds *The Journal of Physical Chemistry C* **118** 16–19

[71] Durajski A P, Skoczylas K M and Szczęśniak R 2019 Superconductivity in bilayer graphene intercalated with alkali and alkaline earth metals *PCCP* **21** 5925–31

[72] Yang S, Li S, Tang S, Dong W, Sun W, Shen D and Wang M 2016 Sodium adsorption and intercalation in bilayer graphene from density functional theory calculations *Theor Chem Acc* **135** 164

[73] Wang Z, Selbach S M and Grande T 2014 Van der Waals density functional study of the energetics of alkali metal intercalation in graphite *Rsc Advances* **4** 4069–79

[74] Do T-N, Lin C-Y, Lin Y-P, Shih P-H and Lin M-F 2015 Configuration-enriched magneto-electronic spectra of AAB-stacked trilayer graphene *Carbon* **94** 619–32

[75] Dimiev A M, Shukhina K, Behabtu N, Pasquali M and Tour J M 2019 Stage transitions in graphite intercalation compounds: Role of the graphite structure *The Journal of Physical Chemistry C* **123** 19246–53

[76] Hafner J 2008 Ab-initio simulations of materials using VASP: Density-functional theory and beyond *J Comput Chem* **29** 2044–78

[77] Hafner J 2007 Materials simulations using VASP—a quantum perspective to materials science *Comput Phys Commun* **177** 6–13

[78] Mattsson T R, Lane J M D, Cochrane K R, Desjarlais M P, Thompson A P, Pierce F and Grest G S 2010 First-principles and classical molecular dynamics simulation of shocked polymers *Physical Review B* **81** 054103

[79] Suzuki S, Bower C and Zhou O 1998 In-situ TEM and EELS studies of alkali—metal intercalation with single-walled carbon nanotubes *Chemical Physics Letters* **285** 230–4

[80] Zhou O, Gao B, Bower C, Fleming L and Shimoda H 2000 Structure and electrochemical properties of carbon nanotube intercalation compounds *Molecular Crystals and Liquid Crystals Science and Technology: Section A: Molecular Crystals and Liquid Crystals* **340** 541–6

[81] Gueorguiev G K, Czigány Z, Furlan A, Stafström S and Hultman L 2011 Intercalation of P atoms in Fullerene-like CPx *Chemical Physics Letters* **501** 400–3

[82] Moradi M, Bagheri Z and Bodaghi A 2017 Li interactions with the B40 fullerene and its application in Li-ion batteries: DFT studies *Physica E: Low-Dimensional Systems and Nanostructures* **89** 148–54

[83] Han N T, Dien V K, Thuy Tran N T, Nguyen D K, Su W-P and Lin M-F 2020 First-principles studies of electronic properties in lithium metasilicate (Li2SiO3) *RSC Advances* **10** 24721–9

[84] Khuong Dien V, Thi Han N, Nguyen T D H, Huynh T M D, Pham H D and Lin M-F 2020 Geometric and electronic properties of Li2GeO3 *Frontiers in Materials* **7**

[85] Nguyen T D H, Pham H D, Lin S-Y and Lin M-F 2020 Featured properties of Li+-based battery anode: Li4Ti5O12 *RSC Advances* **10** 14071–9

[86] Gao Y, Zhu C, Chen Z and Lu G 2017 Understanding ultrafast rechargeable aluminum-ion battery from first-principles *The Journal of Physical Chemistry C* **121** 7131–8

[87] Combelles C and Doublet M-L 2008 Structural, magnetic and redox properties of a new cathode material for Li-ion batteries: The iron-based metal organic framework *Ionics* **14** 279–83

[88] Conder J, Bouchet R, Trabesinger S, Marino C, Gubler L and Villevieille C 2017 Direct observation of lithium polysulfides in lithium—sulfur batteries using operando X-ray diffraction *Nature Energy* **2** 1–7

382 Diverse Quasiparticle Properties

[89] Reimers J N and Dahn J 1992 Electrochemical and in situ X-ray diffraction studies of lithium intercalation in Li x CoO2 *Journal of the Electrochemical Society* **139** 2091

[90] Jung J and MacDonald A H 2014 Accurate tight-binding models for the μ bands of bilayer graphene *Physical Review B* **89** 035405

[91] Yao K P C, Okasinski J S, Kalaga K, Shkrob I A and Abraham D P 2019 Quantifying lithium concentration gradients in the graphite electrode of Li-ion cells using operando energy dispersive X-ray diffraction *Energy & Environmental Science* **12** 656–65

[92] Byles B W, Palapati N K R, Subramanian A and Pomerantseva E 2016 The role of electronic and ionic conductivities in the rate performance of tunnel structured manganese oxides in Li-ion batteries *APL Materials* **4** 046108

[93] Kumar P S, Ayyasamy S, Tok E S, Adams S and Reddy M V 2018 Impact of electrical conductivity on the electrochemical performances of layered structure lithium trivanadate (LiV3—xMxO8, M = Zn/Co/Fe/Sn/Ti/Zr/Nb/Mo, x = 0.01–0.1) as cathode materials for energy storage *ACS Omega* **3** 3036–44

[94] Do T-N, Chang C-P, Shih P-H, Wu J-Y and Lin M-F 2017 Stacking-enriched magneto-transport properties of few-layer graphenes *Physical Chemistry Chemical Physics* **19** 29525–33

[95] Ho Y-H, Chiu Y-H, Lin D-H, Chang C-P and Lin M-F 2010 Magneto-optical selection rules in bilayer Bernal graphene *ACS Nano* **4** 1465–72

[96] Minaev B and Ågren H 2005 Theoretical DFT study of phosphorescence from porphyrins *Chemical Physics* **315** 215–39

[97] Chung J-Y, Sorkin V, Pei Q-X, Chiu C-H and Zhang Y-W 2017 Mechanical properties and failure behaviour of graphene/silicene/graphene heterostructures *J Phys D: Appl Phys* **50** 345302

[98] Li G, Zhang L, Xu W, Pan J, Song S, Zhang Y, Zhou H, Wang Y, Bao L and Zhang Y Y 2018 Stable silicene in graphene/silicene Van der Waals heterostructures *Adv Mater* **30** 1804650

[99] Tran N T T, Nguyen D K, Lin S Y, Gumbs G and Lin M F 2019 Fundamental properties of transition-metals-adsorbed graphene *ChemPhysChem* **20** 2473–81

[100] Pham H D, Nguyen T D H, Vo K D, Huynh T M D and Lin M-F 2020 Rich essential properties of boron, carbon, and nitrogen substituted germanenes *Applied Physics Express* **13** 085502

[101] Li J and Liu C Y 2010 Ag/graphene heterostructures: Synthesis, characterization and optical properties *Eur J Inorg Chem* **2010** 1244–8

[102] Ponomarenko L, Geim A, Zhukov A, Jalil R, Morozov S, Novoselov K, Grigorieva I, Hill E, Cheianov V and Fal'Ko V 2011 Tunable metal—insulator transition in double-layer graphene heterostructures *Nature Physics* **7** 958–61

[103] Ebnonnasir A, Narayanan B, Kodambaka S and Ciobanu C V 2014 Tunable MoS2 bandgap in MoS2-graphene heterostructures *Applied Physics Letters* **105** 031603

[104] Valsaraj A, Register L F, Tutuc E and Banerjee S K 2016 DFT simulations of inter-graphene-layer coupling with rotationally misaligned hBN tunnel barriers in graphene/hBN/graphene tunnel FETs *Journal of Applied Physics* **120** 134310

[105] Wang L, Zhou X, Ma T, Liu D, Gao L, Li X, Zhang J, Hu Y, Wang H and Dai Y 2017 Superlubricity of a graphene/MoS 2 heterostructure: A combined experimental and DFT study *Nanoscale* **9** 10846–53

[106] Baltic R, Donati F, Singha A, Wäckerlin C, Dreiser J, Delley B, Pivetta M, Rusponi S and Brune H 2018 Magnetic properties of single rare-earth atoms on graphene/Ir (111) *Physical Review B* **98** 024412

[107] Liu X, Wang C, Hupalo M, Yao Y, Tringides M, Lu W and Ho K 2010 Adsorption and growth morphology of rare-earth metals on graphene studied by ab initio calculations and scanning tunneling microscopy *Physical Review B* **82** 245408

[108] Mao Y, Yuan J and Zhong J 2008 Density functional calculation of transition metal adatom adsorption on graphene *J Phys: Condens Matter* **20** 115209
[109] Ashour R M, Abdelhamid H N, Abdel-Magied A F, Abdel-Khalek A A, Ali M, Uheida A, Muhammed M, Zou X and Dutta J 2017 Rare earth ions adsorption onto graphene oxide nanosheets *Solvent Extr Ion Exch* **35** 91–103
[110] Sevinçli H, Topsakal M, Durgun E and Ciraci S 2008 Electronic and magnetic properties of 3 d transition-metal atom adsorbed graphene and graphene nanoribbons *Physical Review B* **77** 195434
[111] Pham H D, Su W-P, Nguyen T D H, Tran N T T and Lin M-F 2020 Rich p-type-doping phenomena in boron-substituted silicene systems *Royal Society Open Science* **7** 200723
[112] Sivek J, Sahin H, Partoens B and Peeters F M 2013 Adsorption and absorption of boron, nitrogen, aluminum, and phosphorus on silicene: Stability and electronic and phonon properties *Physical Review B* **87** 085444
[113] Anastas P T and Zimmerman J B 2019 The periodic table of the elements of green and sustainable chemistry *Green Chemistry* **21** 6545–66
[114] Hafner J 2008 Ab-initio simulations of materials using VASP: Density-functional theory and beyond *Journal of Computational Chemistry* **29** 2044–78
[115] McGilly L J, Kerelsky A, Finney N R, Shapovalov K, Shih E-M, Ghiotto A, Zeng Y, Moore S L, Wu W and Bai Y 2020 Visualization of moiré superlattices *Nature Nanotechnology* **15** 580–4
[116] Wu J-Y, Chen S-C, Roslyak O, Gumbs G and Lin M-F 2011 Plasma excitations in graphene: Their spectral intensity and temperature dependence in magnetic field *ACS Nano* **5** 1026–32
[117] Yamada A, Chung S-C and Hinokuma K 2001 Optimized LiFePO4 for lithium battery cathodes *Journal of the Electrochemical Society* **148** A224
[118] Hua A C-C and Syue B Z-W 2010 Charge and discharge characteristics of lead-acid battery and LiFePO4 battery in *The 2010 International Power Electronics Conference-ECCE ASIA* (IEEE) pp 1478–83
[119] Shih P-H, Do T-N, Gumbs G and Lin M-F 2020 Electronic and optical properties of doped graphene *Physica E: Low-Dimensional Systems and Nanostructures* **118** 113894
[120] Tran N T T, Lin S-Y, Glukhova O E and Lin M-F 2016 π-Bonding-dominated energy gaps in graphene oxide *RSC Advances* **6** 24458–63
[121] Lin Y-T, Lin S-Y, Chiu Y-H and Lin M-F 2017 Alkali-created rich properties in grapheme nanoribbons: Chemical bondings *Scientific Reports* **7** 1–14
[122] Azevedo S 2006 Energetic stability of B–C–N monolayer *Physics Letters A* **351** 109–12
[123] Zhang Y-Y, Pei Q-X, Liu H-Y and Wei N 2017 Thermal conductivity of a h-BCN monolayer *PCCP* **19** 27326–31
[124] Chang C-W, Han W-Q and Zettl A 2005 Thermal conductivity of BCN and BN nanotubes *Journal of Vacuum Science & Technology B: Microelectronics and Nanometer Structures Processing, Measurement, and Phenomena* **23** 1883–6
[125] Yap Y K 2009 *BCN Nanotubes and Related Nanostructures* vol 6 (Germany: Springer Science & Business Media)
[126] Shih P-H, Do T-N, Gumbs G, Huang D, Pham T P and Lin M-F 2020 Magneto-transport properties of B-, Si-and N-doped graphene *Carbon* **160** 211–18
[127] Chung H-C, Chang C-P, Lin C-Y and Lin M-F 2016 Electronic and optical properties of graphene nanoribbons in external fields *Physical Chemistry Chemical Physics* **18** 7573–616
[128] Smotlacha J, Pincak R and Pudlak M 2011 Electronic structure of disclinated graphene in a uniform magnetic field *The European Physical Journal B* **84** 255–64
[129] Lin Y-P, Wang J, Lu J-M, Lin C-Y and Lin M-F 2014 Energy spectra of ABC-stacked trilayer graphene in magnetic and electric fields *RSC Advances* **4** 56552–60

[130] Do T-N, Gumbs G, Shih P-H, Huang D, Chiu C-W, Chen C-Y and Lin M-F 2019 Peculiar optical properties of bilayer silicene under the influence of external electric and magnetic fields *Scientific Reports* **9** 1–15

[131] Li T-S and Lin M-F 2006 Electronic properties of carbon nanotubes under external fields *Physical Review B* **73** 075432

[132] Lin C-Y, Wu J-Y, Chiu Y-H, Chang C-P and Lin M-F 2014 Stacking-dependent magnetoelectronic properties in multilayer graphene *Physical Review B* **90** 205434

[133] Chen S-C, Wu J-Y and Lin M-F 2018 Feature-rich magneto-electronic properties of bismuthene *New Journal of Physics* **20** 062001

[134] Tuz V R, Fesenko V I, Fedorin I V, Sun H-B and Shulga V M 2017 Crossing and anticrossing effects of polaritons in a magnetic-semiconductor superlattice influenced by an external magnetic field *Superlattices Microstruct* **103** 285–94

[135] Shih P-H, Do T-N, Gumbs G, Pham H D and Lin M-F 2019 Electric-field-diversified optical properties of bilayer silicene *Optics Letters* **44** 4721–4

[136] Bernevig B A and Zhang S-C 2006 Quantum spin Hall effect *Physical Review Letters* **96** 106802

[137] Chiu C-W, Shyu F-L, Chang C-P, Chen R-B and Lin M-F 2004 Magneto collective excitations of armchair carbon nanotubes *Physica E: Low-Dimensional Systems and Nanostructures* **22** 700–3

[138] Lin C-Y, Wu J-Y, Ou Y-J, Chiu Y-H and Lin M-F 2015 Magneto-electronic properties of multilayer graphenes *PCCP* **17** 26008–35

[139] Lin, Shih-Yang et al. 2020 *Silicene-Based Layered Materials* (Bristol, England: IOP Publishing Limited)

[140] Hanke W and Sham L 1979 Many-particle effects in the optical excitations of a semiconductor *Physical Review Letters* **43** 387

[141] Li Y, Qian F, Xiang J and Lieber C M 2006 Nanowire electronic and optoelectronic devices *Mater Today* **9** 18–27

[142] Piprek J 2013 *Semiconductor Optoelectronic Devices: Introduction to Physics and Simulation* (Amsterdam: Elsevier)

[143] Hsueh H, Guo G and Louie S G 2011 Excitonic effects in the optical properties of a SiC sheet and nanotubes *Physical Review B* **84** 085404

[144] Tiago M L, Northrup J E and Louie S G 2003 Ab initio calculation of the electronic and optical properties of solid pentacene *Physical Review B* **67** 115212

[145] Lin M-F and Shyu F-L 2000 Optical properties of nanographite ribbons *J Phys Soc Jpn* **69** 3529–32

[146] Albrecht S, Reining L, Del Sole R and Onida G 1998 Ab initio calculation of excitonic effects in the optical spectra of semiconductors *Physical Review Letters* **80** 4510

[147] Basera P, Saini S and Bhattacharya S 2019 Self energy and excitonic effect in (un) doped TiO 2 anatase: A comparative study of hybrid DFT, GW and BSE to explore optical properties *Journal of Materials Chemistry C* **7** 14284–93

[148] Huang Y-K, Chen S-C, Ho Y-H, Lin C-Y and Lin M-F 2014 Feature-rich magnetic quantization in sliding bilayer graphenes *Scientific Reports* **4** 1–10

[149] Falkovsky L 2008 Optical properties of graphene in *Journal of Physics: Conference Series* (Bristol, England: IOP Publishing) p 012004

[150] Mak K F, Sfeir M Y, Wu Y, Lui C H, Misewich J A and Heinz T F 2008 Measurement of the optical conductivity of graphene *Physical Review Letters* **101** 196405

[151] Ho Y-H, Wu J-Y, Chen R-B, Chiu Y-H and Lin M-F 2010 Optical transitions between Landau levels: AA-stacked bilayer graphene *Applied Physics Letters* **97** 101905

[152] Chuang Y-C, Wu J-Y and Lin M-F 2013 Electric field dependence of excitation spectra in AB-stacked bilayer graphene *Scientific Reports* **3** 1368

[153] Lin Y-P, Lin C-Y, Ho Y-H, Do T-N and Lin M-F 2015 Magneto-optical properties of ABC-stacked trilayer graphene *Physical Chemistry Chemical Physics* **17** 15921–7

[154] Lin C-Y, Huang B-L, Ho C-H, Gumbs G and Lin M-F 2018 Geometry-diversified Coulomb excitations in trilayer AAB stacking graphene *Physical Review B* **98** 195442

[155] Cao J, Yan X, Ding J and Wang D 2001 Band structures of carbon nanotubes: The sp3s* tight-binding model *J Phys: Condens Matter* **13** L271

[156] Watkins M, Sizochenko N, Moore Q, Golebiowski M, Leszczynska D and Leszczynski J 2017 Chlorophenol sorption on multi-walled carbon nanotubes: DFT modeling and structure—property relationship analysis *J Mol Model* **23** 39

[157] Nanot S, Thompson N A, Kim J-H, Wang X, Rice W D, Hároz E H, Ganesan Y, Pint C L and Kono J 2013 *Springer Handbook of Nanomaterials* ed R Vajtai (Berlin, Heidelberg: Springer) pp 105–46

[158] Capaz R B, Spataru C D, Ismail-Beigi S and Louie S G 2006 Diameter and chirality dependence of exciton properties in carbon nanotubes *Physical Review B* **74** 121401

[159] Lin M-F and Shung K W-K 1994 Plasmons and optical properties of carbon nanotubes *Physical Review B* **50** 17744

[160] Cho T, Su W, Leung T, Ren W and Chan C T 2009 Electronic and optical properties of single-walled carbon nanotubes under a uniform transverse electric field: A first-principles study *Physical Review B* **79** 235123

[161] Guo G, Chu K, Wang D-S and Duan C-G 2004 Linear and nonlinear optical properties of carbon nanotubes from first-principles calculations *Physical Review B* **69** 205416

[162] Nath P, Chowdhury S, Sanyal D and Jana D 2014 Ab-initio calculation of electronic and optical properties of nitrogen and boron doped graphene nanosheet *Carbon* **73** 275–82

[163] Kumagai M and Takagahara T 1989 Excitonic and nonlinear-optical properties of dielectric quantum-well structures *Physical Review B* **40** 12359

[164] Takagahara T and Takeda K 1992 Theory of the quantum confinement effect on excitons in quantum dots of indirect-gap materials *Physical Review B* **46** 15578

[165] Thambidurai M, Muthukumarasamy N, Agilan S, Murugan N, Vasantha S, Balasundaraprabhu R and Senthil T 2010 Strong quantum confinement effect in nanocrystalline CdS *Journal of Materials Science* **45** 3254–8

[166] Klimchitskaya G, Korikov C and Petrov V 2015 Theory of reflectivity properties of graphene-coated material plates *Physical Review B* **92** 125419

[167] Li Q, Lu J, Gupta P and Qiu M 2019 Engineering optical absorption in graphene and other 2D materials: Advances and applications *Advanced Optical Materials* **7** 1900595

[168] Xu F, Yu Z, Ren Y, Wang B, Wei Y and Qiao Z 2016 Transmission spectra and valley processing of graphene and carbon nanotube superlattices with inter-valley coupling *New Journal of Physics* **18** 113011

[169] Gonzalez J M and Oleynik I I 2016 Layer-dependent properties of SnS 2 and SnSe 2 two-dimensional materials *Physical Review B* **94** 125443

[170] Li L, Kim J, Jin C, Ye G J, Qiu D Y, Felipe H, Shi Z, Chen L, Zhang Z and Yang F 2017 Direct observation of the layer-dependent electronic structure in phosphorene *Nature Nanotechnology* **12** 21

[171] Jackson J D 1999 *Classical Electrodynamics* (College Park, Maryland: American Association of Physics Teachers)

[172] Chen X, Meng R, Jiang J, Liang Q, Yang Q, Tan C, Sun X, Zhang S and Ren T 2016 Electronic structure and optical properties of graphene/stanene heterobilayer *PCCP* **18** 16302–9

[173] Huang W L 2009 Electronic structures and optical properties of BiOX (X = F, Cl, Br, I) via DFT calculations *J Comput Chem* **30** 1882–91

[174] Do T-N, Shih P-H, Chang C-P, Lin C-Y and Lin M-F 2016 Rich magneto-absorption spectra of AAB-stacked trilayer graphene *PCCP* **18** 17597–605
[175] Shih P-H, Do T-N, Huang B-L, Gumbs G, Huang D and Lin M-F 2019 Magneto-electronic and optical properties of Si-doped graphene *Carbon* **144** 608–14
[176] Liu Y, Willis R F, Emtsev K and Seyller T 2008 Plasmon dispersion and damping in electrically isolated two-dimensional charge sheets *Physical Review B* **78** 201403
[177] Mahan G D 2013 *Many-Particle Physics* (Berlin/Heidelberg, Germany: Springer Science & Business Media)
[178] Castellani C, Di Castro C and Metzner W 1994 Dimensional crossover from Fermi to Luttinger liquid *Physical Review Letters* **72** 316
[179] Zhu S, Song Y, Wang J, Wan H, Zhang Y, Ning Y and Yang B 2017 Photoluminescence mechanism in graphene quantum dots: Quantum confinement effect and surface/edge state *Nano Today* **13** 10–14
[180] Niaz S, Manzoor T and Pandith A H 2015 Hydrogen storage: Materials, methods and perspectives *Renewable and Sustainable Energy Reviews* **50** 457–69
[181] Schlapbach L and Züttel A 2011 *Materials for Sustainable Energy: A Collection of Peer-Reviewed Research and Review Articles from Nature Publishing Group* (Singapore: World Scientific) pp 265–70
[182] Lobzenko I, Baimova J and Krylova K 2020 Hydrogen on graphene with low amplitude ripples: First-principles calculations *Chemical Physics* **530** 110608
[183] Muhammad R, Shuai Y and Tan H-P 2017 First-principles study on hydrogen adsorption on nitrogen doped graphene *Physica E: Low-Dimensional Systems and Nanostructures* **88** 115–24
[184] Gmitra M, Konschuh S, Ertler C, Ambrosch-Draxl C and Fabian J 2009 Band-structure topologies of graphene: Spin-orbit coupling effects from first principles *Physical Review B* **80** 235431
[185] Irmer S, Frank T, Putz S, Gmitra M, Kochan D and Fabian J 2015 Spin-orbit coupling in fluorinated graphene *Physical Review B* **91** 115141
[186] Gmitra M, Kochan D and Fabian J 2013 Spin-orbit coupling in hydrogenated graphene *Physical Review Letters* **110** 246602
[187] Dien V K, Pham H D, Tran N T T, Han N T, Huynh T M D, Nguyen T D H and Fa-Lin M 2020 Orbital-hybridization-created optical excitations in Li8Ge4O12 *arXiv preprint arXiv:2009.02160*
[188] Yang L, Deslippe J, Park C-H, Cohen M L and Louie S G 2009 Excitonic effects on the optical response of graphene and bilayer graphene *Phys Rev Lett* **103** 186802
[189] Ye Z, Cao T, O'Brien K, Zhu H, Yin X, Wang Y, Louie S G and Zhang X 2014 Probing excitonic dark states in single-layer tungsten disulphide *Nature* **513** 214–8
[190] Song Y, Wang X and Mi W 2017 Spin splitting and reemergence of charge compensation in monolayer WTe 2 by 3d transition-metal adsorption *PCCP* **19** 7721–7
[191] Wang Y, Wang B, Huang R, Gao B, Kong F and Zhang Q 2014 First-principles study of transition-metal atoms adsorption on MoS2 monolayer *Physica E: Low-Dimensional Systems and Nanostructures* **63** 276–82
[192] González J, Guinea F and Vozmediano M 2001 Electron-electron interactions in graphene sheets *Physical Review B* **63** 134421
[193] Kotov V N, Uchoa B, Pereira V M, Guinea F and Neto A C 2012 Electron-electron interactions in graphene: Current status and perspectives *Reviews of Modern Physics* **84** 1067
[194] Kane C, Balents L and Fisher M P 1997 Coulomb interactions and mesoscopic effects in carbon nanotubes *Physical Review Letters* **79** 5086
[195] Tans S J, Devoret M H, Groeneveld R J and Dekker C 1998 Electron—electron correlations in carbon nanotubes *Nature* **394** 761–4

[196] Ihnatsenka S and Kirczenow G 2013 Effect of edge reconstruction and electron-electron interactions on quantum transport in graphene nanoribbons *Physical Review B* **88** 125430

[197] Zarea M and Sandler N 2007 Electron-electron and spin-orbit interactions in armchair graphene ribbons *Physical Review Letters* **99** 256804

[198] Wu J, Chen S and Lin M 2014 Temperature-dependent Coulomb excitations in silicene *New Journal of Physics* **16** 125002

[199] Wu J-Y, Chen S-C, Gumbs G and Lin M-F 2016 Feature-rich electronic excitations of silicene in external fields *Physical Review B* **94** 205427

[200] Cheianov V V and Fal'ko V I 2006 Friedel oscillations, impurity scattering, and temperature dependence of resistivity in graphene *Physical Review Letters* **97** 226801

[201] Egger R and Grabert H 1995 Friedel oscillations for interacting fermions in one dimension *Physical Review Letters* **75** 3505

[202] Lin C-Y, Lee M-H and Lin M-F 2018 Coulomb excitations in ABC-stacked trilayer graphene *Physical Review B* **98** 041408

[203] Lin C-Y, Lee M-H and Lin M-F 2018 Unusual Coulomb excitations in ABC-stacked trilayer graphene *arXiv preprint arXiv:1801.06367*

[204] Dang N D and Zelevinsky V 2001 Improved treatment of ground-state correlations: Modified random phase approximation *Physical Review C* **64** 064319

[205] Potthoff M 2003 Self-energy-functional approach to systems of correlated electrons *The European Physical Journal B-Condensed Matter and Complex Systems* **32** 429–36

[206] Rieger M M, Steinbeck L, White I, Rojas H and Godby R 1999 The GW space-time method for the self-energy of large systems *Comput Phys Commun* **117** 211–28

[207] Lawrence W and Wilkins J 1973 Electron-electron scattering in the transport coefficients of simple metals *Physical Review B* **7** 2317

[208] Zheng L and Sarma S D 1996 Coulomb scattering lifetime of a two-dimensional electron gas *Physical Review B* **53** 9964

[209] Gross E and Kohn W 1990 Time-dependent density-functional theory *Adv. Quantum Chem* **21** 287–323

[210] Parr R G 1983 Density functional theory *Annu Rev Phys Chem* **34** 631–56

[211] Mahan G 2000 *Many-Body Physics* (New York: Kluwer Academic/Plenum Publishers)

[212] Vela A, Moutinho M, Culchac F, Venezuela P and Capaz R B 2018 Electronic structure and optical properties of twisted multilayer graphene *Physical Review B* **98** 155135

[213] Kim S, Ihm J, Choi H J and Son Y-W 2013 Minimal single-particle Hamiltonian for charge carriers in epitaxial graphene on 4H-SiC (0001): Broken-symmetry states at Dirac points *Solid State Communications* **175** 83–9

[214] Pierucci D, Sediri H, Hajlaoui M, Girard J-C, Brumme T, Calandra M, Velez-Fort E, Patriarche G, Silly M G and Ferro G 2015 Evidence for flat bands near the Fermi level in epitaxial rhombohedral multilayer graphene *ACS Nano* **9** 5432–9

[215] Sugawara K, Yamamura N, Matsuda K, Norimatsu W, Kusunoki M, Sato T and Takahashi T 2018 Selective fabrication of free-standing ABA and ABC trilayer graphene with/without Dirac-cone energy bands *NPG Asia Materials* **10** e466-e

[216] Bergman O, Jokela N, Lifschytz G and Lippert M 2010 Quantum Hall effect in a holographic model *Journal of High Energy Physics* **2010** 63

[217] Stormer H L, Tsui D C and Gossard A C 1999 The fractional quantum Hall effect *Reviews of Modern Physics* **71** S298

[218] Wu J-Y, Chen S-C, Do T-N, Su W-P, Gumbs G and Lin M-F 2018 The diverse magneto-optical selection rules in bilayer black phosphorus *Scientific Reports* **8** 1–11

[219] Bena C and Montambaux G 2009 Remarks on the tight-binding model of graphene *New Journal of Physics* **11** 095003

[220] Konschuh S, Gmitra M and Fabian J 2010 Tight-binding theory of the spin-orbit coupling in graphene *Physical Review B* **82** 245412

[221] Balakrishnan V 1998 *Elements of Nonequilibrium Statistical Mechanics* (New York City: Springer) pp 203–22

[222] Ferrier M, Dassonneville B, Guéron S and Bouchiat H 2013 Phase-dependent Andreev spectrum in a diffusive SNS junction: Static and dynamic current response *Physical Review B* **88** 174505

[223] Marsh C, Backx G and Ernst M 1997 Static and dynamic properties of dissipative particle dynamics *Physical Review E* **56** 1676

[224] Parr R G 1980 *Horizons of Quantum Chemistry* (New York City: Springer) pp 5–15

[225] Do T-N, Shih P-H, Gumbs G, Huang D, Chiu C-W and Lin M-F 2018 Diverse magnetic quantization in bilayer silicene *Physical Review B* **97** 125416

[226] Burt M 1992 The justification for applying the effective-mass approximation to microstructures *J Phys: Condens Matter* **4** 6651

[227] Dresselhaus G 1956 Effective mass approximation for excitons *J Phys Chem Solids* **1** 14–22

[228] Menezes M G, Capaz R B and Louie S G 2014 Ab initio quasiparticle band structure of ABA and ABC-stacked graphene trilayers *Physical Review B* **89** 035431

[229] Norimatsu W and Kusunoki M 2010 Selective formation of ABC-stacked graphene layers on SiC (0001) *Physical Review B* **81** 161410

[230] Beniwal S, Hooper J, Miller D P, Costa P S, Chen G, Liu S-Y, Dowben P A, Sykes E C H, Zurek E and Enders A 2017 Graphene-like boron—carbon—nitrogen monolayers *ACS Nano* **11** 2486–93

[231] Pham H D, Gumbs G, Su W-P, Tran N T T and Lin M-F 2020 Unusual features of nitrogen substitutions in silicene *RSC Advances* **10** 32193–201

[232] Dai J, Yuan J and Giannozzi P 2009 Gas adsorption on graphene doped with B, N, Al, and S: A theoretical study *Applied Physics Letters* **95** 232105

[233] Zhang F, Sahu B, Min H and MacDonald A H 2010 Band structure of A B C-stacked graphene trilayers *Physical Review B* **82** 035409

[234] Enoki T, Suzuki M and Endo M 2003 *Graphite Intercalation Compounds and Applications* (Oxford: Oxford University Press)

[235] Rüdorff W 1959 *Advances in Inorganic Chemistry and Radiochemistry* (Amsterdam: Elsevier) pp 223–66

[236] Shih P-H, Chiu C-W, Wu J-Y, Do T-N and Lin M-F 2018 Coulomb scattering rates of excited states in monolayer electron-doped germanene *Physical Review B* **97** 195302

[237] Wu J-Y, Lin C-Y, Gumbs G and Lin M-F 2015 The effect of perpendicular electric field on temperature-induced plasmon excitations for intrinsic silicene *RSC Advances* **5** 51912–18

[238] Wu J-Y, Su W-P and Gumbs G 2020 Anomalous magneto-transport properties of bilayer phosphorene *Scientific Reports* **10** 1–10

[239] Koshino M and Ando T 2008 Magneto-optical properties of multilayer graphene *Physical Review B* **77** 115313

[240] Wang D 2011 Electric-and magnetic-field-tuned Landau levels and Hall conductivity in AA-stacked bilayer graphene *Physics Letters A* **375** 4070–3

[241] Armitage N, Mele E and Vishwanath A 2018 Weyl and Dirac semimetals in three-dimensional solids *Reviews of Modern Physics* **90** 015001

[242] Chiu C-K, Teo J C, Schnyder A P and Ryu S 2016 Classification of topological quantum matter with symmetries *Reviews of Modern Physics* **88** 035005

[243] Pesin D and MacDonald A H 2012 Spintronics and pseudospintronics in graphene and topological insulators *Nature Materials* **11** 409–16

[244] Bernevig B A, Hughes T L and Zhang S-C 2006 Quantum spin Hall effect and topological phase transition in HgTe quantum wells *Science* **314** 1757–61

[245] Hsieh D, Xia Y, Qian D, Wray L, Dil J, Meier F, Osterwalder J, Patthey L, Checkelsky J and Ong N P 2009 A tunable topological insulator in the spin helical Dirac transport regime *Nature* **460** 1101–5

[246] König M, Wiedmann S, Brüne C, Roth A, Buhmann H, Molenkamp L W, Qi X-L and Zhang S-C 2007 Quantum spin Hall insulator state in HgTe quantum wells *Science* **318** 766–70

[247] Liu Z, Zhou B, Zhang Y, Wang Z, Weng H, Prabhakaran D, Mo S-K, Shen Z, Fang Z and Dai X 2014 Discovery of a three-dimensional topological Dirac semimetal, Na3Bi *Science* **343** 864–7

[248] Borisenko S, Gibson Q, Evtushinsky D, Zabolotnyy V, Büchner B and Cava R J 2014 Experimental realization of a three-dimensional Dirac semimetal *Physical Review Letters* **113** 027603

[249] Xu S-Y, Belopolski I, Alidoust N, Neupane M, Bian G, Zhang C, Sankar R, Chang G, Yuan Z and Lee C-C 2015 Discovery of a Weyl fermion semimetal and topological Fermi arcs *Science* **349** 613–17

[250] Martin R M 2020 *Electronic Structure: Basic Theory and Practical Methods* (Cambridge: Cambridge University Press)

[251] Yang L, Liu Z, Sun Y, Peng H, Yang H, Zhang T, Zhou B, Zhang Y, Guo Y and Rahn M 2015 Weyl semimetal phase in the non-centrosymmetric compound TaAs *Nature Physics* **11** 728–32

[252] Wen X-G 2017 Colloquium: Zoo of quantum-topological phases of matter *Reviews of Modern Physics* **89** 041004

[253] Qi X-L and Zhang S-C 2011 Topological insulators and superconductors *Reviews of Modern Physics* **83** 1057

[254] Schnyder A P and Brydon P M 2015 Topological surface states in nodal superconductors *J Phys: Condens Matter* **27** 243201

[255] Ivanovskaya V V, Zobelli A, Teillet-Billy D, Rougeau N, Sidis V and Briddon P R 2010 Hydrogen adsorption on graphene: A first principles study *The European Physical Journal B* **76** 481–6

[256] Elias D C, Nair R R, Mohiuddin T M G, Morozov S V, Blake P, Halsall M P, Ferrari A C, Boukhvalov D W, Katsnelson M I, Geim A K and Novoselov K S 2009 Control of graphene's properties by reversible hydrogenation: Evidence for graphane *Science* **323** 610–13

[257] Ryu S, Han M Y, Maultzsch J, Heinz T F, Kim P, Steigerwald M L and Brus L E 2008 Reversible basal plane hydrogenation of graphene *Nano Lett* **8** 4597–602

[258] Jeloaica L and Sidis V 1999 DFT investigation of the adsorption of atomic hydrogen on a cluster-model graphite surface *Chemical Physics Letters* **300** 157–62

[259] Karmodak N and Andreussi O 2020 Catalytic activity and stability of two-dimensional materials for the hydrogen evolution reaction *ACS Energy Letters* **5** 885–91

[260] Deng D, Novoselov K S, Fu Q, Zheng N, Tian Z and Bao X 2016 Catalysis with two-dimensional materials and their heterostructures *Nature Nanotechnology* **11** 218–30

[261] Hanh T T T, Takimoto Y and Sugino O 2014 First-principles thermodynamic description of hydrogen electroadsorption on the Pt(111) surface *Surf Sci* **625** 104–11

[262] Thi Thu Hanh T, Minh Phi N and Van Hoa N 2020 Hydrogen adsorption on two-dimensional germanene and its structural defects: An ab initio investigation *PCCP* **22** 7210–17

[263] Hanh T T T and Van Hoa N 2020 Zero-point vibration of the adsorbed hydrogen on the Pt(110) surface *Adsorption* **26** 453–9

20 Problems

Nguyen Thi Han, Vo Khuong Dien,
Ching-Hong Ho, Nguyen Thanh Tien,
Phuoc Huu Le, and Ming-Fa Lin

PROBLEM 20.1

The low-energy essential properties of the π-electronic states on graphene-related nanotubes [1] and nanoribbons [2]. Using the single-orbital tight-binding functions [3], the various Hamiltonian matrices could be built for the geometry-enriched single-walled/mono-layer systems. Evaluate the independent matrix elements, only with the nearest-neighbor interactions, in (a) armchair and (b) zigzag carbon nanotubes and (c) armchair and (d) zigzag graphene nanoribbons. (e)/(f)/(g)/(h) Clearly illustrate the main features of electronic energy spectra. It should be noticed that certain electronic states show unique probability distributions. Finally, (i) explore the 1D van Hove singularities and the important differences between these two systems.

PROBLEM 20.2

Carbon nanotubes [4] and graphene nanoribbons [5] can exhibit unusual ballistic transport properties. The Landauer Buttiker formula [6] clearly indicates the quantization behavior of electrical conductance, in which an inverse relation exists between group velocity and density of states in the absence of scattering events within a sample 1D band structures, doping levels, bias-voltage windows, and temperatures will greatly diversify the quantization phenomena. It is worthy of systematic investigations about their strong effects.

PROBLEM 20.3

A uniform electric field in 1D carbon nanotube/graphene nanoribbon along the transverse direction with specific quantum confinement, being created by a gate voltage (Figures 20.1(a) and 20.2(b) [7]), is very suitable in modulating the fundamental properties. Its main effects on the Hamiltonians are to generate the on-site Coulomb potential energies. The diagonal elements of lattice-site-dependent Hermitian matrices are changed from zero to distinct finite values. Apparently, band structures and wave functions would be drastically changed under a sufficiently high electric-field strength. As for four kinds of 1D materials in Figure 20.2, investigate the semiconductor-metal transitions, the asymmetry of hole and electron spectra, the van Hove singularities, and the ballistic conductance.

DOI: 10.1201/9781003322573-20

(a) Armchair   (b) Zigzag   (c) Armchair

(d) Zigzag

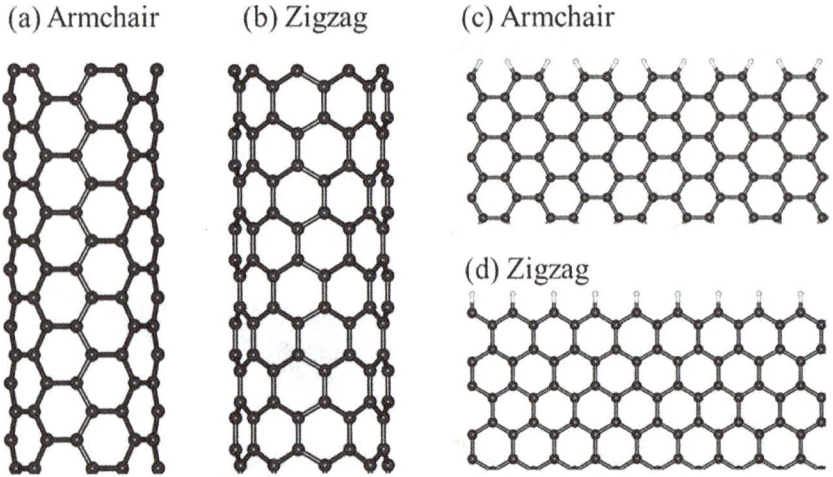

FIGURE 20.1   Geometric structures of 1D honeycomb lattices: (a) armchair and (b) zigzag carbon nanotubes and (c) armchair and (d) zigzag graphene nanoribbons.

PROBLEM 20.4

A cylindrical carbon nanotube in the presence of a longitudinal magnetic field (Figure 20.2(a)). The periodical Aharonov-Bohm effects [8] are responsible for the dramatic changes of the fundamental properties. The clockwise and anti-clockwise angular momenta will experience the different variations through the gradual modulation of magnetic flux, in which the parabolic energy bands keep similar dispersions except for those nearest to the Fermi level. (a) Calculate the magnetic Hamiltonian of a single-walled carbon nanotube by only changing the effective angular momenta. Very interestingly, the semiconductor-metal transitions will occur two times within a magnetic flux quantum. (b) Evaluate the corresponding magnetic fluxes for, respectively, semiconducting and metallic systems. Moreover, there exists a persistent current [an induced magnetic field] in each cylindrical tubule, being characterized by the variation of the total ground state energy with the external magnetic flux [9]. (c) Investigate paramagnetic and diamagnetic configurations of semiconducting and metallic carbon nanotubes [10]. (d) Under a strong enough field, the spin-dependent Zeeman splitting needs to be included in the electronic and magnetic properties.

PROBLEM 20.5

A uniform perpendicular magnetic field can flock the neighboring electronic states with the close energies together and thus create the quantized Landau levels in 2D systems (LLs [3]). A finite-size graphene nanoribbon, as clearly shown in Figure 20.2(b), is expected to exhibit the unusual quantization phenomena, mainly owing to the significant cooperation or competition between the magnetic field and quantum confinement [11]. The vector potential determines the Peierls phases in the hopping integrals of atomic interactions. In general, this leads to an enlarged Moiré super-lattice. However, (a) an armchair/zigzag 1D graphene nanoribbon only presents the same unit cell but the rather different Hamiltonian matrix elements [12]. (b) Explore the magneto-electronic energy spectra in terms of energy dispersions, wave-vector-dependent Landau wave functions (localization centers and zero-point numbers), and width and edge dependences. By using the symmetric and antisymmetric prob-ability distributions of 1D Landau sub-bands, the magneto-optical selection rules could be reached under the electric-dipole perturbations [13]. (c) Discuss the featured magneto-transport properties, as done in Prob. 20.2.

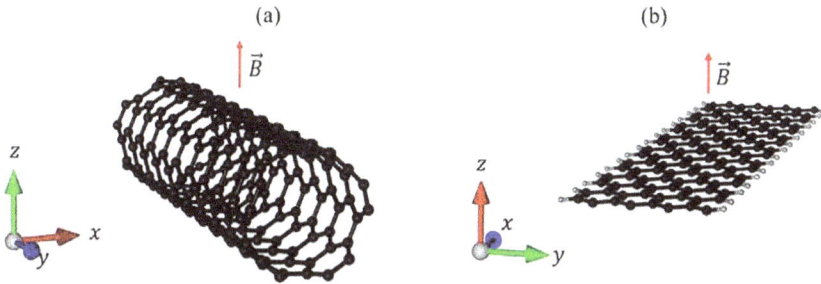

FIGURE 20.2   (a) Carbon nanotube and (b) graphene nanoribbon under electric and mag-netic fields.

PROBLEM 20.6

In general, graphite intercalation compounds exhibit the n- or p-type dopings, in which they are frequently utilized as core materials in current batteries [14]. For example, the Li-atom and $AlCl_4$-molecule intercalations, respectively, come to exist in lithium- [15] and aluminum-ion-related [16] batteries. The former, as shown in Figure 20.3, is chosen for a model study [17]. The different free car-rier densities, stacking configurations [stage-n arrangements], and interlayer dis-tances would play critical roles in the diversified phenomena. Roughly speaking, a superlattice model might be suitable for exploring electronic excitations [18] and optical absorption spectra [19]. First, (a) express a Hamiltonian of a planar honeycomb crystal for the dominating π bonding. Its π-electronic energy spec-trum, with the low-lying Dirac-cone structure, and the unique wave functions are available in investigating the elementary Coulomb excitations [20] by using the random-phase approximation [21]. This model is reliable in understanding the dynamic and static charge screening responses due to the Coulomb-field per-turbations [22] since it has effectively covered the electron-electron scattering amplitudes of the initial and final states, the Pauli exclusion principle, and the conservations of the transferred momentum and energy. (a) Calculate the bare response function of a monolayer system, being characterized by the 2D momen-tum and frequency (q,ω). The low-energy singular structures (the joint van Hove singularities) are worthy of detailed investigations. The induced charge densities of all layers are strongly coupled by the intralayer and interlayer Coulomb interactions; therefore, their close relations are responsible for the kz-dependent dielectric function, as revealed in semiconductor superlattices [23]. (b) Discuss the physical pictures of $\varepsilon\left(q, k_z, \omega\right)$ under the various Fermi energies of the stage-1 LiC_6 [24]. (c) Define the dimensionless energy loss functions corresponding to the EELS measurements [25]. Finally, (d) state the important differences among stage-1 to stage-4 Li-graphite intercalation compounds for the low-frequency plasmon modes and electron-hole excitations of conduction carriers. In addition, (e) the imaginary part of polarization function at long wavelength can provide useful information of the single-particle absorption structures arising from an electromagnetic perturbation.

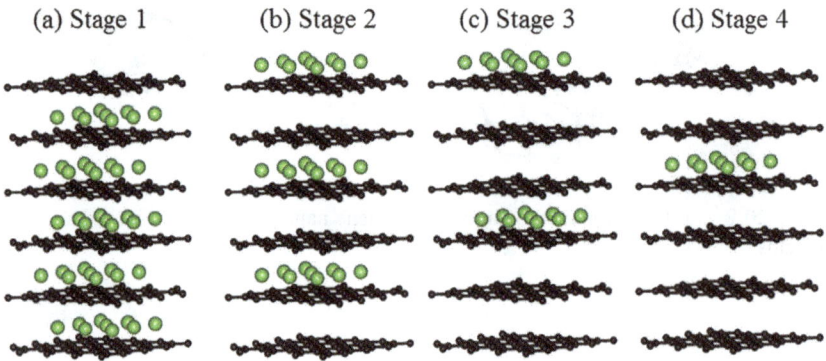

FIGURE 20.3 (a) Stage-1, (b) stage-2, (c) stage-3, and (d) stage-4 lithium-graphite intercalation compounds.

PROBLEM 20.7

The first-principles simulations (details in Chapter 2) are very reliable in achieving the optimal crystal symmetries, especially for the low-dimensional structures. The strong distortions will appear under the active dangling bonds, such as those of the 0D chains/rings/finite-length nanotubes/disks [26]/[27]/[28] and 1D nanoribbons/nanoscrolls [29]/[30]. VASP calculations should be very efficient in comprehending the geometry-diversified crystal structures. (a) Examine the highly non-uniform chemical environments due to the observable C-C bond length fluctuations, which are expected to strongly depend on the total atom number/density. (b) Discuss the main features of 0D and 1D band structures and standing waves. (c) Achieve the significant differences for the van Hove singularities. (d) Plot the spatial charge density distributions and analyze the dominances of the sp, sp^2, and sp^3 chemical bondings in various systems. (e) Investigate the curvature effects on the curved surfaces [31] because of the $2p_z$-orbital misorientations and the hybridizations of π and σ orbitals.

PROBLEM 20.8

Recently, a lot of emergent layered materials have been successfully synthesized by various chemical and physical techniques. Monolayer honeycomb crystal structure is identified to be built from a pure group-IV element; that is, there exists the 2D graphene/silicene/germanene/tinene/plumbene [32–36]. Very interestingly, whether the strong competition/significant coupling of the π and σ bondings survives will be revealed in the planar or buckling [A, B] sublattices (Figure 20.4 [37]). The VASP results are useful in achieving optimal geometries. (a) Characterize the degree of buckling and distinguish the dominances of sp^2 and sp^3 bondings in five systems. And then (b) discuss their composite effects in band structures, spatial charge densities, and van Hove singularities. While the spin-orbital interaction is open during the numerical simulation, the low-lying electronic energy spectrum might have a dramatic change except for the pure carbon systems [38]. Its effects on the electronic properties are greatly enhanced by the increase of atomic number, as observed for the buckling behaviors [38]. (c) The systematic investigations could be finished for direct/indirect band gaps, low-energy K and/or Γ valleys, state degeneracy, π and σ -electronic contributions at low and middle energies, and rich van Hove singularities due to band mixings. They are able to explain the close relations between the intrinsic interactions and the physical/chemical/material properties [39, 40].

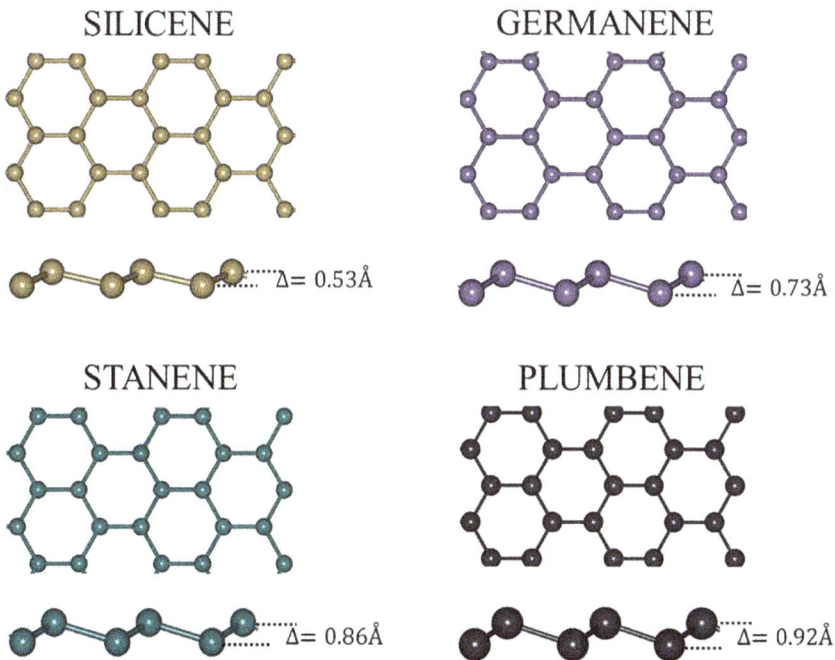

FIGURE 20.4   A buckled silicene/germanene/stanene/plumbene.

PROBLEM 20.9

A zigzag graphene nanoribbon, with/without hydrogen passivation, presents an unusual magnetism, mainly owing to their special edge structure. Use VASP simulations to achieve the optimal geometry in the presence of the spin-dominated on-site Coulomb interactions (the Hubbard-like Hamiltonian in [41]; detailed in Figure 20.5). The strong cooperation of the π bondings and many-particle interactions are expected to induce the rich electronic properties [39, 40]. (a) Explore the low-lying valence and conduction sub-bands, such as the high-symmetry points, energy dispersions, band gaps, localized wave functions, state degeneracy, and width dependences. (b) Explain a vanishing magnetic moment by illustrating the spatial spin density distribution. The dramatic transformation from the anti-ferromagnetic to ferromagnetic configurations will be revealed in the asymmetric chemisorption of guest adatoms [42, 43]. The Li-adatom adsorption on a single edge is suitable for the diversified magnetic configurations. Their positions and concentrations are expected to induce drastic changes in the main features of electronic and magnetic properties [44]. (c) The systematic numerical calculations cover their main features. (d) It is very interesting to fully understand the close relations among the adatom chemisorption, the C-C orbital hybridizations, and the on-site-bi-particle interactions.

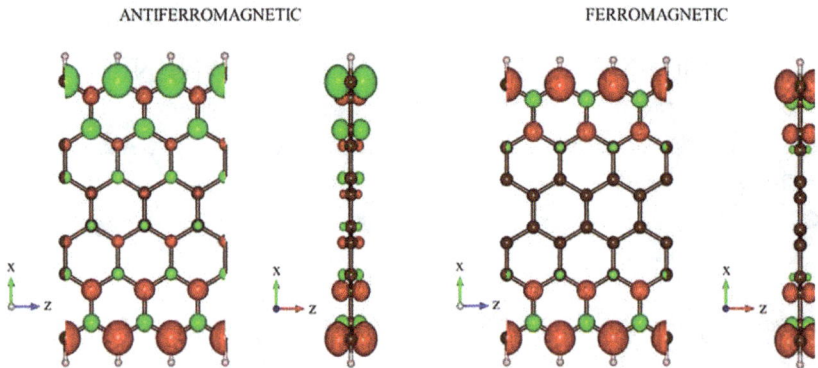

FIGURE 20.5   The anti-ferromagnetic and ferro-magnetic spin configuration of a zigzag graphene nanoribbon is, respectively, revealed at the same edge and across its center.

PROBLEM 20.10

The fundamental properties would become more complicated for 1D silicene/germanene/tinene/plumbene nanoribbons [45–48], compared with graphene ones. As to zigzag systems, the dependence of the edge-induced anti-ferromagnetism on the atomic number is one of the study focuses. The first-principles calculations are able to clarify the combined effects due to the finite-width quantum confinement, the edge-dependent spin Coulomb interactions, the sp^3 multi-orbital hybridizations of a buckled honeycomb crystal, and spin-orbital couplings. (a) The featured crystal symmetries, band structures, charge density distributions, spin arrangements, and van Hove singularities could be thoroughly presented by the delicate analyses. And then (b) the concise physical and chemical pictures are very helpful in comprehending the quasiparticle properties.

PROBLEM 20.11

The spin configurations are easily modulated by the chemical modifications of transition- and rare-earth-metal atoms (Figure 20.6 [49]) on 1D group-IV nanoribbons. Under the chemical adsorptions, such guest adatoms might provide the significant spin density distributions through the active d and f orbitals, belonging to the large on-site Coulomb interactions [50]. Very interestingly, the critical mechanisms mainly arise from the significant sp^3 bondings in a buckled honeycomb crystal, its spin-orbital couplings, the quantum confinement, the magnetic effects of zigzag edge, the multi-orbital mixings of guest and guest/host atoms, and their many-particle spin interactions. At least, there are three kinds of spin-induced significant interactions, and their close relations will dominate the diverse quasiparticle phenomena. Fe/Co/Ni are frequently chosen for model research. Similar calculation, as clearly indicated in Figure 20.6, could be generalized to Fe/Co/Ni-adsorbed 1D silicene/germanene/tinene/plumbene nanoribbons.

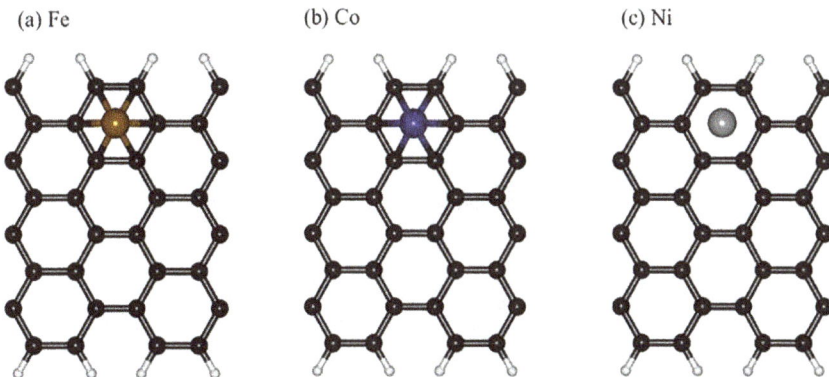

(a) Fe   (b) Co   (c) Ni

FIGURE 20.6 Fe/Co/Ni chemisorption on a 1D graphene nanoribbon. (a) Fe, (b) Co, and (c) Ni chemisorption on a 1D graphene nanoribbon.

PROBLEM 20.12

The single- and many-particle optical excitations of bulk graphite systems with AA, AB, and ABC stackings (Figures 20.7(a)– 20.7(c); Refs [51–53]). (a) The VASP simulations on the perturbations of electromagnetic fields can identify the frequency-dependent dielectric function, reflectance spectrum, absorption coefficient, and energy loss function. (b) Furthermore, they are capable of clarifying the prominent anisotropy of optical properties, the strong sensitivity of stacking configuration, and the existence of excitonic effects. All the 3D graphitic systems belong to semi-metals with the important valance and conduction sub-band overlaps [54, 55], mainly owing to the weak but significant interlayer van der Waals interactions. The low-, middle-, and high-energy electronic states are, respectively, dominated by the π, π and σ bondings. They are expected to exhibit the specific optical absorption structures by examining the imaginary part of the dielectric function. (c) Explain them through the vertical transition channels of the joint van Hove singularities. Most importantly, (d) whether the initial absorption spectrum is drastically changed by the coupled electron-hole pairs during the optical excitation processes is clear evidence of many-body effects. This work represents the first examinations about the stable/quasi-stable excitons in semi-metallic systems.

PROBLEM 20.13

After the chemical modifications, graphite intercalation compounds will display the diverse optical transitions. Obviously, the intercalation-induced Moiré superlattice and charge transfer are responsible for the greatly diversified phenomena, covering the geometric, electronic [17], optical, Coulomb-excitation [24], and transport properties [56]. Certain results, as done for $AlCl_4$ molecule intercalation cases, could be found in this book. These systems might show the metallic behaviors by the p-type or n-type dopings [16, 17]. (a) The combination effects of charge transfer and exciton on optical transitions deserve a closer examination. (b) Similar studies could test the alkali-atom cases and examine the survival of excitons in metals. Propose concise pictures for the calculated results.

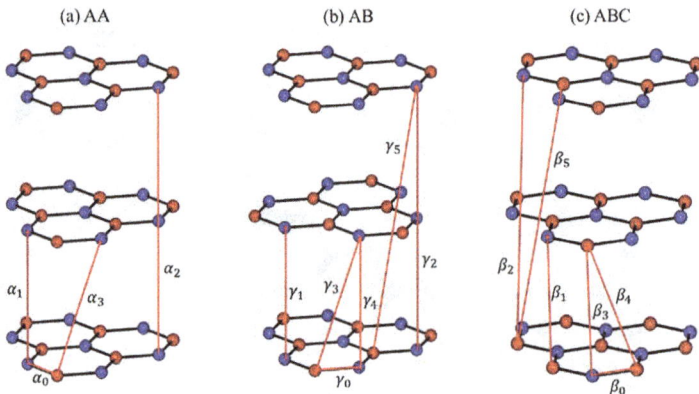

FIGURE 20.7 Optimal crystal structures of (a) AA- (b) AB- and (c) ABC-stacked graphite systems.

(a)   (b)

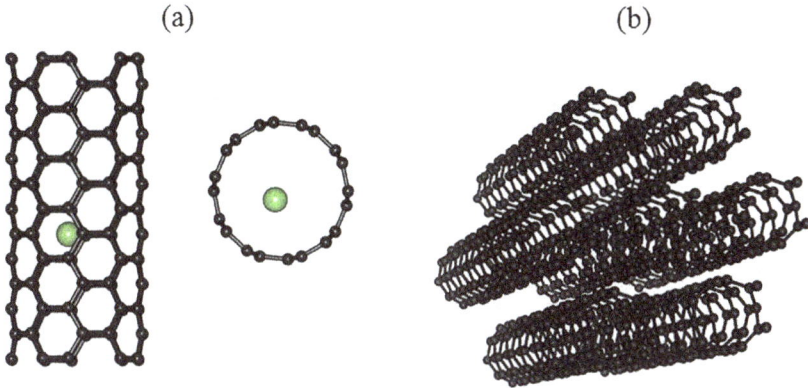

FIGURE 20.8   (a) A 1D alkali-intercalation armchair nanotube and (b) a 3D carbon nano-tube bundle.

PROBLEM 20.14

Excitons in zero-, narrow-, and wide-gap systems are very interesting research topics. It is well known that the band properties across the Fermi level are mainly determined by the intrinsic lattice symmetries [57], hopping integrals [58], many-particle interactions [59], and spin-orbital couplings [60]. Group-IV 2D and 1D honeycomb lattices are outstanding candidates in illustrating the various quasiparticle behaviors. Monolayer graphene/silicene/germanene/tinene plumbene, which have shown the different electronic energy spectra and wave functions, are predicted to exhibit rich and unique optical excitations. (a) The studying focuses principally come from the dimensionality-, orbital-hybridization-, and spin-enriched optical absorption spectra. Discuss the combined effects thoroughly. Similar investigations could be generalized to other layered materials.

PROBLEM 20.15

The simultaneous existence of excitons and spin configurations could be predicted by the VASP simulations. The 2D and 3D ferromagnetic semiconductors/metals are outstanding candidates in revealing strong evidence, e.g., hydrogen-, oxygen-, transition-metal-adsorbed graphene systems [61–63] and LiFe/FePO (ternary/quaternary lithium oxides [64, 65]). (a) Verify that the multi-orbital hybridizations and spin-created Coulomb interactions are able to create the spin-split electronic states [66]. This might lead to the spin-related absorption structures. (b) For both layered and bulk systems, the frequency-dependent absorption coefficient should be a suitable optical property in examining the critical mechanisms (the specific orbital hybridizations, atom- and orbital-decomposed spin configurations, and excitons [67]). The joint van Hove singularities, being due to valence and conduction sub-bands near the Fermi level, will dominate the featured absorption spectra. (c) Discuss their important roles under the single-particle and many-body pictures.

PROBLEM 20.16

The systematic studies on optical excitations of carbon nanotubes could be thoroughly achieved only by the first-principle calculations [68], but not the phenomenological models. The significant couplings of dynamic charge responses and electromagnetic waves are easily diversified by the nanotube radii and chiralities [69–71], coaxial symmetries [72], bundle structures [7], gate voltages [73], magnetic fields [74], guest-atom intercalations [75], and mechanical strains [76]. Apparently, the optical absorption spectra are very sensitive to the critical mechanisms. (a) Explore the initial several absorption structures and investigate the geometric dependences. (b) Examine the strength of excitonic effects by observing the threshold peak and deduce the specific relation with the bare/ screened electron-electron Coulomb interactions. (c) Evaluate the important differences between 1D and 3D nanotube systems by the neighboring van der Waals interactions [77]. Very interestingly, (d) the doping effects on low-frequency absorption structures come to exist after a Li-atom chain intercalation into an armchair carbon nanotube [78]. (f) Similar calculations could be done for a transverse electric field [79].

PROBLEM 20.17

Few- and multi-layer graphene systems, without/with chemical modifications, have shown the diverse phenomena of optical [80, 81] and magneto-optical excitations [82], as revealed in the previous theoretical predictions and experimental measurements. Furthermore, certain important issues need to be further solved by the VASP simulations, e.g., strong effects of excitons [67] and chemisorption/chemical substitutions [62, 63, 66]. That is, (a) this method is available in thoroughly exploring the critical mechanisms, covering the number of layers, stacking configurations (AA, AB, ABC, and AAB ones [20, 53]), gate voltages, chemical modifications, and excitons. On the other side, the generalized tight-binding model and dynamic Kubo formula are capable of dealing with the magneto-optical excitation spectra [82]. (b) How to unify these two approaches in similar studies is the current studying focus, i.e., the phenomenological and numerical methods are consistent with each other under most cases.

REFERENCES

[1] Lin M-F and Shung K W-K 1994 Plasmons and optical properties of carbon nanotubes *PhRvB* **50** 17744
[2] Chen S, Wang T, Lee C and Lin M-F 2008 Magneto-electronic properties of graphene nanoribbons in the spatially modulated electric field *PhLA* **372** 5999–6002
[3] Chiu Y-H, Lai Y, Ho J, Chuu D and Lin M-F 2008 Electronic structure of a two-dimensional graphene monolayer in a spatially modulated magnetic field: Peierls tight-binding model *PhRvB* **77** 045407
[4] Li H J, Lu W, Li J, Bai X and Gu C 2005 Multichannel ballistic transport in multiwall carbon nanotubes *PhRvL* **95** 086601
[5] Calado V, Zhu S-E, Goswami S, Xu Q, Watanabe K, Taniguchi T, Janssen G C and Vandersypen L 2014 Ballistic transport in graphene grown by chemical vapor deposition *ApPhL* **104** 023103

[6] Racec E and Wulf U 2001 Resonant quantum transport in semiconductor nanostructures *PhRvB* **64** 115318

[7] Lin M F and Shung K W K 1997 Optical and magneto-optical properties of carbon nanotube bundles *J Phys Soc Jpn* **66** 3294–302

[8] Zaric S, Ostojic G N, Kono J, Shaver J, Moore V C, Strano M S, Hauge R H, Smalley R E and Wei X 2004 Optical signatures of the Aharonov-Bohm phase in single-walled carbon nanotubes *Sci* **304** 1129–31

[9] Lai P, Chen S and Lin M-F 2008 Electronic properties of single-walled carbon nanotubes under electric and magnetic fields *Physica E: Low-Dimensional Systems and Nanostructures* **40** 2056–8

[10] Lin M and Shung K W-K 1995 Magnetoconductance of carbon nanotubes *PhRvB* **51** 7592

[11] Wu J, Chiu Y-H, Lien J and Lin M-F 2009 The effects of the modulated magnetic fields on electronic structures of graphene nanoribbons *Journal of Nanoscience and Nanotechnology* **9** 3193–200

[12] Hancock Y, Uppstu A, Saloriutta K, Harju A and Puska M J 2010 Generalized tight-binding transport model for graphene nanoribbon-based systems *PhRvB* **81** 245402

[13] Gopalan S, Furdyna J and Rodriguez S 1985 Inversion asymmetry and magneto-optical selection rules in n-type zinc-blende semiconductors *PhRvB* **32** 903

[14] Yang C, Chen J, Ji X, Pollard T P, Lü X, Sun C-J, Hou S, Liu Q, Liu C and Qing T 2019 Aqueous Li-ion battery enabled by halogen conversion—intercalation chemistry in graphite *Nature* **569** 245–50

[15] Ji K, Han J, Hirata A, Fujita T, Shen Y, Ning S, Liu P, Kashani H, Tian Y and Ito Y 2019 Lithium intercalation into bilayer graphene *Nature Communications* **10** 1–10

[16] Bhauriyal P, Mahata A and Pathak B 2017 The staging mechanism of AlCl 4 intercalation in a graphite electrode for an aluminium-ion battery *PCCP* **19** 7980–9

[17] Li W-B, Lin S-Y, Tran N T T, Lin M-F and Lin K-I 2020 Essential geometric and electronic properties in stage-n graphite alkali-metal-intercalation compounds *RSC Advances* **10** 23573–81

[18] Shyu F-L, Lin M-F and Lu Y-T 1999 Electronic excitations in cylinder superlattices *J Phys Soc Jpn* **68** 3352–9

[19] Chiu C-W, Huang Y-C, Shyu F-L and Lin M-F 2012 Optical absorption spectra in ABC-stacked graphene superlattice *Synth Met* **162** 800–4

[20] Ho J-H, Lu C, Hwang C, Chang C and Lin M-F 2006 Coulomb excitations in AA-and AB-stacked bilayer graphites *PhRvB* **74** 085406

[21] Ren X, Rinke P, Joas C and Scheffler M 2012 Random-phase approximation and its applications in computational chemistry and materials science *JMatS* **47** 7447–71

[22] Mahan G 2000 *Many-Body Physics* (New York: Kluwer Academic/Plenum Publishers)

[23] Ho J, Chang C and Lin M-F 2006 Electronic excitations of the multilayered graphite *PhLA* **352** 446–50

[24] Lin M, Huang C and Chuu D 1997 Plasmons in graphite and stage-1 graphite intercalation compounds *PhRvB* **55** 13961

[25] Vig S, Kogar A, Mitrano M, Husain A A, Mishra V, Rak M S, Venema L, Johnson P D, Gu G D and Fradkin E 2017 Measurement of the dynamic charge response of materials using low-energy, momentum-resolved electron energy-loss spectroscopy (M-EELS) *SciPost Phys* **3** 026

[26] Milani A, Tommasini M, Russo V, Bassi A L, Lucotti A, Cataldo F and Casari C S 2015 Raman spectroscopy as a tool to investigate the structure and electronic properties of carbon-atom wires *Beilstein Journal of Nanotechnology* **6** 480–91

[27] Pérez-Guardiola A, Ortiz-Cano R, Sandoval-Salinas M E, Fernández-Rossier J, Casanova D, Pérez-Jiménez A and Sancho-Garcia J-C 2019 From cyclic nanorings to single-walled carbon nanotubes: Disclosing the evolution of their electronic structure with the help of theoretical methods *PCCP* **21** 2547–57

[28] Chen X, Hu R and Bai F 2017 DFT study of the oxygen reduction reaction activity on Fe– N4-patched carbon nanotubes: The influence of the diameter and length *Materials* **10** 549

[29] Tien N T, Phuc V T and Ahuja R 2018 Tuning electronic transport properties of zigzag graphene nanoribbons with silicon doping and phosphorus passivation *AIP Advances* **8** 085123

[30] Chang S-L, Wu B-R, Yang P-H and Lin M-F 2016 Geometric, magnetic and electronic properties of folded graphene nanoribbons *RSC Advances* **6** 64852–60

[31] Chang S-L, Wu B-R, Yang P-H and Lin M-F 2012 Curvature effects on electronic properties of armchair graphene nanoribbons without passivation *PCCP* **14** 16409–14

[32] Geim A K and Novoselov K S 2010 *Nanoscience and Technology: A Collection of Reviews from Nature Journals* (Cambridge: World Scientific) pp 11–19

[33] Vogt P, De Padova P, Quaresima C, Avila J, Frantzeskakis E, Asensio M C, Resta A, Ealet B and Le Lay G 2012 Silicene: Compelling experimental evidence for graphene-like two-dimensional silicon *PhRvL* **108** 155501

[34] Dávila M, Xian L, Cahangirov S, Rubio A and Le Lay G 2014 Germanene: A novel two-dimensional germanium allotrope akin to graphene and silicene *NJPh* **16** 095002

[35] Zhu F-F, Chen W-J, Xu Y, Gao C-L, Guan D-D, Liu C-H, Qian D, Zhang S-C and Jia J-F 2015 Epitaxial growth of two-dimensional stanene *Nature Materials* **14** 1020–5

[36] Yuhara J, He B, Matsunami N, Nakatake M and Le Lay G 2019 Graphene's latest cousin: Plumbene epitaxial growth on a "nano watercube" *Adv Mater* **31** 1901017

[37] Martinez-Guerra E, Hernández K, Cifuentes-Quintal E and de Coss R 2013 Ab initio study of the buckling on silicene and germanene *APS* **2013** F5.004

[38] Krasovskii E 2015 Spin—orbit coupling at surfaces and 2D materials *J Phys: Condens Matter* **27** 493001

[39] Lin S-Y, Lin Y-T, Tran N T T, Su W-P and Lin M-F 2017 Feature-rich electronic properties of aluminum-adsorbed graphenes *Carbon* **120** 209–18

[40] Nguyen D K, Tran N T T, Chiu Y-H, Gumbs G and Lin M-F 2020 Rich essential properties of Si-doped graphene *Sci Rep* **10** 1–16

[41] Coury M, Dudarev S, Foulkes W, Horsfield A, Ma P-W and Spencer J 2016 Hubbard-like Hamiltonians for interacting electrons in s, p, and d orbitals *PhRvB* **93** 075101

[42] Nguyen D K, Lin Y-T, Lin S-Y, Chiu Y-H, Tran N T T and Fa-Lin M 2017 Fluorination-enriched electronic and magnetic properties in graphene nanoribbons *PCCP* **19** 20667–76

[43] Tran N T T, Nguyen D K and Lin M-F 2017 Diversified essential properties in halogenated graphenes *arXiv preprint arXiv:170602169*

[44] Lin Y-T, Lin S-Y, Chiu Y-H and Lin M-F 2017 Alkali-created rich properties in grapheme nanoribbons: Chemical bondings *Sci Rep* **7** 1–14

[45] Liu W, Zheng J, Zhao P, Cheng S and Guo C 2017 Magnetic properties of silicene nanoribbons: A DFT study *AIP Advances* **7** 065004

[46] Pang Q, Zhang Y, Zhang J-M, Ji V and Xu K-W 2011 Electronic and magnetic properties of pristine and chemically functionalized germanene nanoribbons *Nanoscale* **3** 4330–8

[47] Mahmud S and Alam M K 2019 Large bandgap quantum spin Hall insulator in methyl decorated plumbene monolayer: A first-principles study *RSC Advances* **9** 42194–203

[48] Zhang J, Lang X and Jiang Q 2018 Density functional theory calculations for armchair stanene nanoribbons with fluorine and sulfur functionalization *Physica E: Low-Dimensional Systems and Nanostructures* **101** 71–7

[49] Lin S and Lin M 2018 Metal-adsorbed graphene nanoribbons *arXiv preprint arXiv:180605290*

[50] Badrtdinov D I and Nikolaev S A 2020 Localised magnetism in 2D electrides *Journal of Materials Chemistry C* **8** 7858–65

[51] Chiu C, Lee S, Chen S, Shyu F and Lin M-F 2010 Absorption spectra of AA-stacked graphite *NJPh* **12** 083060

[52] Chiu C-W, Ho Y-H, Shyu F-L and Lin M-F 2014 Layer-enriched optical spectra of AB-stacked multilayer graphene *Applied Physics Express* **7** 115102

[53] Lin C Y and Lin M-F 2018 Unusual Coulomb excitations in tri-layer ABC-stacked graphene *APS* **2018** H36.013

[54] Huang B-L, Chuu C-P and Lin M-F 2019 Asymmetry-enriched electronic and optical properties of bilayer graphene *Sci Rep* **9** 1–12

[55] Lin C-Y, Ho C-H, Wu J-Y and Lin M-F 2020 Unusual electronic excitations in ABA trilayer graphene *Sci Rep* **10** 1–9

[56] Piraux L, Amine K, Bayot V, Issi J-P, Tressaud A and Fujimoto H 1992 Transport properties in graphite intercalation compounds with transition metal fluorides *SSCom* **82** 371–5

[57] Vasseur G, Fagot-Revurat Y, Kierren B, Sicot M and Malterre D 2013 Effect of symmetry breaking on electronic band structure: Gap opening at the high symmetry points *Symmetry* **5** 344–54

[58] Shyu F-L, Lin M F, Chang C, Chen R-B, Shyu J, Wang Y and Liao C 2001 Tight-binding band structures of nanographite multiribbons *J Phys Soc Jpn* **70** 3348–55

[59] Mo W-C, Lien J-Y and Lin M-F 2008 Competition between intertube interactions and external electric fields on a pair of narrow-gap carbon nanotubes *Physica E: Low-Dimensional Systems and Nanostructures* **40** 1674–6

[60] Kurpas M, Junior P E F, Gmitra M and Fabian J 2019 Spin-orbit coupling in elemental two-dimensional materials *PhRvB* **100** 125422

[61] Huang H-C, Lin S-Y, Wu C-L and Lin M-F 2016 Configuration-and concentration-dependent electronic properties of hy   drogenated graphene *Carbon* **103** 84–93

[62] Thanh Thuy Tran N, Lin S-Y, Lin Y-T and Lin M-F 2015 Chemical bondings induced rich electronic properties of oxygen absorbed few-layer graphenes *arXiv arXiv:1507.07057*

[63] Tran N T T, Nguyen D K, Lin S Y, Gumbs G and Lin M F 2019 Fundamental properties of transition-metals-adsorbed graphene *ChemPhysChem* **20** 2473–81

[64] Boufelfel A 2013 Electronic structure and magnetism in the layered LiFeO2: DFT+ U calculations *J Magn Magn Mater* **343** 92–8

[65] Nakayama M, Yamada S, Jalem R and Kasuga T 2016 Density functional studies of olivine-type LiFePO4 and NaFePO4 as positive electrode materials for rechargeable lithium and sodium ion batteries *Solid State Ionics* **286** 40–4

[66] Nguyen D K, Tran N T T, Nguyen T T and Lin M-F 2018 Diverse electronic and magnetic properties of chlorination-related graphene nanoribbons *Sci Rep* **8** 1–12

[67] Khuong Dien V, Duong Pham H, Thanh Thuy Tran N, Han N T, Duyen Huynh T M, Dieu Hien Nguyen T and Fa-Lin M 2020 Orbital-hybridization-created optical excitations in Li8Ge4O12 *arXiv e-prints arXiv:2009.02160*

[68] Spataru C D, Ismail-Beigi S, Benedict L X and Louie S G 2004 Quasiparticle energies, excitonic effects and optical absorption spectra of small-diameter single-walled carbon nanotubes *Appl Phys A* **78** 1129–36

[69] Lin M, Chuu D, Huang C, Lin Y and Shung K-K 1996 Collective excitations in a single-layer carbon nanotube *PhRvB* **53** 15493

[70] Lin M and Shyu F 1999 Electronic excitations in coupled armchair carbon nanotubes *PhLA* **259** 158–63

[71] Lin M F and Chuu D S 1998 Electronic states of toroidal carbon nanotubes *J Phys Soc Jpn* **67** 259–63

[72] Yannouleas C, Bogachek E N and Landman U 1996 Collective excitations of multishell carbon microstructures: Multishell fullerenes and coaxial nanotubes *PhRvB* **53** 10225

[73] Zhong D, Zhao C, Liu L, Zhang Z and Peng L-M 2018 Continuous adjustment of threshold voltage in carbon nanotube field-effect transistors through gate engineering *ApPhL* **112** 153109

[74] Li T-S and Lin M-F 2006 Conductance of carbon nanotubes in a transverse electric field and an arbitrary magnetic field *Nanotechnology* **17** 5632

[75] Chacón-Torres J, Dzsaber S, Vega-Diaz S, Akbarzadeh J, Peterlik H, Kotakoski J, Argentero G, Meyer J, Pichler T and Simon F 2016 Potassium intercalated multiwalled carbon nanotubes *Carbon* **105** 90–5

[76] Khandoker N, Hawkins S, Ibrahim R, Huynh C and Deng F 2011 Tensile strength of spinnable multiwall carbon nanotubes *Procedia Engineering* **10** 2572–8

[77] Ansari R, Daliri M and Hosseinzadeh M 2013 On the van der Waals interaction of carbon nanotubes as electromechanical nanothermometers *AcMSn* **29** 622–32

[78] Chiu C-W, Lee S-H and Lin M-F 2010 Inelastic Coulomb scatterings of doped armchair carbon nanotubes *Journal of Nanoscience and Nanotechnology* **10** 2401–8

[79] Chen S, Hseih W and Lin M-F 2005 Charge screening of single-walled carbon nanotubes in a uniform transverse electric field *PhRvB* **72** 193412

[80] Demetriou G, Bookey H T, Biancalana F, Abraham E, Wang Y, Ji W and Kar A K 2016 Nonlinear optical properties of multilayer graphene in the infrared *Opt Express* **24** 13033–43

[81] Lee S, Ho Y, Chiu C and Lin M-F 2010 Optical properties of deformed few-layer graphenes with AB stacking *JAP* **108** 043509

[82] Lin C-Y, Wu J-Y, Ou Y-J, Chiu Y-H and Lin M-F 2015 Magneto-electronic properties of multilayer graphenes *PCCP* **17** 26008–35

Index

Note: **Boldface** page references indicate tables. *Italic* references indicate figures.

405

For Product Safety Concerns and Information please contact our EU
representative GPSR@taylorandfrancis.com
Taylor & Francis Verlag GmbH, Kaufingerstraße 24, 80331 München, Germany